Advances in
INORGANIC CHEMISTRY
AND
RADIOCHEMISTRY

Volume 15

CONTRIBUTORS TO THIS VOLUME

N. W. Alcock
G. M. Bancroft
D. C. Bradley
W. R. Cullen
H. G. Heal
R. H. Platt

Advances in
INORGANIC CHEMISTRY
AND
RADIOCHEMISTRY

EDITORS

H. J. EMELÉUS

A. G. SHARPE

University Chemical Laboratory
Cambridge, England

VOLUME 15

1972

COPYRIGHT © 1972, BY ACADEMIC PRESS, INC.
ALL RIGHTS RESERVED.
NO PART OF THIS PUBLICATION MAY BE REPRODUCED OR
TRANSMITTED IN ANY FORM OR BY ANY MEANS, ELECTRONIC
OR MECHANICAL, INCLUDING PHOTOCOPY, RECORDING, OR ANY
INFORMATION STORAGE AND RETRIEVAL SYSTEM, WITHOUT
PERMISSION IN WRITING FROM THE PUBLISHER.

ACADEMIC PRESS, INC.
111 Fifth Avenue, New York, New York 10003

United Kingdom Edition published by
ACADEMIC PRESS, INC. (LONDON) LTD.
24/28 Oval Road, London NW1

LIBRARY OF CONGRESS CATALOG CARD NUMBER: 59-7692

PRINTED IN THE UNITED STATES OF AMERICA

CONTENTS

LIST OF CONTRIBUTORS vii

Secondary Bonding to Nonmetallic Elements
N. W. ALCOCK

I.	Introduction	2
II.	Group VIII	6
III.	Group VII	9
IV.	Group VI	18
V.	Group V	29
VI.	Group IV	33
VII.	Cyanides	36
VIII.	Conclusion	41
IX.	Appendix	50
	References	53

Mössbauer Spectra of Inorganic Compounds: Bonding and Structure
G. M. BANCROFT AND R. H. PLATT

I.	Preface	59
II.	Introduction	60
III.	Fingerprint Uses	89
IV.	Bonding and Structure	103
	References	241

Metal Alkoxides and Dialkylamides
D. C. BRADLEY

I.	Introduction	259
II.	Metal Alkoxides	260
III.	Metal Dialkylamides	298
	References	316

Fluoroalicyclic Derivatives of Metals and Metalloids
W. R. CULLEN

I.	Introduction	323
II.	Preparative Methods	324
III.	Other Chemical Properties	342
IV.	Physical Properties	345
V.	Coordination Complexes	346
	References	368

The Sulfur Nitrides

H. G. HEAL

I. Introduction	375
II. Thiazyl, SN, and Its Polymers	376
III. Tetrasulfur Dinitride, S_4N_2	396
IV. Saturated Sulfur–Nitrogen Frameworks, Coupled and Fused Rings, and Polymers	399
V. Conclusion	408
References	409

AUTHOR INDEX	413
SUBJECT INDEX	441
CONTENTS OF PREVIOUS VOLUMES	445

LIST OF CONTRIBUTORS

Numbers in parentheses indicate the pages on which the authors' contributions begin.

N. W. ALCOCK (1), *Department of Molecular Sciences, University of Warwick, Coventry, England*

G. M. BANCROFT (59), *Chemistry Department, University of Western Ontario, London, Canada*

D. C. BRADLEY (259), *Department of Chemistry, Queen Mary College, London, England*

W. R. CULLEN (323), *Chemistry Department, University of British Columbia, Vancouver, British Columbia, Canada*

H. G. HEAL (375), *Queen's University of Belfast, Belfast, Northern Ireland*

R. H. PLATT (59), *University Chemical Laboratory, University of Cambridge, Cambridge, England*

SECONDARY BONDING TO NONMETALLIC ELEMENTS

N. W. Alcock

Department of Molecular Sciences, University of Warwick, Coventry, England

I. Introduction	2
A. Thesis	2
B. Evidence	2
C. Van der Waals Distances	3
D. Bridge Bonds	5
E. Charge Transfer Adducts	5
F. Classification	5
II. Group VIII	6
Xenon with Fluorine	6
III. Group VII	9
A. Chlorine and Bromine	9
B. Iodine with Oxygen	13
C. Iodine with Halogens	15
IV. Group VI	18
A. With Oxygen and Sulfur	18
B. With Halogens	23
V. Group V	29
A. With Oxygen and Sulfur	30
B. With Halogens	31
VI. Group IV	33
VII. Cyanides	36
VIII. Conclusion	41
A. Angles	41
B. Geometry	42
C. Distances	44
D. Bonding	46
E. Theoretical Calculations	48
F. Significance	49
IX. Appendix	50
A. Group VII	50
B. Group VI	51
C. Group V	52
D. Group IV	52
E. Cyanides	53
References	53

I. Introduction

A. Thesis

A number of recent crystal structure determinations on compounds of the nonmetals have discovered intramolecular distances that are much longer than normal bonds and intermolecular distances that are much shorter than van der Waals distances. In this chapter, these interactions are examined and a qualitative explanation attempted. It will become clear that in most of them an approximately linear arrangement is found,

$$Y-A\cdots X$$

where Y–A is a normal bond and A---X is a short intermolecular distance.* It is with these approximately linear interactions that we are particularly concerned, and it will be our contention that they are the result of directed forces and that their behavior is sufficiently regular and understandable for the name *secondary bond* to be appropriate.†

B. Evidence

The only conclusive method of establishing the presence of secondary interactions is by crystal structure determinations. An intermolecular interaction can be recognized as being significant by being shorter than the expected intermolecular (van der Waals) distance, but if it is the result of directed forces, i.e., bonds, rather than electrostatic or nondirectional van der Waals forces, then it must satisfy one or both of the following criteria: (*a*) the interacting neighbor(s) are not in the most favorable positions for nondirected forces, and/or (*b*) the interacting neighbor(s) are in stereochemically significant positions. These are, of course, the same criteria that distinguish covalent from electrostatic primary bonding. For virtually all the short interactions that have been found in compounds of the nonmetallic elements, criterion (*b*) is satisfied, with the stereochemically significant arrangement being a linear one.

---

* Throughout the review, a single dash is used for normal bonds and triple dashes for short secondary interactions.

† The principal sources have been searched to July 1971, but in many structure determinations the existence of secondary interactions is only mentioned in passing, if at all, and this means that some earlier examples have probably been missed. Additional information to May 1972 is in the Appendix.

The physical evidence of melting points and boiling points can indicate association, but it cannot demonstrate the presence of directed interactions. The most it can do is to suggest that one compound shows similar intermolecular interactions to those in some similar compound whose structure is known. Infrared spectra can occasionally go further than this when the interaction is strong enough for either the secondary A---X bond to produce identifiable absorptions, or for other molecular frequencies to be shifted because of the interaction.

C. Van der Waals Distances

Before an intermolecular distance can be considered short, there must be some idea of the van der Waals distance that would be expected. If such distances are to be obtained as the sum of atomic van der Waals radii, with each element having a standard radius, then these radii must not be affected substantially by the atomic environment. This has not been proved, and what evidence there is shows that van der Waals radii are much more variable than, say, covalent bond radii. However, tables of van der Waals radii have been presented by Pauling (*129*, p. 260) and Bondi (*21*). Those of Pauling are simply the ionic radii of the atoms as anions; the equating of these to van der Waals radii is only tested for one or two elements. Those of Bondi are derived for the purpose of calculating molecular volumes and may not be valid for determining contact distances. He also gives compilations of intermolecular distances from which the following radii (in Å) are deduced: N, 1.55; O, 1.50; S, 1.83; F, 1.50; Cl, 1.78; Br, 1.85; and I, 2.00. These values differ systematically by up to 0.15 Å from those of Pauling, being larger for N, O, and F and smaller for Br and I; the use of ionic radii is therefore not a good approximation. However, they do correspond, to within 0.03 Å, with the mean van der Waals radii for volume calculations of Bondi. These will, therefore, be taken as the best available standard values of van der Waals radii, and are given in Table I. For four elements, Ge, Sn, Sb, and Si, values have been estimated by extrapolation, and for Xe, the figure has been revised as noted below.

Because of the problems concerning the constancy of van der Waals radii, comparison of a particular intermolecular distance with the standard value may still be unreliable, and estimation of the relative strengths of interactions between different pairs of atoms from such comparisons will be even more doubtful. An alternative approach stems from Pauling's observation (*129*, p. 263) that van der Waals radii are all approximately 0.8 Å longer than the corresponding covalent radii. Bondi (*21*) gives figures that support this, but that show that the best value of the increment varies slightly from period to period. For the elements N to F, the van der Waals radii are approximately 0.85 Å larger than the covalent radii, for Si to Cl and Ge to As, the differences are about 0.75 Å, and for Te and I, about 0.70 Å.* Therefore, a comparison of a secondary bond length with the standard single bond length may

TABLE I

VAN DER WAALS AND SINGLE BOND RADII[a]

		N	O	F	
r_w		1.55	1.52	1.47	
r_c		0.70	0.66	0.64	
	Si	P	S	Cl	
r_w	2.10	1.80	1.80	1.75	
r_c	1.17	1.10	1.04	0.99	
	Ge	As	Se	Br	
r_w	1.95	1.85	1.90	1.85	
r_c	1.22	1.18	1.14	1.11	
	Sn	Sb	Te	I	Xe
r_w	2.10	2.05	2.06	1.98	2.00
r_c	1.40	1.36	1.32	1.28	1.29
		Bi			
r_w		2.15			
r_c		1.55			

[a] r_w = van der Waals radius in Å, after Bondi (21); See text p. 3. r_c = covalent single bond radius in Å, derived by subtracting 0.76 Å from the r_b figures of Bondi (21, Table I); bismuth from Bi–Bi in the element.

give a useful estimate of the strength of the secondary bond, and these comparisons are included in Tables II–V.

One general statement can be made about van der Waals distances that is perhaps most valuable because it is most likely to be correct. Any element in periods 3, 4, and 5 (K to Rn) will have a van der Waals radius that in corresponding compounds is *at least as large* as that of the element of the same group in the preceding period. When the later

* This leads to the suggestion embodied in Table I that for Xe *in compounds*, the van der Waals radius should be 2.00 Å, rather than 2.16 Å derived from the packing of free atoms.

element forms an intermolecular contact that is *shorter* than that formed by the earlier element, then the intermolecular attraction must be stronger.

D. BRIDGE BONDS

The atom X forming a secondary interaction with atom A is often itself attached to another atom of type A, i.e., it forms a bridge

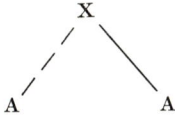

Such bridges occur in which the A---X interaction covers the whole range from being essentially nonexistent to being as strong as the normal X–A bond, and it can be very difficult to decide whether it should be called a primary or a secondary interaction. In principle, if the bridge is at all asymmetric, the weaker bond could be called secondary, but in practice it is better to reserve the term for cases where there is a difference of at least several tenths of an Ångström between the two bond lengths. Conversely, when the bridge is symmetrical but both bonds are much longer than normal, the atom X could be regarded as forming secondary bonds to both atoms A. This situation is less common, but the bonding in one or two molecules is more comprehensible if viewed in this way.

E. CHARGE TRANSFER ADDUCTS

A number of structures of addition compounds have been determined that show short interactions, mainly between halogens and O, S, Se, or N of heteroatom-containing organic molecules. These interactions have been identified as charge-transfer bonds, and the subject has been reviewed (*10, 86*). They show the structural feature of bond linearity Y–A---X, and it is almost certain that the bonding is similar to that in the compounds to be described. There is no clear dividing line, but the present chapter will concentrate on systems containing a single molecular species (or pair of ions) rather than on addition complexes.

F. CLASSIFICATION

In the main sections, compounds are grouped under each element A, the "central" atom of the linear arrangement. Usually, if the overall atomic arrangement is Y–A---X–Z, there is no difficulty deciding whether A or X is "central" because only the angle of Y–A---X is approximately 180° and the angle of A---X–Z is in the range 90°–120°. The only exceptions are the compounds in which X–Z is the N–C of a covalent cyanide, which often form long linear chains. Although there is good evidence that the noncyanide atom A corresponds to the "central" atom in other compounds, these cyanides have been collected separately. For secondary interactions to occur, the central atom must not already be surrounded by atoms forming primary covalent bonds; this means that the main area of chemistry involved is that of nonmetals in their lower oxidation states, when they have incomplete coordination polyhedra. In most of the compounds concerned, the primary geometries have been explained by the electron pair repulsion model (81).

Quantitative information about bond lengths and angles is collected in Tables II-V for the short interactions that are approximately linear. In the figures illustrating the geometries, bond distances are generally mean values, but in the tables the individual values are given. In some cases the figures only show the immediate environment of the central atom.

II. Group VIII*

Xenon with Fluorine

The melting points of the xenon fluorides decrease from XeF_2 (140°C) to XeF_4 (114°C) to XcF_6 (48°C), and from this it should follow that the intermolecular forces decrease in the same order. Unexpectedly, the crystal structures reveal that the Xe–F intermolecular distances actually decrease from XeF_2 to XeF_4, while XeF_6 contains no extended Xe---F intermolecular interactions. This example illustrates the danger of using melting point evidence unsupported by structure determinations. In XeF_2 (118) the xenon atom has eight nonbonded contacts, with Xe---F equal to 3.41 Å, directed at the corners of a cube; in XeF_4 (34) there are four interactions with a length of 3.23 Å, arranged approximately tetrahedrally, and in the mixed crystal $XeF_2 \cdot XeF_4$ (35), there are a series of contacts of lengths between 3.28 and 3.42 Å. The van der Waals distance is 3.47 Å. Presumably the melting point behavior is caused by larger numbers of weaker forces producing greater binding than smaller numbers of stronger forces.

* Summarized in Table II.

The investigation of XeF_6 (3, 32, 33) has proved difficult and it shows at least four phases, but it seems that all contain tetramers or hexamers. The intermolecular forces are between fluorine atoms only, and this presumably accounts for the low melting point. In one phase, the tetramers can be approximately described as pyramidal XeF_5^+ units linked by F^-; some Xe–F bridges are symmetrical with Xe–F, 2.5 Å, some asymmetrical with distances of 2.23 and 2.60 Å. There are either two or three interactions with the xenon through the base face of the pyramid, and none of these appear to be aligned with the apical F–Xe bond.

The coordination of xenon in $XeF_5^+PtF_6^-$ (16) is very similar to that in XeF_6. The pyramidal XeF_5^+ has mean bond length of 1.89 Å, and there are four additional fluorines (Fig. 1) below the base of the

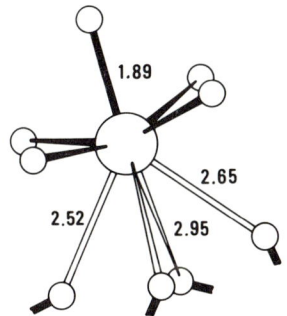

FIG. 1. XeF_5^+ in $XeF_5^+PtF_6^-$. Large circle, Xe; small circles, F. Redrawn from Bartlett et al. (16).

pyramid with Xe---F equal to 2.52, 2.65, 2.95, and 2.95 Å. These interactions are staggered relative to the primary bonds forming the base of the pyramid. Very similar again is the coordination found in the addition compound $XeF_2 \cdot IF_5$ (108), in which the iodine in the pyramidal IF_5 molecule has four close contacts with I---F of 3.14 Å in the same staggered configuration. In none of these compounds do the secondary interactions show directional properties.

By contrast, in $XeF^+Sb_2F_{11}^-$ (126) the XeF^+ has a bond length of 1.84 Å and the next neighbor is at 2.35 Å, forming an almost linear bridge to the $Sb_2F_{11}^-$ ion (Fig. 2). A very similar arrangement is found in $Xe_2F_3^+AsF_6^-$ (142), with one fluorine linking two F–Xe groups (Fig. 3). The terminal distances are 1.90 Å and both bridge distances are 2.14 Å. Both these compounds can be thought of as containing straightforward

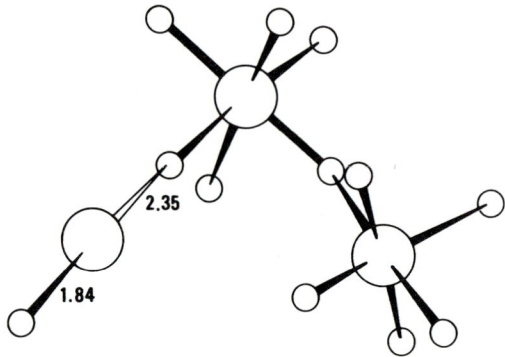

FIG. 2. XeF+Sb$_2$F$_{11}$-. Large circles, Xe and Sb; small circles, F. Redrawn from McRae et al. (126).

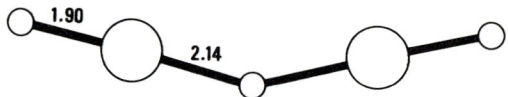

FIG. 3. Xe$_2$F$_3$+ in Xe$_2$F$_3$+AsF$_6$-. Large circles, Xe; small circles, F. Redrawn from Sladky et al. (142).

fluorine bridges, but for the first, the asymmetry makes a description involving a secondary interaction equally valid.

For the polymeric (XeO$_3$F$^-$)$_n$ ion (97) (Fig. 4) two explanations are also possible. The Xe–F distances of 2.36 and 2.48 Å are only slightly different, and so the choice is between standard fluorine bridges, donating electron pairs, or secondary interactions between XeO$_3$ molecules and F$^-$ ions. In both cases, pseudo-octahedral coordination would be expected, but in the first there should be six sterically active electron

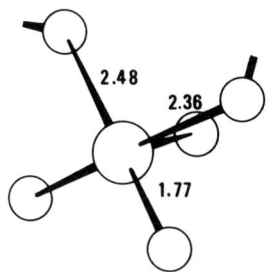

FIG. 4. (XeO$_3$F$^-$)$_n$ in K+XeO$_3$F$^-$. Large circle, Xe; small circles, F (bridging) and O (terminal). Redrawn from Hodgson and Ibers (97).

pairs and in the second four. One piece of evidence supports the second view. This is that the XeO_3 moiety is very similar in dimensions to XeO_3 itself with O–Xe–O angles averaging 100°. For six electron pairs with one vacant position lone pair–bonded pair repulsion would be expected to give angles rather less than 90°. It is also interesting that the shortest Xe–O, 1.75 Å, is opposite the vacant position, while the longest, 1.79 Å, is opposite the shorter Xe---F bond.

XeO_3 is considered below, with HIO_3.

III. Group VII*

A. Chlorine and Bromine

Although short secondary bonds are most prominent with iodine, there are some examples among chlorine and bromine compounds. In ClF_3 (30) the shortest intermolecular contacts have Cl–F equal to 3.06

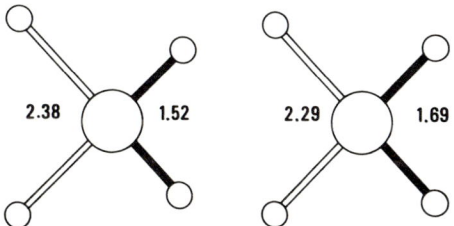

Fig. 5. ClF_2^+ (left) in $ClF_2^+SbF_6^-$ and BrF_2^+ (right) in $BrF_2^+SbF_6^-$. Large circles, Cl and Br; small circles, F. Redrawn from Edwards and Sills (62).

(twice) and 3.17 Å (twice) (with others rather longer) fairly evenly arranged on the unobstructed sides of the T-shaped molecule, and these are close to the expected van der Waals distance of 3.22 Å. BrF_3 is similar (31), although the structure has not been fully refined. However, $ClF_2^+SbF_6^-$ (62) and $BrF_2^+SbF_6^-$ (57) are very different (Fig. 5) and show directed secondary interactions. In each, the halogen atom has a square-planar environment, with two fluorines at normal bonding distances and two further away, 2.38 Å for ClF_2^+ and 2.29 Å for BrF_2^+.

In N-chlorosuccinimide (28) there is a linear N–Cl---O grouping with Cl---O equal to 2.88 Å, compared to a van der Waals distance of 3.27 Å.

* Summarized in Table II.

TABLE II

Secondary Bonds to Group VIII and VII Elements[a,b]

Compound	Ref.	Fig.	Linear group Y–A---X		Angle at A	Length Y–A	Length A---X	Single A–X	van der Waals A---X	Difference secondary minus single	Accuracy and notes
XeF$^+$Sb$_2$F$_{11}^-$	126	2	F	Xe F	(180)	1.84	2.35	1.93	3.47	0.51	c
Xe$_2$F$_3^+$AsF$_6^-$	142	3	F	Xe F	178	1.90	2.14	1.93	3.47	0.24	c
K$^+$XeO$_3$F$^-$	97	4	O	Xe F	171	1.76	2.48	1.93	3.47	0.55	c
					172	1.79	2.36			0.43	
XeO$_3$	145	—	O	Xe O	159	1.74	2.89	1.95	3.52	0.94	c
					163	1.76	2.80			0.85	
					172	1.77	2.89			0.94	
ClF$_2^+$SbF$_6^-$	62	5	F	Cl F	176	1.57	2.33	1.63	3.22	0.74	c
					176	1.58	2.43			0.85	
BrF$_2^+$SbF$_6^-$	57	5	F	Br F	178	1.69	2.29	1.75	3.32	0.60	c
n-Chlorosuccinimide	28	—	N	Cl O	170	1.69	2.88	1.65	3.27	1.23	c
POBr$_3$	128	—	P	Br O	169	2.14	3.08	1.77	3.37	1.31	c
Oxalyl bromide	84	—	C	Br O	169	1.84	3.27	1.77	3.37	1.41	Original results inconsistent
NH$_4^+$IO$_3^-$	111	6	O	I O	168	1.77	2.83	1.94	3.50	0.89	d
					172	1.81	2.78			0.84	
					174	1.84	2.82			0.88	
Li$^+$IO$_3^-$	52a	—	O	I O	165	1.81	2.89	1.94	3.50	0.95	Three times[d]
Na$^+$IO$_3^-$	125	—	O	I O	(166)	(1.83)	(3.01)	1.94	3.50	(1.07)	—
Rb$^+$IO$_3$	5	—	O	I O	168	1.80	2.76	1.94	3.50	0.82	Three times[d]
Ce^{4+}(IO$_3^-$)$_4$·H$_2$O	105	—	O	I O	160	1.81	2.93	1.94	3.50	0.99	I1[e]
					160	1.83	3.00			1.06	—
					164	1.84	2.99			1.05	—
					170	1.82	2.78			0.84	I2
					171	1.82	2.56			0.62	—

SECONDARY BONDING TO NONMETALLIC ELEMENTS

Compound										
Cu²⁺IO₃⁻OH⁻	80	—	O	I	152	1.83	2.99	—	—	1.05	
					169	1.82	2.75	—	—	0.81	I3
					161	1.83	3.07	—	—	1.13	
					175	1.86	2.52	—	—	0.58	I4
					167	1.77	3.05	—	—	1.11	
					172	1.82	2.65	—	—	0.71	
					174	1.82	2.55	—	—	0.61	c
α-HIO₃	78	—	O	I	172	1.80	2.50	1.94	3.50	0.56	Twice
					172	1.83	2.70	—	—	0.76	a
					163	1.78	2.88	1.94	—	0.94	
					166	1.82	2.78	—	—	0.84	
HI₃O₈	70	7	O	I	174	1.90	2.50	—	3.50	0.56	
					171	1.78	2.71	1.94	—	0.77	I1ᶜ
					161	1.80	2.62	—	—	0.68	
					168	1.97	2.58	—	—	0.64	
					168	1.78	2.56	—	—	0.62	I2
					169	1.79	2.83	—	—	0.89	
					174	1.95	2.38	—	—	0.44	
					178	1.81	2.59	—	—	0.65	I3
					177	1.90	2.54	—	—	0.60	
I₂O₅	140	—	O	I	169	1.76	2.93	1.94	3.50	0.99	I1ᶜ
					176	1.78	2.72	—	—	0.78	
					173	1.92	2.44	—	—	0.50	
					172	1.82	2.53	—	—	0.59	
					171	1.94	2.24	—	—	0.30	I2
p-ClC₆H₄IO₂	11	—	O	I	160	1.60	2.95	1.94	3.50	1.01	
					172	1.69	2.85	—	—	0.91	
K⁺IO₂F₂⁻	131	—	C	I	163	1.92	2.71	—	—	0.77	
					170	1.92	2.89	1.94	3.50	0.95	
					161	1.94	2.81	—	—	0.87	
Cs⁺₂I₈²⁻	88	8	I	I	175	2.80	3.42	2.56	3.96	0.86	Twiceᶜ
[N(CH₃)₄]⁺I₉⁻	107	—	I	I	168	2.67	3.43	2.56	3.96	0.85	
					178	2.67	3.24	—	—	0.68	
					169	2.91	3.24	—	—	0.68	

continued

TABLE II—continued

Compound	Ref.	Fig.	Linear group Y–A···X		Angle At A	Length Y–A	Length A···X	Single A–X	van der Waals A···X	Difference secondary minus single	Accuracy and notes
I_2	20		I	I	170	2.72	3.50	2.56	3.96	0.94	a
α-ICl	23	9	Cl	I	179	2.44	3.08	2.56	3.96	0.52	c
			Cl	I	179	2.37	3.00	2.27	3.73	0.73	—
β-ICl	41		Cl	I	180	2.44	3.06	2.56	3.96	0.50	c
			Cl	I	175	2.35	2.94	2.27	3.73	0.67	—
$ICl_2^+SbCl_6^-$	150		Cl	I	174	2.29	3.00	2.27	3.73	0.73	c
					177	2.33	2.85	—	—	0.58	—
$ICl_2^+AlCl_4^-$	150	10	Cl	I	174	2.26	2.86	2.27	3.73	0.59	c
					176	2.29	2.88	—	—	0.61	—
I_2Cl_6	24		Cl	I	175	2.38	2.68	2.27	3.73	0.41	Twice each[c]
					175	2.39	2.72	—	—	0.45	c
$(C_6H_5)_2I^+Cl^-$	112	11	C	I	179	2.08	3.08	2.27	3.73	0.81	c
					179	2.08	3.24	—	—	0.97	—
$(C_6H_5)_2I^+I^-$	112		C	I	(180)	(2.08)	3.34	2.56	3.96	0.78	c
$C_6H_5ICl_2$	12	12	C	I	164	2.02	3.45	2.27	3.73	1.13	c

[a] Notes to Tables II–V: The standard single bond lengths and the van der Waals distances are derived from the figures in Table I. The *Difference* column gives the difference between the A···X interaction length and an A–X single bond length. The latter is taken as the standard A–X single bond length unless (a) Y is the same as X, (b) Y–A is formally a single bond, and (c) Y–A is shorter than the standard A–X single bond length. In this situation, the estimate of the single bond length found within the molecule is likely to be more valid than the standard value; for example, see $XeF^+Sb_2F_{11}^-$ in Table II.

[b] Parentheses around a figure indicate either that it cannot be computed from published information or that it appears particularly unreliable.

[c] Standard deviation of A···X is 0.04 Å or better, corresponding approximately to a standard deviation for the angle Y–A···X of 2° or better (if X and Y are the same).

[d] Standard deviation of A···X is 0.01 Å or better, corresponding approximately to a standard deviation of 0.5° for the angle Y–A···X (if X and Y are the same).

Two compounds show bromine–oxygen interactions. In phosphorus oxybromide, the molecules are linked into infinite chains by almost linear P–Br---O interactions with Br---O distance of 3.08 Å (*128*), whereas in oxalyl bromide (*84*) there is an O---Br interaction having a length of 3.27 Å. This is not much shorter than the van der Waals distance of 3.37 Å, but again there is a linear grouping, C–Br---O.

B. IODINE WITH OXYGEN

The short contacts formed by iodine fall into two groups, those with oxygen and those with halogens. With one exception oxygen contacts have only been found for I^V (iodic acid and related compounds), but this is probably an accidental restriction. Periodate almost always occurs

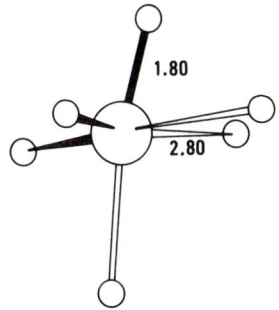

FIG. 6. IO_3^- in $NH_4^+IO_3^-$. Large circle, I; small circles, O. Redrawn from Keve et al. (*111*).

as the IO_6 group in which the regular octahedron will not allow extra iodine contacts, while I^{III} oxygen compounds are generally unstable, and crystal structures have not been determined. There is some spectroscopic evidence that $(IO)_2SO_4$, I_2O_4, and related compounds contain iodine–oxygen secondary bonds similar to those in iodates (*51*).

The recently determined structure of $NH_4^+IO_3^-$ (*111*) (Fig. 6) can be taken as a typical example of the I^V coordination. The structure is related to that of perovskite in its cell dimensions, but the octahedron of oxygen atoms around each iodine atom is very distorted, so that there are three near and three distant neighbors (1.80 and 2.80 Å). An equivalent description is as a pyramidal IO_3^- ion with three close contacts. The expected van der Waals distance is 3.50 Å.

The same arrangement occurs in $LiIO_3$ (*52a, 132*), $NaIO_3$ (*125*), and $RbIO_3$ (*5*) and is very likely in KIO_3, $CsIO_3$, and $TlIO_3$, which are almost isomorphous with NH_4IO_3 (6). In zirconium iodate (116), however, the iodate has five neighbors, at an average distance of 2.87 Å, forming with the three oxygens of the iodate group, a crude antiprism around the iodines, presumably with nondirected interactions. Ceric iodate (49) is similar, the iodate having five neighbors, three nearer (2.88 Å) and two further (3.27 Å). The packing in these two may well be controlled by the coordination around the metal atom because in $Ce(IO_3)_4 \cdot H_2O$ (105) all four iodines are surrounded by distorted octahedra. The iodine atom in $Cu(IO_3)OH$ (80) also has this coordination.

Perhaps the most interesting of this group is α-HIO_3 (78), where, although there are some hydrogen bonds, the molecular packing is controlled by the intermolecular interactions. This is conclusively

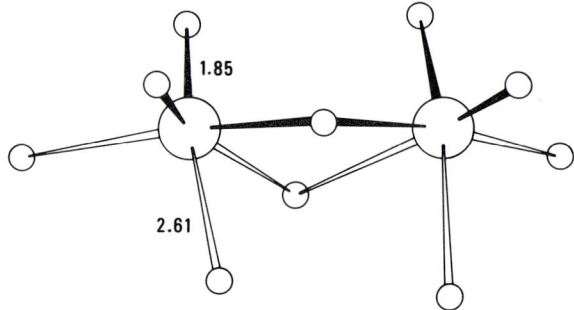

FIG. 7. The I_2O_5 moiety in HI_3O_8. Large circles, I; small circles, O. Redrawn from Feikema and Vos (70).

proved because XeO_3 (145), though lacking hydrogen bonds, is isostructural with α-HIO_3. In both compounds the central atom is surrounded by a distorted octahedron of oxygen atoms. The short Xe–O is slightly shorter than the short I–O (1.76 vs. 1.83 Å) and the long Xe---O is slightly longer than the long I---O (2.87 vs. 2.75 Å). In the addition compound, $HIO_3 \cdot I_2O_5$ (70), the iodine of the HIO_3 molecule has only two short intermolecular contacts, but each iodine of the I_2O_5 molecule has three contacts (Fig. 7), completing its octahedron. In I_2O_5 itself (140), one of the iodine atoms forms three secondary bonds, but the other has only two, its pseudo-octahedral coordination consisting of three short and two long bonds and one vacant position. If the hydroxy group of iodic acid is replaced by an aryl group, as in p-chloroiodoxybenzene (11), the iodine still makes a short oxygen contact (2.77 Å). Similarly, in one compound of iodine(VII), KIO_2F_2, where the iodine is linked to only four bonded atoms, there are two oxygen atoms (I---O, 2.81, 2.89 Å) which complete an octahedron (131).

C. Iodine with Halogens

The structures of the polyhalogens and polyhalides were reviewed in 1961 (*153*), and recently their bonding has been examined (*154*). This will be considered below. These compounds show great variations in bond lengths, and it is perhaps more difficult for secondary bonds to be distinguished from primary bonds than from van der Waals interactions. Only typical examples are included here and details of other structures can be found in Refs. (*153*) and (*154*).

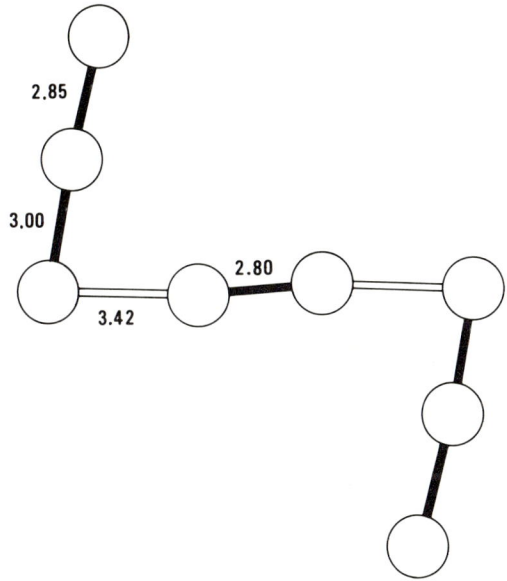

FIG. 8. I_8^{2-} in Cs_2I_8. Redrawn from Havinga et al. (*88*).

In Cs_2I_8 (*88*) the I_8^{2-} anion (Fig. 8) consists of an I_2 molecule which interacts at each end with I_3^- ions, with I---I equal to 3.42 Å. The van der Waals distance is 3.96 Å. A very similar structure is found in $N(CH_3)_4I_9$ (*107*), in which three I_2 molecules are attached to one I_3^- at distances of 3.24 (twice) and 3.43 Å. In solid I_2 (*20*) the intermolecular distance is 3.496 Å, and the intramolecular distance, 2.715 Å, is 0.05 Å longer than in the gas phase (*109*), whereas in I_8^{2-} it is longer still, 2.80 Å. Two modifications of ICl (Fig. 9) show similar interactions (*23*, *41*) with a very short I---I of 3.06 or 3.08 Å linking the molecules into zigzag chains; this is hardly longer than the bond distance in iodine itself.

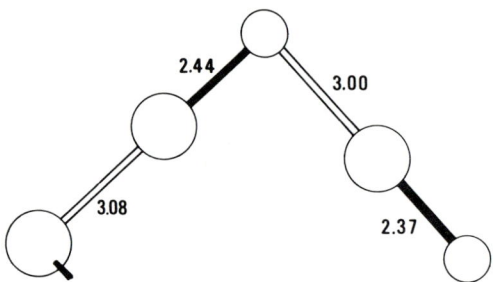

FIG. 9. α-ICl. Large circles, I; small circles, Cl (23).

There are also significant I---Cl interactions of 3.00 or 2.94 Å compared to a van der Waals distance of 3.73 Å.

Rather shorter I---Cl interactions occur in two compounds formally containing ICl_2^+, i.e. $ISbCl_8$ and $IAlCl_6$ (150). In both, the iodine atoms have two close chlorine atoms (2.29 Å, mean) and two completing a square plane at 2.90 Å (mean), which form part of the $SbCl_6^-$ or $AlCl_4^-$ groups (Fig. 10).

In the dimeric I_2Cl_6 (24) the bridges are symmetrical, but the I–Cl distances are 2.70 Å, only slightly shorter than the interaction distances in $IAlCl_6$ and $ISbCl_8$; the terminal I–Cl distances are rather long, 2.38 Å. The length of the bridge bonds suggests that it is equally valid to describe I_2Cl_6 either as a bridged molecule or as two ICl_2^+ groups interacting with two Cl⁻ ions. Both diphenyl iodonium chloride, $[(C_6H_5)_2ICl]_2$, and the corresponding bromide (112) have bridged structures (Fig. 11) in which the terminal iodine atoms are four-coordinated by two phenyl groups and two bridge atoms. For the chloride, the bridge is asymmetric with

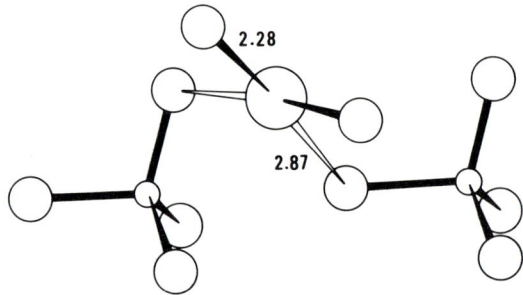

FIG. 10. $ICl_2^+AlCl_4^-$. Large circle, I; medium circles, Cl; small circle, Al. Redrawn from Vonk and Wiebanga (150).

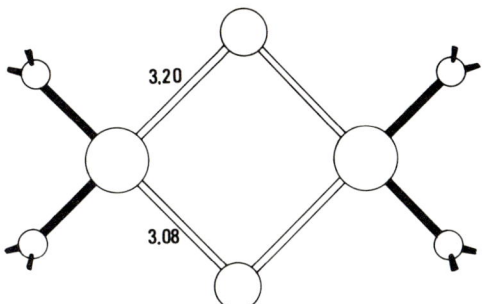

FIG. 11. Two central atoms of $[(C_6H_5)_2ICl]_2$. Large circles, I; medium circles, Cl; small circles, C (112).

I–Cl equal to 3.08 and 3.20 Å; but for the iodide it is symmetric, with I–I equal to 3.34 Å. For both, the distances are much longer than the normal single bond lengths (I–Cl, 2.27 Å; I–I, 2.56 Å) and the bonds can clearly be called secondary. In iodobenzene dichloride, $C_6H_5ICl_2$ (12) (Fig. 12), the iodine is linked to a phenyl group and two chlorine atoms at normal distances (I–Cl, 2.45 Å) and a square is completed by a weak I---Cl interaction of 3.40 Å.

Two compounds may show I---F interactions. Diphenyl iodonium borofluoride (144) has three I–F interactions between 2.94 and 3.00 Å (compared to the van der Waals distance of 3.45 Å), but only one C–I---F angle is close to 180°. In IOF_3 (149) many short I–F distances have been reported, but details have not been published.

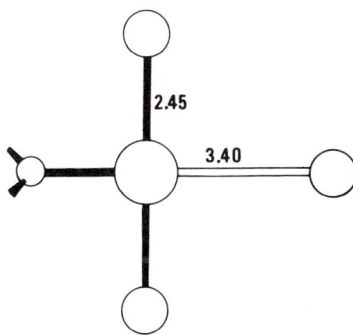

FIG. 12. The central atoms of $C_6H_5ICl_2$. Large circles, I; medium circles, Cl; small circles, C (12).

IV. Group VI*

The great majority of interactions of elements in this group are formed by selenium and tellurium. Sulfur is found in several adducts with charge-transfer interactions (*86*), while both it and selenium form thiothiophthene-like compounds containing bonds of order less than unity, which have been reviewed recently (*113*). Neither of these groups are considered further here.

A. With Oxygen and Sulfur

Both selenium and tellurium dioxide show secondary interactions leading to octahedral environments. In selenium dioxide (*122*) there are three primary Se–O bonds in a pyramid, length 1.78 Å (mean), and three secondary bonds, length 2.72 Å. Tellurium dioxide exists in two modifications, both related to TiO_2 structures, and both with four short and two long *cis* Te–O bonds. In the tetragonal modification (*117*) (Fig. 13),

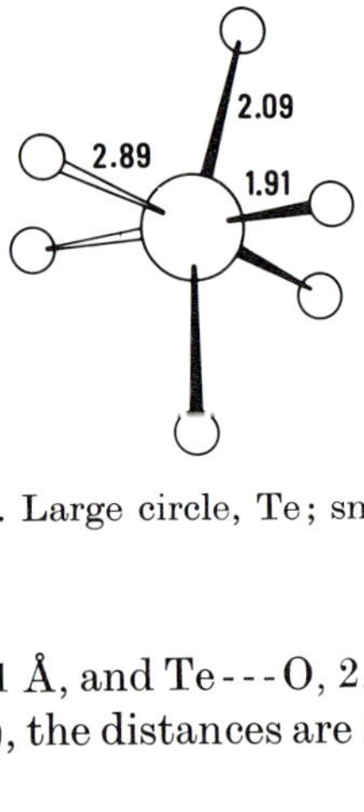

Fig. 13. TeO_2 (tetragonal). Large circle, Te; small circles, O. Redrawn from Leciejecwitz (*117*).

the distances are Te–O, 1.91 Å, and Te---O, 2.89 Å; in the orthorhombic modification, tellurite (*106*), the distances are 2.11 and 2.73 Å. Tellurium catecholate (*121*)

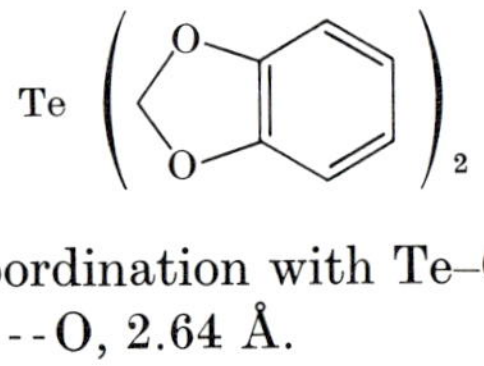

has the same primary coordination with Te–O, 2.01 Å, but forms only one secondary bond, Te---O, 2.64 Å.

* Summarized in Table III.

Selenous acid (*152*), H_2SeO_3, is similar to iodic acid in having three primary and three secondary bonds, with lengths of 1.74, and 3.06 Å (mean), while benzeneselinic acid (*29*), $C_6H_5SeO_2H$, forms one secondary interaction, with a C–Se---O linear system and Se---O equal to 3.16 Å, and also a weak O–Se---O interaction. Unlike iodates, salts of selenium and tellurium oxyacids have not been found to show short interactions, although rather few have been examined.*

In contrast to the four similar Te–O distances in tellurium catecholate, tellurium diethylxanthate (*104*) has two shorter Te–S, 2.49 Å, and two longer distances, 2.88 Å; these could be regarded as long dative

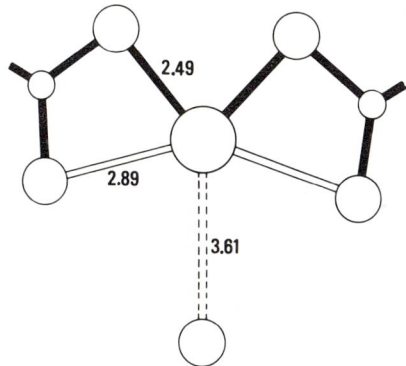

FIG. 14. The central atoms of $Te(S_2COC_2H_5)_2$. Large circle, Te; medium cicles, S; small circles, C. Redrawn from Husebye (*104*).

bonds or as short secondary interactions. The atoms are coplanar, but the S–Te---S groups are rather far from linear, with bond angles of 149° (Fig. 14). In the largest gap, a further Te---S interaction occurs of a length of 3.61 Å, still rather short compared to the van der Waals distance of 3.86 Å.

In tellurium bis(dimethyldithiophosphate), $Te[S_2P(OCH_3)_2]_2$ (*103*) (Fig. 15), the four sulfur atoms form a much distorted square around the tellurium with short Te–S, 2.44 Å, and long (cis), 3.31 Å, distances. Two of the selenium atoms in the dimeric anion $[Se(SeCN)_3]_3^{2-}$ (*87*) have similar environments (Fig. 16). The planar ion can be regarded as two $Se(SeCN)_2$ molecules bridged by two $SeCN^-$ ions with weak symmetrical interactions, and the mean distances are: short, 2.41 Å; long, 3.18 Å. By contrast, an asymmetric planar Te_2S_6-bridged system occurs in trithiourea tellurium hydrogen difluoride (*74*); the terminal Te–S bonds

* But see Appendix.

TABLE III

Secondary Bonds to Group VI Elements[a,b]

Compound	Ref.	Fig.	Linear group Y–A---X	Angle at A	Length Y–A	Length A---X	Single A–X	van der Waals A---X	Difference secondary minus single	Accuracy and notes
SeO$_2$	122	—	O Se O	(167)	(1.75)	(2.61)	1.80	3.42	(0.81)	—
				(156)	(1.79)	(2.78)			(0.99)	Twice
TeO$_2$, tetragonal	117	13	O Te O	180	1.91	2.89	1.98	3.58	0.98	Twice
TeO$_2$, tellurite	106	—	O Te O	165	2.19	2.88	1.98	3.58	0.90	—
				166	2.25	2.79			0.81	—
Te(O$_2$C$_6$H$_4$)$_2$	121	—	O Te O	154	1.98	2.65	1.98	3.58	0.67	c
H$_2$SeO$_3$	152	—	O Se O	172	1.73	3.01	1.80	3.42	1.21	—
				162	1.75	3.16			1.36	—
				150	1.76	3.00			1.20	—
C$_6$H$_5$SeO$_2$H	29	—	C Se O	178	1.90	3.16	1.80	3.42	1.36	c
			O Se O	170	1.71	3.32			1.52	—
Te(S$_2$COC$_2$H$_5$)$_2$	104	14	S Te S	150	2.49	2.86	2.36	3.86	0.50	Twice[c]
Te[S$_2$P(OCH$_3$)$_2$]$_2$	103	15	S Te S	173	2.44	3.31	2.36	3.86	0.95	Twice[d]
[Se(SeCN)$_3$]$_2^{2-}$	87	16	Se Se Se	175	2.40	3.21	2.28	3.80	0.93	d
				177	2.40	3.24			0.96	—
				173	2.41	3.17			0.89	—
				173	2.42	3.02			0.74	—
[TeSC(NH$_2$)$_2$]$_2^{2+}$(HF$_2^-$)$_2$	74	—	S Te S	173	2.47	3.02	2.36	3.86	0.66	c
TeF$_4$	56	—	F Te F	165	1.87	2.26	1.96	3.53	0.39	c
TeCl$_4$	36	—	Cl Te Cl	(180)	2.31	2.91	2.31	3.81	0.60	Te1[a]
				—	2.32	2.90	—	—	0.59	—
				—	2.32	2.95	—	—	0.64	—
				—	2.30	2.95	—	—	0.64	Te2
				—	2.32	2.96	—	—	0.65	—
				—	2.32	2.92	—	—	0.61	—

SECONDARY BONDING TO NONMETALLIC ELEMENTS

Compound	Ref		Atoms									Notes
			F	Se	F	170	1.08	2.69	1.78	3.37	1.01	Twice[c]
SeF$_3$+Nb$_2$F$_{11}$–	59	18	O	Se	F	166	1.60	2.88	—	—	1.10	c
			F	Se	F	174	1.64	2.42	1.78	3.37	0.78	—
						172	1.66	2.47	—	—	0.83	—
SeF$_3$+NbF$_6$–	60	—				171	1.67	2.40	—	—	0.76	—
			F	Se	F	176	1.69	2.33	1.78	3.37	0.55	Sel[c]
						176	1.72	2.41	—	—	0.63	—
						174	1.77	2.24	—	—	0.46	—
						177	1.73	2.43	—	—	0.65	Se2 three times[c]
SeCl$_3$+AlCl$_4$–	143	—	Cl	Se	Cl	171	2.07	3.03	2.13	3.65	0.96	c
						165	2.11	3.05	—	—	0.94	—
						166	2.13	3.11	—	—	0.98	—
SeOCl$_2$·SbCl$_5$	93	—	Cl	Se	Cl	170	2.13	3.05	2.13	3.65	0.92	c
2SeOCl$_2$·SnCl$_4$	92	—	Cl	Se	Cl	170	2.13	3.01	2.13	3.65	0.88	d
SeOCl$_2$·2C$_5$H$_5$N	119	—	O	Se	Cl	152	1.73	3.34	—	—	1.21	—
(C$_9$H$_8$NO)+SeOCl$_3$–	48	19	O	Se	Cl	170	1.59	3.65	2.13	3.65	1.52	c
			O	Se	Cl	161	1.59	3.38	2.13	3.65	1.25	c
			Cl	Se	Cl	162	2.27	2.96	—	—	0.83	—
			Cl	Se	Cl	165	2.23	2.99	—	—	0.86	—
(C$_5$H$_6$N+)$_2$ SeOCl$_3$–Cl–	151	20	Cl	Se	Cl	168	2.25	2.99	2.13	3.65	0.86	a
[(CH$_3$)$_4$N+]$_2$ Cl$_2$(SeOCl$_2$)$_{10}^{2-}$	94	21	Cl	Se	Cl	170	2.20	3.09	2.13	3.65	0.96	Sel[d]
			Cl	Se	O	143	2.18	3.14	1.80	3.42	1.34	—
			Cl	Se	Cl	162	2.20	3.05	2.13	3.65	0.92	Se2
			O	Se	O	157	1.57	3.60	—	—	1.47	—
			Cl	Se	Cl	156	2.20	2.85	1.80	3.42	1.05	—
			Cl	Se	O	161	2.22	2.97	2.13	3.65	0.84	Se3
			O	Se	Cl	165	1.58	3.44	—	—	1.31	—
			Cl	Se	Cl	157	2.20	2.89	1.80	3.42	1.09	—
			Cl	Se	O	171	2.17	3.66	2.13	3.65	1.53	Se4
			Cl	Se	O	170	2.17	2.87	1.80	3.42	1.07	—
			Cl	Se	Cl	168	2.20	3.05	2.13	3.65	0.92	Se5

continued

TABLE III—continued

Compound	Ref.	Fig.	Linear group Y-A···X			Angle at A	Length			Single A-X	van der Waals A···X	Difference secondary minus single	Accuracy and notes
							Y-A	A···X					
			O	Se	O	163	2.24	2.93		—	—	0.80	—
$(CH_3)_3S^+I^-$	156	—	C	S	I	158	1.62	3.12		1.80	3.42	1.32	—
$(CH_3)_3Se^+I^-$	98	—	C	Se	I	176	1.77	3.95		2.32	3.78	1.63	c
						179	1.96	3.78		2.42	3.88	1.46	d
$(CH_3)_3Te^+CH_3TeI_4^-$	64	22	C	Te	I	149	2.01	3.97		2.60	4.04	1.37	Cation[a]
						170	2.08	4.00		—	—	1.40	—
						173	2.13	3.84		—	—	1.24	—
						166	2.15	3.88		—	—	1.28	Anion
Thiuret hydriodide	77	—	S	S	I	167	2.09	3.62		2.32	3.78	1.30	c
$C_4H_8SeI_2$	99	—	I	Se	I	167	2.76	3.64		2.42	3.88	1.22	d
$C_6H_5TeCl \cdot SC(NH_2)_2$	75	23	C	Te	Cl	164	2.12	3.71		2.31	3.81	1.40	c
$C_6H_5TeBr \cdot SC(NH_2)_2$	76	—	C	Te	Br	164	2.12	3.77		2.43	3.91	1.34	c
$C_6H_5Te[SC(NH_2)_2]_2^+Cl^-$	76	—	C	Te	Cl	163	2.11	3.61		2.31	3.81	1.30	c
$(CH_3)_2TeCl_2$	46	24	C	Te	Cl	163	2.08	3.52		2.31	3.81	1.21	d
						171	2.10	3.46		—	—	1.15	—
$[p\text{-}ClC_6H_4Se]_2$	115	—	Se	Se	Cl	166	2.33	3.66		2.13	3.65	1.53	—
$[p\text{-}ClC_6H_4Te]_2$	115	—	Te	Te	Cl	(180)	2.70	(3.79)		2.31	3.81	1.48	—

[a] See footnote a, Table II.

[b] Parentheses around a figure indicate either that it cannot be computed from published information or that it appears particularly unreliable.

[c] Standard deviation of A···X is 0.04 Å or better, corresponding approximately to a standard deviation for the angle Y-A···X of 2° or better (if X and Y are the same).

[d] Standard deviation of A···X is 0.01 Å or better, corresponding approximately to a standard deviation of 0.5° for the angle Y-A···X (if X and Y are the same).

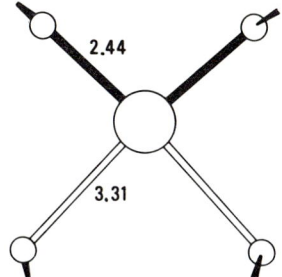

Fig. 15. The central atoms of Te[S$_2$P(OCH$_3$)$_2$]$_2$. Large circle, Te; small circles, S. Redrawn from Husebye (103).

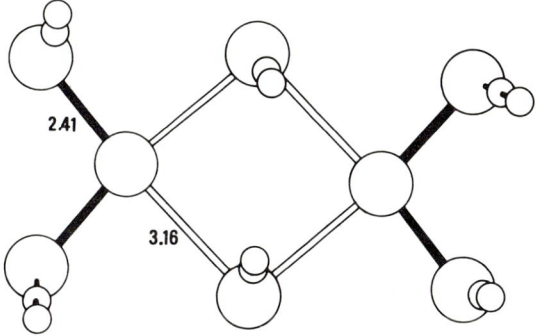

Fig. 16. [Se(SeCN)$_3$]$_2^{2-}$. Large circles, Se; small circles, C and N. Redrawn from Hauge (87).

are 2.47 and 2.53 Å, with bridge bonds of 2.86 Å opposite the long terminal bonds and 3.02 Å opposite the short terminal bonds.

B. With Halogens

Tellurium tetrafluoride (56) also shows asymmetric bridges, having endless chains

$$F-TeF_3-F-TeF_3-$$

with bridge distances of 2.08 and 2.26 Å (terminal Te–F of 1.89 Å); the weaker bridge bond can be regarded as a strong secondary interaction. The coordination is octahedral, with one vacant (apex) position and the two bridge atoms are cis-equatorial. This can also be regarded, very approximately, as TeF$_3^+$F$^-$, with strong interaction between the ions.

The same structure type has been proposed from spectroscopic evidence for selenium and tellurium tetrachloride, tetrabromide, and tetraiodide (2, 83, 89) and has recently been proved for $TeCl_4$ (36) (Fig. 17). Tetrameric units are built up of $TeCl_3^+$ ions symmetrically bridged by Cl^- ions. Each tellurium has three short Te–Cl bonds, 2.32 Å, and three long bonds, 2.92 Å. $SeCl_4$ and $TeBr_4$ are isomorphous with this compound.

Both SF_4 and SeF_4 are monomeric in the vapor phase, but the Raman spectra of the solids suggest the presence of fluorine bridging (13).

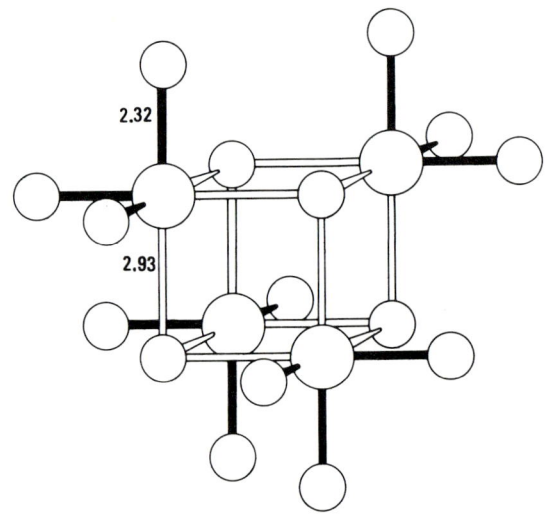

FIG. 17. $TeCl_4$. Large circles, Te; small circles, Cl. Redrawn from Buss and Krebs (36).

However, TeF_5^-, has been shown to be monomeric, by spectroscopic (1a) and crystallographic evidence (61).

Weak interactions occur in the adduct of $SeOF_2$ with NbF_5 (58) where the pyramidal $SeOF_2$ molecule (Se–F, 1.68 Å) makes three further contacts [Se---F, 2.75 Å (mean)] to complete an octahedron. The cation SeF_3^+ (Fig. 18) in both $SeF_3^+Nb_2F_{11}^-$ (59) and $SeF_3^+NbF_6^-$ (60) is very similar, but has rather shorter distances: Se–F, 1.66 and 1.73 Å; Se---F, 2.43 and 2.35 Å, respectively. The corresponding chlorine-containing cation in $SeCl_3^+AlCl_4^-$ (143) has Se–Cl equal to 2.18 Å and Se---Cl equal to 3.08 Å (mean) with the same geometry.

The structures of six compounds containing selenium oxychloride or its derivatives have been determined. All show secondary bonding resulting in distorted octahedra, sometimes with one or two vacant

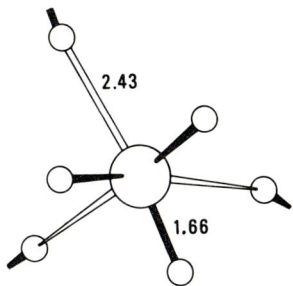

FIG. 18. SeF$_3^+$ in SeF$_3^+$Nb$_2$F$_{11}^-$. Large circle, Se; small circles, F. Redrawn from Edwards and Jones (59).

positions. In two compounds, SeOCl$_2$·SbCl$_5$ (93) and 2SeOCl$_2$·SnCl$_4$ (92), the oxygen atom of the SeOCl$_2$ molecule acts as a donor. Each contains a Cl–Se---Cl interaction to a Cl of another molecule, with Se---Cl of 3.05 or 3.01 Å, and the second also has a weaker O–Se---Cl. Both also contain intramolecular interactions involving four-membered rings, such as

```
    O————Sb
    |     |
    Se————Cl
```

but it is not clear whether these are significant. In SeOCl$_2$·2C$_5$H$_5$N (119) the selenium accepts two electron pairs from pyridine molecules becoming five-coordinate, and the sixth position is occupied by a long Se---Cl interaction, 3.65 Å. The SeOCl$_3^-$ ion has been examined in two compounds. When the cation is 8-hydroxyquinolinium (48), the anion can be viewed as polymerized into chains by symmetrical chlorine bridges (Fig. 19).

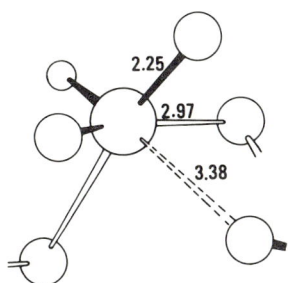

FIG. 19. [SeOCl$_3^-$]$_n$ in C$_9$H$_8$NO$^+$SeOCl$_3^-$. Large circle, Se; medium circles, Cl; small circle, O. Redrawn from Cordes (48).

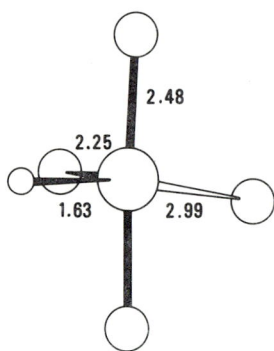

Fig. 20. SeOCl₃⁻ in [C₅H₆N⁺]₂SeOCl₃⁻Cl⁻. Large circle, Se; medium circles, Cl; small circle, O. Redrawn from Wang and Cordes (*151*).

This gives a five-coordinate selenium, and the sixth position is filled by a weak Se---Cl intramolecular interaction. However, the Se–Cl bridge bonds are also long, 2.97 Å, and it is valid to consider the complex as being built up of SeOCl₂ molecules bound together by interactions with chloride ions. In (C₅H₆N⁺)₂SeOCl₃⁻Cl⁻ (*151*) there are discrete SeOCl₃⁻ ions, each of which interacts with one chloride ion, with Se---Cl equal to 2.99 Å, to give a five-coordinate selenium atom (Fig. 20). As a culmination of the secondary interactions of SeOCl₂, the complex (CH₃)₄NCl·5SeOCl₂ (*94*) contains a [Cl₂(SeOCl₂)₁₀]²⁻ anion, a centrosymmetrical species held together entirely by secondary bonds (Fig. 21).

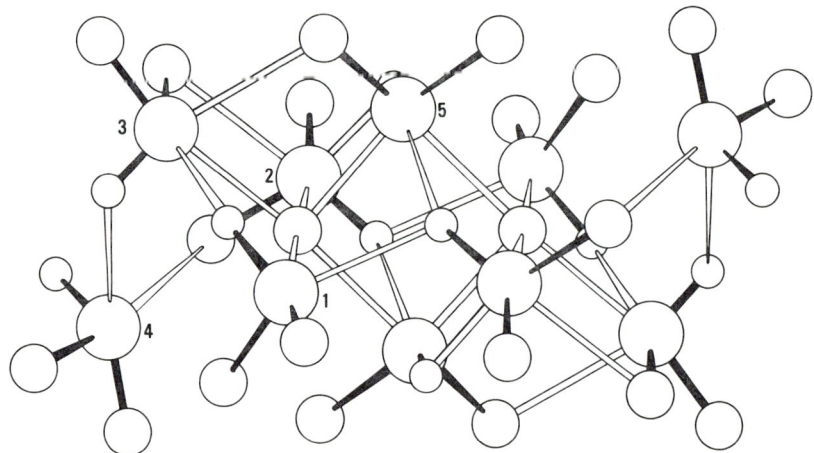

Fig. 21. Cl₂(SeOCl₂)₁₀²⁻, in (CH₃)₄NCl·5SeOCl₂. Large circles, Se; medium circles, Cl; small circles, O. The five different selenium atoms are numbered. Redrawn from Hermodsson (*94*).

These are mainly Se---Cl distances of about 3.20 Å, but there are some Se---O. Six of the selenium atoms achieve six-coordination, and four five-coordination, although some of the interactions are rather long.

An interesting group of compounds for which S, Se, and Te (=X) can all be compared contain the ion $(CH_3)_3X^+$, interacting with iodine. These are pyramidal, and the first two have one linear C–X---I grouping, while the third has three, completing an octahedron. The interacting distances are S---I, 3.89 Å (156); Se---I, 3.78 Å (98); Te---I, 3.95 Å (mean) (64). Iodide is the anion for the first two, but the third occurs in the compound $(CH_3)_3Te^+CH_3TeI_4^-$ (Fig. 22); the anion is octahedral,

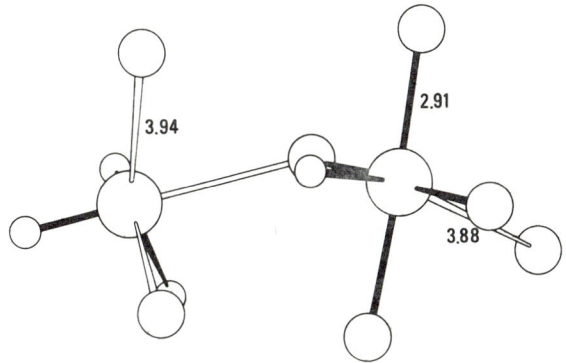

FIG. 22. $(CH_3)_3Te^+CH_3TeI_4^-$. Large circles, Te; medium circles, I; small circles, C. Redrawn from Einstein et al. (64).

with one empty position and this position is filled with a further Te---I interaction of length 3.88 Å. Although the S---I contact is no shorter than the van der Waals distance of 3.78 Å, the linearity and the similarity to the other compounds suggests that there is a directed interaction, even though it is very weak. A rather shorter S---I interaction is known, 3.62 Å, in thiuret hydriodide (77) with an approximately linear S–S–I grouping. However, in 2,2'-diiododiethyl trisulfide (54), a S---I contact of 3.74 Å without directional properties may well be due to van der Waals forces only. An Se---I interaction occurs between molecules of the tetrahydroselenophene·iodine adduct (99) with the linear grouping I–I–Se---I and distances Se–I, 2.76 Å and Se---I, 3.64 Å.

Tellurium to chlorine and bromine interactions are found in several compounds of Te(II). Typical of these is the complex of thiourea with benzene tellurium chloride (75). This is a T-shaped molecule, with Te–C as the stem, Te–Cl (3.00 Å) and Te–S as the arms, and a Te---Cl interaction forming a distorted square with a length of 3.71 Å (Fig. 23).

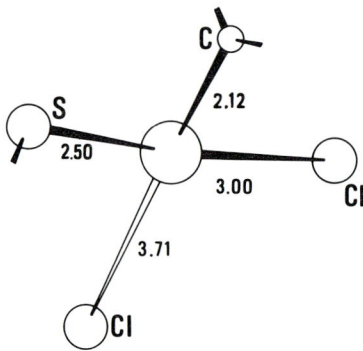

Fig. 23. The central atoms in $C_6H_5TeSC(NH_2)Cl$. Large circle, Te; medium circles, Cl and S; small circle, C. Redrawn from Foss and Husebye (75).

The corresponding bromide (76) has Te–Br, 3.11 Å, and Te---Br, 3.77 Å. If two molecules of thiourea (76) are complexed, a chloride ion is released to give $PhTe(tu)_2{}^+Cl^-$, but the chlorine is only 3.61 Å away, in line with the carbon atom, completing a distorted square. A similar C–Te---Cl is found in a Te(IV) compound, dimethyl tellurium dichloride (46), which has two short Te–Cl distances, 2.51 Å, and two longer ones, 3.49 Å, which complete an octahedron (Fig. 24). In p,p'-dichlorodiphenyl ditelluride (115), the grouping Cl---Te–Te---Cl is approximately linear, with Cl---Te somewhat longer, 3.79 Å (mean). The corresponding selenium compound (115) is isomorphous; its Cl---Se distances are slightly shorter (3.72 Å). Neither of these are significantly shorter than the expected van der Waals distances of 3.81 and 3.65 Å, but the geometry suggests directed interactions.

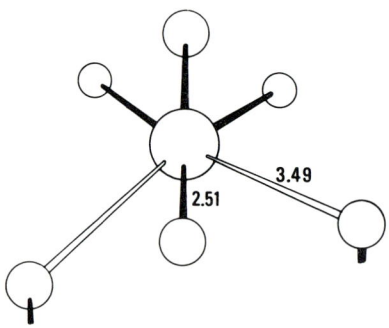

Fig. 24. $(CH_3)_2TeCl_2$. Large circle, Te; medium circles, Cl; small circles, C. Redrawn from Christofferson et al. (46).

A number of tellurium(II)–thiourea complexes show Te–hal bonds rather longer than the standard single bond length. These have been reviewed by Foss (73) and explained in terms of trans interactions; in these compounds, as with the polyhalides, the distinction between primary and secondary bonds becomes blurred.

V. Group V*

Nitrogen is concerned in some charge-transfer bonded adducts (86) and in the interactions formed by cyanides which are considered below, but apart from this, neither it nor phosphorus appears to be the principal

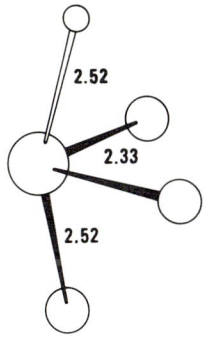

Fig. 25. The central atoms in $SbCl_3 \cdot C_6H_5NH_2$. Large circle, Sb; medium circles, Cl; small circle, N. Redrawn from Hulme (101).

atom of directed secondary bonds, while for bismuth, only two examples have been found. Arsenic and antimony interact mainly with chalcogens and halogens, but an Sb---N interaction of 2.53 Å occurs in $SbCl_3 \cdot C_6H_5NH_2$ (101) (Fig. 25). The Cl–Sb bond in the group Cl–Sb---N is 2.52 Å, considerably longer than in isolated $SbCl_3$ (2.37 Å) or in the other two bonds in the adduct (2.33 Å). There is an As---As interaction of 3.24 Å in dimethylarsinodimethyldithioarsinate (39) (Fig. 26); as could be expected, this occurs between pyramidal As^{III} groups rather than tetrahedral As^V groups, and the atoms S–As---As–S are approximately linear. In one compound, the adduct of antimony trichloride and

* Summarized in Table IV.

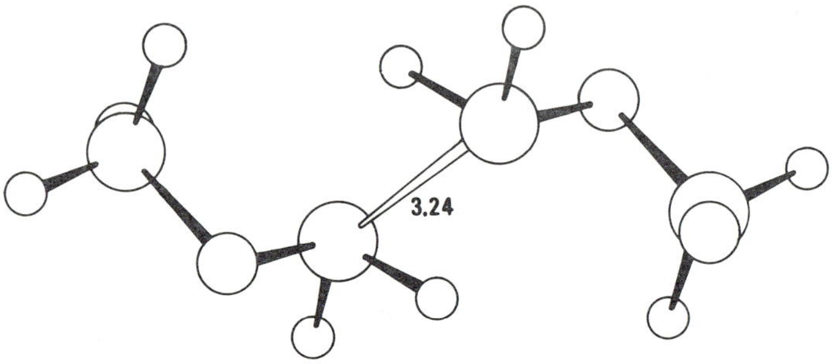

FIG. 26. (CH$_3$)$_2$AsSAsS(CH$_3$)$_2$. Large circles, As; medium circles, S; small circles, C. Redrawn from Camerman and Trotter (39).

naphthalene (102), there is an interaction with the π electrons of the naphthalene. In the group

$$Cl-Sb----\overset{C}{\underset{C}{\|}}$$

the Cl–Sb bond, 2.367 Å, is 0.02 Å longer than the other two Cl–Sb bonds; the distance to the π bond is 3.2 Å which is long, but the structure clearly shows that this is a directed interaction.

A. WITH OXYGEN AND SULFUR

Both arsenious and antimonous oxides exist as molecular and polymerized species in the solid state; in the cubic (molecular) form (9), the central atoms have three near oxygens in a pyramid and three further completing an octahedron: As–O, 1.8 Å; As---O, 3.0 Å; Sb–O, 2.0 Å, and Sb---O, 2.9 Å. The monoclinic form (17, 18) of As$_2$O$_3$ does not show similar contacts, while the structure of valentinite (155), although apparently containing Sb$_2$O$_3$ chains, has not been accurately determined. Bismuth in bismuth silicate (eulytite) has the same environment with Bi–O, 2.15 Å, and Bi---O, 2.63 Å (139). By contrast, several other arsenic and antimony oxy compounds do not show secondary contacts, one example being potassium di-o-phenylene dioxyarsenate (141 and see references therein).

Compounds with thio acids have quite different structures, and two xanthates and two diethyl(dithiocarbamates) contain distorted octahedra. In As(S$_2$COC$_2$H$_5$)$_3$ (42) the distances are As–S, 2.28 Å, and As---S, 2.94 Å, while the antimony compound (82) has Sb–S, 2.52 Å, and Sb---S, 3.00 Å. As [S$_2$CN(C$_2$H$_5$)$_2$]$_3$ has As–S, 2.35 Å, and As---S, 2.85 Å (47) (Fig. 27); phenylarsine bis(diethyl dithiocarbamate) (15) is

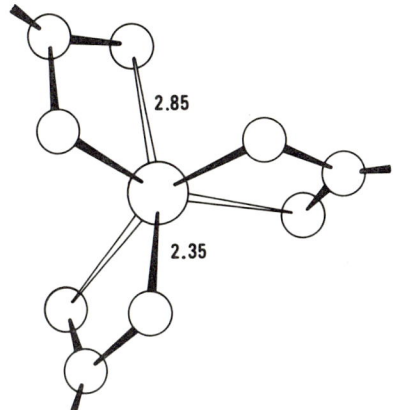

FIG. 27. The central atoms in As[S₂CN(C₂H₅)₂]₃. Large circle, As; small circles, S (next to As) and C. Redrawn from Colapietro et al. (47).

very similar with two S–As---S groupings and distances of 2.33 and 2.87 Å. Antimony triiodide forms an adduct with 1,4-dithiane (19), containing Sb---S links, 3.30 Å; the Sb-I bonds opposite these interactions are 0.02 Å longer than the third Sb–I bond.

B. WITH HALOGENS

The structure of solid arsenic trifluoride has not been determined, but in antimony trifluoride (55) there are three short, pyramidal Sb–F bonds of 1.92 Å and three longer bonds of 2.61 Å, completing an octahedron (Fig. 28). In the adduct, AsF₃·SbF₅ (63), the arsenic is also in a distorted octahedron, but the fluorine atoms show several degrees of interaction (Fig. 29). Two near, 1.64 Å, with one at 2.01 Å forming a bridge to a SbF₅ group, make up a distinct adduct molecule, but there is

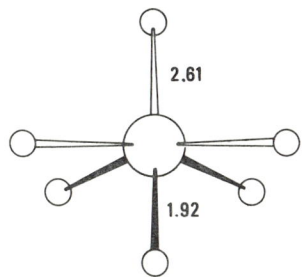

FIG. 28. SbF₃. Large circle, Sb; small circles, F. Redrawn from Edwards (55).

a second bridge through a fluorine at 2.39 Å and two longer contacts of 2.73 Å; the van der Waals distance is 3.32 Å. Infrared evidence shows that the anion SbF_4^- exists both as monomeric and polymeric species, with large cation size favoring the former (*1a*); in $NaSbF_4$ (*38*), one fluorine of each unit forms an asymmetric bridge, with Sb–F of 2.19 and 2.51 Å, and there is a weaker Sb---F interaction of 2.84 Å as well, giving overall six-coordination. Unexpectedly, in view of the infrared work, $KSbF_4$ (*37*) has been reported to have more symmetric bridges with Sb–F of 2.18 and 2.29 Å; it has secondary interactions of 2.90 and 2.98 Å, but there are no linear groups.

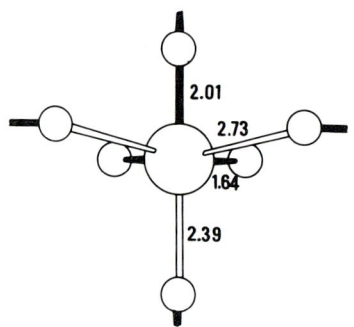

FIG. 29. $AsF_3 \cdot SbF_5$. Large circle, As; small circles, F. Redrawn from Edwards and Sills (*63*).

Although SbF_3 and SbI_3 (below) show secondary links, it seems that neither $AsBr_3$ (*25*), $SbCl_3$ (*120*), nor $SbBr_3$ (*25, 50*) do so. In $SbCl_3$, the bond distance is 2.35 Å, whereas the next nearest neighbors are at 3.5 Å. However, in $Cs_3As_2Cl_9$ (*96*), the arsenic is reported to be in a distorted octahedron with As–Cl, 2.25 Å, and As---Cl, 2.76 Å; there is a series of isomorphous compounds containing either As or Sb with Cl, Br, or I. This was an early determination in two dimensions only, but the results are not dissimilar to more recent work on pyridinium tetrachloroantimonite (*130*), where the $SbCl_4^-$ ions are linked by asymmetric bridges with Sb–Cl, 2.38 Å, and Sb---Cl, 3.12 Å.

The triiodides of arsenic, antimony, and bismuth form an interesting series (*146, 148*). All are isomorphous, with octahedra of iodine atoms containing the central atoms. Arsenic and antimony are significantly displaced from the centers of the octahedra, to give three near and three far X–I distances, but for Bi no irregularity can be detected. The bond distances are as tabulated below.

	Short (Å)	Long (Å)
AsI$_3$	2.556	3.56
SbI$_3$	2.868	3.32
BiI$_3$	3.1	3.1

VI. Group IV*

There are very few relevant interactions involving group IV elements because (a) the earlier members almost always have regular tetrahedral coordination, and (b) for the later members, the description *nonmetal* is doubtfully appropriate. One example is GeF$_2$ (*147*) (Fig. 30) which is

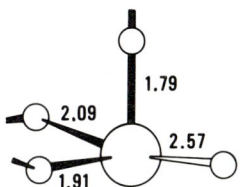

Fig. 30. GeF$_2$. Large circle, Ge; small circles, F. Redrawn from Trotter et al. (*147*).

not dissimilar to SbF$_3$ and SeF$_3^+$. It is polymerized into a chain with almost symmetrical bridges (Ge–F, 1.91 and 2.09 Å) and each germanium is coordinated to a pyramid of fluorine atoms, but in a fourth position opposite the longer Ge–F bridge bond, there is another fluorine at 2.57 Å. A number of Sn(IV) compounds are polymerized, and the bridges range from symmetrical, as in dimethyltin fluorine (*138*) with a regular octahedron, to highly unsymmetrical, as in dimethyltin chloride (*52*) with Sn–Cl bonds of 2.40 and 3.54 Å. Here, the tin coordination is still basically tetrahedral, but the CH$_3$–Sn–CH$_3$ angle has widened to 123° toward the octahedral angle of 180° (Fig. 31). (CH$_3$)$_2$Sn(NCS)$_2$ (*43*, *71*) is similar to the chloride, having two Sn---S interactions, and the distortion toward octahedral is greater, with CH$_3$–Sn–CH$_3$ being 148°. (CH$_3$)$_3$Sn(NCS) (*43*, *85*) is reported to show a similar Sn---S interaction leading to five-coordinate tin, but details have not yet been published. In tribenzyltin acetate (*8*) (Fig. 32) the bridge is moderately asymmetrical, with Sn–O equal to 2.14 and 2.65 Å. The coordination is almost

---

* Summarized in Table IV.

TABLE IV

Secondary Bonds to Group V and IV Elements[a,b]

Compound	Ref.	Fig.	Linear group Y-A---X	Angle at A	Length Y-A	Length A---X	Single A-X	van der Waals A---X	Difference secondary minus single	Accuracy and notes
SbCl$_3$·C$_6$H$_5$NH$_2$	101	25	Cl Sb N	166	2.52	2.53	2.06	3.60	0.47	c
(CH$_3$)$_2$AsSAsS(CH$_3$)$_2$	39	26	S As As	162	2.28	3.24	2.36	3.70	0.88	d
As$_2$O$_3$, cubic	9	—	O As O	—	1.8	3.0	1.84	3.37	1.2	Three times
Sb$_2$O$_3$, cubic	9	—	O Sb O	—	2.0	2.9	2.02	3.57	0.9	Three times
Bi$_4$(SiO$_4$)$_3$	139	—	O Bi O	156	2.15	2.62	2.21	3.67	0.47	Three times
As(S$_2$COC$_2$H$_5$)$_3$	42	—	S As S	160	2.28	2.94	2.22	3.65	0.72	Three times[a]
Sb(S$_2$COC$_2$H$_5$)$_3$	82	—	S Sb S	152	2.52	3.00	2.40	3.85	0.60	Three times[a]
As[S$_2$CN(C$_2$H$_5$)$_2$]$_3$	47	27	S As S	(177)	2.34	2.90	2.22	3.65	0.68	d
				—	2.35	2.81	—	—	0.59	—
				—	2.36	2.82	—	—	0.60	—
C$_6$H$_5$As[S$_2$CN(C$_2$H$_5$)$_2$]$_2$	15	—	S As S	152	2.32	2.91	2.22	3.65	0.69	d
				153	2.33	2.84	—	—	0.62	—
SbI$_3$·S$_2$C$_4$H$_4$	19	—	I Sb S	169	2.77	3.27	2.40	3.85	0.87	d
				171	2.77	3.34	—	—	0.94	—

Compound			Y	A	X							Notes
SbF$_3$	*55*	28	F	Sb	F	162	1.90	2.63	2.00	3.52	0.63	*c*
AsF$_3$·SbF$_5$	*63*	29	F	As	F	156	1.94	2.60	—	—	0.60	Twice
			F	Sb	F	164	1.64	2.73	1.82	3.32	1.09	Twice*c*
NaSbF$_4$	*38*	—	F	Sb	F	163	2.01	2.39	—	—	0.57	—
						152	1.93	2.51	2.00	3.52	0.58	—
Cs$_3$As$_2$Cl$_9$	*96*	—	Cl	As	Cl	170	2.03	2.84	—	—	0.84	—
						172	2.25	2.75	2.17	3.60	0.57	Three times
(C$_5$H$_6$N)$^+$SbCl$_4^-$	*130*	—	Cl	Sb	Cl	(180)	2.38	3.12	2.35	3.80	0.72	Published coordinates inconsistent[a]
AsI$_3$	*146*	—	I	As	I	161	2.56	3.50	2.46	3.83	1.04	Three times
SbI$_3$	*148*	—	I	Sb	I	171	2.87	3.32	2.64	4.03	0.68	Three times
BiI$_3$	*148*	—	I	Bi	I	(180)	(3.1)	(3.1)	2.73	4.13	(0.37)	Three times
GeF$_2$	*147*	30	F	Ge	F	162	2.09	2.57	1.86	3.42	0.71	*c*
(CH$_3$)$_2$SnCl$_2$	*52*	31	Cl	Sn	Cl	164	2.40	3.54	2.39	3.85	1.15	*c*
(CH$_3$)$_2$Sn(NCS)$_2$	*71*	—	N	Sn	S	163	2.14	3.20	2.44	3.90	0.74	*c*
(C$_7$H$_7$)$_3$SnO$_2$CCH$_3$	*8*	32	O	Sn	O	169	2.14	2.65	2.06	3.62	0.59	*c*

[a] See footnote *a*, Table II.
[b] Parentheses around a figure indicate either that it cannot be computed from published information or that it appears particularly unreliable.
[c] Standard deviation of A---X is 0.04 Å or better, corresponding approximately to a standard deviation for the angle Y-A---X of 2° or better (if X and Y are the same).
[d] Standard deviation of A---X is 0.01 Å or better, corresponding approximately to a standard deviation of 0.5° for the angle Y-A---X (if X and Y are the same).

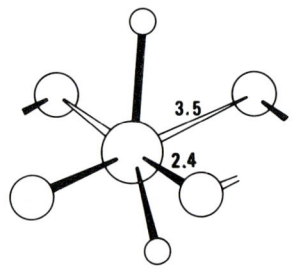

FIG. 31. (CH₃)₂SnCl₂. Large circle, Sn; medium circles, Cl; small circles, C. Redrawn from Davies *et al.* (*52*).

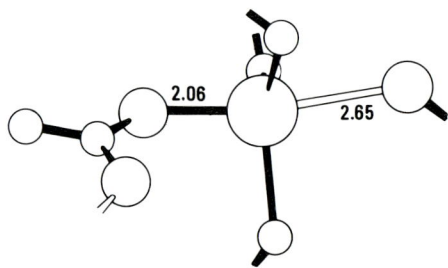

FIG. 32. The central atoms in (C₆H₅CH₂)₃SnO₂CCH₃. Large circle, Sn; medium circles, O; small circles, C (*8*).

trigonal-bipyramidal, but the average C–Sn–O (short) is 95° rather than 90° for a trigonal bipyramid or 109° for a tetrahedron. In other compounds, such as Na₂SnF₆ (*123*), which shows Sn–F bonds of various lengths, it is hardly valid to treat the tin as a nonmetal.

VII. Cyanides*

Virtually all nonmetal cyanides that have been investigated show interactions between the nonmetal and the nitrogen of the cyanide, often giving a long linear group

$$\cdots X\text{–}C\text{–}N\cdots X\text{–}C\text{–}N\cdots X\text{–}$$

If the group is bent, it is bent at the nitrogen, showing that it is X that is the "principal atom." Despite this, these compounds have been

* Summarized in Table V.

collected here because of their general similarity. They have been reviewed recently (27) with special reference to their intermolecular interactions.

Three halogen cyanides (79, 91, 110) have been examined, and for all, a linear infinite chain is required by the space group symmetry, but only for ClCN have the atomic positions been determined accurately. This has an interaction distance Cl---N of 3.01 Å. The approximate bond lengths suggest that the degree of interaction increases for I > Br > Cl [Table 3 in Hassel and Römming (86)]. Cyanuric chloride also shows a Cl---N interaction, somewhat longer than that in cyanogen chloride (100). Inclusion of an organic moiety between the halogen and the cyanide does not disturb the chain structure, as shown by IC≡C–CN

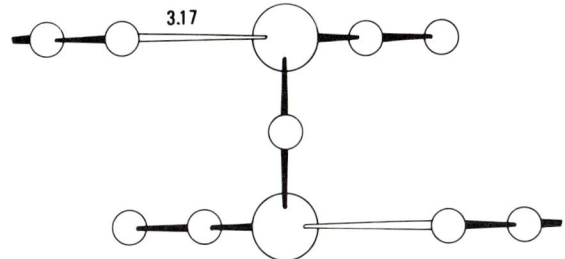

FIG. 33. CH$_2$(SCN)$_2$. Large circle, S; small circles, C and N. Redrawn from Konnert and Britton (114).

(22) and a series of halogenophenyl cyanides and isocyanides (27, pp. 147, 134). These and several related compounds have been examined by Britton and co-workers, but the only complete publication is for p-iodobenzonitrile (134), in which the I---N distance is 3.18 Å, substantially longer than the probable 2.8 Å for ICN.

With sulfur and selenium, some dozen compounds have been examined (114) of which ten show interaction between S or Se and CN and are summarized in Table V; in an eleventh, tetracyano-1,4-dithiin (53), there is a preferential interaction between the CN groups and the double bonds. The sulfur or selenium atoms generally show distorted square-planar coordination, with two near neighbors and two far S---N interactions. With CH$_2$(SCN)$_2$ (114) (Fig. 33) there is only one S---N interaction, whereas with (CH$_3$)$_2$SC(CN)$_2$ (45) the sulfur forms three primary bonds and so has the ability to form three secondary bonds completing a distorted octahedron. The shortest interactions are 2.95 Å for S---N in S(CN)$_2$ (65) and 2.98 Å for Se---N in Se(SCN)$_2$ (127). Most of the

TABLE V

SECONDARY BONDS FORMED BY CYANIDES[a,b]

Compound	Ref.	Fig.	Linear group Y–A...N			Angle CN...A	Angle YA...N	Length Y–A	Length A...N	Standard single A–N	van der Waals A...N	Difference A...N minus single	Accuracy and notes
			Y	A	N								
ClCN	91	—	C	Cl	N	180	180	1.57	3.01	1.69	3.30	1.32	c
BrCN	79	—	C	Br	N	180	180	(1.79)	(2.86)	1.82	3.40	(1.04)	—
ICN	110	—	C	I	N	180	180	(2.03)	(2.79)	1.98	3.53	(0.81)	—
IC≡CN	22, 27	—	C	I	N	(180)	(180)	(2.03)	(2.81)	1.98	3.53	(0.83)	—
p-IC6H4CN	134	—	C	I	N	180	180	2.06	3.18	1.98	3.53	1.20	a
CH2(SCN)2	114	33	C	S	N	132	176	1.68	3.17	1.74	3.35	1.43	a
(CH3)2SC(CN)2	45	—	C	S	N	140	171	1.73	3.29	1.74	3.35	1.55	c
						114	162	1.80	3.35			1.61	
						119	174	1.82	3.45			1.71	
S(CN)2	65	—	C	S	N	138	176	1.72	2.95	1.74	3.35	1.21	c
						148	170	1.74	2.98			1.24	
Se(CN)2	65, 90	—	C	Se	N	(145)	(180)	(1.8)	(2.4)	1.84	3.45	(0.6)	Very poor. Space group doubtful
S(SCN)2	69	—	S	S	N	139	166	2.12	3.12	1.74	3.35	1.38	S1 twice
			S	S	N	119	162	2.12	3.25			1.51	S2
			C	S	N	115	169	1.69	3.12			1.38	
Se(SCN)2	127	—	S	Se	N	133	168	2.21	2.98	1.84	3.45	1.14	Se1 twice; Atom 1 is central, atom 2 attached to cyanide
Se(SeCN)2	4, 124	34	Se	S	N	122	167	2.21	3.32	1.74	3.35	1.58	Se1 twice
			C	S	N	115	172	1.69	3.03			1.29	S2
			Se	Se	N	129	168	2.30	3.16	1.84	3.45	1.32	Se1 twice[c]
			Se	Se	N	126	164	2.30	3.27			1.43	Se2
			C	Se	N	122	172	1.92	3.07			1.23	

Compound	Ref		Y	A	X							Notes
p-C₆H₄(SeCN)₂	124	—	C	Se	N	162	175	1.84	3.06	1.84	3.45	1.22	C of CN^c
						113	168	1.91	3.32	—	—	1.48	C of C₆H₄
(CH₂SCN)₂	26	—	C	S	N	140	169	1.63	3.28	1.74	3.35	1.54	C of CN^c
						119	177	1.80	3.39	—	—	1.65	C of CH₂
C₄(CN)₄S	133	—	C	S	N	127	174	1.69	3.22	1.74	3.35	1.48	^c
						124	171	1.72	3.26	—	—	1.52	
P(CN)₃	67	—	C	P	N	148	165	1.77	2.99	1.80	3.35	1.19	^c
						154	164	1.79	2.97	—	—	1.17	
						155	166	1.80	2.85	—	—	1.05	
As(CN)₃	66, 135	35	C	As	N	166	162	1.97	2.74	1.88	3.40	0.86	
CH₃As(CN)₂	135	—	C	As	N	149	165	1.99	2.93	1.88	3.40	1.05	C of CN^c
						150	160	1.97	3.32	—	—	1.44	
						106	165	2.00	3.32	—	—	1.44	C of CH₃
(CH₃)₂AsCN	40	—	C	As	N	168	168	2.00	3.18	1.88	3.40	1.30	C of CN^c
SiH₃CN	7	—	C	Si	N	(180)	(180)	—	—	—	—	—	Distances not known accurately
(CH₃)₃GeCN	137	—	C	Ge	N	180	180	1.98	3.57	1.92	3.50	1.65	—
(CH₃)₂Ge(CN)₂	27	—	C	Ge	N	(180)	(180)	—	3.20	1.92	3.52	1.28	—
(CH₃)₂Si(CN)₂	27	—	C	Si	N	(180)	(180)	—	3.53	1.67	3.65	1.86	—
(CH₃)₃SnCN	136	—	X	Sn	X	180	180	2.49	2.49	2.10	3.65	0.39	X = either C or N^e

^a See footnote a, Table II.

^b Parentheses around a figure indicate either that it cannot be computed from published information or that it appears particularly unreliable.

^c Standard deviation of A---X is 0.04 Å or better, corresponding approximately to a standard deviation for the angle Y–A---X of 2° or better (if X and Y are the same).

^d Standard deviation of A---X is 0.01 Å or better, corresponding approximately to a standard deviation of 0.5° for the angle Y–A---X (if X and Y are the same).

others are weak with an S---N of about 3.2 Å, compared to the van der Waals distance of 3.35 Å, but the directional properties suggest that these are real secondary bonds. In the isomorphous series $S(SCN)_2$ (69), $Se(SCN)_2$ (127), and $Se(SeCN)_2$ (4, 124), all the S and Se atoms achieve distorted square coordination, and Fig. 34 shows the molecular packing

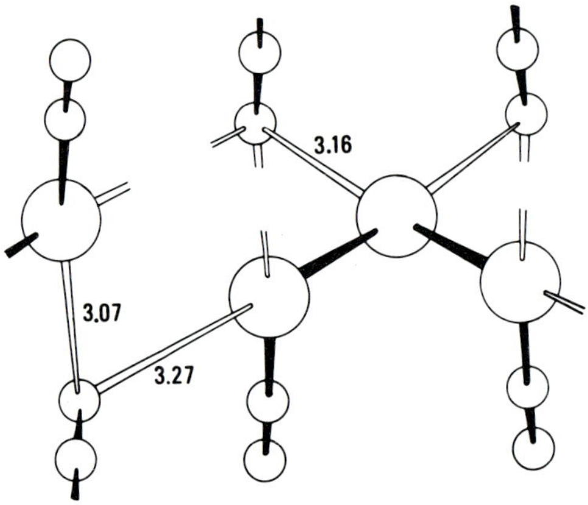

FIG. 34. $Se(SeCN)_2$. Large circles, Se; small circles, C and N. Redrawn from McDonald and Pettit (124).

in $Se(SeCN)_2$. This group of compounds is sufficiently extensive for two effects on bond lengths to appear. The shortest interactions are of the type NC–X---NC forming a linear chain, with cyanide opposite the secondary bond. If the chain is bent at the nitrogen, i.e.,

$$\text{NC}-\text{X}---\text{N} \diagup \text{C}$$

the interaction is longer, and it is also longer in compounds of type R–X---NC with alkyl or aryl groups opposite the secondary bond.

Several group V cyanides have been examined. $P(CN)_3$ (67) has three interactions completing an octahedron with P---N equal to 2.93 Å. By contrast, in $As(CN)_3$ (66, 135) (Fig. 35), there is only one interaction, but it is much stronger, 2.74 Å. On substitution with first one and then a second methyl group (40, 135), the interaction weakens to 2.94 and 3.18 Å. In $CH_3As(CN)_2$ (135) there are also weaker H_3C–As---N and NC–As---N interactions, both with As---N of 3.32 Å.

With group IV, an incomplete study (7) on silyl cyanide, which has an unexpectedly high melting point, has shown a linear chain, ---Si–C–N---Si, but the bond lengths were not determined accurately.

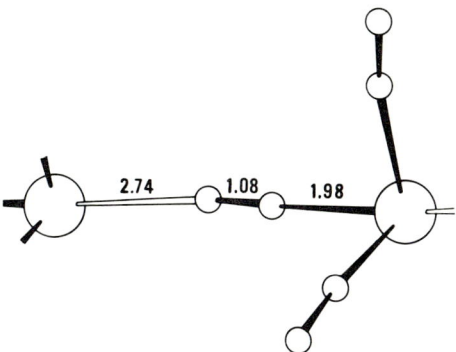

FIG. 35. As(CN)₃. Large circles, As; small circles, C and N. Redrawn from Emerson and Britton (*66*).

Trimethylgermyl cyanide (*137*) again has an abnormal melting point and the crystal structure determination shows a linear chain. The interaction is weak (3.57 Å), but probably real, as the C–Ge–C angles are slightly wider than tetrahedral (average 115°). Further evidence that this is real comes from the structure of $(CH_3)_2Ge(CN)_2$, (*27*, p. 115 f.), where the interaction is stronger, 3.20 Å; the change parallels that found for As(CN)₃ and CH₃As(CN)₂. In $(CH_3)_2Si(CN)_2$ (*27*, p. 155 f.), the overall structure is retained, but the interaction is very weak, with Si---N of 3.53 Å. Trimethyltin cyanide (*136*) has approximately the same structure as the germanium compound, but with one great difference. The CN group is now detached and is equidistant between two tin atoms, which are coordinated by three methyl groups in a plane; the cyanide group is disordered, but a linear group Sn---C–N---Sn is retained. It appears that trimethyllead cyanide also has this structure (*44*).

VIII. Conclusion

A. ANGLES

The crystallographic results given above have been interpreted in terms of linear interactions, but some workers (*140*, *154*) have suggested that the relevant features are right-angled interactions,

$$\begin{array}{c} A\text{---}X \\ | \\ Y \end{array}$$

The strongest evidence for the linear group comes from those compounds where there is only one primary bond from the central atom, when the secondary bond is always found to be in line with, rather than at right angles to it.

Tables II–V include as many angles as are known, and taking their accuracy into consideration, it can be seen that deviations of up to 15° from the straight line are common, but that few exceed this. When they do, of course, it becomes difficult to recognize a particular linear Y–A---X group, and so the preceding statement is equivalent to saying that only in very few compounds with secondary bonds are linear groups absent or doubtful.

B. Geometry

Whatever the nature of the bonding in the linear interactions, the resulting geometries can be analyzed, and with a set of simple rules, can be understood. Considering both primary and secondary bonds (and sometimes vacant positions), the central atom may have one of four coordinations: the centrosymmetric line, square, and octahedron, or the pseudo-centrosymmetric trigonal-bipyramid (with an axial secondary bond). Using this classification, the observed geometries are given in Table VI.

The rules that explain these structures are: (1) The geometry of the primary bonds of the central atom is governed in the usual way by the number of lone and bonded electron pairs; (2) secondary bonds can form in any direction in line with primary bonds, but (3) not in the same direction as a lone pair on the central atom.

As riders to (2) can be added (a) that secondary bonds are rare with the elements P, S, or Cl, and (b) that the number of secondary bonds may be restricted by packing considerations.

There are a very few exceptions to (1), involving group IV elements, in which the overall geometry has changed, partly or completely, from that predicted for the primary bonds (tetrahedral) toward either trigonal-bipyramidal or octahedral. These can be seen as special cases, because they are the only examples where the primary coordination contains no vacant positions. Thus, the introduction of bonds to extra atoms can be expected to distort the original tetrahedron. As an example, in $(CH_3)_2SnCl_2$ where the secondary bonds to chlorine are rather long, the angles CH_3–Sn–CH_3 and Cl–Sn–Cl are 123.5° and 93°, rather than 109° for the original tetrahedron or 180° for a perfect octahedron. A similar situation might be found with an AB_5 molecule with five bonding pairs

TABLE VI

GEOMETRY OF SECONDARY INTERACTIONS[a]

Line

Primary: 1   Electron pairs: 4   Secondary: 1

XeF^+ (II, 2); $Xe_2F_3^+$ (II, 3); $POBr_3$ (II); I_8^{2-} (II, 8); ICl (II, 9); n-Chlorosuccinimide (II); $ClCN$ (V)

Square

Primary: 2   Electron pairs: 4   Secondary: 1   Vacant: 1

$CH_2(SCN)_2$ (V, 33)

Primary: 2   Electron pairs: 4   Secondary: 2

ClF_2^+ (II, 5); BrF_2^+ (II, 5); ICl_2^+ (II, 10); $(C_6H_5)_2ICl_2$ (II, 12); $Te[S_2P(OCH_3)_2]_2$ (III, 15); $[Se(SeCN)_3]_2^{2-}$ (III, 16); $Se(SeCN)_2$ (V, 34)

Primary: 3   Electron pairs: 5   Secondary: 1

$C_6H_5ICl_2$ (II, 11); $C_6H_5Te(SC(NH_2)_2Cl$ (III, 23)

Trigonal bipyramid

Primary: 4   Electron pairs: 4   Secondary: 1

$(C_6H_5CH_2)_3Sn(OCOCH_3)$ (IV, 32); $(CH_3)_3GeCN$ (V)

Octahedron

Primary: 3   Electron pairs: 4   Secondary: 1   Vacant: 2

$SeOCl_2 \cdot SbCl_5$ (III); $(CH_3)_3Se^+$ (III); $C_6H_5SeO_2H$ (III); $SbCl_3 \cdot C_6H_5NH_2$ (IV, 25); $(CH_3)_2AsSAsS(CH_3)_2$ (IV, 26); GeF_2 (IV, 30); $As(CN)_3$ (V, 35)

Primary: 3   Electron pairs: 4   Secondary: 2   Vacant: 1

XeO_3F^- (II, 4); I_2O_5 (II); $Cl_2(SeOCl_2)_{10}^{2-}$ (III, 21)

Primary: 3   Electron pairs: 4   Secondary: 3

IO_3^- (II, 6); I_2O_5 (in HI_3O_8; II, 7); H_2SeO_3 (III); $TeCl_4$ (III, 17); SeF_3^+ (III, 18); $SeCl_3^+$ (III); $SeOF_2$ (III); $(SeOCl_3^-)_n$ (III, 19); $Cl_2(SeOCl_2)_{10}^{2-}$ (III, 21); $(CH_3)_3Te^+$ (III, 22); $Bi_4(SiO_4)_3$ (IV); $As[S_2CN(C_2H_5)_2]_3$ (IV, 27); SbF_3 (IV, 28); $AsF_3 \cdot SbF_5$ (IV, 29); AsI_3 (IV); $P(CN)_3$ (V)

Primary: 4   Electron pairs: 5   Secondary: 1   Vacant: 1

$Te(O_2C_6H_4)_2$ (III); TeF_4 (III); $SeOCl_3^-$ (III, 20).

Primary: 4   Electron pairs: 5   Secondary: 2

$IO_2F_2^-$ (II); TeO_2 (III, 13); $(CH_3)_2TeCl_2$ (III, 24)

Primary: 5   Electron pairs: 6   Secondary: 1

$CH_3TeI_4^-$ (III, 22); $SeOCl_2 \cdot 2C_5H_5N$ (III)

Distorted octahedron

Primary: 4   Electron pairs: 4   Secondary: 2

$(CH_3)_2SnCl_2$ (IV, 31); $(CH_3)_2Sn(NCS)_2$ (IV)

[a] Typical examples only are included. After each is given the table number and, if it is illustrated, the figure number.

forming one secondary bond and becoming octahedral, but no example has been reported.

The effect of (3) is demonstrated by considering the geometry defined by six electron pairs (5 bonding and 1 lone pair) giving a square-pyramidal molecule. Three species containing this arrangement, XeF_6, XeF_5^+, and IF_5, do not have a secondary bond in line with the apical atom. With AB_4 molecules having six electron pairs no secondary bonding is possible because the geometry is a square, e.g., XeF_4. One can predict that an AB_3 species with six electron pairs would not form secondary bonds, but the only likely species of this type, XeF_3^-, has not been prepared. The underlying reason is that the octahedron defined by the original six electron pairs is centrosymmetrical.

There are two exceptions to (3) and the preceding discussion; they are CH_3TeI_4 and $SeOCl_2 \cdot 2C_5H_5N$. Both form five primary bonds with square-pyramidal geometry, but *do* have a secondary interaction occupying the sixth position of an octahedron. It is possible that their bonding should be attributed to p electrons only, in which case there will not be six sterically active electron pairs in the original molecules. A parallel can be drawn with $TeCl_6^{2-}$ *(68)* and $TeBr_6^{2-}$ *(14)*, having regular octahedral geometry but formally seven electron pairs.

C. Distances

The problems of definition of van der Waals radii, etc., discussed in the Introduction make it difficult to use secondary bond distances as a measure of interaction strength, but there are two possible approaches to this.

At least one worker *(62)* has used the *ratio* (secondary bond length)/(primary bond length) as a measure of the strength of the secondary bond, but this must almost certainly be invalid. For a very weak interaction, almost at the van der Waals distance (taken to be 1.50 Å longer than a single bond) the ratio will be (primary bond length + 1.50)/(primary bond length), and this can clearly vary, for example, from 1.3 for Te–Br to 2.1 for Cl–F interactions. For a stronger interaction, the ratio will be just as dependent on the participating elements.

The alternative approach uses the *difference* (secondary bond length minus single bond length). This is supported by the calculations (discussed below) of Wiebenga and Kracht *(154)* of bond orders in polyhalogens. Their plot of this difference ($R - R_{cov}$) against bond order is shown in Fig. 36; their bond orders are in the range 0.4 to 1.0 corresponding to length differences 0.7 to 0.0 Å. Part of the shortening from the van der Waals distance is assumed to be due to electrostatic attraction

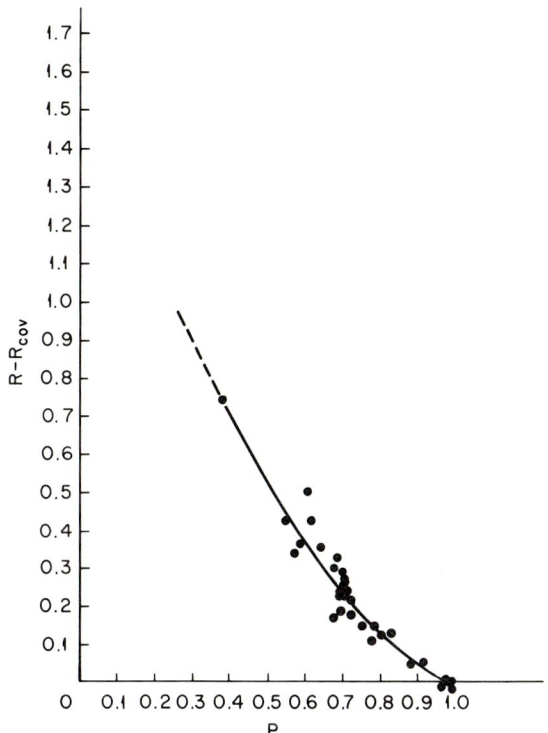

FIG. 36. The relationship between bond order and bond length minus standard single-bond length for polyhalogens. From Wiebenga and Kracht (*154*).

and a calculated correction is made for this (mostly small, but as much as 0.3 Å in one case). This correction presents a problem for the estimation of secondary bond orders from the bond length differences, because it uses both the charges on the individual atoms and the force constant of the bond being corrected, information which is not normally available. In the future with this correction, reliable estimates of bond order can probably be obtained, but at present they can only be semiquantitative.

From the data two trends can be established for the strength of secondary bonds. In the system Y–A---X, Y and X are often identical and there is not much evidence of the result of changing Y alone. However, if A is changed, the effect is clear. When A becomes more electropositive, the interaction becomes stronger. Good examples of this are F–Br---F and F–Cl---F in BrF_2^+ and ClF_2^+, with secondary distances of 2.29 and 2.38 Å, respectively, or I–X---I in AsI_3, SbI_3, and BiI_3 with distances of 3.50, 3.32, and 3.1 Å, and there are many others. Similarly, if the atom A is positively charged, it forms stronger interactions than the same atom when neutral.

One might expect that making X (or X and Y together) more electronegative would also strengthen the bond, but this is not necessarily so. There are few examples where compounds with similar geometries can be compared, but one case is O–As---O in As_2O_3, 3.0 Å, S–As---S in $As[S_2CN(C_2H_5)_2]_3$, 2.84 Å (mean), and corresponding bond length differences of 1.16 and 0.82 Å. Although the first determination is inaccurate, the large decrease make it almost certain that here As---S is stronger than As---O. Similarly, in $C_6H_5Te(tu)Cl$ and $C_6H_5Te(tu)Br$, the bond length differences are 0.40 and 0.34 Å, and it is again likely that Te---Cl is weaker than Te---Br. Further evidence can be obtained by considering the effect of making all of A, X, and Y more electropositive without changing their relative electronegativity. This clearly strengthens the interaction in some cases. Cl–I---Cl in ICl_2^+ can be compared with F–Cl---F and F–Br---F in ClF_2^+ and BrF_2^+, with bond length differences of 0.63 (mean), 0.80 (mean), and 0.60 Å, where I---Cl is stronger than Cl---F and no weaker than Br---F. Taking the first trend into consideration, it appears from this that the interaction tends to be strengthened if X becomes *less*, not more electronegative.

D. BONDING

With as much empirical information available as possible, the nature of the secondary bonding itself can be considered. One simple explanation is that nothing more than electrostatic attraction is responsible, with the lone pair(s) of the central atom taking up some space. This is not unattractive for octahedra with three primary and three secondary bonds, but it does not explain the pseudo-octahedral geometry when one or two secondary bonds are not formed. Also, with two primary bonds there is no obvious reason why only two secondary attractions should be strong. This explanation, then, is plausible for SeF_3^+ forming an octahedron, but not useful for BrF_2^+ forming a square, two compounds whose bonding should be very similar.

What follows is put forward as a theory for secondary bonds that may be more satisfactory than the electrostatic view. In general terms, a dative interaction is suggested between a lone pair on the "outer" atom (X) and an empty orbital on the "cental" atom (A). Further, reasoning from the linearity of the bond, it is likely that the empty orbital is the σ^* orbital of the primary bond, Y–A; this can be expected to project furthest in space in the direction away from Y. Considering both primary and secondary bonds together, the three atomic orbitals involved will give three molecular orbitals, occupied by two electron pairs. If the three atoms are arranged symmetrically, there are occupied bonding and nonbonding orbitals and an unoccupied antibonding orbital, and the result is a weak bond between each pair of atoms. With an unsymmetrical arrangement, the bonding orbital is concentrated between the close pair, giving a strong bond, and the nonbonding one is concentrated on the other side and (with a small contribution from the bonding orbital) gives a weak bond. The overall scheme is identical to that put forward for the hydrogen bond, which has the same possibility of asymmetry, O–H---O, or symmetry, (F–H–F)$^-$.

In theory, the formation of a secondary bond should produce a weakening of the primary bond (effectively one bonding electron pair spread over two bonds). But this will only be true if there is no change at all in the orbitals involved; there are generally d orbitals on the central atom of the correct symmetry to mix with the bonding or nonbonding three-center orbitals, and the formation of the secondary bond might lead to greater interaction with these d orbitals. In this case, the primary bond might be unchanged, or even strengthened by forming the secondary bond. The best evidence of either effect is found with iodine and the polyiodides, where the intramolecular I–I is significantly longer than in the gas phase; some similar examples have been noted during the main part of the chapter. There are also cases in which a primary bond that is long because of the nature of the structure, e.g., the bridge I–O in I_2O_5, has the shortest secondary bond opposite it. Hassel and Römming (86a) have shown that for iodine in charge-transfer bonded adducts and trihalides, there is a linear correlation with Y–I lengthening as I---X shortens.

The two trends noted for the strength of the secondary bonds can be understood qualitatively from this model. When the central atom (A) becomes more electropositive, the bonding and antibonding orbitals of the primary bond will both become less compact, leading to better overlap with the distant lone-pair orbital of the outer atom (X); there will also be a greater electrostatic force between the central and outer atoms. Similarly, when the outer atom becomes more electropositive, the lone-pair orbital will become less compact and will overlap better with the σ^* orbital of the central atom. In both cases, the greater diffuseness of the orbitals is the first step toward full delocalization of electrons, the first step toward metallic bonding; one or two compounds with secondary bonds do have a metallic luster.

This model corresponds closely to that suggested for charge-transfer bonded species (10), and it has already been noted that the two types of bond are geometrically similar. The electron pair donation is effectively a charge transfer, and it would be possible to give the whole class of secondary bonds this label. However, as presently recognized, charge-transfer interactions are always thought of as weak. It is therefore better to introduce the new term *secondary bond* which can cover a whole range of interactions including charge-transfer bonds, than to extend that term far beyond its present use.

E. Theoretical Calculations

In the work of Wiebenga and Kracht (*154*) on the bonding of polyhalogens, a Hückel procedure was used, modified to allow for charges on the atoms and based only on p orbitals. The overlap integral was only given a nonzero value for two adjacent p orbitals directed at each other, and its variation with bond distance (R) was taken to be a linear function of $(R - R_{covalent})$, fitted to the known values for I_2 and I_3^-. It was assumed that all angles were 90° or 180°.

All the species examined have bond lengths longer than the standard covalent distances, and the bond orders were calculated corresponding to the observed geometries. These agreed well with the known distances (Fig. 36), which is not surprising in view of the assumed form of the overlap integral. A more independent test was a comparison of the calculated atomic charges with those obtained from nuclear quadrupole resonance and Mössbauer measurements, and the agreement was satisfactory; other comparisons also suggest that this theoretical model accounts rather well for the bonding in these compounds.

This model is in essence the same as that proposed above for secondary bonds, but is more sophisticated because it allows full delocalization rather than delocalization over a three-atom system only. In two species, ICl (solid) and I_8^{2-}, bonds were examined that have been termed secondary in the present study; the bond orders were 0.60 for the I---I–Cl interaction and 0.39 for the I---I–I interaction. This indicates very strongly that the secondary bond model can account satisfactorily for the observed interactions.

These two polyhalogen secondary bonds, although weak, are part of a continuous range of bond strength (Fig. 36), and to distinguish them as secondary might seem unnecessary. In principle, this is true, but in practice in most systems the possible continuous range of bond lengths has not been observed. The bonding is usually very asymmetric, and it is then valuable to distinguish the longer one as *secondary*, while recognizing that the bonding is of the same type in both the primary and secondary bond, apart from a possibly greater electrostatic component in the latter. The alternative in the past has been to ignore or discard the secondary bond, rather than, ideally, to treat it as equivalent to the primary one.

The predominance of asymmetric systems points to a problem that cannot yet be resolved. Above, the primary geometry was explained by electron pair repulsion, and in this model the electrons are in hybrid orbitals which include d orbitals. The asymmetry then appears as a consequence of the lone pairs. From the model of Wiebenga and Kracht, the lone pairs are in s or p orbitals directed at right angles to the molecular plane (if there is one). It is not clear whether asymmetry can arise naturally in this case, as calculations of geometries with minimum energy were not made. With I_3^-, the immediate cause of asymmetry seems to be an asymmetric crystal environment (154), and the same may be true of other systems. This makes it quite likely that d orbitals are not necessary for asymmetry; their contribution may only be in second-order effects, e.g., partial multiple bonding in I–O bonds of iodates. The trans effects proposed by Foss (73) can produce long bonds and asymmetric systems, but only when the atoms X and Y in the Y–A---X group are not the same.

The way to further understanding of secondary bonds may well be through calculations like those of Wiebenga and Kracht on more general systems, which might establish bond orders and polarities. Examination of electronic spectra might also give useful information.

Some much more detailed calculations have been made (95) on the interactions in solid halogens, which demonstrate the importance of charge-transfer interactions in these systems.

F. Significance

The facts collected in this chapter are clearly very much at variance with bonding models that allow only single bonds at standard bond distances and interactions at van der Waals distances. In reality, in some types of compounds, secondary bonds occur in virtually every example examined, and the understanding of these compounds must be incomplete if secondary bonding is ignored. Its more general importance is as part of a continuum of bonding types, of particular importance in bridging the gap between covalent and metallic bonding. By including secondary bonding, we can go one stage further in bringing the whole range of chemical interaction into one conceptual group, where the nature of the bonding is controlled by the electronegativity of the participating atoms. This can be illustrated for bridging interactions –B–A–B– in Fig. 37.

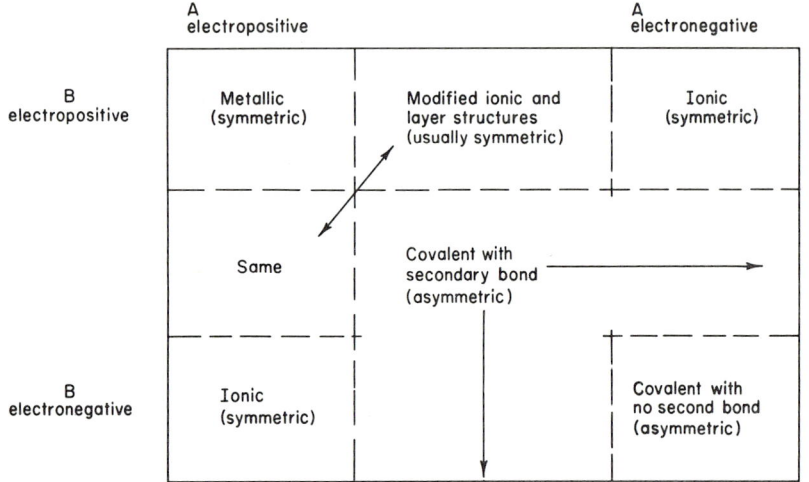

FIG. 37. The variation of bond type with electronegativity in a system B–A–B.

IX. Appendix

The following further structure determinations have been published up to the end of May 1972. Their results have not been incorporated into Tables II–V. The bond lengths are almost always accurate to better than 0.04 Å.

A. GROUP VII

An important discovery is that some 15 steroid molecules whose structures were determined as bromine derivatives have linear interactions C–Br---O and these appear to dominate the molecular packing. The Br---O distances are in the range 2.9–3.2 Å (129a). A chlorine oxygen interaction occurs in $POCl_3$. Its crystals are isomorphous with those of $POBr_3$ and contain a P–Cl---O grouping with P–Cl of 1.97 Å, Cl---O of 3.05 Å, and the angle at Cl of 165° (127a). The structure of $POBr_3$ has been refined in an alternative space group without significant change in the atomic positions (145a). In $BrF_4{}^+Sb_2F_{11}{}^-$ the bromine is surrounded by a distorted octahedron of fluorine atoms with four at 1.81 Å (avg.) and two (cis) at 2.24 and 2.49 Å. The F–Br---F angles are 168° and 175° (118a).

Four further iodate structures have been reported. In $Ca(IO_3)_2 \cdot 6H_2O$ the iodine has the usual distorted octahedral coordination (25a) while in $Sr(IO_3)_2 \cdot H_2O$, there are two linear O–I---O interactions but instead of a third I---O, there are two oxygen atoms at 3.2 Å (*126b*). In $KIO_3 \cdot HIO_3$ there are four independent iodine atoms. All have distorted octahedral environments but one has an extra iodine oxygen contact through the center of the widest face, between the three I---O interactions (*42b*). In $KIO_3 \cdot HIO_3 \cdot KCl$, there are I---Cl as well as I---O interactions. The octahedron around one iodine atom contains two unusually short I---O distances, 2.59 and 2.47 Å and one I---Cl of 3.03 Å. The octahedron of the second iodine has one I---O of 2.61 Å, one I---Cl of 3.07 Å and one vacant position (*25b*).

B. Group VI

The most important information is that a number of selenites and tellurites have distorted octahedra around Se or Te as would be expected by comparison with iodates. In $LiH_3(SeO_3)_2$, the three Se---O interactions range from 2.80 to 3.29 Å while the bonded Se–O vary between 1.65 and 1.77 Å (*126e*). The same coordination is reported for $NaD_3(SeO_3)_2$ (*126d*) and $NaH_3(SeO_3)_2$ (*108a*) while in $KH_3(SeO_3)_2$ two of the Se---O interactions are relatively short (3.00 Å, 3.22 Å) but the third (3.45 Å) is close to the van der Waals distance of 3.42 Å (*85a*). In $BaTeO_3 \cdot H_2O$, the Te–O distance is 1.85 Å, there are two linear O–Te---O interactions of 3.03 and 3.33 Å and one position of the octahedron is vacant (*126f*). The structure of H_2SeO_3 has been refined by neutron diffraction (*114b*) confirming the previous results. It is worth noting that although it has the same space group ($P2_12_12_1$) and similar cell dimensions to HIO_3, the two compounds are not isomorphous, unlike HIO_3 and XeO_3.

The short intramolecular S---S and S---O distances in thiothiophthenes and desaurins are reviewed (*121a*). Both Cl---S–S---Cl and Cl---S–C interactions (with Cl---S of 3.23 Å and 3.19 Å respectively) occur in 4-phenyl-1,2-dithiolium chloride monohydrate (*84a*) and there are similar interactions in the corresponding bromide (*100a*) and iodide (*100b*). The structure of thiuret hydriodide has been refined (*130a*) and the same paper discusses a number of other compounds containing S–S---Hal interactions.

In $TeCl_3^+AlCl_4^-$, the tellurium atom has 3 short Te–Cl bonds, 2.28 Å and 3 long ones, 3.06 Å in a distorted octahedron (*36b*), very similar to that in $TeCl_4$, whose structure determination has been published in detail (*36a*). In the compound $ClC_2H_4TeCl_3$, bridging chlorine atoms link the molecules into infinite chains. The tellurium has octahedral coordination with one vacant position, and in the Cl–Te---Cl groups, Cl–Te is 2.39 Å, Te---Cl is 2.72 Å (*113a*).

In Te(CH$_3$)$_2$I$_2$, each of the three independent tellurium atoms has a distorted octahedral environment with two Te–C of 2.10–2.16 Å, two short Te–I of 2.85–2.99 Å and two C–Te---I interactions with Te---I of 3.65–4.03 Å (*42c*).

C. GROUP V

The structures of two further adducts of SbCl$_3$ have been determined. In SbCl$_3$·2C$_6$H$_5$NH$_2$, the antimony atom is in an octahedron with one position vacant. It forms two long interactions with nitrogen atoms of aniline molecules with Sb---N of 2.64 Å. The bond angle Cl–Sb---N is 166° and this Sb–Cl is 2.50 Å, substantially longer than the apical Sb–Cl, 2.36 Å (*101a*). A complex with phenanthrene, C$_{14}$H$_{10}$·2SbCl$_3$, is similar to the corresponding naphthalene complex (*52b*). In KSb$_2$F$_7$, SbF$_3$ and SbF$_4^-$ units are linked into infinite chains by long interactions. The antimony atom in the SbF$_4^-$ ion remains four-coordinate but the other one has distorted octahedral coordination with three short Sb–F, 1.92–1.96 Å, two long Sb---F, 2.41 Å and 2.57 Å and one vacant position (*126c*). In BiCl$_3$, which might be expected to show directed interactions, the bismuth has three near chlorines (2.46–2.52 Å) forming a pyramidal BiCl$_3$ group and five at 3.22–3.45 Å giving in all a bicapped trigonal prism (*126g*). Although two Cl–Bi---Cl groups are approximately linear (155°) it is clearly not valid to label these as directed to the exclusion of the others. If the bonding of the five chlorines is partly covalent, as a comparison with the van der Waals distance (3.90 Å) would suggest, then it must be of a more complex type than for most of the compounds described here. There is some support for this from a comparison (*126g*) with two molecules probably isomorphous with BiCl$_3$: SbCl$_3$ and β-SbBr$_3$. In these and in α-SbBr$_3$, the same bicapped trigonal prismatic coordination occurs and it is very significant that the Sb---Cl interaction (3.63 Å, avg.) is longer than the Bi---Cl interaction (3.31 Å, avg.).

In α-Bi$_2$O$_3$, one bismuth atom has distorted octahedral coordination with three short Bi–O, 2.14–2.29 Å and three longer, 2.48–2.80 Å; the second bismuth atom has three short, 2.08–2.21 Å, and two longer distances, 2.54, 2.63 Å, and the shortest Bi–O is opposite the vacant position (*126a*). Similar octahedra with one vacant position occur in the oxides Bi$_{12}$GeO$_{20}$ (*1*) and Bi$_2$Ti$_4$O$_{11}$ (*107a*) while a complete octahedron is found in bismuth formate (*142a*) with Bi–O, 2.34 Å and Bi---O, 2.50 Å.

D. GROUP IV

In trimethyl tin isothiocyanate, (CH$_3$)$_3$SnNCS, there is an Sn---S interaction, 3.13 Å, which links the molecules into chains; at the same time the C–Sn–C angle increases to 119° (avg.) and N–Sn–C is 95° (avg.), i.e., the molecule is close to a trigonal bipyramid (72).

E. CYANIDES

Further examples of halogen–cyanide interactions have been found: Cl---N of 3.22 Å in 2,4,6-trichlorobenzonitrile, Br---N of 3.06 Å in the corresponding bromo compound (42a) and Br---N of 3.03 Å in bromotricyanomethane (154a). Interestingly, chlorotricyanomethane forms no interactions (154a). In the two salts $K(SeCN)_3 \cdot \frac{1}{2}H_2O$ (87b) and $Rb(SeCN)_3 \cdot \frac{1}{2}H_2O$ (87a), Se---N interactions occur of length 3.47–3.60 Å, which provide square planar environments around each selenium with either three short and one long bond (central Se) or two short, one long bond and one vacant position (terminal Se). The whole series of group IV cyanides, $(CH_3)_2M(CN)_2$, M = Si–Pb, has been examined. With M = Si and Ge, there are weak interactions, N---Si, 3.48 Å, N---Ge, 3.28 Å, and the molecules are somewhat distorted from tetrahedral, H_3C–Si–CH_3, 120°, H_3C–Ge–CH_3, 121°. There are also very weak interactions, N---Si, 3.97 Å and N---Ge, 3.84 Å, about 0.3 Å *longer* than the expected van der Waals distances but which are linear and which appear to be significant. With M = Sn, both interactions are of equal length, Sn---N, 2.48 Å and the molecules are closer to octahedral, H_3C–Sn–CH_3, 149°. The compound with M = Pb appears to be similar to this, but single crystals could not be obtained (114a).

REFERENCES

1. Abrahams, S. C., Jamieson, P. B., and Bernstein, J. L., *J. Chem. Phys.* **47**, 4034 (1967).
1a. Adams, C. J., and Downs, A. J., *J. Chem. Soc. A* 1534 (1971).
2. Adams, D. M., and Lock, P. J., *J. Chem. Soc. A* 145 (1967).
3. Agron, P. A., Johnson, C. K., and Levy, H. A., *Inorg. Nucl. Chem. Lett.* **1**, 145 (1965).
4. Aksnes, O., and Foss, O., *Acta. Chem. Scand.* **8**, 702, 1787 (1954).
5. Alcock, N. W., *Acta Crystallogr. Sect. B*, in press.
6. Alcock, N. W., Ph.D. Thesis, Cambridge Univ., 1963.
7. Alcock, N. W., and Sheldrick, G. M., unpublished results.
8. Alcock, N. W., and Timms, R. E., *J. Chem. Soc. A* 1873 (1968).
9. Almin, K. E., and Westgren, A., *Ark. Kemi Mineral. Geol.* **15B**, No. 22, 1 (1942).
10. Andrews, L. J., and Keefer, R. M., *Advan. Inorg. Chem. Radiochem.* **3**, 91 (1961).
11. Archer, E. M., *Acta Crystallogr.* **1**, 64 (1948).
12. Archer, E. M., and van Schalkwyk, T. G. D., *Acta Crystallogr.* **6**, 88 (1953).
13. Aynsley, E. E., Dodd, R. E., and Little, R., *Spectrochim. Acta* **18**, 1005 (1962).

14. Bagnall, K. W., d'Eye, R. W. M., and Freeman, J. H., *J. Chem. Soc.* 3959 (1955).
15. Bally, R., *Acta Crystallogr.* **23**, 295 (1967).
16. Bartlett, N., Einstein, F., Stewart, D. F., and Trotter, J., *J. Chem. Soc. A* 1190 (1967).
17. Becker, K. A., Plieth, K., and Stranski, I. N., *Z. Anorg. Allg. Chem.* **266**, 293 (1951).
18. Becker, K. A., Plieth, K., and Stranski, I. N., *Z. Anorg. Allg. Chem.* **269**, 92 (1952).
19. Bjorvatten, T., *Acta Chem. Scand.* **20**, 1863 (1966).
20. Bolhuis, F. V., Koster, P. B., and Migchelsen, T., *Acta Crystallogr.* **23**, 90 (1967).
21. Bondi, A., *J. Phys. Chem.* **68**, 441 (1964).
22. Borgen, B., Hassel, O., and Römming, C., *Acta Chem. Scand.* **16**, 2469 (1962).
23. Boswijk, K. H., van der Heide, J., Vos, A., and Wiebenga, E. H., *Acta Crystallogr.* **9**, 274 (1956).
24. Boswijk, K. H., and Wiebenga, E. H., *Acta. Crystallogr.* **7**, 417 (1954).
25. Braekken, H., *Kgl. Nor. Vidensk. Selsk. Forh.* **8**, No. 10, 1 (1935).
25a. Braibanti, A., Manotti Lanfredi, A. M., Pellinghelli, M. A., and Tiripicchio, A., *Inorg. Chim. Acta* **5**, 590 (1971).
25b. Braibanti, A., Tiripicchio, A., and Manotti Lanfredi, A. M., *Chem. Commun.* 1128 (1967).
26. Bringeland, R., and Foss, O., *Acta Chem. Scand.* **12**, 79 (1958).
27. Britton, D., in "Perspectives in Structural Chemistry" (J. D. Dunitz and J. A. Ibers, eds.), Vol. I, pp. 109–171. Wiley, New York, 1967.
28. Brown, R. N. *Acta Crystallogr.* **14**, 711 (1961).
29. Bryden, J. H., and McCullough, J. D., *Acta Crystallogr.* **7**, 833 (1954).
30. Burbank, R. D., and Bensey, F. N., *J. Chem. Phys.* **21**, 602 (1953).
31. Burbank, R. D., and Bensey, F. N., *J. Chem. Phys.* **27**, 982 (1957).
32. Burbank, R. D., and Jones, G. R., *Science* **168**, 248 (1970).
33. Burbank, R. D., and Jones, G. R., *Science* **171**, 485 (1971).
34. Burns, J. H., Agron, P. A., and Levy, H. A., in "Noble-Gas Compounds" (H. H. Hyman, ed.), p. 211. Univ. of Chicago Press, Chicago, Illinois, 1963; see also pp. 195 f, 203 f.
35. Burns, J. H., Ellison, R. D., and Levy, H. A., in "Noble-Gas Compounds" (H. H. Hyman, ed.), p. 226. Univ. of Chicago Press, Chicago, Illinois, 1963;
36. Buss, B., and Krebs, B., *Angew. Chem.* **82**, 446 (1970); *Angew Chem. Int. Ed. Eng.* **9**, 463 (1970).
36a. Buss, B., and Krebs, B., *Inorg. Chem.* **10**, 2795 (1971).
36b. Buss, B., Krebs, B., and Altena, D., *Z. Anorg. Allg. Chem.* **386**, 257 (1971).
37. Byström, A., Bäcklund, S., and Wilhelmi, K.-A., *Ark. Kemi* **4**, 175 (1952).
38. Byström, A., Bäcklund, S., and Wilhelmi, K.-A., *Ark. Kemi* **6**, 77 (1953).
39. Camerman, N., and Trotter, J., *J. Chem. Soc.* 219 (1964).
40. Camerman, N., and Trotter, J., *Can. J. Chem.* **41**, 460 (1963).
41. Carpenter, G. B., and Richards, S. M. *Acta Crystallogr.* **15**, 360 (1962).
42. Carrai, G., and Gottardi, G., *Z. Kristallogr.* **113**, 373 (1960).
42a. Carter, V. B., and Britton, D., *Acta Crystallogr. Sect. B* **28**, 945 (1972).
42b. Chan, L. Y. Y., and Einstein, F. W. B., *Canad. J. Chem.* **49**, 468 (1971).

42c. Chan, L. Y. Y., and Einstein, F. W. B., *J. Chem. Soc. Dalton Trans.* 316 (1972).
43. Chow, Y. M., *Inorg. Chem.* **9**, 794 (1970).
44. Chow, Y. M., and Britton, D., *Acta Crystallogr. Sect. B* **27**, 856 (1971).
45. Christensen, A. T., and Witmore, W. G., *Acta Crystallogr Sect. B* **25**, 73 (1969).
46. Christofferson, G. D., Sparks, R. A., and McCullough, J. D., *Acta Crystallogr.* **11**, 782 (1958).
47. Colapietro, M., Domenicano, A., Scaramuzza, L., and Vaciago, A., *Chem. Commun.* 302 (1968).
48. Cordes, A. W., *Inorg. Chem.* **6**, 1204 (1967).
49. Cromer, D. T., and Larson, A. C., *Acta Crystallogr.* **9**, 1015 (1956).
50. Cushen, D. W., and Hulme, R., *J. Chem. Soc.* 2218 (1962).
51. Dasent, W. E., and Waddington, T. C., *J. Chem. Soc.* 3350 (1960).
52. Davies, A. G., Milledge, H. J., Puxley, D. C., and Smith, P. J., *J. Chem. Soc. A* 2862 (1970).
52a. de Boer, J. L., Bolhuis, F. V., Olthof-Hazekamp, R., and Vos, A., *Acta Crystallogr.* **21**, 841 (1966).
52b. Delmaldé, A., Mangia, A., Nardelli, M., Pelizzi, G., and Vidoni Tani, M. E., *Acta Crystallogr. Sect. B* **28**, 147 (1972).
53. Dollase, W. A., *J. Amer. Chem. Soc.* **87**, 979 (1965).
54. Donohue, J., *J. Amer. Chem. Soc.* **72**, 2701 (1950).
55. Edwards, A. J., *J. Chem. Soc. A* 2751 (1970).
56. Edwards, A. J., and Haiwaidy, F. I., *J. Chem. Soc. A* 2977 (1968).
57. Edwards, A. J., and Jones, G. R., *J. Chem. Soc. A* 1467 (1969).
58. Edwards, A. J., and Jones, G. R., *J. Chem. Soc. A* 2858 (1969).
59. Edwards, A. J., and Jones, G. R., *J. Chem. Soc. A* 1491 (1970).
60. Edwards, A. J., and Jones, G. R., *J. Chem. Soc. A* 1891 (1970).
61. Edwards, A. J., and Monty, M. A., *J. Chem. Soc. A* 703 (1969).
62. Edwards, A. J., and Sills, R. J. C., *J. Chem. Soc. A* 2697 (1970).
63. Edwards, A. J., and Sills, R. J. C., *J. Chem. Soc. A* 942 (1971).
64. Einstein, F., Trotter, J., and Williston, C., *J. Chem. Soc. A* 2018 (1967).
65. Emerson, K., *Acta Crystallogr.* **21**, 970 (1966).
66. Emerson, K., and Britton, D., *Acta Crystallogr.* **16**, 113 (1963).
67. Emerson, K., and Britton, D., *Acta Crystallogr.* **17**, 1134 (1964).
68. Engel, G., *Z. Kristallogr.* **90**, 341 (1935).
69. Fehler, F., and Linke, K. H., *Z. Anorg. Allg. Chem.* **327**, 151 (1964).
70. Feikema, Y. D., and Vos. A., *Acta Crystallogr.* **20**, 769 (1966).
71. Forder, R. A., and Sheldrick, G. M., *J. Organometal. Chem.* **22**, 611 (1970).
72. Forder, R. A., and Sheldrick, G. M., *J. Organometal. Chem.* **21**, 115 (1970).
73. Foss, O., *in* "Selected Topics in Structure Chemistry" (P. Andersen, O. Bastiansen, and S. Furberg, eds.), pp. 145–173. Universitets forlag, Oslo, 1967.
74. Foss, O., and Hauge, S., *Acta Chem. Scand.* **19**, 2395 (1965).
75. Foss, O., and Husebye, S., *Acta Chem. Scand.* **20**, 132 (1966).
76. Foss, O., and Maroy, K., *Acta Chem. Scand.* **20**, 123 (1966).
77. Foss, O., and Tjomsland, O., *Acta Chem. Scand.* **12**, 1799 (1958).
78. Garrett, B. S., U.S. At. Energy Comm. Rep., ORNL-1745 (1954); abstracted in *Struct. Rep.* **18**, 393 (1954).
79. Geller, S., and Schawlow, A. L., *J. Chem. Phys.* **23**, 779 (1955).

80. Ghose, S., *Acta Crystallogr.* **15**, 1105 (1962).
81. Gillespie, R. J., *Angew Chem. Int. Ed. Engl.* **6**, 819 (1967).
82. Gottardi, G., *Z. Krystallogr.* **115**, 451 (1961).
83. Greenwood, N. N., Straughan, B. P., and Wilson, A. E., *J. Chem. Soc. A* 1479 (1966).
84. Groth, P., and Hassel, O., *Acta Chem. Scand.* **16**, 2311 (1962).
84a. Grundtvig, F., and Hordvik, A., *Acta Chem. Scand.* **25**, 1567 (1971).
85. Hall, J. B., *Inorg. Chem.* In press.
85a. Hansen, F., Hazell, R. G., and Rasmussen, S. E., *Acta Chem. Scand.* **23**, 2561 (1969).
86. Hassel, O., and Römming, C., *Quart. Rev., Chem. Soc.* **16**, 1 (1962).
86a. Hassel, O., and Römming, C., *Acta Chem. Scand.* **21**, 2659 (1967).
87. Hauge, S., *Acta Chem. Scand.* **25**, 1135 (1971).
87a. Hauge, S., *Acta Chem. Scand.* **25**, 3103 (1971).
87b. Hauge, S., and Sletten, J., *Acta Chem. Scand.* **25**, 3094 (1971).
88. Havinga, E., Boswijk, K. H., and Wiebenga, E. H., *Acta Crystallogr.* **7**, 487 (1954).
89. Hayward, G. C., and Hendra, P. J., *J. Chem. Soc. A* 643 (1967).
90. Hazell, A. C., *Acta Crystallogr.* **16**, 843 (1963).
91. Heiart, R. B., and Carpenter, G. B., *Acta Crystallogr.* **9**, 889 (1956).
92. Hermodsson, Y., *Acta Crystallogr.* **13**, 656 (1960).
93. Hermodsson, Y., *Acta Chem. Scand.* **21**, 1313 (1967).
94. Hermodsson, Y., *Acta Chem. Scand.* **21**, 1328 (1967).
95. Hillier, I. H., and Rice, S. A., *J. Chem. Phys.* **46**, 3881 (1967).
96. Hoard, J. L., and Goldstein, L., *J. Chem. Phys.* **3**, 117 (1935).
97. Hodgson, D. J., and Ibers, J. A., *Inorg. Chem.* **8**, 326 (1969).
98. Hope, H., *Acta Crystallogr.* **20**, 610 (1966).
99. Hope, H., and McCullough, J. D., *Acta Crystallogr.* **17**, 712 (1964).
100. Hoppe, W., Lenné, H. U., and Morandi, G., *Z. Kristallogr.* **108**, 321 (1957).
100a. Hordvik, A., and Baxter, R. M., *Acta Chem. Scand.* **23**, 1082 (1969).
100b. Hordvik, A., and Sletten, E., *Acta Chem. Scand.* **20**, 1874 (1966).
101. Hulme, R., *J. Chem. Soc. A* 2448 (1968).
101a. Hulme, R., private communication.
102. Hulme, R., and Szymański, J. T., *Acta Crystallogr. Sect. B* **25**, 753 (1969).
103. Husebye, S., *Acta Chem. Scand.* **20**, 24 (1966).
104. Husebye, S., *Acta Chem. Scand.* **21**, 42 (1967).
105. Ibers, J. A., *Acta Crystallogr.* **9**, 225 (1956).
106. Ito, T., and Sawada, H., *Z. Kristallogr.* **102**, 13 (1939).
107. James, W. J., Hach, R. J., French, D., and Rundle, R. E., *Acta Crystallogr.* **8**, 814 (1955).
107a. Jensen, G., *Lab. Insulation Res. Mass. Inst. Tech.*, Tech. Rep. 198 (see ref. *126a* and *Chem. Abs*, **63**, 15650h).
108. Jones, G. R., Burbank, R. D., and Bartlett, N., *Inorg. Chem.* **9**, 2264 (1970).
108a. Kaplan, S. F., Kay, M. I., and Morosin, B., *Ferroelectrics* **1**, 31 (1970).
109. Karle, I. L., *J. Chem. Phys.* **23**, 1739 (1955).
110. Ketelaar, J. A. A., and Zwartsenberg, J. W., *Rec. Trav. Chim. Pays-Bas* **58**, 448 (1939).
111. Keve, E. T., Abrahams, S. C., and Bernstein, J. L., *J. Chem. Phys.* **54**, 2556 (1971).

112. Khotsyanova, T. L., *Kristallografiya* 1, 524 (1956); *Chem. Abstr.* 51, 4793 (1957); 52, 4282 (1958).
113. Klinsberg, E., *Quart. Rev., Chem. Soc.* 23, 537 (1969).
113a. Kobelt, D., and Paulus, E. F., *Angew. Chem. Int. Ed. Engl.* 10, 74 (1971).
114. Konnert, J. H., and Britton, D., *Acta Crystallogr. Sect. B* 27, 781 (1971).
114a. Konnert, J., Britton, D., and Chow, Y. M., *Acta Crystallogr. Sect. B* 28, 180 (1972).
114b. Krebs Larsen, F., Lehmann, M. S., and Søtofte, I., *Acta Chem. Scand.* 25, 1233 (1971).
115. Kruse, F. H., Marsh, R. E., and McCullough, J. D., *Acta Crystallogr.* 10, 201 (1957).
116. Larson, A. C., and Cromer, D. T., *Acta Crystallogr.* 14, 128 (1961).
117. Leciejewicz, J., *Z. Kristallogr.* 116, 345 (1961).
118. Levy, H. A., and Agron, P. A., in "Noble-Gas Compounds" (H. H. Hyman, ed.), p. 221. Univ. of Chicago Press, Chicago, Illinois, 1963.
118a. Lind, M. D., and Christe, K. O., *Inorg. Chem.* 11, 608 (1972).
119. Lindqvist, I., and Nahringbauer, G., *Acta Crystallogr.* 12, 638 (1959).
120. Lindqvist, I., and Niggli, A., *J. Inorg. Nucl. Chem.* 2, 345 (1956).
121. Lindqvist, O., *Acta Chem. Scand.* 21, 1473 (1967).
121a. Lynch, T. R., Mellor, I. P., and Nyburg, S. C., *Acta Crystallogr. Sect. B* 27, 1948 (1971).
122. McCullough, J. D., *J. Amer. Chem. Soc.* 59, 789 (1937).
123. McDonald, R. R., Larson, A. C., and Cromer, D. T., *Acta Crystallogr.* 17, 1104 (1964).
124. McDonald, W. S., and Pettit, L. D., *J. Chem. Soc. A* 2044 (1970).
125. MacGillivray, C. H., and van Eck, C. L. P., *Rec. Trav. Chim. Pays-Bas* 62, 729 (1943).
126. McRae, V. M., Peacock, R. D., and Russell, D. R., *Chem. Commun.* 62 (1969).
126a. Malmros, G., *Acta Chem. Scand.* 24, 384 (1970).
126b. Manotti Lanfredi, A. M., Pellinghelli, M. A., Tiripicchio, A., and Tiripicchio Camellini, M., *Acta Crystallogr. Sect. B* 28, 679 (1972).
126c. Mastin, S. H., and Ryan, R. R., *Inorg. Chem.* 10, 1757 (1971).
126d. Mohana Rao, J. K., to be published (see ref. *126e*).
126e. Mohana Rao, J. K., and Viswamitra, M. A., *Acta Crystallogr. Sect. B* 27, 1765 (1971).
126f. Nielsen, B. R., Hazell, R. G., and Rasmussen, S. E., *Acta Chem. Scand.* 25, 3037 (1971).
126g. Nyburg, S. C., Ozin, G. A., and Szymański, J. T., *Acta Crystallogr. Sect. B* 27, 2298 (1971).
127. Ohlberg, S. A., and Vaughan, P. A., *J. Amer. Chem. Soc.* 76, 2649 (1954).
127a. Olie, K., *Acta Crystallogr. Sect. B.* 27, 1459 (1971).
128. Olie, K., and Mijlhoff, F. C., *Acta Crystallogr. Sect. B* 25, 974 (1969).
129. Pauling, L., "The Nature of the Chemical Bond." Cornell Univ. Press, Ithaca, New York, 1960.
129a. Peck, D. N., Duax, W. L., Eger, C., and Norton, D. A., *American Crystallographic Association Abstracts*, Summer 1970, 71 (L6).
130. Porter, S. K., and Jacobson, R. A., *J Chem. Soc. A* 1356 (1970)
130a. Rodesiler, P. F., and Amma, E. L., *Acta Crystallogr. Sect. B* 27, 1687 (1971).
131. Rogers, M. T., and Helmholtz, L., *J. Amer. Chem. Soc.* 62, 1537 (1940).
132. Rosenzweig, A. and Morosin, B., *Acta Crystallogr.* 20, 758 (1966).

133. Rychnovsky, V., and Britton, D., *Acta Crystallogr. Sect. B* **24**, 725 (1968).
134. Schlemper, E. O., and Britton, D., *Acta Crystallogr.* **18**, 419 (1965).
135. Schlemper, E. O., and Britton, D., *Acta Crystallogr.* **20**, 777 (1966).
136. Schlemper, E. O., and Britton, D., *Inorg. Chem.* **5**, 507 (1966).
137. Schlemper, E. O., and Britton, D., *Inorg. Chem.* **5**, 511 (1966).
138. Schlemper, E. O., and Hamilton, W. C., *Inorg. Chem.* **5**, 995 (1966).
139. Segal, D. J., Santoro, R. P., and Newnham, R. E., *Z. Kristallogr.* **123**, 73 (1966).
140. Selte, K., and Kjekshus, A., *Acta Chem. Scand.* **24**, 1912 (1970).
141. Skapski, A. C., *Chem. Commun.* 10 (1966).
142. Sladky, F. O., Bulliner, P. A., Bartlett, N., de Boer, B. G., and Zalkin, A., *Chem. Commun.* 1048 (1968).
142a. Stålhandske, C.-I., *Acta Chem. Scand.* **23**, 1525 (1969).
143. Stork-Blaisse, B. A., and Romers, C., *Acta Crystallogr. Sect. B* **27**, 386 (1971).
144. Struchkov, Yu. T., and Khotsyanova, T. L., *Izv. Akad. Nauk. SSSR Ser. Khim.* **11**, 821 (1960); *Struct. Rep.* **24**, 661 (1960).
145. Templeton, D. H., Zalkin, A., Forrester, J. D., and Williamson, S. M., in "Noble-Gas Compounds" (H. H. Hyman, ed.), p. 229. Univ. of Chicago Press, Chicago, Illinois, 1963.
145a. Templeton, L. K., and Templeton, D. H., *Acta Crystallogr. Sect. B.* **72**, 1678 (1971).
146. Trotter, J., *Z. Kristallogr.* **121**, 81 (1965).
147. Trotter, J., Akhtar, M., and Bartlett, N., *J. Chem. Soc. A* 30 (1966).
148. Trotter, J., and Zobel, T., *Z. Kristallogr.* **123**, 67 (1966).
149. Viers, J. W., and Baird, H. W., *Chem. Commun.* 1093 (1967).
150. Vonk, C. G., and Wiebenga, E. H., *Acta Crystallogr.* **12**, 859 (1959).
151. Wang, B.-C., and Cordes, A. W., *Inorg. Chem.* **9**, 1643 (1970).
152. Wells, A. F., and Bailey, M., *J. Chem. Soc.* 1282 (1949).
153. Wiebenga, E. H., Havinga, E. E., and Boswijk, K. H., *Advan. Inorg. Chem. Radiochem.* **3**, 133 (1961).
154. Wiebenga, E. H., and Kracht, D., *Inorg. Chem.* **8**, 738 (1969).
154a. Witt, J. R., Britton, D., and Mahon, C., *Acta Crystallogr. Sect. B* **28**, 950 (1972).
155. Wyckoff, R. G., "Crystal Structures," Vol II, p. 20. Wiley (Interscience), New York, 1964.
156. Zuccaro, D. E., and McCullough, J. D., *Z. Kristallogr.* **112**, 401 (1959).

MÖSSBAUER SPECTRA OF INORGANIC COMPOUNDS: BONDING AND STRUCTURE

G. M. Bancroft

Chemistry Department, University of Western Ontario, London, Canada

and

R. H. Platt

University Chemical Laboratory, University of Cambridge, Cambridge, England

I. Preface 59
II. Introduction 60
   A. The Mössbauer Effect 60
   B. Isomer and Center Shift 61
   C. Quadrupole Splitting 63
   D. The Additivity Model for Quadrupole Splittings 71
III. Fingerprint Uses 89
   A. Oxidation States and Inequivalent Mössbauer Atoms . . . 89
   B. Decomposition Reactions. 94
   C. The Effect of Temperature and Pressure on the Electronic Structure of Iron Compounds 95
   D. Site Populations in Silicate Minerals 97
   E. Preparation of Novel Compounds 100
   F. Frozen Solution Studies 101
IV. Bonding and Structure 103
   A. Sn^{IV} Compounds 103
   B. Fe^{II} Low-Spin Compounds 166
   C. Ru^{II} and Ir^{III} Compounds 184
   D. Iodine Compounds 187
   E. Sn^{II} Compounds 201
   F. Other Oxidation States of Iron 208
   G. Other Mössbauer Isotopes 227
References 241

I. Preface

The Mössbauer effect, discovered by R. L. Mössbauer in 1957 (*413*) began to be widely applied to chemical problems after it was shown in 1960 that ^{57}Fe exhibited this phenomenon (*369*). In the last ten years, Mössbauer spectroscopy has become a powerful and versatile probe in many areas of chemistry. The great fraction of Mössbauer research has been carried out using ^{57}Fe and ^{119}Sn. However, over thirty other isotopes exhibit the effect (*411*) and chemically interesting information has been obtained on compounds of Ni, Kr, Ru, Sb, Te, I, Xe, W, Ir, Au,

Np, and many of the rare earths, although the use of several of the above isotopes presents considerable experimental difficulties and inconvenience.

Despite the great growth of Mössbauer research in inorganic chemistry in the last ten years and the number of books (*14, 110, 247, 270, 274, 412, 536, 559*) and reviews (*175, 198, 235, 277, 297, 324, 437, 516, 577*), the technique has largely remained in the hands of the specialists. It is our intention in this chapter to review critically recent Mössbauer studies which illustrate the use of this technique first, in rather simple fingerprint applications, and second, in the determination of the structure and bonding in inorganic compounds. Hopefully, this will hasten the advance of Mössbauer spectroscopy as a routine method in inorganic chemistry departments and industry.

We severely limit the scope of this chapter by neglecting not only the basic physics of the experiment and how it arises (*247*), but also experimental details (*110*). We will also just discuss the use of two Mössbauer parameters, the isomer or center shift and quadrupole splitting, and neglect magnetic splitting and magnetic effects. Useful bonding information can be obtained from magnetic splittings (*357*), but the overwhelming majority of inorganic compounds do not show magnetic splitting at temperatures of 77°K (liquid N_2) and above.

In a chapter of this scope it is neither feasible nor desirable to attempt to give a comprehensive account of every pertinent paper or to include a catalog of all Mössbauer data. Rather we have tried to present a critical account of the major areas in which Mössbauer spectroscopy has been applied recently to the problems of structure and bonding for all isotopes except the rare earths and actinides and to include enough data for the reader to judge the scope and limitations of this spectroscopy. Annual surveys of the Mössbauer literature are available in the excellent series of articles by Greenwood (*298*). For ^{119}Sn Mössbauer spectroscopy, two very recent compilations of data have been published (*437, 577*).

II. Introduction

A. THE MÖSSBAUER EFFECT

Mössbauer spectroscopy can be likened to ultraviolet spectroscopy. Both techniques employ a source of radiation, an absorber and a detector. In Mössbauer spectroscopy, we consider transitions between nuclear energy levels with the emission and absorption of γ rays; whereas, in ultraviolet spectroscopy, we consider the transitions between electronic energy levels with the emission and absorption of ultraviolet radiation.

To observe resonance, a range of source photon energies is scanned: in ultraviolet by the use of a prism or grating, and in Mössbauer by employing the Döppler effect. The energy of the γ ray (E_γ) is varied by the well-known Döppler formula

$$\Delta E = (v/c)E_\gamma$$

where ΔE = change in energy of γ photon, v = velocity of source relative to absorber, and c = velocity of light. As in ultraviolet, absorption is plotted versus the energy of source photon (usually in velocity units for Mössbauer). Different compounds of one isotope give different spectra, i.e., the nuclear energy levels are sensitive to the extranuclear environment. These differences in spectra can be attributed to the hyperfine interactions—the interactions between the nuclear charge distribution and the extranuclear electric and magnetic fields. These hyperfine interactions give rise to the isomer shift (I.S.), the quadrupole splitting (Q.S.), and the magnetic Zeeman splitting. As mentioned previously, we will be concerned with the first two parameters.

B. Isomer and Center Shift

The isomer shift results from the electrostatic interaction between the charge distribution of the nucleus and those electrons which have a finite probability of being found in the region of the nucleus. The above interaction results in a slight shift of both ground and excited nuclear energy levels in a compound relative to those in the free atom. The shifts will be different in source and absorber and a Döppler velocity will have to be supplied to the source or absorber to observe resonance, i.e., for resonance

$$^sE_\gamma \pm (v/c)\,^sE_\gamma = {}^aE_\gamma,$$

where $^sE_\gamma$ and $^aE_\gamma$ are the source and absorber transition energies, respectively.

However, a similar shift in energy levels can arise from the second-order Döppler (S.O.D.) shift which arises from the thermal motion of the Mössbauer atoms (*319*, *360*). The observed center or chemical shift (C.S.) is a resultant of both the isomer shift and S.O.D. shift; although the latter is usually much smaller than the isomer shift and is usually neglected. The center shift (C.S.) will be used in this chapter, although other terms such as chemical or chemical isomer shift are also commonly used.

The isomer shift can be computed classically by considering the effect of the overlap of s electron density with the nuclear charge density.

First-order perturbation theory (75, 247) gives the isomer shift δ as:

$$\delta = K \frac{\delta R}{R} \{[\Psi(O)_s]_a{}^2 - [\Psi(O)_s]_s{}^2\} \quad (1)$$

where $[\Psi(O)_s]_a{}^2$ and $[\Psi(O)_s]_s{}^2$ are the total s electron densities at the absorber and source atoms, respectively, K is a constant for a given isotope, and $\delta R = R_{ex} - R_{gr}$, where R_{ex} and R_{gr} are the radius of the excited and ground nuclear states, respectively. The isomer shift depends on a nuclear factor $\delta R/R$ and an extranuclear factor—the s electron densities. For some isotopes such as ^{57}Fe, δR is negative and, thus, if $[\Psi(O)_s]_a{}^2 < [\Psi(O)_s]_s{}^2$, then a positive shift is observed. For other isotopes,

TABLE I

s Electron Densities at the Iron Nucleus for Different d Configurations[a]

d Electron configuration	Electron density at iron nucleus (atomic units)
d^8	11,878.6
d^7	11,879.1
d^6	11,879.5
d^5	11,881.3
d^4	11,885.2
d^3	11,892.0

[a] From Refs. (160, 555).

such as ^{119}Sn, δR is positive and, thus, if $[\Psi(O)_s]_a{}^2 > [\Psi(O)_s]_s{}^2$, a positive shift is observed. For a given source, e.g., ^{57}Co in Pd, an increase in I.S. is observed when the s electron density in absorbers decreases.

Although changes in isomer shift are due to variations in s electron density, differences in isomer shifts are observed on addition or removal of p or d electrons which do not themselves interact with the nuclear charge density. Hartree–Fock calculations for different d^n configurations by Watson (Table I) (160, 555) show that a decrease in the number of Fe $3d$ electrons causes a marked increase in the $3s$ electron density at the nucleus. This trend arises indirectly via the $3s$ electrons which spend a fraction of their time further from the nucleus than the $3d$ electrons, causing the $3s$ electrons to expand, and thus reduce the s electron density at the nucleus. On this basis, one would expect that a d^6 ion (Fe^{2+}) would have a larger isomer shift than a d^5 ion (Fe^{3+}).

In a molecule, the picture becomes more complex because the s, p, and d electron densities will be modified by covalent bonding (553). For example, in ^{57}Fe, the two important bonding interactions, σ bonding and π back-bonding, contribute both to a $4s$ population and a change in the d-orbital population from the free ion value. Isomer shifts have been calculated for a variety of $3d$ and $4s$ configurations for ^{57}Fe, but for a variety of reasons, these calculations are not entirely satisfactory (216). For our purposes, it is usually extremely useful just to obtain *relative* changes in valency orbital populations from *relative* isomer shifts, keeping in mind that an increase in $4s$ density decreases the isomer shift, whereas an increase in $3d$ density increases the isomer shift.

For ^{119}Sn, the sign of $\delta R/R$ was in dispute for some time. However, this is now known to be positive (381), and the isomer shift will thus increase with an increase in s electron density, but decrease with an increase in p electron density. Lees and Flinn have derived an equation for the isomer shift relative to Mg_2Sn as a function of the number of $5s$ and $5p$ electrons (designated n_s and n_p, respectively) (381):

$$\text{I.S.} = -2.36 + 3.01 n_s - 0.20 n_s^2 - 0.17 n_s n_p \tag{2}$$

Similar equations have been derived for other isotopes (498, 510). These equations indicate that the isomer shift should be much more sensitive to the s electron density than the p electron density, and could be used in conjunction with Q.S. data to obtain both n_s and n_p. It would be expected that covalent tin ($5s\,5p^3$) would have a more positive shift than "ionic" Sn^{IV} compounds, which can, for convenience, be classified as having the electronic configuration $4d^{10}$, and that the isomer shift in Sn compounds will vary with the covalent character of its bonds.

C. Quadrupole Splitting

If the nucleus is not spherical, as assumed above, and $I > \frac{1}{2}$, the interaction of noncubic extranuclear electric fields with the nuclear charge density results in a splitting of the nuclear energy levels. For example, for ^{57}Fe and ^{119}Sn, the $I = \frac{3}{2}$ level splits into two, while the $I = \frac{1}{2}$ level remains degenerate (Fig. 1). A characteristic two-line spectrum is obtained (Fig. 1) from the two allowed transitions. The separation of the peaks is the quadrupole splitting (Q.S.), and the centroid of the two peaks relative to the source is the center shift. For noninteger nuclear spins the quadrupole interaction results in $I + \frac{1}{2}$ levels, but for integer spins, it is possible to obtain $2I + 1$ levels, and the degeneracy is completely removed. If I is large for ground and excited states, then complex spectra can be observed, and it becomes more difficult to obtain the hyperfine parameters (252, 345, 568).

The magnitude of the quadrupole splitting is proportional to the electric field gradient (EFG) tensor which interacts with the quadrupole moment of the nucleus. While the isomer shift measures the s electron density at the nucleus, the EFG tensor measures the distortion from

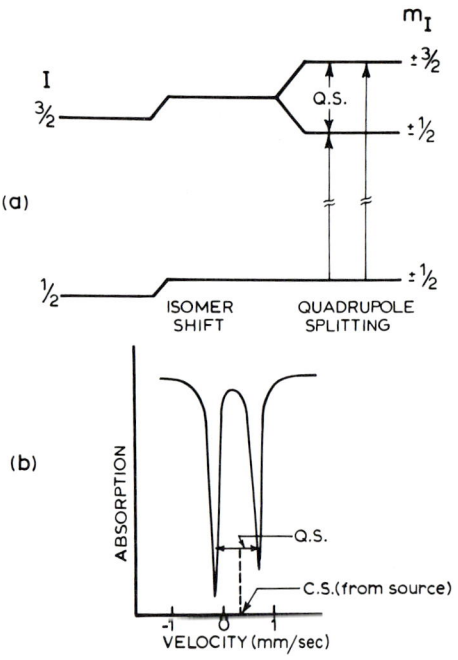

FIG. 1. (a). Nuclear energy levels, the isomer shift, and quadrupole splitting for $I_{gr} = \frac{1}{2}$, $I_{ex} = \frac{3}{2}$. (b) Resultant Mössbauer spectrum.

cubic symmetry of the electron distribution and ligands about the Mössbauer atom.

The EFG tensor has nine components which arise in the following way (136). The electric field at the Mössbauer nucleus is the negative gradient of the potential V and the EFG is the gradient of the electric field E.

$$\text{EFG} = \nabla E = - \begin{bmatrix} V_{xx} & V_{xy} & V_{xz} \\ V_{yx} & V_{yy} & V_{yz} \\ V_{zx} & V_{zy} & V_{zz} \end{bmatrix} \quad (3)$$

where

$$V_{xx} = \frac{\partial^2 V}{\partial x^2}, \qquad V_{xy} = \frac{\partial^2 V}{\partial x \partial y}, \text{ etc.}$$

If we assume that the EFG is set up by point charges, e, the contribution of one point charge to each component of the EFG tensor is given by the expressions in Table II (*136*). The polar coordinates are defined in the

TABLE II

COMPONENTS OF ELECTRIC FIELD GRADIENT TENSOR FOR
A POINT CHARGE OF 1 PROTONIC CHARGE e

Components	Coordinates
$V_{xx} = er^{-3}(3\sin^2\theta\cos^2\phi - 1)$ $V_{yy} = er^{-3}(3\sin^2\theta\sin^2\phi - 1)$ $V_{zz} = er^{-3}(3\cos^2\theta - 1)$ $V_{xy} = V_{yx} = er^{-3}(3\sin^2\theta\sin\phi\cos\phi)$ $V_{xz} = V_{zx} = er^{-3}(3\sin\theta\cos\theta\cos\phi)$ $V_{yz} = V_{zy} = er^{-3}(3\sin\theta\cos\theta\sin\phi)$	

normal fashion. The above tensor can be reduced to diagonal form if the coordinate axes are properly chosen so that the EFG can be completely specified by the three components V_{XX}, V_{YY}, and V_{ZZ}.* However, even these three are not independent, since

$$V_{XX} + V_{YY} + V_{ZZ} = 0 \qquad (4)$$

There are, then, only two independent parameters and these are normally chosen to be V_{ZZ} and η, where $\eta = (V_{XX} - V_{YY})/V_{ZZ}$. The EFG axes are chosen so that $|V_{ZZ}| \geq |V_{YY}| \geq |V_{XX}|$, thus constraining η to have values between 0 and 1. Many of the properties of the EFG tensor can be deduced from the symmetry properties of the crystal. For example, if a molecule has a fourfold axis of symmetry (as in *trans*-MA_2B_4) (where M is a Mössbauer isotope), then this axis is chosen to be the Z direction of the EFG tensor. A rotation of 90° about the Z axis produces no change in the EFG tensor, and it follows that $V_{XX} = V_{YY}$ and $\eta \equiv 0$. However, in some cases of interest (e.g. *cis*-MA_2B_4), the Z EFG axes do *not* coincide with the highest symmetry axes of the molecule. The proper

* The initial axes will be distinguished from the principal EFG axes in this article by using small letters x, y, z, and V_{xx}, V_{yy}, V_{zz}, etc., for the former, but X, Y, Z and V_{XX}, V_{YY}, V_{ZZ} for the latter.

choice of EFG axes is illustrated later in this article for many cases of interest (Table IV).

For the $I = \frac{3}{2}$ case (^{57}Fe and ^{119}Sn), the Q.S. can be expressed as:

$$\Delta \equiv \text{Q.S.} = \tfrac{1}{2} e^2 qQ(1 + \eta^2/3)^{1/2} \tag{5}$$

where Q is the quadrupole moment of the nucleus, $eq = V_{ZZ} = -$ the Z component of the EFG, and $e = $ protonic charge $= 4.80 \times 10^{-10}$ e.s.u. Either q or V_{ZZ} is normally referred to as the field gradient. It should be kept in mind that the signs of q and the Z component of the EFG are different.

Ideally, we would like to obtain three pieces of information from a measurement of the quadrupole splitting: the magnitude of q and η, and the sign of q. It should be noted that like $\delta R/R$, Q can be either positive or negative, and thus q and the Q.S. may not have the same sign. If the $\pm\frac{3}{2}$ state is at high energy (as in Fig. 1), the sign of the Q.S. is positive. For Mössbauer nuclei such as ^{57}Fe and ^{119}Sn, a two-line spectrum is obtained, and the Q.S. easily measured. By the nature of Eq. (5), it is obvious that both q and η cannot be calculated from a measurement of the Q.S. Also, the sign of the Q.S. cannot generally be determined from a powder spectrum since the $\pm\frac{3}{2}$ and $\pm\frac{1}{2}$ lines are of very similar intensity and cannot be distinguished. Techniques for measuring the sign of Q (and q) and estimating η for ^{57}Fe and ^{119}Sn will be discussed later in this section.

For convenience, assume that $\eta = 0$ and express q as*:

$$q = (1 - \gamma_\infty)q_{\text{lat}} + (1 - R)q_{\text{val}} \quad (351) \tag{6}$$

where (a) R and γ_∞ are the Sternheimer antishielding factors, (b) q_{lat}, the contribution from the external ligand charges Z_i, is given by

$$q_{\text{lat}} = \sum_i \frac{Z_i(3\cos^2\theta_i - 1)}{r_i^3} \tag{7}$$

and (c) q_{val}, the valence contribution from the valence electrons equals

$$-\sum_{\substack{\text{valence}\\\text{electrons}}} \langle 3\cos^2\theta_i - 1\rangle \langle r^{-3}\rangle \tag{8}$$

Table III gives the contribution to q_{val} from p and d electrons. The Sternheimer factors take into account the effect of q_{val} or q_{lat} on the inner

---

* The maximum value of $(1 + \eta^2/3)^{1/2} = 1.15$ so that a large η makes little difference to the magnitude of the Q.S.

nonvalence electrons and result in magnifying any contribution to q from the valence electrons or lattice, without changing the sign.

Both q_{lat} and q_{val} terms can, in principle, be calculated using Eqs. (7) and (8) if the crystal structure and the valence orbital populations are known. However, it is usually difficult to assign charges to different atoms in the structure (17) and valence orbital populations are usually not known and are difficult to calculate. In addition, the value of Q is often not accurately known for most isotopes making it difficult to compare calculated and observed splittings.

TABLE III

VALUES OF $V_{zz}/e(=q)$ FOR p and d ELECTRONS

Wavefunction	V_{zz}/e	Wavefunction	V_{zz}/e
p_x	$+\frac{2}{5}\langle r^{-3}\rangle_p$	d_{z^2}	$-\frac{4}{7}\langle r^{-3}\rangle_d$
p_y	$+\frac{2}{5}\langle r^{-3}\rangle_p$	$d_{x^2-y^2}$	$+\frac{4}{7}\langle r^{-3}\rangle_d$
p_z	$-\frac{4}{5}\langle r^{-3}\rangle_p$	d_{xy}	$+\frac{4}{7}\langle r^{-3}\rangle_d$
		d_{xz}	$-\frac{2}{7}\langle r^{-3}\rangle_d$
		d_{yz}	$-\frac{2}{7}\langle r^{-3}\rangle_d$

Because of these (and other) difficulties, more semiempirical methods such as the additivity model are used to rationalize Q.S. values, but before discussing the additivity model, it seems appropriate to examine features of the field gradient more closely.

If the electron and ligand charge distributions have cubic symmetry, both q_{val} and q_{lat}, respectively, will be zero. For example, the q_{val} contribution from p electrons can be written using Eq. (8) and Table III:

$$q_{val} = K_p[-N_{p_z} + \tfrac{1}{2}(N_{p_x} + N_{p_y})]^* \qquad (9)$$

where $K_p = +\tfrac{4}{5}\langle r^{-3}\rangle_p$ and N = orbital populations. It is evident from Eq. (9) that if the three p orbitals are equally populated, then $q_{val} = 0$. If $N_{p_z} > \tfrac{1}{2}(N_{p_x} + N_{p_y})$, then q_{val} is negative; if $N_{p_z} < \tfrac{1}{2}(N_{p_x} + N_{p_y})$, q_{val} will then be positive. A concentration of negative charge along the Z EFG axis gives a negative q. Similarly for d electrons

$$q_{val} = K_d[-N_{d_{z^2}} + N_{d_{x^2-y^2}} + N_{d_{xy}} - \tfrac{1}{2}(N_{d_{xz}} + N_{d_{yz}})] \qquad (10)$$

* This is the result of the Townes–Dailey treatment for quadrupole splittings. For a discussion of the assumptions and approximations see Lucken (391).

where $K_d = +\tfrac{4}{7}\langle r^{-3}\rangle_d$. If the component orbitals of the t_{2g} and/or the e_g levels have equal populations, then $q_{\text{val}} = 0$.

Similarly for q_{lat}, it can easily be shown using Eq. (7) that an octahedral or tetrahedral array of charges of equal magnitude gives $q_{\text{lat}} = 0$. Compressing the axial ligands in the octahedral case gives a negative q_{lat}; compressing the equatorial ligands gives a positive q_{lat}. Again a concentration of negative charge along the Z EFG axis gives a negative q.

It is convenient now to divide q_{val} into two contributions:

$$q_{\text{val}} = q_{\text{C.F.}} + q_{\text{M.O.}} \tag{11}$$

where $q_{\text{C.F.}}$ is the valence contribution considering a crystal field model with no overlap of ligand and metal orbitals, and $q_{\text{M.O.}}$ is the valence contribution considering bonding between metal and ligands.

The $q_{\text{C.F.}}$ term is dominant in transition metal ions such as Fe^{2+} high spin or Fe^{3+} low spin in which the t_{2g} and/or e_g levels are not fully or half populated. In many other cases of interest, I and Te compounds, Sn^{IV}, Fe^{II} low spin, Ru^{II}, and Ir^{III}, the Q.S. can be attributed to $q_{\text{M.O.}}$ (and possibly q_{lat}). To illustrate the value of the above separation, consider Fe^{II} low spin, where the major part of the Q.S. is due to $q_{\text{M.O.}}$, and Fe^{II} high spin, where the major part of the Q.S. is due to $q_{\text{C.F.}}$. For Fe^{II} low spin (t_{2g}^6) (44), if there is no covalent bonding, $N_{d_{xy}} = \tfrac{1}{2}(N_{d_{xz}} + N_{d_{yz}})$ and $q_{\text{C.F.}} = 0$. Consider a hypothetical species $trans$-$[FeA_2B_4]^{2+}$, where A and B are neutral ligands, and suppose that any Q.S. is due to the differences in π back-bonding capacity of A and B. If A is a better π acceptor than B, then more electron density will be withdrawn along the Z axes than along the X or Y axes and $N_{d_{xy}} > \tfrac{1}{2}(N_{d_{xz}} + N_{d_{yz}})$ and $q_{\text{M.O.}}$ is positive; if A is a poorer π acceptor than B, then $N_{d_{xy}} < \tfrac{1}{2}(N_{d_{xz}} + N_{d_{yz}})$ and $q_{\text{M.O.}}$ is negative. The quadrupole splittings for many other isotopes can also be rationalized by considering the electron imbalance about the Mössbauer isotope given by Eqs. (9) and (10).

Quadrupole splittings which arise from a $q_{\text{M.O.}}$ term normally vary little with temperature. Any small variation with temperature is due to small changes in bond lengths.

In contrast to the above situation, the fourth electron in the t_{2g} level in Fe^{2+} high spin $(t_{2g}^4 e_g^2)$ normally gives rise to a large $q_{\text{C.F.}}$ term which is more temperature dependent (351). If the Fe^{2+} is surrounded by a perfect octahedron of point charges, then the degeneracy of the t_{2g} levels is not removed, and the extra electron spends an equal time in all three of the t_{2g} orbitals and $q_{\text{val}} = 0$, because $N_{d_{xy}} = \tfrac{1}{2}(N_{d_{xz}} + N_{d_{yz}}) = 1.33$. However, this system is inherently subject to a Jahn–Teller distortion which removes the degeneracy of the t_{2g} and e_g levels. If the axial ligands are compressed slightly, then the sixth electron preferentially occupies the d_{xy} orbital and a large positive $q_{C.F.}$ results. This field gradient is normally very temperature dependent because the splitting of the t_{2g} levels is usually of the order of kT. Thus, the d_{xz} and d_{yz} orbitals are Boltzmann populated, and this Boltzmann population decreases on lowering the temperature. For the axial compression, $q_{C.F.}$ can be expressed as

$$q_{C.F.} = \tfrac{4}{7}\langle r^{-3}\rangle \left[\frac{1 - e^{-\Delta_3/kT}}{1 + 2e^{-\Delta_3/kT}}\right] \quad (12)$$

where Δ_3 is the energy separation between d_{xy} and (d_{xz}, d_{yz}), and k is Boltzmann's constant. Thus, the maximum value of $q_{C.F.}$ due to one d electron ($\tfrac{4}{7}\langle r^{-3}\rangle$) is decreased owing to thermal population; and as T decreases, $q_{C.F.}$ increases ($e^{-\Delta_3/kT}$ decreases).

Spin-orbit coupling decreases the Q.S. from the above value (*351*). The $q_{M.O.}$ and q_{lat} terms are usually much smaller than $q_{C.F.}$, but both usually *decrease* the observed quadrupole splitting from that expected just from the $q_{C.F.}$ term.

A similar type of treatment is applicable to Fe^{III} (*281, 283*) low spin. In other cases of interest, such as Fe^0 (d^8), and Fe^{-I} (d^9), it is usually not possible to separate contributions from $q_{M.O.}$ and $q_{C.F.}$. Quadrupole splittings for a large number of compounds in the above valency states have remained largely unexplained since detailed molecular orbital calculations are needed.

From the above discussion, it is apparent that more bonding information can be obtained on ions having symmetric ground states (Fe^{II} and Sn^{IV}), since the major part of the Q.S. is due to $q_{M.O.}$ and "complications" from $q_{C.F.}$ do not enter the picture. However, information of considerable structural and bonding interest can still be obtained for ions such as Fe^{2+} high spin, and these will be discussed in later sections.

The Sign of the Quadrupole Splitting

As we have discussed previously, it is often desirable to determine the sign of the Q.S. and q, but for ^{57}Fe and ^{119}Sn it is not possible by taking the usual powder spectrum to obtain the sign of the Q.S. and q. For ^{57}Fe, Q is positive, whereas for ^{119}Sn, Q is negative. Thus, a measured positive Q.S. would correspond to a positive q for ^{57}Fe, but a negative q for ^{119}Sn.

There are two methods which are normally used to obtain the sign of q. The first involves obtaining spectra of a sample with all crystallites oriented in one known way relative to the direction of the gamma beam

(*5, 121*); the second involves measuring the spectrum of a polycrystalline sample at 4°K in a large magnetic field (*132, 136*). Because of the difficulties in obtaining single crystals in a known orientation, the second method is usually the most useful and will be briefly described here.

On application of a large magnetic field to a powdered sample, the degeneracy of the nuclear levels is completely removed. The EFG axes take all orientations with respect to the magnetic field, and a large

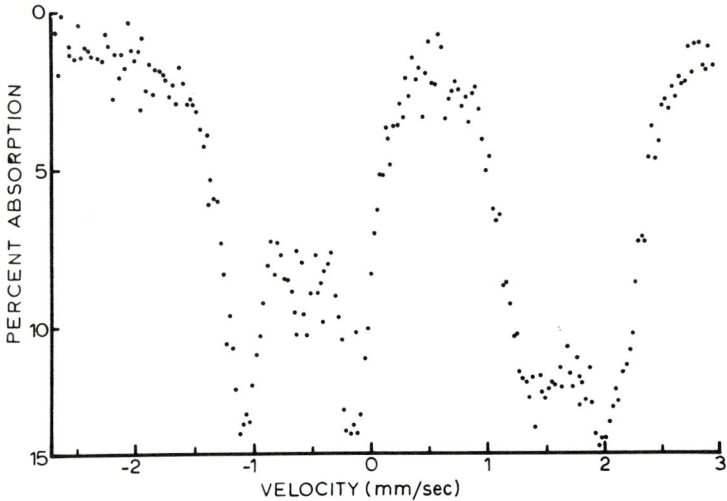

FIG. 2. Effect of a magnetic field on an ^{57}Fe powder spectrum—the Mössbauer spectrum of ferrocene at 4.2°K in an applied longitudinal magnetic field of 40 kgauss (*132*). The doublet lies to positive velocities and V_{zz} and q are positive.

number of superimposed spectra are observed. For $\eta = 0$ or a small value, the two-line zero field spectrum splits into a doublet and a triplet (Fig. 2), with the doublet being due to the $+\frac{1}{2} \rightarrow +\frac{3}{2}$ and $-\frac{1}{2} \rightarrow -\frac{3}{2}$ transitions. For ^{57}Fe, then, if the doublet lies to positive velocity, the sign of both Q.S. and q is positive. If η approaches 1, the spectrum goes from the doublet–triplet structure to a symmetric triplet–triplet structure. Using detailed computation, an estimate of η can be made (*136*). Orientation of the crystallites or an anisotropic f factor can markedly alter the spectrum and lead to difficulties in detailed interpretation, especially for small quadrupole splittings (<0.50 mm/sec), but the sign can still usually be obtained.

For ^{119}Sn, a doublet–quartet structure is observed (*265, 438*) in an applied magnetic field, and the sign of the Q.S. is positive (q negative) if the doublet is at positive velocities.

D. The Additivity Model for Quadrupole Splittings

The interpretation of the quadrupole splittings of Sn^{IV}, Fe^{II} low spin, and Fe^{-II} compounds has been greatly facilitated by application of the additivity model [for example, see Refs. (*36, 44, 122, 406, 440, 442*)] and it should prove to be useful for Fe^{III} high spin, Ir^{III}, Ru^{II}, W^0, W^{VI}, and a number of other Mössbauer atoms. In this model, use is made of the basic premise that the quadrupole splitting can be regarded to a first approximation as the sum of independent contributions, one for each ligand bound to the metal atom. The additivity model is expected to apply for compounds of transition metal ions whose t_{2g} and/or e_g subshells are filled or half-filled, e.g., Fe^{II} (t_{2g}^6) or $Ir^{III}(t_{2g}^6)$, or for compounds of main group ions whose s and p shells are empty [e.g., $Sn^{IV}(4d^{10})$ or $Te^{VI}(4d^{10})$] in the free ion.

In the above cases, $q_{C.F.} = 0$, and quadrupole splittings will be purely a function of the nature and distribution of the ligand bonds. For other compounds (e.g., Fe^{II} high spin, Fe^{III} low spin, Fe^0), the $q_{C.F.}$ term obscures the dependence of the quadrupole splitting on the nature of the M–L bonds. In other cases (Sn^{II} compounds), the inherent asymmetric occupation of the valence orbitals ($Sn^{II} = 5s^{2-x}5p^x$) obscures the dependence of the Q.S. on the M–L bond types.

In this section, we will explore the consequences of additivity and attempt to relate the various formulations of the additivity model. The agreement between theory and experiment will be assessed later using the extensive Q.S. data for Sn^{IV} and Fe^{II} low-spin compounds. The use of the additivity treatment in predicting structures of compounds and bonding properties of ligands will be discussed.

1. Point Charge Formalism

The simplest formulation of an additive EFG is the point charge model, in which each ligand is assigned a charge, the magnitude of which *represents* the contribution of that ligand to the EFG. Since the EFG is expressed in terms of separate contributions, q_{val} and q_{lat} (Section II,C). these two contributions are usually *represented* by separate charges.

Thus, for a compound containing n ligands bound to a metal atom, M(A,B,C, ... N), the nine components of the EFG tensor, V_{rs}, written in terms of the axes defined in Table II, are given by the equations:

$$V_{xx} = e \sum_L [L](3\sin^2\theta_L \cos^2\phi_L - 1) \qquad (13.1)$$

$$V_{yy} = e \sum_L [L](3\sin^2\theta_L \sin^2\phi_L - 1) \tag{13.2}$$

$$V_{zz} = e \sum_L [L](3\cos^2\theta_L - 1) \tag{13.3}$$

$$V_{xy} = V_{yx} = e \sum_L 3[L]\sin^2\theta_L \sin\phi_L \cos\phi_L \tag{13.4}$$

$$V_{xz} = V_{zx} = e \sum_L 3[L]\sin\theta_L \cos\theta_L \cos\phi_L \tag{13.5}$$

$$V_{yz} = V_{zy} = e \sum_L 3[L]\sin\theta_L \cos\theta_L \sin\phi_L \tag{13.6}$$

where

$$[L] = \frac{C_1(1-R)}{r_1^3} + \frac{C_L(1-\gamma)}{r_L^3} \tag{14}$$

In Eqs. (13.1)→(13.6) C_1 and C_L are the equivalent charges at distances r_1 and r_L, *representing* the valence and lattice contributions of the generic ligand L, respectively. The angles θ_L and ϕ_L, together with r_1 and r_L, form the spherical polar coordinates of the charges C_1 and C_L relative to the Z axis of the EFG tensor. The summation is made over all ligands.

Expressions for the relative values of V_{ZZ} and η expected for various idealized structural types may be derived in terms of the parameters [L]. The axes are chosen so as to diagonalize the EFG tensor and preserve the ordering convention $|V_{ZZ}| \geq |V_{YY}| \geq |V_{XX}|$. Expressions obtained for common structural types are summarized in Table IV.

A more elegant treatment of the point charge model has been described by Clark (*119*), who calculated the contribution of a ligand (L), to the total EFG tensor assuming that the Z axis is directed along the metal ligand bond. If the bond has cylindrical symmetry, a "$C_{\infty v}$ bond," it may be represented by a point, and the contribution to the EFG at the nucleus is given by the tensor

$$\begin{bmatrix} -\tfrac{1}{2}q_L & 0 & 0 \\ 0 & -\tfrac{1}{2}q_L & 0 \\ 0 & 0 & q_L \end{bmatrix} \tag{15}$$

This tensor is the partial field gradient (p.f.g.)$_L$ of ligand L. The parameter q_L, in point charge formalism is equal to

$$\frac{2C_1(1-R)}{r_1^3} + \frac{2C_L(1-\gamma)}{r_L^3} \tag{16}$$

and, hence, comparing Eqs. (13.3), (14), and (16), the Z component of (p.f.g.)$_L = 2[L]e$. The total EFG tensor of a molecule may be calculated by placing each ligand, in turn, on the Z axis, writing down its (p.f.g.)$_L$, and then rotating the tensor, (p.f.g.)$_L$, to its true position. The contributions from each ligand are then summed. For example, the EFG tensors calculated for octahedral *trans-* and *cis-* MA_4B_2 species are

and

$$\begin{bmatrix} -(q_B - q_A) & 0 & 0 \\ 0 & -(q_B - q_A) & 0 \\ 0 & 0 & 2(q_B - q_A) \end{bmatrix} \quad (17.1)$$

$$\begin{bmatrix} \tfrac{1}{2}(q_B - q_A) & 0 & 0 \\ 0 & \tfrac{1}{2}(q_B - q_A) & 0 \\ 0 & 0 & -(q_B - q_A) \end{bmatrix} \quad (17.2)$$

respectively, which give the same 2 : −1 ratio found in Table IV.

The expressions in Table IV allow the relative quadrupole splittings for various structural types to be compared using the expression:

$$\Delta = \tfrac{1}{2} e V_{ZZ} Q (1 + \eta^2/3)^{1/2} \quad (18)$$

In any such comparison, we tacitly assume that the [L] of a ligand is a constant in all compounds of a given valence state, e.g., Sn^{IV}. This, of course, would not be expected to hold exactly, but the expressions in Table IV are found to give a good semiquantitative guide to relative quadrupole splittings. As discussed later in this section, it is more convenient and accurate to assign separate [L] values to a ligand for different coordination numbers.

2. Molecular Orbital Approaches

Although the point charge model may be anticipated to be a reasonably accurate guide to lattice effects, it is obviously rather a crude and unrealistic approximation to the asymmetries of metal–ligand bonding interactions. Attempts have been made, therefore, to gain a clear understanding of additive electron field gradients arising from bonding interactions.

Bancroft *et al.* (*44*), in considering the quadrupole splittings of low-spin Fe^{II} six-coordinate compounds, have described a model based on a suggestion by McClure (*393*) for the interpretation of electronic spectra. In this treatment, instead of representing the valence contribution of a ligand to the EFG as a point charge, the effect of the ligand upon the populations of the iron atomic orbitals is considered. Thus, σ donation will populate the $4s$, $4p$, $3d_{z^2}$, and $3d_{x^2-y^2}$ orbitals, while

TABLE IV

Point Charge Model Expressions for the Components of the EFG Tensor for Some Common Structures[a]

Code No.[b]	Structure	Components of EFG[c]
1	Z | B | M---X / | \\ B B B	$V_{ZZ} = V_{XX} = V_{YY} = 0$
2	Z | A | M---X / | \\ B B B	$V_{ZZ} = \{2[A] - 2[B]\}e$ $V_{YY} = \{-[B] + [A]\}e$ $V_{XX} = \{-[B] + [A]\}e$ $\eta = 0$
3[d]	Z | A | M / | \\ A B B	$V_{ZZ} = \{2[A] - 2[B]\}e$ $V_{YY} = \{2[B] - 2[A]\}e$ $V_{XX} = 0$ $\eta = 1$
4[e]	Z | A | M / | \\ B C C	$V_{zz} = \{2[A] - \tfrac{2}{3}([B] + 2[C])\}e$ $V_{yy} = \{-[A] - [B] + 2[C]\}e$ $V_{xx} = \{-[A] + \tfrac{5}{3}[B] - \tfrac{2}{3}[C]\}e$ $V_{xz} = V_{zx} = \left\{\dfrac{\sqrt{2}}{3}(-2[B] + 2[C])\right\}e$ $V_{xy} = V_{yx} = V_{yz} = V_{zy} = 0$ $\eta \neq 0$ $\Delta = \tfrac{1}{2}e^2\, Q(\tfrac{4}{3}P^2 + 2Q^2)^{1/2}$ $P = [A] + [B] - 2[C]$ $Q = [A] - [B]$ Sign = sign of P
5[g]	Z | A | M / | \\ D B C	$V_{zz} = \{2[A] - \tfrac{2}{3}([B] + [C] + [D])\}e$ $V_{yy} = \{-[A] - [B] + [C] + [D]\}e$ $V_{xx} = \{-[A] + \tfrac{5}{3}[B] - \tfrac{1}{3}([C] + [D])\}e$ $V_{yz} = V_{zy} = \left\{\dfrac{\sqrt{2}}{\sqrt{3}}([C] - [D])\right\}e$ $V_{xz} = V_{zx} = \left\{\dfrac{\sqrt{2}}{3}(-2[B] + [C] + [D])\right\}e$ $V_{xy} = V_{yx} = \left\{\dfrac{2}{\sqrt{3}}([C] - [B])\right\}e$ $\eta \neq 0$

TABLE IV—continued

Code No.[b]	Structure	Components of EFG[c]
6[h]	Z B B—M—B X B— B	$V_{ZZ} = \{4[B]^{tba} - 3[B]^{tbe}\} e$ $V_{YY} = \{\tfrac{3}{2}[B]^{tbe} - 2[B]^{tba}\} e$ $V_{XX} = \{\tfrac{3}{2}[B]^{tbe} - 2[B]^{tba}\} e$ $\eta = 0$
7[h]	X B B—M—A Z B— B	$V_{ZZ} = \{2[A]^{tbe} - 2[B]^{tba} - \tfrac{1}{2}[B]^{tbe}\} e$ $V_{YY} = \{-[A]^{tbe} - 2[B]^{tba} + \tfrac{5}{2}[B]^{tbe}\} e$ $V_{XX} = \{-[A]^{tbe} + 4[B]^{tba} - 2[B]^{tbe}\} e$ $\eta \neq 0$
8[h]	Z B B—M—B X B— A	$V_{ZZ} = \{2[A]^{tba} - 3[B]^{tbe} + 2[B]^{tba}\} e$ $V_{YY} = \{-[A]^{tba} + \tfrac{3}{2}[B]^{tbe} - [B]^{tba}\} e$ $V_{XX} = \{-[A]^{tba} + \tfrac{3}{2}[B]^{tbe} - [B]^{tba}\} e$ $\eta = 0$
9[h]	Z B A—M—B X A— B	$V_{ZZ} = \{-2[A]^{tbe} - [B]^{tbe} + 4[B]^{tba}\} e$ $V_{YY} = \{\tfrac{5}{2}[A]^{tbe} - [B]^{tbe} - 2[B]^{tba}\} e$ $V_{XX} = \{-\tfrac{1}{2}[A]^{tbe} + 2[B]^{tbe} - 2[B]^{tba}\} e$ $\eta \neq 0$
10[h]	Z A B—M—B X B— A	$V_{ZZ} = \{4[A]^{tba} - 3[B]^{tbe}\} e$ $V_{YY} = \{-2[A]^{tba} + \tfrac{3}{2}[B]^{tbe}\} e$ $V_{XX} = \{-2[A]^{tba} + \tfrac{3}{2}[B]^{tbe}\} e$ $\eta = 0$
11	Z Y B B—M—B B— —B B X	$V_{ZZ} = V_{XX} = V_{YY} = 0$
12	Z A Y B—M—B B— —B A X	$V_{ZZ} = \{4[A] - 4[B]\} e$ $V_{YY} = \{2[B] - 2[A]\} e$ $V_{XX} = \{2[B] - 2[A]\} e$ $\eta = 0$

continued

TABLE IV—*continued*

POINT CHARGE MODEL EXPRESSIONS FOR THE COMPONENTS OF THE EFG TENSOR FOR SOME COMMON STRUCTURES[a]

Code No.[b]	Structure	Components of EFG[c]
13	Z, B-M-B axial B's, A-M-A equatorial (2A, 4B: trans A's in equatorial, 2B axial, 2B equatorial)	$V_{ZZ} = \{2[B] - 2[A]\}e$ $V_{YY} = \{[A] - [B]\}e$ $V_{XX} = \{[A] - [B]\}e$ $\eta = 0$
14	Z, A axial, 4B equatorial, B axial	$V_{ZZ} = \{2[A] - 2[B]\}e$ $V_{YY} = \{[B] - [A]\}e$ $V_{XX} = \{[B] - [A]\}e$ $\eta = 0$
15	Z, A axial, 4B equatorial, C axial	$V_{ZZ} = \{2[A] + 2[C] - 4[B]\}e$ $V_{YY} = \{2[B] - [A] - [C]\}e$ $V_{XX} = \{2[B] - [A] - [C]\}e$ $\eta = 0$
16	Z, A axial, C,B,B,B equatorial, B axial	$V_{ZZ} = \{2[A] - [B] - [C]\}e$ $V_{XX} = \{-[B] + 2[C] - [A]\}e$ $V_{YY} = \{2[B] - [C] - [A]\}e$ $\eta \neq 0$
17	Z, A axial, B,C,B,B equatorial, C axial	$V_{ZZ} = \{2[A] + [C] - 3[B]\}e$ $V_{XX} = \{3[B] - 2[C] - [A]\}e$ $V_{YY} = \{[C] - [A]\}e$ $\eta \neq 0$
18	Z, A axial, C,B,C,B equatorial, B axial	$V_{ZZ} = \{2[A] - 2[C]\}e$ $V_{YY} = \{[C] - [A]\}e$ $V_{XX} = \{[C] - [A]\}e$ $\eta = 0$
19	Z, A axial, C,B,B,C equatorial, B axial	$V_{ZZ} = \{2[A] - 2[C]\}e$ $V_{XX} = \{4[C] - 3[B] - [A]\}e$ $V_{YY} = \{3[B] - 2[C] - [A]\}e$ $\eta \neq 0$

TABLE IV—continued

Code No.[b]	Structure	Components of EFG[c]
20	Z A Y B\\|/C M B/|\\C A X	$V_{ZZ} = \{4[A] - 2[B] - 2[C]\}e$ $V_{YY} = \{[C] + [B] - 2[A]\}e$ $V_{XX} = \{[C] + [B] - 2[A]\}e$ $\eta = 0$
21	Z A Y B\\|/C M C/|\\B A X	$V_{ZZ} = \{4[A] - 2[B] - 2[C]\}e$ $V_{XX} = \{4[B] - 2[C] - 2[A]\}e$ $V_{YY} = \{4[C] - 2[B] - 2[A]\}e$ $\eta \neq 0$
22	Z B Y A\\|/B M A/|\\C C X	$V_{ZZ} = \{[B] + [C] - 2[A]\}e$ $V_{XX} = \{[C] + [A] - 2[B]\}e$ $V_{YY} = \{[B] + [A] - 2[C]\}e$ $\eta \neq 0$
23	Z A Y A\\|/B M A/|\\B B X	$V_{ZZ} = V_{YY} = V_{XX} = 0$
24	Z A Y B\\|/A M B/|\\B A X	$V_{ZZ} = \{3[A] - 3[B]\}e$ $V_{XX} = \{3[B] - 3[A]\}e$ $V_{YY} = 0$ $\eta = 1$

[a] These expressions are taken from Refs. (29, 39, 44, 55, 122, 234, 440, 442).

[b] When referring to a structure in the text, the code number will be prefixed by table number.

[c] The choice of axes is indicated on the diagram of the structure or in a footnote. In all cases, except 4 and 5 this choice of axes serves to diagonalize the EFG tensor. The ordering of the axes to preserve the convention $|V_{ZZ}| \geqslant |V_{YY}| \geqslant |V_{XX}|$ will depend on the [L] values. Thus, the final choice of axes may not be the same as given here, i.e., V_{ZZ} may become V_{XX} or V_{YY}, etc.

[d] The X axes coincide with the C_2 symmetry axis, and the Y and Z axes lie in the symmetry planes.

[e] The Y axis is perpendicular to the symmetry plane, while the X and Z axes lie in the plane. The orientation of the X and Z axes depends on the relative magnitudes of [A], [B], and [C], and the tensor must be diagonalized separately for each case considered.

[f] This expression gives the magnitude of the quadrupole splitting and is obtained from the symmetrized parameters of Clark (120).

[g] The EFG tensor must be diagonalized for each example considered.

[h] The superscripts tbe and tba refer to trigonal-bipyramidal equatorial and trigonal-bipyramidal axial bonds, respectively.

π bonding will reduce the populations of the d_{xy}, d_{xz}, and d_{yz} orbitals. The effect of a particular ligand, L, on the iron atomic orbital populations will depend on (a) the σ-donating (σ_L) and π-withdrawing (π_L) capacities of the ligands and (b) the relative involvement of the atomic orbitals in the Fe–L bond which for σ bonding is proportional to the squares of the coefficients of the hybrid orbitals:

$$h_{1,2} = 1/\sqrt{6}\, s \pm 1/\sqrt{2}\, p_z + 1/\sqrt{3}\, d_{z^2} \qquad (19.1)$$

$$h_{3,4} = 1/\sqrt{6}\, s \pm 1/\sqrt{2}\, p_x - 1/\sqrt{12}\, d_{z^2} + \tfrac{1}{2} d_{x^2-y^2} \qquad (19.2)$$

$$h_{5,6} = 1/\sqrt{6}\, s \pm 1/\sqrt{2}\, p_y - 1/\sqrt{12}\, d_{z^2} - \tfrac{1}{2} d_{x^2-y^2} \qquad (19.3)$$

The π orbitals (d_{xy}, d_{yz}, d_{xz}) have equal π-bonding power in the three principal directions of the EFG tensor which lie along the M–L bonds.

TABLE V

RELATIVE CHARGE DENSITIES IN p AND d ORBITALS OF THE SIX OCTAHEDRAL σ BONDING HYBRIDS[a]

Ligand	p_x	p_y	p_z	$d_{x^2-y^2}$	d_{z^2}
L_Z	0	0	$\tfrac{1}{2}$	0	$\tfrac{1}{3}$
L_Z	0	0	$\tfrac{1}{2}$	0	$\tfrac{1}{3}$
L_X	$\tfrac{1}{2}$	0	0	$\tfrac{1}{4}$	$\tfrac{1}{12}$
L_X	$\tfrac{1}{2}$	0	0	$\tfrac{1}{4}$	$\tfrac{1}{12}$
L_Y	0	$\tfrac{1}{2}$	0	$\tfrac{1}{4}$	$\tfrac{1}{12}$
L_Y	0	$\tfrac{1}{2}$	0	$\tfrac{1}{4}$	$\tfrac{1}{12}$

[a] From Ref. (44).

The normalized relative amounts of charge density in a given bonding direction are summarized in Table V. We neglect the $4p$ contributions, since the dependence of V_{rs} upon $\langle r^{-3} \rangle$ will result in a much smaller contribution to the EFG from $4p$ orbitals.

The relative populations of the $3d$ orbitals may be computed as σ_L or π_L multiplied by the appropriate coefficient in Table V. For example, a ligand, L, on the Z axis will increase the population of the $3d_{z^2}$ orbital by an amount proportional to $\tfrac{1}{3}\sigma_L$ and decrease the population of the d_{xz} and d_{yz} orbitals by an amount proportional to π_L. As an illustration of the method, Table VI contains the relative changes in effective population of each $3d$ orbital from t_{2g}^6 configuration for *trans-* and *cis-* FeA_2B_4 species. Substitution into Eq. (10) gives the expressions

$$q_{trans} \alpha [-2(\pi_B + \pi_A) + (\tfrac{2}{3}\sigma_B - \tfrac{2}{3}\sigma_A)]* \qquad (20.1)$$

$$q_{cis} \alpha [+\pi_B - \pi_A - \tfrac{1}{3}\sigma_B + \tfrac{1}{3}\sigma_A] \qquad (20.2)$$

TABLE VI

RELATIVE CHANGES IN ORBITAL POPULATIONS FOR *trans-* AND *cis-*MA_2B_4

Ligand	p_x	p_y	p_z	$d_{x^2-y^2}$	d_{z^2}
		trans-$MA_2B_4{}^a$			
L_Z	0	0	$\tfrac{1}{2}\sigma_A$	0	$\tfrac{1}{3}\sigma_A$
L_Z	0	0	$\tfrac{1}{2}\sigma_A$	0	$\tfrac{1}{3}\sigma_A$
L_X	$\tfrac{1}{2}\sigma_B$	0	0	$\tfrac{1}{4}\sigma_B$	$\tfrac{1}{12}\sigma_B$
L_X	$\tfrac{1}{2}\sigma_B$	0	0	$\tfrac{1}{4}\sigma_B$	$\tfrac{1}{12}\sigma_B$
L_Y	0	$\tfrac{1}{2}\sigma_B$	0	$\tfrac{1}{4}\sigma_B$	$\tfrac{1}{12}\sigma_B$
L_Y	0	$\tfrac{1}{2}\sigma_B$	0	$\tfrac{1}{4}\sigma_B$	$\tfrac{1}{12}\sigma_B$
Total	σ_B	σ_B	σ_A	σ_B	$\tfrac{2}{3}\sigma_A + \tfrac{1}{3}\sigma_B$
		cis-$MA_2B_4{}^a$			
L_Z	0	0	$\tfrac{1}{2}\sigma_B$	0	$\tfrac{1}{3}\sigma_B$
L_Z	0	0	$\tfrac{1}{2}\sigma_B$	0	$\tfrac{1}{3}\sigma_B$
L_X	$\tfrac{1}{2}\sigma_A$	0	0	$\tfrac{1}{4}\sigma_A$	$\tfrac{1}{12}\sigma_A$
L_X	$\tfrac{1}{2}\sigma_B$	0	0	$\tfrac{1}{4}\sigma_B$	$\tfrac{1}{12}\sigma_B$
L_Y	0	$\tfrac{1}{2}\sigma_A$	0	$\tfrac{1}{4}\sigma_A$	$\tfrac{1}{12}\sigma_A$
L_Y	0	$\tfrac{1}{2}\sigma_B$	0	$\tfrac{1}{4}\sigma_B$	$\tfrac{1}{12}\sigma_B$
Total	$\tfrac{1}{2}(\sigma_A + \sigma_B)$	$\tfrac{1}{2}(\sigma_A + \sigma_B)$	σ_B	$\tfrac{1}{2}(\sigma_A + \sigma_B)$	$\tfrac{1}{6}\sigma_A + \tfrac{1}{2}\sigma_B$

[a] From Ref (44). See Table IV, species 12 and 13 for the correct assignment of axes.

and a similar treatment for $FeAB_5$ gives

$$q_{AB_5} \alpha [-\pi_B + \pi_A - \tfrac{1}{3}\sigma_B + \tfrac{1}{3}\sigma_A] \qquad (20.3)$$

Clearly Eqs. (20.1)–(20.3), give the same $q_{trans}:q_{cis}:q_{AB_5}$ ratios calculated from the point charge model (Table IV) and lend confidence to the application of the simple additivity relationship in Table IV to the quadrupole splitting of low-spin Fe^{II} compounds.

* In the original equations in Bancroft et al. (44), the above σ_L values were multiplied by four. This constant does not affect the relative values.

For Sn^{IV} species, the situation is more complex as the range of structures to be considered spans four-, five-, and six-coordinate compounds. A simple but general molecular orbital theory for an additive EFG has been described and applied to Sn^{IV} compounds (122). The model is closely related to the ideas used in the interpretation of NQR spectroscopy (391) and to the model of Bancroft et al. (44) described above. In the first instance only σ bonding will be considered as π-bonding effects are not thought to make an important contribution to the EFG of Sn^{IV} compounds (see Section IV, A, 1a).

For the general closed shell molecule with n ligands $M(A, B, \ldots N)$, the total wavefunction of the valence electrons can be written

$$\Psi = |\Psi_1^\alpha \Psi_1^\beta \Psi_2^\alpha \Psi_2^\beta \ldots \Psi_n^\alpha \Psi_n^\beta| \quad (21)$$

where $\Psi_1, \Psi_2, \ldots \Psi_n$ are n valence molecular orbitals containing $2n$ valence electrons. The symbol $|\ )$ denotes a Slater determinant and α, β denote $m_S = +\frac{1}{2}, -\frac{1}{2}$ respectively.

The components V_{rs} of the EFG tensor at M are given by the diagonal matrix elements of the EFG tensor operator acting upon Ψ and may be written in terms of one-electron matrix elements.

$$V_{rs} = (\Psi_1^\alpha \Psi_1^\beta \ldots \Psi_n^\beta | \sum_{el} \mathscr{V}_{rs} | \Psi_1^\alpha \Psi_1^\beta \ldots \Psi_n^\beta)$$

$$= 2 \sum_{L=A}^{N} (1 - R_L)(\Psi_L | \mathscr{V}_{rs} | \Psi_L) \quad (22)$$

where $\mathscr{V}_{rs} = -er^{-5}(3x_r x_s - r^2 \delta_{rs})$.

In Eq. (22), $\sum_{el}$ denotes the summation over all electrons, e is the protonic charge, and R_L is the appropriate Sternheimer factor, $x_r x_s = x, y, z$ and $\delta_{rs} =$ Kronecker delta (i.e., $\delta_{xy} = 0; \delta_{xx} = 1$).

The molecular orbital may be transformed into a set of localized orbitals α_L (L = A, B, ... N), chosen so that each α_L is, so far as possible, localized in the region of the M–L axis. The Slater determinant [Eq. (21)] and V_{rs} remain unchanged by such a unimodular transformation and, hence,

$$V_{rs} = 2 \sum_{L=A}^{N} (1 - R_L)(\alpha_L | \mathscr{V}_{rs} | \alpha_L)$$

$$= \sum_{L=A}^{N} V_{rs}(L) \quad (23)$$

If the orbital α_L is to a large extent localized in the M–L bond axis, $V_{rs}(L)$ will depend mainly on the properties of the ligand, L. In this case the total EFG is simply the tensor sum of approximately independent contributions and these localized orbitals provide a natural framework for a discussion of additive electric field gradients. If the tensor $V_{rs}(L)$ is written in terms of the local axes (i.e., axes referring to the M–L bond) with Z directed along the M–L bond, then V_{rs} (L) form the elements of the "partial field gradient" due to L, $(p.f.g.)_L$, encountered earlier in Clark's treatment of the point charge model.

It is also convenient if the localized orbitals are equivalent orbitals. The members of a set of equivalent orbitals can be permuted among themselves by the operations of the point group or one of its subgroups. The set of localized orbitals α_L may span one or more sets of equivalent orbitals. In the first case, the orbitals are unique and it follows from the definition of equivalent orbitals that *different members of the same set of equivalent orbitals give rise to the same* $(p.f.g.)_L$.

As a further approximation the localized orbitals may be considered as a linear combination of a metal orbital h_L and a ligand orbital χ_L.

$$\alpha_L = c_1 h_L + c_2 \chi_L \qquad (24)$$

The orbitals α_L, h_L, and χ_L may all be taken as real, so that c_1 and c_2 are also real. The metal orbital is an appropriate equivalent orbital formed from the metal atomic orbitals and is called a "hybrid orbital." The matrix element in Eq. (23) may be written

$$\langle \alpha_L | \mathscr{V}_{rs} | \alpha_L \rangle = c_1^2 \, \langle h_L | \mathscr{V}_{rs} | h_L \rangle + 2 c_1 c_2 \langle h_L | \mathscr{V}_{rs} | \chi_L \rangle + c_2^2 \langle \chi_L | \mathscr{V}_{rs} | \chi_L \rangle \qquad (25)$$

The first and third terms of Eq. (25) may be identified with the valence and lattice contributions to the EFG represented in the point charge model by the equivalent charges C_1 and C_L [Eq. (14)], respectively. As V_{rs} depends on $\langle r^{-3} \rangle$ (*vide infra*), the three terms will fall off roughly as $1 : 10^{-1} : 10^{-2}$. Hence, only a small contribution from the lattice is expected (although this may be inflated by an order of magnitude due to Sternheimer effects) and the second and third terms of Eq. (25) may be neglected. In this case, using the local axes (X, Y, and Z) and taking the Z axis along the M–L bond

$$V_{ZZ}(L) = 2(1 - R_L)\langle h_L | \mathscr{V}_{ZZ} | h_L \rangle c_1^2 = 2[L]e \qquad (26)$$

The above expression is sufficient for a $C_{\infty V}$ (*119*) bond, but for a bond without axial symmetry

$$\eta_L = \frac{\langle h_L | \mathscr{V}_{XX} - \mathscr{V}_{YY} | h_L \rangle}{\langle h_L | \mathscr{V}_{ZZ} | h_L \rangle} \qquad (27)$$

where X and Y are chosen so as to diagonalize $V(L)$.

For tin(IV) compounds the structures of interest are tetrahedral, octahedral, and trigonal-bipyramidal. For tetrahedral and octahedral systems, the appropriate metal hybrids h_L ($5sp^3$ and $5sp^3d^2$, respectively) span a single equivalent set, whereas the trigonal-bipyramidal hybrids span two sets (apical and equatorial) as the $5s$ and $5d_{z^2}$ may participate in both apical and equatorial bonds. In Table VII are given typical

TABLE VII
CALCULATION OF PARTIAL FIELD GRADIENTS IN IMPORTANT STRUCTURAL TYPES[a,b]

Tetrahedral

$$h_z^{tet} = \tfrac{1}{2}s + \frac{\sqrt{3}}{2}p_z$$

$$[L]^{tet} = -\tfrac{3}{10}\langle r^{-3}\rangle_p \, \sigma_L^{tet}$$

Octahedral

$$h_z^{oct} = \frac{1}{\sqrt{6}}s + \frac{1}{\sqrt{2}}p_z + \frac{1}{\sqrt{3}}d_{z^2}$$

$$[L]^{oct} = \left(-\tfrac{1}{5}\langle r^{-3}\rangle_p - \tfrac{2}{21}\langle r^{-3}\rangle_d - \frac{\sqrt{2}}{3\sqrt{5}}\langle r^{-3}\rangle_{sd}\right)\sigma_L^{oct}$$

Trigonal-bipyramidal
Apical

$$h_z^{tba} = \frac{1}{\sqrt{2}}\cos\theta\, s + \frac{1}{\sqrt{2}}p_z + \frac{1}{\sqrt{2}}\sin\theta\, d_{z^2}$$

$$[L]^{tba} = \left(-\tfrac{1}{5}\langle r^{-3}\rangle_p - \tfrac{1}{7}\sin^2\theta\langle r^{-3}\rangle_d - \frac{1}{\sqrt{5}}\sin\theta\cos\theta\langle r^{-3}\rangle_{sd}\right)\sigma_L^{tba}$$

Equatorial

$$h_z^{tbe} = \frac{1}{\sqrt{3}}\sin\theta\, s + \frac{\sqrt{2}}{\sqrt{3}}p_z + \frac{1}{2\sqrt{3}}\cos\theta\, d_{z^2} - \tfrac{1}{2}\cos\theta\, d_{x^2-y^2}$$

$$[L]^{tbe} = \left(-\tfrac{4}{15}\langle r^{-3}\rangle_p + \tfrac{1}{21}\cos^2\theta\langle r^{-3}\rangle_d - \frac{1}{3\sqrt{5}}\cos\theta\sin\theta\langle r^{-3}\rangle_{sd}\right)\sigma_L^{tbe}$$

$$\eta_L^{tbe} = \left|\frac{15\cos^2\theta\langle r^{-3}\rangle_d - 21\sqrt{5}\cos\theta\sin\theta\langle r^{-3}\rangle_{sd}}{28\langle r^{-3}\rangle_p - 5\cos^2\theta\langle r^{-3}\rangle_d + 7\sqrt{5}\cos\theta\sin\theta\langle r^{-3}\rangle_{sd}}\right|$$

[a] tet = tetrahedral, oct = octahedral, tba = trigonal-bipyramidal-apical, tbe = trigonal-bipyramidal-equatorial.
[b] From Ref. (121).

hybrids for each structural type; the nonequivalence of the trigonal-bipyramidal hybrids is accommodated by use of the parameter θ ($0 < \theta < 2\pi$) to describe the distribution of s and d_{z^2} character between apical and equatorial hybrids.

Also in Table VII are given the partial field gradient parameters, [L] [N.B., the Z component of (p.f.g.)$_L$ = 2[L]e], obtained by substitution of the expressions h_L into Eqs. (26) and (27). The radial averages $\langle r^{-3} \rangle_p$, $\langle r^{-3} \rangle_d$, and $\langle r^{-3} \rangle_{sd}$ are equal to the appropriate radial integrals, corrected for Sternheimer effects, for example,

$$\langle r^{-3} \rangle_{sd} = (1 - R_{sd}) \int_0^\infty \rho_s(r)(r^{-3})\rho_d(r)r^2 \, dr \quad (28)$$

where $\rho_s(r)$ and $\rho_d(r)$ are the radial parts of the 5s and 5d wave functions. The empirical parameter $\sigma_L^{\text{superscript}}$ is proportional to $2c_1^2$. Assuming that Sternheimer effects are constant, calculations of the integrals $\langle r^{-3} \rangle_p$, $\langle r^{-3} \rangle_d$, and $\langle r^{-3} \rangle_{sd}$ for the valence orbitals of tin (using Herman Skillman wave functions) indicate that the ratios $\langle r^{-3} \rangle_{sd} : \langle r^{-3} \rangle_p$ and $\langle r^{-3} \rangle_d : \langle r^{-3} \rangle_p$ are of order of magnitude 10^{-3} and 10^{-2}, respectively (122). Thus, terms in $\langle r^{-3} \rangle_{sd}$ and $\langle r^{-3} \rangle_d$ can be neglected. The [L] values in Table VII then reduce to

$$[L]^{\text{tet}} = -\tfrac{3}{10} \langle r^{-3} \rangle_p \sigma_L^{\text{tet}} \quad (29.1)$$

$$[L]^{\text{oct}} = -\tfrac{1}{5} \langle r^{-3} \rangle_p \sigma_L^{\text{oct}} \quad (29.2)$$

$$[L]^{\text{tba}} = -\tfrac{1}{5} \langle r^{-3} \rangle_p \sigma_L^{\text{tba}} \quad (29.3)$$

$$[L]^{\text{tbe}} = -\tfrac{4}{15} \langle r^{-3} \rangle_p \sigma_L^{\text{tbe}} \quad (\eta_L^{\text{tbe}} = 0) \quad (29.4)$$

Equations (29.1)–(29.4), together with the properties of equivalent orbitals, lead to some important conclusions. First, the equivalence of the octahedral and tetrahedral hybrid orbitals means that the additivity model should provide a good rationalization of the relative quadrupole splittings found for compounds restricted to *one* of these coordination numbers. In these cases, the relative quadrupole splittings can be compared using the equations in Table IV and Eq. (18), where the appropriate [L] value is defined in Eq. (29). However, it is clear from Eqs. (29.1)–(29.4) that different values must be used for [L]^{oct} and [L]^{tet} and, hence, quadrupole splittings for four- and six-coordinate compounds cannot be directly compared. Equations (29.1)–(29.4) imply that the ratio [L]^{oct}/[L]^{tet} is equal to 0.67 ($\sigma_L^{\text{oct}}/\sigma_L^{\text{tet}}$). If, as might be expected, ($\sigma_L^{\text{oct}}/\sigma_L^{\text{tet}}$) does not differ greatly from unity, [L]^{oct} should be approximately 70% of [L]^{tet}. This ratio is close to that found experimentally. For five-coordinate compounds, the equations in Table IV may be used, but separate $[L]^{tba}$ and $[L]^{tbe}$ values must be employed; this is often neglected in a simple point charge approach (*442*).

From Eqs. (29.1)–(29.4) it is clear that tabulation of quantities proportional to $[L]^{superscript}$ for a set of ligands is equivalent to tabulation of corresponding $\sigma_L^{superscript}$ values. But it is also evident from Eqs. (29.1)–(29.4) and Table IV, that a constant quantity may be added to or subtracted from $\sigma_L^{superscript}$ without altering the magnitudes of the calculated quadrupole splittings. For example, consider a tetrahedral R_3SnX species. Substitution of expressions for $[R]^{tet}$ and $[X]^{tet}$ into the appropriate additivity expressions in Table IV gives

$$V_{ZZ} = \tfrac{3}{5} \langle r^{-3} \rangle_p (\sigma_R^{tet} - \sigma_X^{tet}) e \tag{30}$$

V_{ZZ} depends on the difference of $\sigma_R^{tet} - \sigma_X^{tet}$ and it is, therefore, not possible to determine absolute $[L]^{superscript}$ values direct from quadrupole splitting data. The best that can be done for tetrahedral and octahedral compounds is to evaluate parameters related to

$$[L]^{tet} - [X]^{tet} \quad \text{and} \quad [L]^{oct} - [X]^{oct} \tag{31}$$

where X is some standard ligand. For five-coordinate compounds one may evaluate parameters related only to

$$[L]^{tba} - [X]^{tba} \quad \text{and} \quad [L]^{tbe} - \tfrac{4}{3}[X]^{tba} \tag{32.1}$$

or, alternatively,

$$[L]^{tba} - \tfrac{3}{4}[X]^{tbe} \quad \text{and} \quad [L]^{tbe} - [X]^{tbe} \tag{32.2}$$

The above five-coordinate conditions are perhaps more meaningful if written in terms of σ_L, in which cases the quantities which may be evaluated are related to

$$-\tfrac{1}{5}\langle r^{-3} \rangle_p (\sigma_L^{tba} - \sigma_X^{tba}) \quad \text{and} \quad -\tfrac{4}{15}\langle r^{-3} \rangle_p (\sigma_L^{tbe} - \sigma_X^{tba}) \tag{33.1}$$

or, alternatively,

$$-\tfrac{1}{5}\langle r^{-3} \rangle_p (\sigma_L^{tba} - \sigma_X^{tbe}) \quad \text{and} \quad -\tfrac{4}{15}\langle r^{-3} \rangle_p (\sigma_L^{tbe} - \sigma_X^{tbe}) \tag{33.2}$$

It is perhaps appropriate to point out the relationship of this treatment to that of Bancroft *et al.* (*44*). For low-spin Fe^{II} compounds, the equivalent orbitals for the σ bonds are the d^2sp^3 hybrids formed from the $3d_{z^2}$, $3d_{x^2-y^2}$, $4s$, and $4p$ iron atomic orbitals. Neglecting contributions from $4p$ orbitals or cross-terms

$$[L]^\sigma = \frac{-2}{21} \langle r^{-3} \rangle_{3d} \, \sigma_L^{oct} \tag{34}$$

where $[L]^\sigma$ is related to the parameter σ_L of Bancroft (44) by the relationship

$$[L]^\sigma = -\tfrac{2}{3}\sigma_L \tag{35}$$

In order to consider π bonding in low-spin Fe^{II} using this model, it is necessary to form 12 localized π orbitals (2 per M–L axis). Symmetry considerations, however, show that for the case of low-spin Fe^{II} octahedrally coordinated by ligands with empty π orbitals, it is not possible to write a general set of localized orbitals. Although this result does not preclude additivity of π contributions in special cases, it does suggest that there is no theoretical justification for writing a π-bonding contribution to the total EFG as a partial field gradient parameter (π_L), which is completely independent of the other ligands. In other words, some deviations from strict additivity might be expected for the π-bonding contribution to the EFG of low-spin Fe^{II} compounds.

The calculations so far described in this section have referred to regular geometries, whereas in practice, distortions from a regular geometry will be anticipated. Attempts have been made to calculate the effect of distortions using a point charge model (29, 64, 288, 402). In this case, changes in V_{rs} arising from changes in the orientation of the M–L axis are calculated assuming that [L] values remain constant. This method yields equations for V_{ZZ} which differ in the relative magnitudes of the coefficients of [L]. For example, for a tetrahedral MA_3B molecule distorted as in Fig. 3, but preserving C_{3v} symmetry,

$$V_{ZZ} = \{2[B]^{tet} - 3(1 - 3\cos^2\alpha)[A]^{tet}\}e \tag{36}$$

where α is defined in Fig. 3. Similar results are obtained for MA_2B_2 systems.

The application of the point charge model to distorted systems does not seem realistic as the expression for a given bond orbital α_L will be expected to change with orientation of the bond. The equivalence of the orbitals will, therefore, be lost and constant [L] values can no longer be assigned.

The molecular orbital model of Clark et al. [122] has been extended to consider distortions of SnA_3B (Fig. 3) and SnA_2B_2 (Fig. 3) species. The distortion of SnA_3B preserves C_{3v} symmetry, whereas that of SnA_2B_2 preserves C_{2v} symmetry, and the angles α and β are related by the equations

$$\text{for } SnA_3B: 2.19 \geqslant \alpha \geqslant \tfrac{1}{2}\pi \qquad \cos\beta = \tfrac{1}{2}(3\cos^2\alpha - 1) \tag{37.1}$$

$$\text{for } SnA_2B_2: \tfrac{1}{2}\pi \geqslant \alpha \geqslant \tfrac{1}{4}\pi \qquad \cos^2\beta = 1 - \cot^2\alpha \tag{37.2}$$

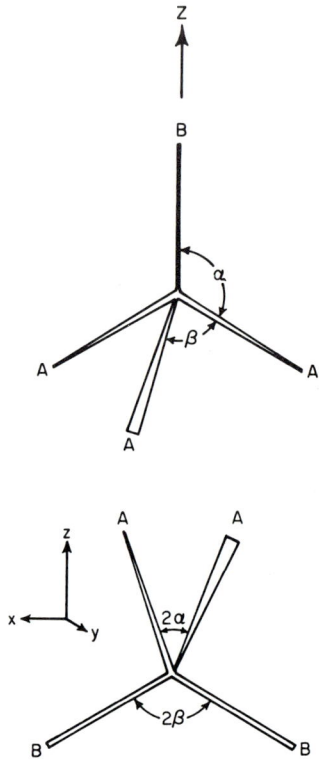

FIG. 3. Distortions of the SnA$_3$B and SnA$_2$B$_2$ compounds from tetrahedral symmetry (122). See Eqs. (37.1) and (37.2).

The calculated values of Δ are

$$\text{for SnA}_3\text{B}: \Delta_1(\alpha) = f_1(\alpha)\,\Delta_1^{\text{tet}} \tag{38.1}$$

$$\text{for SnA}_2\text{B}_2: \Delta_2(\alpha) = f_2(\alpha)|\Delta_2^{\text{tet}}|\,\text{sgn}\,\{Q\} \tag{38.2}$$

where $f_1(\alpha)$ and $f_2(\alpha)$ are functions of angle α, plotted in Figs. 4a and 4b. For SnA$_2$B$_2$, η is also a function of α and it is plotted as a broken line in Fig. 4b.

Equations (38.1) and (38.2) show that in contrast to the point charge model, Δ remains proportional to $[\text{B}]^{\text{tet}} - [\text{A}]^{\text{tet}}$. Thus, for [L] values calculated from observed quadrupole splittings on the basis of an idealized geometry, the values obtained will be *apparent* values equal to $|f(\alpha)|$ times the value for exact tetrahedral geometry. The *true* partial field gradient parameters will be different from these apparent values.

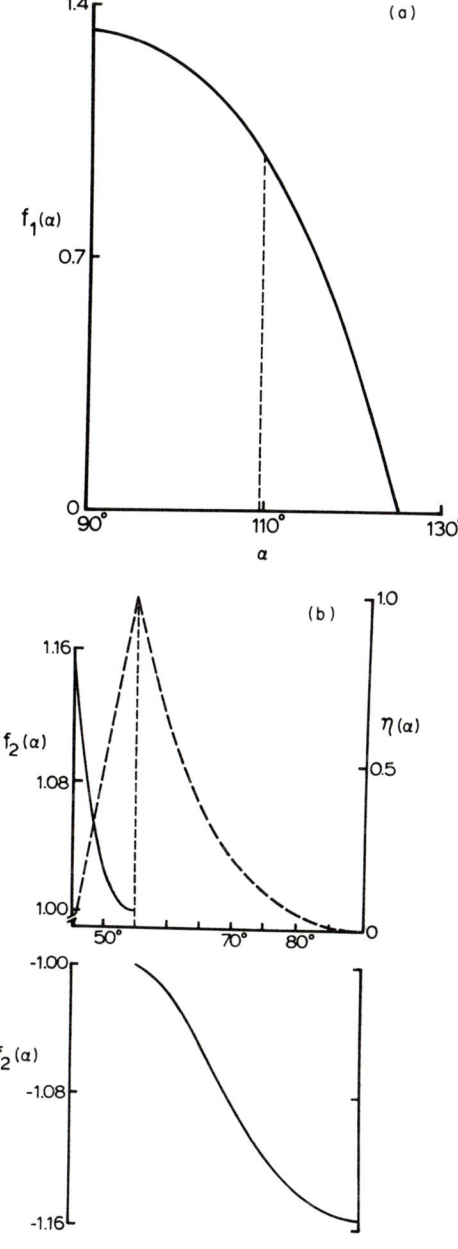

Fig. 4. $f(\alpha)$ [Eqs. (38.1) and (38.2)] plotted for distorted SnA_3B and SnA_2B_2 structures (122). (a) $f_1(\alpha)$ for SnA_3B species versus α. (b) $f_2(\alpha)$ for SnA_2B_2 species versus α. η is also a function of α, and it is plotted as broken line.

The compounds in Table VIII provide a test for this treatment as structural and Mössbauer information are available. Further, the values for α and β satisfy the appropriate Eq. (37) almost exactly. The ratios of Q.S. (ii): Q.S. (i) and Q.S. (iii) : Q.S. (i) calculated directly from observed quadrupole splittings are 0.77 and 0.68. If the quadrupole splittings are corrected to "tetrahedral values," the ratios become 1.67 and 1.23, respectively. However, the predicted ratio is 0.87 (assuming $[Cl]^{tet} = [Br]^{tet}$), which is in much closer agreement with the uncorrected value.

TABLE VIII

Mössbauer Quadrupole Splitting and Structural Data for some Tetrahedral Compounds Showing Distortions from Idealized Geometry

Code	Compound[a]	α^b	β^b	Ref.	Δ^c (mm/sec)	Ref.
(i)	$[cpFe(CO)_2]_2SnCl_2$	47.0	64.3	(426)	+2.38	(64, 288, 331)
(ii)	$[cpFe(CO)_2]SnCl_3$	119.2	98.3	(296)	+1.81	(29, 64, 288, 331)
(iii)	$[cpFe(CO)_2]SnBr_3$	117.7	100.2	(407)	1.60	(64)
(iv)	$Me_3SnMn(CO)_5$	111.6	107.3	(80)	0.75	(29, 431, 569)
(v)	$[Mn(CO)_5]_3SnCl$	101.0	116.5	(543)	1.55	(364)

[a] cp = π-Cyclopentadienyl.
[b] Average of nominally equal angles; see Fig. 3 for definition.
[c] An average of the data has been taken.

Similarly a value of $\frac{1}{2}e^2|Q|$ ($[Me]^{tet} - [Cl]^{tet}$) may be calculated from (iv) and (v) (Table VIII), to be equal to -1.13 mm/sec if no corrections for distortions are made or -1.02 mm/sec if corrections are applied. The uncorrected value is close to the value of -1.37 mm/sec obtained in Section IV,A,1,b.

The overall conclusions therefore are that small distortions from regular geometry are best ignored when applying the additivity model. One possible explanation is that whereas the quadrupole splitting weights the portion of the bond close to the metal nucleus, bond angle data weights the portion roughly midway between.

Finally, it should be noted that additive electric field gradients are, in fact, manifestations of underlying special symmetry features, which have been elegantly elucidated by Clark (119).

III. Fingerprint Uses

Any spectroscopic parameter such as the center shift and quadrupole splitting which is sensitive to the electronic or molecular structure of a compound is capable of providing the chemist with two general types of information: the characterization of the type of atom or molecule by comparison of the spectroscopic parameter with those of known species (the so-called fingerprint technique), and the elucidation of bonding and structure. The first of these will be reviewed quite briefly in this section, while perhaps the more important application—structure and bonding—will be discussed in some detail in Section IV.

The fingerprint application is very useful in the following ways: to indicate the purity of a compound; to characterize the oxidation state of, and the coordination environment about the Mössbauer atom; to detect inequivalent Mössbauer atoms in compounds or minerals, and to estimate the amount of each Mössbauer atom present; and to identify the compounds or species in a complex mixture.

Unlike many other spectroscopic experiments, the line shapes in Mössbauer spectra are normally very well-behaved Lorentzians making the spectra amenable to detailed computer processing. Thus, if the peak or peaks are slightly asymmetric, or if there is a slight shoulder on one of them, one can usually be confident that there is a Mössbauer atom present which is causing this asymmetry or shoulder.

A. Oxidation States and Inequivalent Mössbauer Atoms

Mössbauer spectra are often very useful to characterize the oxidation state of the Mössbauer atom, especially in chemically difficult situations. The very different C.S. values for some of the different oxidation states of iron indicate that these should be diagnostic (Table IX). Any gross deviation of this parameter from the expected value might indicate

TABLE IX

Center Shifts (C.S.) for High-Spin Iron Compounds Relative to Nitroprusside[a,b]

Oxidation state	+1	+2	+3	+4	+6
Center shift	~+2.2	~+1.4	~+0.7	~+0.2	~−0.6

[a] From Ref. (297).
[b] Data in mm/sec at room temperature.

undesirable reactions. An excellent example concerns the Mössbauer spectra of high-spin Fe^{II} chelates of salicylaldoxime, salicylaldehyde, and others. The first and subsequent reported spectra of these compounds gave C.S. (~0.60 mm/sec) and small Q.S. values characteristic of Fe^{III} compounds (236, 376, 532), but these parameters were still attributed to the expected Fe^{II} species with a large degree of π backbonding. However, recent work (177) conclusively shows that the above compounds are, in fact, oxidation products of Fe^{II} compounds, which give normal Fe^{II} C.S. of about 1.4 and 2.5 mm/sec, respectively. Preparation of the Fe^{II} compounds must be carried out in a vacuum system.

TABLE X

CENTER SHIFTS OF Sn^{II} AND Sn^{IV} COMPOUNDS

Compound	C.S.[a]	Ref.
1. α-Tin	+2.10	
2. (catecholate Sn)	+2.95	(50)
3. (catecholate Sn)	+3.13	(50)
4. (naphthalenediolate Sn)	+3.08	(50)
5. $(C_5H_5)_2Sn$	+3.73	(317)
6. $(Ph_2Sn)_n$	+1.42	(280)
7. $(Bu_2Sn)_n$	+1.55	(280)
8. $[(C_5H_5)_2Sn]_n$	+0.72	(317)

[a] Data in mm/sec, relative to SnO_2 or $BaSnO_3$.

Mössbauer spectra are very useful here, because the Fe^{II} and Fe^{III} compounds cannot be readily distinguished by magnetic measurements or chemical analyses.

For a number of other Mössbauer isotopes, $\delta R/R$ [Eq. (1)] is large enough so that different oxidation states give measurably different center shifts. For example, the difference in center shift for the +3 and +5 oxidation states of ^{121}Sb is over 10 mm/sec (62, 496, 522), and this large difference has been used to estimate the amount of Sb^{III} and Sb^V

in nonstoichiometric oxides and sulfides of antimony, such as Sb_2O_4 and Sb_2S_3 (62, 71, 522). Similarly, for ^{119}Sn Mössbauer, Sn^{II} compounds invariably give higher C.S. values than α-tin, whereas Sn^{IV} compounds give lower C.S. values than α-tin. For example, in Table X, compounds 2–5 are clearly Sn^{II} species (50, 317), while compounds 6, 7, and 8 are Sn^{IV} species with polymeric structures.

For ^{129}I (90, 310, 445), ^{127}I (205, 361, 452), ^{129}Xe (451, 453, 454), and ^{125}Te (219, 266, 361, 545) the C.S. is usually sensitive enough to determine oxidation states in chemically difficult situations [for a summary of results, see Shenoy and Ruby (510)]. Similarly, ^{197}Au (49, 108, 222), ^{193}Ir (495, 552), ^{237}Np (458), ^{99}Ru (128, 130, 362, 470, 471), and ^{151}Eu (197, 272), to name a few, often give diagnostic C.S. values. The above isotopes, partly because of sensitivity of the C.S., are among the most widely studied species at the present time.

However, there are a number of situations where Mössbauer cannot distinguish between oxidation states readily. For low-spin iron compounds containing strong π-acceptor groups such as CO, the C.S. values have a fairly small range for oxidation states from +2 to −2 (110). Thus, $K_3Fe^{III}(CN)_6$ and $K_4Fe^{II}(CN)_6$ have remarkably similar center shifts (76, 238, 239). Similarly, in a study of Ir compounds (562) the center shift values for $IrCl(CO)[P(C_6H_5)_3]_2$ and $XYIrCl(CO)[P(C_6H_5)_3]_2$, where

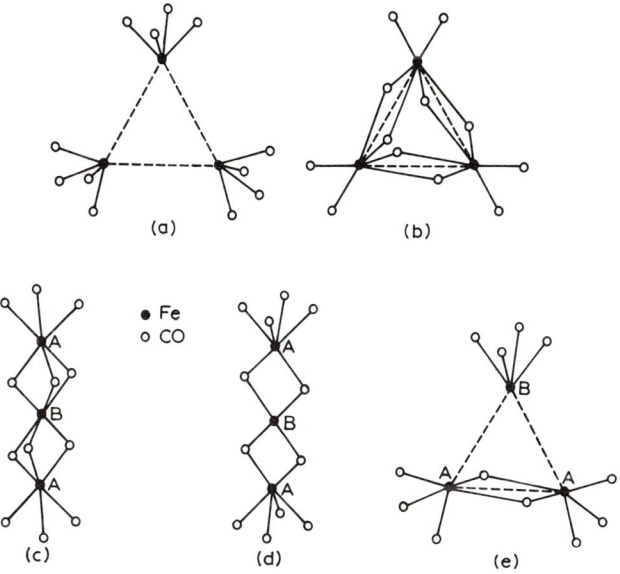

FIG. 5. Possible structures for $Fe_3(CO)_{12}$ (218).

X,Y = Cl, H, Br, etc., were remarkably similar despite the formal Ir oxidation states of +1 and +3, respectively.

The Mössbauer effect has been widely used for compounds containing two or more Mössbauer atoms to determine whether they are equivalent. This information is usually very useful for structural predictions. The first and most notable contribution in this area is given by the Mössbauer spectrum of $Fe_3(CO)_{12}$ (218). Until a definitive crystal structure was published recently (557), several different structures were proposed on the basis of incomplete X-ray work and infrared data. These structures were based either on a triangle of Fe atoms or a linear array of Fe atoms (Fig. 5). The Mössbauer spectrum of $Fe_3(CO)_{12}$ (Fig. 6) clearly

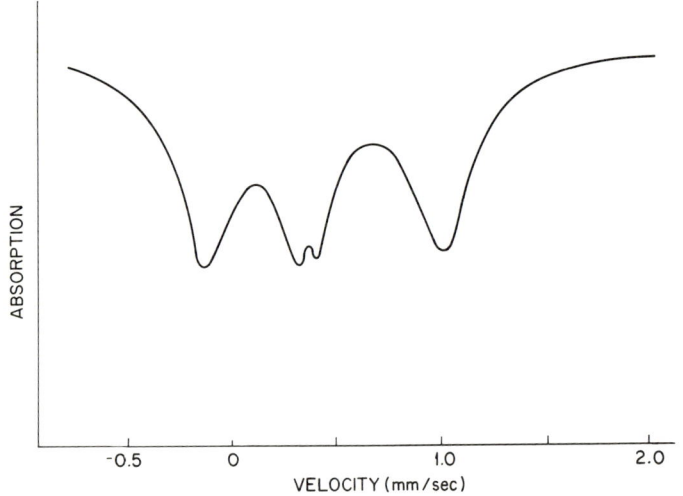

FIG. 6. Mössbauer spectrum of $Fe_3(CO)_{12}$ (293).

indicates that there are two distinguishable (to Mössbauer) iron atoms present, with the outer peaks due to one type of iron atom (Fe_A) and a narrow doublet due to Fe_B. From the areas, it is apparent that Fe_A : Fe_B ~2 : 1. Linear structures of the type shown in Fig. 5c,d would be consistent with this spectrum, but the spectrum of the $Fe_3(CO)_{11}H^-$ anion, which is very similar to that of $Fe_3(CO)_{12}$ (Table XI), rules out a linear structure. If there are any bridging carbonyls in a linear structure, it is not possible to replace any CO group by H and leave two iron atoms equivalent. The above evidence strongly suggests that $Fe_3(CO)_{12}$ has an unsymmetrical triangular structure, with one iron atom coordinated to the same atoms in both $Fe_3(CO)_{12}$ and the hydride, since the parameters for iron atom A in both the parent molecule and the anion are very similar. One structure which is consistent with this evidence is given in Fig. 5e. The H⁻ substitutes for a bridging carbonyl, leaving atoms A equivalent, but the C.S. is decreased somewhat from that in $Fe_3(CO)_{12}$. The very similar parameters for Fe_B in both structures are consistent with the identical nearest neighbors for Fe_B in both structures. There are other unsymmetrical triangular structures which would fit the above evidence, but the recent X-ray structure (557) confirms that $Fe_3(CO)_{12}$ has the structure shown in Fig. 5e.

TABLE XI

MÖSSBAUER PARAMETERS FOR $Fe_3(CO)_{12}$ AND $NaFe_3(CO)_{11}H$ AT 298°K[a]

Compound	Fe atom	C.S.[b]	Q.S.
$Fe_3(CO)_{12}$	A	+0.34	1.05
	B	+0.32	<0.20
$Na[Fe_3(CO)_{11}H]$	A	+0.26	1.32
	B	+0.28	<0.20

[a] Relative to nitroprusside (mm/sec).
[b] From Ref. (218).

There are a number of other examples, where the observation of more than the expected number of lines for one Mössbauer atom has been very useful in structural elucidation: e.g., $I_2Cl_4Br_2$ (445), and a host of polynuclear iron compounds (223, 224, 295, 370) to name a few. However, it should be emphasized that if a polynuclear compound gives rise to apparently one set of absorptions, this negative evidence cannot be taken as proof of equivalence. For example, the crystal structure of $[(\pi\text{-}C_5H_5)_2Fe_2(CO)_3]_2DPPA$ (102) shows that the two iron atoms are inequivalent with a DPPA molecule ($Ph_2PC\equiv CPPh_2$) linking two $Fe_2(CO)_3(\pi\text{-}C_5H_5)_2$ units. One iron atom is bound to a C_5H_5, a terminal CO, and two bridging CO, whereas the other is bound to a C_5H_5 and the two bridging CO, but a terminal P on the DPPA. However, the Mössbauer spectrum (102, 104) shows only one narrow doublet with line widths of 0.24 and 0.26 mm/sec, which might suggest that the two irons are equivalent. Similarly, a previous Mössbauer prediction of the equivalence of iron atoms in $C_8H_{10}Fe_2(CO)_6$ (206) has recently been shown incorrect by an X-ray study (143). For many such spectra, detailed computation becomes important for detecting any line broadening or asymmetry which could be due to an overlap of Lorentzians.

B. Decomposition Reactions

Mössbauer spectroscopy has been particularly valuable in several studies for following reactions in the solid state. The information from Mössbauer spectra can often not be obtained by other techniques, but it is usually important to use Mössbauer in conjunction with other techniques.

The classic example in this area of research is the elucidation of the complex decomposition scheme of $Sr_3[Fe^{III}(C_2O_4)_3]_2 \cdot 2H_2O$ using mainly C.S. values as a guide to the oxidation state of iron (Table IX) and weight losses from thermal analyses (253). Table XII shows that the C.S. varies from the initial value of 0.65 mm/sec (Fe^{III}), to 1.44 mm/sec (Fe^{II}) after heating to 200°C, to 0.60 mm/sec (Fe^{III}) at 400°C, to 0.27 (Fe^{IV}) and 0.82 mm/sec (Fe^{III}) at temperatures of 700°C and above. A mechanism which is consistent with these oxidation states of iron and the weight loss data is:

$$Sr_3[Fe^{III}(C_2O_4)_3]_2 \cdot 2H_2O \xrightarrow{100°-175°} Sr_3[Fe^{III}(C_2O_4)_3]_2 + 2H_2O$$

$$Sr_3[Fe^{III}(C_2O_4)_3]_2 \xrightarrow{175°-350°} Sr_3[Fe_2^{II}(C_2O_4)_5] + 2CO_2$$

$$Sr_3[Fe_2^{II}(C_2O_4)_5] \xrightarrow{300°-450°} 3SrCO_3 \cdot Fe_2^{III}O_3 + 5CO + 2CO_2$$

$$3SrCO_3 \cdot Fe_2O_3 \xrightarrow{500°-650°} SrCO_3 \cdot 2SrFe^{IV}O_{2.8} + 0.6CO + 1.4CO_2$$

$$SrCO_3 \cdot 2SrFe^{IV}O_{2.8} \xrightarrow{700°-1000°} Sr_3Fe_2^{III}O_6(+\text{---})$$

In a similar fashion, using Mössbauer to distinguish between the oxidation states of iron, the thermal decomposition of $K_3Fe^{III}(C_2O_4)_3 \cdot 3H_2O$ (31) and $Fe(C_2O_4) \cdot 2H_2O$ (314) have been followed. The $FeSO_4 + KCN$ reaction has also been studied using Mössbauer (308).

The effect of external radiation on compounds containing Mössbauer atoms has been recently studied. The radiation-induced decomposition of $K_3Fe(C_2O_4)_3 \cdot 3H_2O$ proceeds stoichiometrically to either $K_2Fe^{II}(C_2O_4)_2(H_2O)_2$ (30) or $K_6Fe_2^{II}(C_2O_4)_5$ in vacuum and air, respectively. Other studies (309, 560) indicate that Fe^{III} is formed in Fe^{II} compounds during irradiation. In $Fe(acac)_3$ (32), the large line width is reduced substantially on irradiation, and this has been attributed to a relaxation mechanism.

TABLE XII

ROOM TEMPERATURE MÖSSBAUER PARAMETERS FOR THE
THERMAL DECOMPOSITION OF $Sr_3[Fe(C_2O_4)_3]_2 \cdot 2H_2O$ [a]

Temperature of heating (°C)	C.S.[b]	Q.S.	Assignment
25	+0.65	0.44	$Sr_3[Fe^{III}(C_2O_4)_3]_2 \cdot 2H_2O$
200	+1.44	2.3	$Sr_3Fe_2^{II}(C_2O_4)_5$
400	+0.60	0.70	$SrCO_3 \cdot Fe_2^{III}O_3$
600	+0.44	0.74	
700	+0.27	~0	$SrCO_3 \cdot 2SrFe^{IV}O_{2.8}$(A)
	+0.82	~0	$Sr_3Fe_2^{III}O_6$(B)
1000	+0.23	~0	(A)
	+0.75	~0	(B)

[a] Relative to nitroprusside (mm/sec).
[b] From Ref. (253).

C. THE EFFECT OF TEMPERATURE AND PRESSURE ON THE ELECTRONIC STRUCTURE OF IRON COMPOUNDS

Mössbauer spectroscopy has been used to study a number of very interesting changes in the electronic state of the iron atom in Fe^{II} and Fe^{III} compounds at low temperatures and high pressures. A large number of Mössbauer and magnetic studies of high spin–low spin equilibria in Fe^{II} and Fe^{III} compounds have been undertaken (73, 145, 195, 287, 356, 373–375, 482). For example, in the series of Fe^{II} bisphenanthroline complexes, $Fe(phen)_2X_2$, if $X = Cl^-$, Br^-, I^-, or other ligands low in the spectrochemical series, high-spin compounds are obtained having, first, the large C.S. and Q.S. (Table XIII) characteristic of Fe^{II} high-spin compounds and, second, large magnetic moments ($\mu = $ 5.0–5.3 B.M.) slightly higher than those expected for four unpaired electrons. In contrast, if X is CN^- or NO_2^-, then low-spin compounds are obtained with the small C.S. and Q.S. characteristics of Fe^{II} low-spin compounds (Table XIII) and the very small magnetic moments expected for no unpaired electrons. For intermediate ligands such as NCS or NCSe (374, 375), the high- and low-spin configurations are of very similar energy and in the thermally accessible range. Thus, $Fe(phen)_2(NCS)_2$ is high spin at room temperature, but low spin at liquid N_2 temperatures. Using Mössbauer spectra, magnetic susceptibilities, and other spectral techniques, Konig and Madeja (375) have presented evidence that for Fe(phen)$_2$mal . 7H$_2$O and Fe(phen)$_2$F$_2$. 4H$_2$O, Fe is surprisingly in the intermediate $S = 1$ state.

Other, more subtle changes in the ligand environment lead to a similar high spin–low spin equilibria. In a study of octahedral Fe^{II} chelates based on the hydrotris(1-pyrazolyl)borate ligand, Jesson et al. (356) have shown that the H,H- and CH$_3$,CH$_3$-substituted complexes are low spin and high spin, respectively, over the temperature range 295 to

TABLE XIII

Mössbauer Parameters for Fe(phen)$_2$X$_2$ and Related Compounds[a]

Compound	Temperature (°K)	C.S.[b]	Q.S.	Type of spin
Fe(phen)$_2$(NO$_2$)$_2$	293	0.53	0.38	Low
Fe(phen)$_2$Cl$_2$	293	1.21	3.00	High
Fe(phen)$_2$(NCS)$_2$	293	1.23	2.67	High
	77	0.62	0.34	Low
Fe(phen)$_2$(NCSe)$_2$	293	1.28	2.52	High
	77	0.60	0.18	Low
Fe(phen)$_2$mal · 7H$_2$O	293	0.59	0.18	Intermediate
	77	0.52	0.18	
Fe(phen)$_2$F$_2$ · 4H$_2$O	293	0.58	0.21	Intermediate
	77	0.55	0.16	

[a] Relative to nitroprusside (mm/sec).
[b] From Refs. (374, 375).

4°K, whereas the CH$_3$, H-substituted complex is high spin at 295°K, but reverts to low spin near 150°. Goodgame and Machado (287) have shown that Fe(pyim)$_3$(ClO$_4$)$_2$ forms two geometrical isomers, one low spin and the other giving high spin–low spin equilibria over the temperature range studied.

In many of these studies, the energy separation between the $^5T_2$ and $^1A_1$ states has been determined, although recent evidence (373) suggests that this energy separation is strongly temperature dependent.

Drickamer and co-workers (48, 107, 196, 230, 250, 251, 385, 386, 435) have done a great deal of interesting work on the effect of high pressure on a wide variety of iron-containing chemicals, and these studies again indicate the great diagnostic use of Mössbauer for detecting different types of iron atoms. Many of these studies indicate that Fe^{III} reduces to

Fe^{II} with increasing pressure. The phenomenon is reversible with some hysteresis. Typical spectra of $Fe(acac)_3$ are shown in Fig. 7 (107) indicating that the initial Fe^{III} species is reduced partially to Fe^{II} on application of large pressures. On releasing the pressure, the Fe^{II} peaks disappear. The Fe^{III} to Fe^{II} reduction involves an electron transfer from a nonbonding ligand level to an antibonding $3d$ level on the iron. The $3d$ orbitals spread with increasing pressure, lowering their energy relative to the ligand levels, thus permitting the thermal transfer of an electron from ligand to metal.

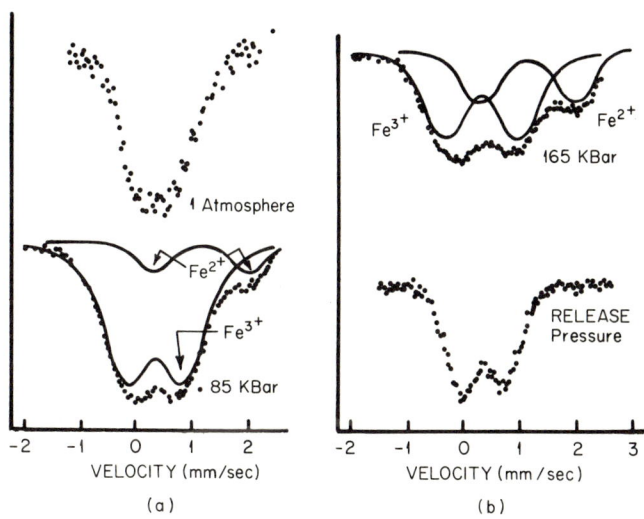

Fig. 7. Mössbauer spectra of $Fe(acac)_3$ (107). Note the Fe^{II} peaks after pressure is applied, and the return to a pure Fe^{III} spectrum after release of pressure.

Similarly, Fe^{IV} in $SrFeO_{2.86}$ is reduced reversibly to Fe^{III}, whereas the Fe^{VI} in $BaFeO_{4-x}$ is reduced irreversibly to Fe^{IV} with perhaps some Fe^{III} (435), presumably again by ligand to metal charge transfer.

These studies should be of the greatest importance in understanding not only the electronic structure of iron compounds, but also the mechanism of reduction of iron in the earth's interior.

D. SITE POPULATIONS IN SILICATE MINERALS

Mössbauer spectroscopy has been applied as a fingerprint technique to a wide range of iron-containing minerals, and a great deal of useful information has been obtained. The most significant and important studies center around the determination of Fe^{2+} site populations and Fe^{3+}/Fe^{2+} ratios in silicate minerals (20, 21, 23–26, 40, 181, 200, 221, 264, 311, 343, 474, 503, 549, 550 and references]. Most silicate minerals such as orthopyroxenes [$(FeMg)_2Si_2O_6$] have two or more cation sites into which Fe^{2+} and Mg^{2+} can enter (172). In orthopyroxenes, there are two such sites; the distorted M2 site, and the more regularly octahedral M1 site. X-Ray diffraction measurements (263) first showed that Fe^{2+} orders in the M2 position, i.e., it prefers the M2 position over the M1

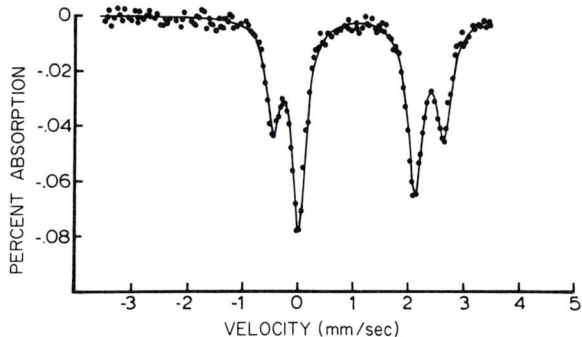

FIG. 8. Mössbauer spectrum of an orthopyroxene of approximate composition $(Fe^{2+}_{0.57}Mg_{0.43})Si_2O_6$ at 80°K (549). The outer two peaks are due to Fe^{2+} in M1, while the inner two peaks are due to Fe^{2+} in M2.

position. Thus, in an orthopyroxene of composition $(Fe_{0.5}Mg_{0.5})_2Si_2O_6$, about 85% of the Fe^{2+} enters the M2 position. Since the amount of ordering should be temperature dependent, the determination of site populations has become of great interest not only for determining crystal chemical relationships, but also as a potential geothermometer.

Mössbauer spectroscopy offers a convenient, rapid, and accurate method for obtaining Fe^{2+} site populations, since structurally different Fe^{2+} ions (such as Fe^{2+} in M2 and M1 above) often give distinctive spectra (Fig. 8). The area (A) under a peak is proportional to the amount (n) of Fe^{2+} in each site, i.e., $A_{M2}/A_{M1} = n_2/n_1$. The first quantitative results reported in 1966 for orthopyroxenes, anthophyllites, and cummingtonites (23, 40) have led to the determination of site populations in a large range of iron-containing minerals such as pyroxenes (25, 311, 503, 549, 550), biotites (343), and other amphiboles (24). Although there are difficulties in obtaining accurate site populations for some minerals (21), the excellent agreement between Mössbauer and X-ray results

(Table XIV) suggests that accurate values can be obtained for most minerals.

In addition to Fe^{2+} ratios, Fe^{3+}/Fe^{2+} ratios in minerals can easily be obtained (24, 181), and these are, in general, in good agreement with the corresponding ratios from chemical analysis (Table XV). Wet chemical methods for obtaining these ratios are often inaccurate, and the nondestructive Mössbauer method should become a standard method for determining this ratio.

Virgo and Hafner (311, 549, 550) and Saxena and Ghose (503) have obtained Fe^{2+} site populations for a large number of natural and syn-

TABLE XIV

COMPARISON OF X-RAY AND MÖSSBAUER SITE POPULATIONS IN SILICATE MINERALS

Mineral	Total Fe^{2+} per formula unit	Fe^{2+} site populations (per formula unit)			
		Mössbauer	Ref.	X-ray	Ref.
Orthopyroxene	1.06	M2 = 0.87	(550)	M2 = 0.90	(263)
		M1 = 0.19		M1 = 0.15	
Glaucophane	0.61	M3 = 0.28	(24)	M3 = 0.29	(436)
		M1 = 0.33		M1 = 0.32	
Grunerite	6.13	M4 = 1.96	(26)	M4 = 1.97	(228)
		M123^a = 4.17		M123 = 4.13	
Anthophyllites	1.61	M4 = 1.39	(40)	—	
		M123 = 0.22		—	
	1.47	—		M4 = 1.30	(229)
		—		M123 = 0.17	
Cummingtonites	2.48	M4 = 1.65	(26)	—	
		M123 = 0.83		—	
	2.50	—		⎧M4 = 1.50 ⎨ ⎩M123 = 1.29	(231)
		—		⎧M4 = 1.74 ⎨ ⎩M123 = 0.58	(262)

^{a} M123 refers to Fe^{2+} per formula unit in M1 + M2 + M3.

thetic pyroxenes and determined both preliminary thermodynamic and kinetic data for the exchange reaction:

$$Mg(M1) + Fe^{2+} (M2) \rightleftarrows Fe^{2+}(M1) + Mg(M2)$$

A number of papers have appeared recently which discuss the kinetics and thermodynamics of the above type of exchange reaction (*305, 414, 503, 538* and references). These papers and the experimental papers mentioned above indicate the great potential of site population determinations for determining the highest temperature of formation of orthopyroxenes and kinetics of cooling, although the low activation energies for forward and reverse reactions may make such determinations very difficult.

TABLE XV

COMPARISON OF MÖSSBAUER AND CHEMICAL ANALYSES VALUES FOR % Fe^{3+}/TOTAL IRON[a]

Mineral	% $Fe^{3+}/(Fe^{2+} + Fe^{3+})$	
	Mössbauer	Chemical analyses
Howieite	24	20
Deerite	37	37
Crocidolite	41	41
Glaucophane	28	32
Crossite	37	40

[a] From Refs. (*24, 27*).

E. PREPARATION OF NOVEL COMPOUNDS

It is apparent from the previous discussion that the vast majority of Mössbauer spectra have been taken with a constant standard single-line source with the absorber as the compound under study. However, a number of interesting papers have appeared in which a standard single-line absorber has been used with the source compound being of interest. Because the source is radioactive, extensive rearrangement of the electrons and bond breaking may occur on the Mössbauer time scale (the half-life of the excited state). As a result, the observed Mössbauer spectrum may not reflect the initial electronic and/or ligand environment about the Mössbauer atom. The differences in spectra can be very useful to the radiochemist in investigating the effect of the radioactive process [for example, see (*225, 248, 249, 272, 352, 358, 398, 541*)], but we are concerned here with the situations where the spectrum does reflect the initial electronic and ligand environment on the Mössbauer lifetime.

The classic studies in this area have been undertaken by the Perlows (*450, 451, 453, 454*), who have produced a number of ^{129}Xe compounds by the β decay of iodine compounds prepared with ^{129}I. For example, the spectra of $XeCl_4$ (*451*), $XeCl_2$ (*453*), and $XeBr_2$ (*454*) were obtained after the decay of $KICl_4$, $KICl_2$, and $KIBr_2$, respectively, using a standard Xe absorber such as a xenon clathrate. These compounds have to be stable for longer than the lifetime of the nuclear excited state ($\sim 10^{-9}$ sec) in order to be "observed."

Somewhat similar experiments have been undertaken using ^{83}Kr Mössbauer (*322, 444, 497*). Using ^{83}SeO$_2$ as a source (^{83}Se$\rightarrow$^{83}Kr) and a krypton clathrate HQ–Kr as absorber, evidence has been presented (*322*) for the existence of a bona fide Kr–O bond. However, using K^{83}BrO$_3$ as source (*444*) (^{83}Br$\rightarrow$^{83}Kr), no evidence was found for the existence of a KrO$_3$ species, in contrast to the production of XeO$_3$ from KIO$_3$ and its observation by Mössbauer (*450*).

Some source work (*495*) using ^{193}Os compounds as sources and single-line ^{193}Ir absorbers have also provided interesting results. The 73-keV excited nuclear energy level of ^{193}Ir is populated in the β decay of ^{193}Os. As in the ^{129}I$\rightarrow$^{129}Xe transition, the majority of the Os atoms should end up in a valence state which is higher by unity than the parent Os atom. The spectra of OsO$_4$, K$_2$OsO$_4 \cdot$2H$_2$O, and Os(C$_5$H$_5$)$_2$ sources appear to be characteristic of IrO$_4^+$, IrO$_4^-$, and Ir(C$_5$H$_5$)$_2^+$, respectively, and provide valuable information on unknown and highly unstable iridium species.

In other forms of radioactive decay, more substantial electronic rearrangement takes place. For example, ^{57}Co captures an orbital electron to give ^{57}Fe. Co^{III}(d^6) should yield Fe^{III}(d^5) if the resulting Auger process does not disturb the molecular structure. In most cases, a mixture of charge states are obtained (*225, 248, 249, 352, 541*), and in other cases, the absorptions are not characteristic of Fe^{II} or Fe^{III} species and very broad lines are observed. Yet, for interesting Co compounds such as vitamin B$_{12}$ and Co^{II}phthalocyanine (*422, 423*), the spectra are characteristic of Fe^{II} with apparently very little, if any, fragmentation. The potential uses and advantages of such emission spectra have been discussed (*423*).

F. Frozen Solution Studies

The Mössbauer effect cannot be observed in a liquid, but Mössbauer spectra of glasses and smectic liquid crystals have been observed. Thus, it is possible to obtain valuable information on the structure of a molecule in solution, or the structure of the solvent, by taking a spectrum of a glass at, for example, 80°K. There are, however, several difficulties which should be emphasized. First, one is often not sure whether a true glass has been obtained, or whether the solid has crystallized. Microcrystalline domains are very difficult to detect. Second, it is sometimes possible that the solvent may coordinate to the Mössbauer atom in unexpected situations. For example, Sn often coordinates to solvent molecules which are normally very "weak" ligands. In other cases, Mössbauer spectra are used to study solvent coordination. Mössbauer spectra of five-coordinate bis(N,N-diethyldithiocarbamato)iron(III) chloride [Fe(dtc)$_2$Cl] and related compounds show a large Q.S. of about 2.7 mm/sec in the solid state, but a Q.S. value of about 0.8 mm/sec in solvents such as dimethylformamide (*178*). By contrast the six-coordinate Fe(dtc)$_3$ compound gives a small Q.S. of about 0.6 mm/sec both in solution and solid. These results, and other spectroscopic evidence, strongly indicate that the coordination about the Fe^{III} changes from five in the solid to six in solution.

Solution Mössbauer spectra of Sn compounds of the type R′SnR$_3$ (R = C$_2$H$_5$, C$_6$H$_5$; R′ = 4-thiopyridone, thiophenol) have also been very helpful in determining the coordination number about the Sn atom—both in solid and solution (*424*).

In an interesting Mössbauer study of the pH dependence of the species in an aqueous solution of NH$_4$Fe^{III} EDTA (*15*), the Mössbauer spectrum changes markedly from pH 6.8 to 3.9. By comparison with solid spectra, the species at pH 6.8 and 3.9 have been assigned to [Fe^{III}(EDTA(OH))]$^{2-}$ and Fe^{III}(EDTA)$^-$, respectively.

In other studies, the use of frozen solution spectra has not provided as much useful information. In a study of [π-C$_5$H$_5$Fe(CO)$_2$]SnCl$_3$ and related compounds (*331*), reasonably conclusive evidence was not obtained as to whether the short Fe–Sn bond was due to intermolecular stacking forces or the specific nature of the chemical bonds. In another area, Fe$_3$(CO)$_{12}$ supposedly has a different structure in solution than in the solid (*461*), but recent attempts to observe a different spectrum in solution than in the solid have not met with success (*38*).

A number of workers have studied the temperature dependence of the recoil free fraction of Fe^{2+} doped water (*100, 180, 425, 447*) and methanol (*515*). Here the Fe^{2+} ions are used as a probe to study the change in frozen solvent properties over a range of temperature. In all cases, the f value drops off sharply, and sometimes approaches zero at a point variously ascribed to a cubic-hexagonal phase change in ice (*180, 425*), a two-phase model, with the ferrous ions associated with a glassy or amorphous variety of ice (*447*), and a transition from a rigid glass to a supercooled state (*515*). Recent results (*100*) on Fe^{2+} ions in the cubic phase of ice indicate that the Fe^{2+} ions are unaffected by this phase and that the Fe^{2+} ions are associated with the glassy fraction of the absorber.

IV. Bonding and Structure

A. Sn^{IV} Compounds

1. Quadrupole Splittings

Until very recently, two major problems have hindered any systematic interpretation of Sn^{IV} quadrupole splittings. First, there has been considerable debate as to the origin of the EFG, i.e., whether π-bonding asymmetries, σ-bonding asymmetries, or q_{lat} effects are predominantly responsible for the observed quadrupole splittings. Second, the influence of changing coordination number on the magnitude of the quadrupole splitting was not known, mainly because sufficient structural data was not available. The effect of structure on quadrupole splittings is of particular interest, as many compounds with a four-coordinate stoichiometry do, in fact, possess associated structures in the solid state, involving five- and six-coordinate tin atoms. In this section we shall first deal with the origins of the quadrupole splittings of tin(IV) compounds and then consider in detail the correlation between quadrupole splitting and structure. In the first part of the discussion, it will be necessary to make some structural assumptions, which will be justified later.

In the following discussion we will employ the general symbols R and X to denote an organic group or an electron-withdrawing ligand, respectively. This convention will also be used in Section IV,A,2.

a. σ or π Bonding? Early ^{119}Sn Mössbauer work (*5, 6, 8, 9, 84, 138, 279*) revealed large quadrupole splittings (in the range 2–4 mm/sec) for a wide variety of substituted organotin compounds of the type R_nSnX_{4-n} ($n = 1$–3) with X = F, Cl, Br, I, O, and S. It was suggested (*6, 138*) that the EFG's in these compounds were produced by differences in the polarities of the tin–ligand σ bonds, which could give rise to both a q_{lat} and a q_{val} contribution to the EFG [Eqs. (7) and (9)]. Later work revealed that the compounds $(Ph_3Sn)_2$ (*269*), $(Ph_3Sn)_4M$ (M = Sn, Ge, Pb) (*269*), Ph_3SnX (X = H, Li) (*9*), R_nSnH_{4-n} ($n = 1$–3, R = Me, Bu, Ph) (*9, 334*), and Me_3SnNa (*141*) showed no resolvable quadrupole splitting. If the difference in electronegativity between the donor atoms of the ligands ($\Delta\Psi$) is used as a criterion of differences in bond polarity (*138, 269*), then from Table XVI it is clear that $\Delta\Psi$ values for compounds which show quadrupole splitting are in the same range as those for which single-line spectra are observed (*269*). From this observation it was suggested (*269*) that EFG's produced by inequalities in the tin–ligand σ bonds were too small to result in resolvable quadrupole splitting and that some other factor was responsible for the observed quadrupole splittings for organotin halides and chalcogenides.

Gibb and Greenwood noted (*269*) that quadrupole splitting seemed to be confined to those organotin compounds in which the X ligand contained nonbonding lone-pair electrons. This observation led to the

TABLE XVI

ELECTRONEGATIVITY DIFFERENCES, $\Delta \Psi$, BETWEEN CARBON AND OTHER ELEMENTS[a]

No quadrupole splitting		Quadrupole splitting	
Element	$\Delta \Psi_L$[b]	Element	$\Delta \Psi_L$[b]
H	−0.30	F	+1.60
Li	−1.53	Cl	+0.33
Ge	−0.48	Br	+0.24
Sn	−0.78	I	−0.29
Pb	−0.95	O	+1.00
		S	−0.06
		N	+0.57

[a] Allred Rochow values from F. A. Cotton and G. Wilkinson, "Advanced Inorganic Chemistry," 2nd ed., p. 103. Wiley (Interscience), New York, 1966.
[b] $\Delta \Psi_L = \Psi_L - \Psi_C$.

suggestion that these lone-pair electrons formed dative $p_\pi \to d_\pi$ (*269*) or $p_\pi \to p_\pi$ (*334*) bonds with the empty 5d and 6p orbitals of the tin atom and that this asymmetric introduction of π-electron density was responsible for the observed quadrupole splittings. A similar explanation was proposed to account for the quadrupole splitting trends of a range of six-coordinate tin species (*300*). The data available for these compounds at that time are summarized in Table XVII, and it can be seen that although organotin complexes show large quadrupole splittings, in the absence of a tin–carbon bond, no quadrupole splittings could be observed, even when there is a noncubic arrangement of the ligands. It was proposed (*300*) that for compounds in which all six atoms directly bonded to the Sn atom have filled nonbonding shells available for π bonding, the σ-bonding asymmetries are insufficient to give quadrupole splitting.

However, for organotin derivatives, the lack of π bonding in the Sn–R bond results in a large degree of π-electron asymmetry and, hence, in large quadrupole splitting.

Parish and Platt (440) have presented data for a series of substituted four-coordinate organotin compounds of the type R_3SnX (in which R=Me or Ph and $X = C_6F_5$, C_6Cl_5, C≡CPh, CH=CH$_2$, and CCl=CCl$_2$) for which the observed quadrupole splitting trends are incompatible with

TABLE XVII

SELECTION OF EARLY MÖSSBAUER DATA FOR SIX-COORDINATE Sn^{IV} COMPLEXES[a]

Code No.[b]	Compound	C.S.[c,d]	Q.S.[c]	Γ[c,e]	Refs.
1	SnCl$_4$bipy	0.42	—	1.31	(300)
2	SnBr$_4$bipy	0.66	—	1.17	(300)
3	SnI$_4$bipy	0.95	—	1.42	(300)
4	SnCl$_4$(en)$_2$	0.50	—	—	(278)
5	SnBr$_4$(en)$_2$	0.43	—	—	(278)
6	SnCl$_4$(py)$_2$	0.00	—	—	(278)
7	SnCl$_4$(oxH)$_2$	0.45	—	1.22	(300)
8	SnCl$_4$[(NH$_2$)$_2$CS]$_2$	0.95	—	—	(278)
9	SnBr$_4$[(NH$_2$)$_2$CS]$_2$	0.80	—	—	(278)
10	SnCl$_2$(ox)$_2$	0.30	—	1.50	(300)
11	SnCl$_2$(acac)$_2$	0.25	—	1.15	(300)
12	Me$_2$Sn(ox)$_2$	0.85	1.93	1.05, 1.21	(300)
13	Me$_2$Sn(H$_2$O)$_n$OH$^+$	1.37	3.90	1.31, 1.50	(300)

[a] More data for similar compounds may be found in Refs. (300) and (278)
[b] Code number will be preceded by table number in text.
[c] Data in mm/sec at liquid N$_2$ temperature.
[d] Relative to SnO$_2$, values from Ref. (278) converted assuming the center shift of Sn = 2.70.
[e] Full width at half height; data taken with SnO$_2$ source.

the π-bonding theory of quadrupole splitting, but which are fully consistent with a σ-bonding interpretation. Sams et al. (114, 115, 150) have reported further data for halogen-substituted organotin phenyl, vinyl, and acetylene derivatives which confirm and further clarify the dependence of quadrupole splitting on σ-bond polarities. Data for these types of compounds are collected in Table XVIII.

The compounds $Me_3SnC_nF_{2n+1}$ ($n = 1$–3) (Table XVIII, compounds 5–7) contain no lone-pair electrons suitable for π bonding and, hence, the only type of π interaction which is possible is hyperconjugation. In contrast, the polarity of the tin–fluorocarbon bond is expected to be

TABLE XVIII

Mössbauer Data for Some Tetrahedral Organotin (IV) Compounds

Code No.[a]	Compound	C.S.[a,b]	Q.S.[a]	Ref.
1	Me_4Sn	1.31	—	(516)
2	$Et_3SnCH_2X^c$	1.29–1.43	—	(368)
3	$Me_3SnCH_2X^d$	1.32–1.38	—	(368)
4	Me_3SnCHF_2	1.28	0.94	(150)
5	Me_3SnCF_3	1.31	1.48	(150, 440)
6	$Me_3SnCF_2CF_3$	1.30	1.63	(150)
7	$Me_3SnCF(CF_3)_2$	1.32	1.89	(150)
8	$Me_3SnCH(CF_3)_2$	1.30	1.57	(150)
9	$Me_3SnCH=CH_2$	1.30	—	(9)
10	$Me_3SnCCl=CCl_2$	1.31	1.24	(440)
11	$Me_3SnCF=CF_2$	1.30	1.41	(150)
12	$Ph_3SnCH=CH_2$	1.28	—	(440)
13	$PhSn(CH=CH_2)_3$	1.25	—	(440)
14	$Me_3SnC{\equiv}C{\cdot}CHMe_2$	1.15	1.06	(494)
15	$Me_3SnC{\equiv}CPh$	1.22	1.29	(65, 440, 494)
16	$Me_3SnC{\equiv}C{\cdot}C{=}CEt$	1.20	1.80	(494)
17	$Me_3SnC{\equiv}CCF_3$	1.25	1.77	(150)
18	$Me_2Sn(C{\equiv}CCF_3)_2$	1.19	1.95	(150)
19	$Et_3SnC{\equiv}CH$	1.44	1.42	(456, 494)
20	$Et_3SnC{\equiv}CMe$	1.37	1.22	(456, 494)
21	$Et_3SnC{\equiv}CEt$	1.35	1.05	(456, 494)
22	$Et_3SnC{\equiv}C{\cdot}CHMe_2$	1.36	1.09	(494)
23	$Et_3SnC{\equiv}C\left(\begin{smallmatrix}CH{-}CH_2\\ \diagdown\diagup\\ CH_2\end{smallmatrix}\right)$	1.38	1.25	(494)
24	$Et_3SnC{\equiv}CPh$	1.38	1.48	(456, 494)
25	$Et_3SnC{\equiv}CCl$	1.39	1.75	(456, 494)
26	$Et_3SnC{\equiv}CBr$	1.40	1.70	(494)
27	$Et_3SnC{\equiv}C{\cdot}P(O)(OEt)_2$	1.42	2.40	(456, 494)
28	$Et_3SnC{\equiv}C{\cdot}SnEt_3$	1.38	1.18	(494)
29	$Pr_3SnC{\equiv}CH$	1.42	1.37	(494)
30	$Pr_2Sn(C{\equiv}CPr)_2$	1.27	1.60	(494)
31	$Bu_3SnC{\equiv}CH$	1.40	1.42	(456, 494)
32	$BuSn(C{\equiv}CPh)_3$	0.81	1.72	(440)
33	Me_3SnPh	1.21	—	(9, 141)
34	$MeSnPh_3$	1.19	—	(367)
35	$1\text{-}R_3Sn{\cdot}C_6H_4{\cdot}X\text{-}4^e$	1.16–1.43	0–0.48	(115, 141, 554)
36	$1\text{-}Me_3Sn{\cdot}C_6H_4{\cdot}R\text{-}x^f$	1.18–1.24	—	(115)
37	$1,2\text{-}(Me_3Sn)_2C_6Ph_4$	1.25	—	(115)

TABLE XVIII—continued

Code No.[a]	Compound	C.S.[a,b]	Q.S.[a]	Ref.
38	1-$Me_3Sn \cdot C_6H_4 \cdot I$-2	1.18	—	(115)
39	1-$Me_3Sn \cdot C_6Ph_4 \cdot I$-2	1.21	0.74	(115)
40	1-$Me_3Sn \cdot C_6H_4 \cdot CF_3$-2	1.21	0.66	(115)
41	1-n-$Bu_3SnC_6H_4 \cdot CF_3$-2	1.37	0.77	(115)
42	$Me_3SnC_6Cl_5$	1.32	1.09	(440)
43	1-$Me_3SnC_6Cl_4 \cdot H$-2	1.24	0.83	(115)
44	1,2-$(Me_3Sn)_2C_6Cl_4$	1.25	0.78	(115)
45	1,4-$(Me_3Sn)_2C_6Cl_4$	1.26	1.10	(115)
46	$Me_3SnC_6F_5$	1.27	1.31	(440)
47	1,2-$(Me_3Sn)_2C_6F_4$	1.26	0.85	(115)
48	1,4-$(Me_3Sn)_2C_6F_4$	1.20	1.20	(115)
49	1-$Me_3Sn \cdot C_6F_4 \cdot H$-4	1.24	1.08	(115)
50	$Me_2Sn(C_6F_5)_2$	1.25	1.51	(115, 440, 528)
51	1-$Me_2Sn(C_6F_5 \cdot Br$-2$)_2$	1.25	1.41	(115)
52	$MeSn(C_6F_5)_3$	1.19	1.14	(528)
53	1-$Ph_3Sn \cdot C_6H_4 \cdot X$-4[g]	1.30	—	(368)
54	1-$Ph_3Sn \cdot C_6H_4 \cdot Me$-2	1.30	—	(368)
55	$Ph_3SnC_6Cl_5$	1.27	0.84	(440)
56	$Ph_2Sn(C_6Cl_5)_2$	1.43	1.14	(140, 440)
57	$PhSn(C_6Cl_5)_3$	1.11	0.80	(140)
58	$Ph_3SnC_6F_5$	1.28	0.99	(138, 440, 528)
59	$Ph_2Sn(C_6F_5)_2$	1.22	1.11	(528)
60	$PhSn(C_6F_5)_3$	1.16	0.92	(528)
61	(4-$Me \cdot C_6H_4)_2Sn(C_6F_5)_2$	1.22	1.18	(528)
62	(4-$Me \cdot C_6H_4)Sn(C_6F_5)_3$	1.18	1.02	(528)
63	$Me_3SnC_5H_5$	1.05	1.20	(317)
64	$Me_3SnCH_2C_6H_5[Cr(CO)_3]$	1.67	0.59	(460)
65	$Me_2(Ph)SnPh[Cr(CO)_3]$	1.74	0.64	(460)
66	(4-$Me_3Sn)_2C_6H_4Cr(CO)_3$	1.69	0.72	(460)
67	$Me_3SnPhCr(CO)_3$	1.67	0.72	(460)
68	$Me_3SnPhMo(CO)_3$	1.43	0.84	(460)
69	$Me_2Sn[C_6H_5Cr(CO)_3]_2$	1.75	0.89	(460)

[a] Code number will be preceded by table number in text. Data given in mm/sec at liquid nitrogen temperature; when appropriate an unweighted average has been taken.

[b] Relative to SnO_2, assuming an identical shift for $BaSnO_3$, and a shift of 2.1 mm/sec for α-tin.

[c] Range of data for X = Cl, OMe, NMe_2, CN, or pyridine.

[d] Range of data for X = F, Cl, OMe, NMe_2, or pyridine.

[e] Range of data for R = Me or Et, X = F, Cl, OMe, NMe_2, t-Bu, Me.

[f] R = Me, Ph, x = 2, 3.

[g] X = Me, Cl, Br.

much greater than the tin–methyl bond. For example, the Taft σ^* constants (535) of substituents (X), which give a measure of the polarities of the C–X bonds, should provide a reasonable guide to the relative polarities of the Sn–X bonds. The Taft σ^* constants of CF_3 (2.58) and Me (0.00) indicate considerable inequalities in the polarities of the tin σ bonds in Me_3SnCF_3. The compounds Me_3SnCH_2F, Me_3SnCHF_2, and Me_3SnCF_3 show (150) a trend toward increasing quadrupole splitting with successive fluorine substitution as expected from a σ-bonding interpretation. Further, the order of quadrupole splittings $Me_3SnCF(CF_3)_2 > Me_3SnC_2F_5 > Me_3SnCF_3$ parallels the effective electronegativities of the fluorocarbon groups (150) as calculated from NMR data (151), although it is surprising that the compound $Me_3SnCH(CF_3)_2$ has such a large quadrupole splitting.

Of the compounds Me_nSnPh_{4-n} ($n = 1, 3$; Table XVIII, compounds 33 and 34) $R_nSn(CH=CH_2)_{4-n}$ ($n = 1, 3$; Table XVIII, compounds 9, 12, 13) and $R_nSn(C\equiv C-R')_{4-n}$ ($n = 1-3$; Table XVIII, compounds 14–16, 19–24, etc.), only the acetylene derivatives show quadrupole splitting even though phenyl, vinyl, and acetylene groups have $p\pi$ electrons suitable for π bonding. The evidence for the existence of π bonding in these systems is not conclusive (18), and there is certainly no indication that tin-alkynyl bonds have a greater π-bond order than tin–phenyl or tin-vinyl bonds. However, it is generally recognized that sp hybridized carbon atoms are considerably more electron-withdrawing than sp^2 hybridized carbon atoms as shown, for example, by the Taft σ^* constants of 1.35 for a $C\equiv CR$ group compared with 0.36 and 0.6 for vinyl and phenyl groups, respectively. These data strongly indicate that the quadrupole splitting in the alkynyl derivatives is produced by the polarity of the tin–alkynyl bond. Similarly, the acidic cyclopentadienyl group produces a small quadrupole splitting in $Me_3SnC_5H_5$ (Table XVIII, compound 63).

The substitution of fluorine or chlorine atoms into the phenyl or vinyl group produces quadrupole splitting. It would be anticipated that the electron-withdrawing fluorine and chlorine atoms would increase the electron-withdrawing power of the phenyl or vinyl group and, thus, increase the polarity of the tin–phenyl and tin–vinyl bond. This expectation is confirmed by dipole moment data (346), which show that for the compounds $Me_3Sn \cdot C_6H_4 \cdot p$-X (X = F, Cl, Br), the Me_3Sn group is electron-releasing with respect to the ring. These data, therefore, provide further strong evidence for a σ-orbital imbalance interpretation of quadrupole splitting which is also supported by the observation (114, 115, 440) that the quadrupole splittings of chlorocarbon compounds are invariably lower than those of their fluorocarbon analogs; for example, compare the pairs of compounds in Table XVIII: 42 and 46, 44 and 47. 45 and 48, 55 and 58, 56 and 59, 57 and 60, 10 and 11. A suggestion (528) that the quadrupole splittings of fluorocarbon derivatives may arise from π-bonding effects is not consistent with NMR evidence (342) which shows no donation of π electrons to the tin atom in $Me_3SnC_6F_5$. It is also interesting that the coordination of a $Cr(CO)_3$ or $Mo(CO)_3$ group to the phenyl ring of the species Me_3SnCH_2Ph, Me_2SnPh_2, $(1,4-Me_3Sn)_2C_6H_4$, or Me_3SnPh (Table XVIII, compounds 64–67) results in a small quadrupole splitting, probably owing to the increased polarity of the tin–phenyl bond (460).

In the series of acetylene derivatives $R_3SnC\equiv C-X$ (Table XVIII, compounds 14–26 and 28–31), the largest quadrupole splittings are found for those compounds in which X is an electron-withdrawing halogen atom or CF_3 group. The series of triethyltin derivatives (Table XVIII, compounds 19–26 and 28) also show (494) good correlations with dipole moment and the inductive, σ^*, (576) and inductive plus mesomeric, σ_n, (576) constants of the X group.

The data discussed above are clearly fully consistent with a σ-bonding interpretation of quadrupole splitting, but provide no indication of π-bonding effects. It was, therefore, suggested (440) that the quadrupole splittings of tin(IV) compounds *in general* arise from a σ-orbital imbalance owing to inequalities in bond polarities. A similar conclusion was reached by Drago et al. (338).

In attempting to make a general assessment of the dependence of quadrupole splitting on the nature of the tin–ligand bonds, it is important to realize that the major variations of quadrupole splitting which are observed arise from structural rather than bonding changes (*vide infra*) (Table XIX). Variations in quadrupole splittings which can be attributed to differences in bond character will only be revealed in the quadrupole splittings of isostructural series, such as the tetrahedral compounds discussed above. For example, it has been shown (*vide infra*) that triethyl- and trimethyltin halides are probably isostructural with associated five-coordinate structures, whereas the triphenyl- and trineophyltin halides (with the exception of Ph_3SnF) have monomeric structures with tetrahedral coordination of the tin atom. Data for these compounds are contained in Table XX. Although the four- and five-coordinate species have very different quadrupole splittings due to the structural differences, both series of compounds show a small variation of quadrupole splitting with bond polarity as illustrated by the straight-line relationship between quadrupole splitting and Taft σ^* (400, 440). Further, the lowering of the quadrupole splittings of the triphenyl compared with the trineophyl derivatives is consistent with the greater polarity of the tin–phenyl bond. For other series of five- or six-coordinate compounds, quadrupole splitting changes are small. For example in the series of five-coordinate compounds in Table XXXVI, variations are difficult to interpret in terms of bond polarities. In cases when significant variations are observed, for example, the five-coordinate carboxylate species Me_3SnO_2CR' ($R' = CH_2X$, CX_3, $X = F$, Cl, Br, I) (*170, 459*) and the six-coordinate complexes Ph_2SnX_2bipy (*465*) and $Bu_2SnX_2 \cdot$ phen ($X = $ Cl, Br, I) (*417*), the trends are usually those expected from a σ-bonding interpretation.

TABLE XIX

RANGES OF QUADRUPOLE SPLITTINGS FOR COMPOUNDS WITH DIFFERING STRUCTURES[a]

Structural type	Range of quadrupole splitting[b]
Tetrahedral $R_n SnX_{4-n}$ ($n = 1$–3)	0.00–2.31
Octahedral $RSnX_5$	1.92
Octahedral *cis*-R_2SnX_4	1.63–2.34
Trigonal-bipyramidal R_3SnX_2 (X axial)	2.76–3.86
Octahedral *trans*-R_2SnX_4	3.37–4.32

[a] Summary of data considered in Refs. (*234*) and (*440*), in which papers full details of data with references are given.
[b] Data given in mm/sec at liquid nitrogen temperature.

From the above discussion, it is clear that although there is much positive evidence that quadrupole splittings are produced by σ-bonding inequalities, there is little or no evidence for a significant contribution from π-bonding effects. This is not unexpected as, from the dependence of q_{val} on $\langle r^{-3} \rangle$ (Section II,D), asymmetries in the population of the $5d$ orbitals will make a much smaller contribution to q_{val} than an equivalent asymmetry in the $5p$ electron density (*122, 141*). Calculations of the relative $\langle r^{-3} \rangle$ values of $5p$ and $5d$ orbitals show that the ratio $\langle r^{-3} \rangle_{5d}$: $\langle r^{-3} \rangle_{5p}$ is of the order 0.01, and this has been confirmed by comparison of the relative quadrupole splittings of octahedral and tetrahedral compounds (*vide infra*). Hence, even if a significant degree of π bonding is present in the tin–ligand bonds, this will be expected to have a minimal effect on the quadrupole splitting.

It is now worthwhile to reexamine the data which led to the postulation of π-bonding effects, i.e., the lack of quadrupole splitting in the

TABLE XX

MÖSSBAUER DATA FOR TRIALKYL- AND TRIPHENYLTIN HALIDES[a]

R	R$_3$SnF C.S.[b,c]	R$_3$SnF Q.S.[b]	R$_3$SnCl C.S.[b,c]	R$_3$SnCl Q.S.[b]	R$_3$SnBr C.S.[b,c]	R$_3$SnBr Q.S.[b]	R$_3$SnI C.S.[b,c]	R$_3$SnI Q.S.[b]	M[R$_3$SnCl$_2$] C.S.[b,c]	M[R$_3$SnCl$_2$] Q.S.[b]	M[R$_3$SnBr$_2$] C.S.[b,c]	M[R$_3$SnBr$_2$] Q.S.[b]
Me	1.26[d]	3.82[d]	1.42[d]	3.44[d]	1.45[d]	3.40[d]	1.49[d]	3.10[d]	1.33[j,k]	3.28[j,k]	1.43[j,l]	3.45[j,l]
Et	1.48[e]	3.94[e]	1.60[e]	3.70[e]	1.61[e]	3.29[e]	1.56[e]	3.05[e]	1.50[l,m]	3.44[l,m]	—	—
n-Pr	1.46[f]	4.01[f]	1.62[f]	3.66[f]	1.66[f]	3.52[f]	1.59[f]	2.90[f]	—	—	—	—
n-Bu	1.34[f]	3.73[f]	1.46[f]	3.48[f]	1.44[f]	3.29[f]	1.43[f]	2.63[f]	—	—	—	—
i-Bu	1.47[f]	3.82[f]	1.61[f]	3.36[f]	1.60[f]	3.20[f]	1.63[f]	2.73[f]	—	—	—	—
Neo[h]	1.33[g]	2.79[g]	1.39[g]	2.65[g]	1.42[g]	2.65[g]	1.41[g]	2.40[g]	—	—	—	—
Ph	1.22[i]	3.58[i]	1.34[i]	2.54[i]	1.29[i]	2.50[i]	1.20[i]	2.15[i]	1.32[l,n]	3.00[l,n]	1.29[o]	2.88[o]
									1.23[o]	2.87[o]		

[a] Data have been selected from references reporting full series of compounds. Other data for these and related compounds may be found in Ref. (516).
[b] Data in mm/sec at liquid nitrogen temperature assuming a shift of 2.1 mm/sec for α-tin.
[c] Relative to SnO$_2$.
[d] An average of data from Refs. (141), 258).
[e] An average of data from Refs. (176, 258, 442).
[f] An average of data from Refs. (176, 258).
[g] Reference (336).
[h] Neo = Me$_2$(Ph)CCH$_2$-.
[i] An average of data from Refs. (258, 442).
[j] M = (Et$_4$N)$^+$.
[k] An average of data from Refs. (440, 442).
[l] Reference (442).
[m] M = (Ph$_3$CCH$_2$Ph)$^+$.
[n] M = (Me$_4$N)$^+$.
[o] Reference (207). M = (Ph$_3$PC$_{10}$H$_{21}$)$^+$.

compounds $(Ph_3Sn)_2$, $(Ph_3Sn)_4M$ (M = Sn, Ge, Pb), Ph_3SnX (X = H, Li), R_nSnH_{4-n} (n = 1-3, R = Me, Bu, Ph), Me_3SnNa, SnX_4Y_2, and SnX_2Y_4. The original observation (278, 300) of zero quadrupole splittings for all SnX_4Y_2 and SnX_2Y_4 species, in which X is a halogen atom and Y is an electronegative donor ligand, has been superseded by more recent data. Quadrupole splittings have been reported for many SnX_4Y_2 compounds in which both X and Y have π-donor electrons (see data in Table XXXVII) and it has also been observed that some compounds in which the Y group has no suitable π electrons [e.g. $SnCl_4 \cdot$ en (457) and $SnCl_4$-$[Ph_2P \cdot (CH_2)_2 \cdot PPh_2]$ (103)] give only small or zero quadrupole splitting. The differences in the bond polarity for organotin hydrides are not expected to be large, as illustrated by the Taft σ^* constants of Me(0.00) and H(0.49) and molecular orbital calculations for Me_3SnH (299); the lack of quadrupole splitting for the compounds is, therefore, not unexpected. Similarly, large variations in bond polarity could not be expected for the Sn–M(M = C, Si, Ge, Sn, Pb) bonds and recent work (65) has shown some evidence of unresolved quadrupole splitting in the species R_3Sn-MR_3 (R = Me or Ph, M = C, Si, Ge, or Sn), which is at a maximum for M = Sn. Finally it is difficult to evaluate the significance of the reported data for Ph_3SnLi and Me_3SnNa, as these compounds do not exist in the solid state (558), and Goldanskii (57) has shown that frozen solution spectra vary markedly with solvent.

b. Influence of Coordination Number and Structural Determination.
Tin(IV) compounds are found with a wide variety of structures including four-, five-, and six-coordinate tin atoms. Many compounds with a nominally tetrahedral stoichiometry have associated solid state structures involving five- or six-coordinate tin atoms. For example, X-ray diffraction studies of Me_3SnX [X = F (122), CN (506), OH (366), or NCS (243)] show polymeric structures with five-coordination of the tin atom, whereas Me_2SnF_2 (507) has a polymeric structure with trans-octahedral coordination of the tin atom. The quadrupole splitting observed in the Mössbauer spectra of tin(IV) compounds provide a very powerful means of studying these structural variations.

The first attempt to correlate quadrupole splitting with structure was made by Herber et al. (336), who noted that R_3SnX and R_2SnX_2 compounds which probably have associated structures (e.g., Me_2SnF_2, Me_3SnOH, Me_3Sn-imidazole) appeared to give larger quadrupole splitting than other R_3SnX and R_2SnX_2 species. A parameter ρ (defined as the quadrupole splitting divided by the center shift relative to SnO_2) was introduced, and it was postulated that R_3SnX and R_2SnX_2 compounds for which ρ was greater than 2.1 had associated structures.

Later the additivity model, using a point charge formalism, was used

TABLE XXI

Quadrupole Splitting Data for Some Organotin Halides and Halide Anions

Compound	V_{ZZ}/e^a	Q.S.(obs.)[b]	Ref.	Q.S.(cal.)[b,c]
$K_2[R_2SnF_4]^d$	$4[R]^{oct} - 4[F]^{oct}$	4.28^d	(442)	4.16
$M_2[R_2SnCl_4]^e$	$4[R]^{oct} - 4[Cl]^{oct}$	4.21^e	(234, 442)	3.76
$Cs_2[Me_2SnBr_4]$	$4[R]^{oct} - 4[Br]^{oct}$	4.22	(442)	3.64
$(pyH)_2[Ph_2SnCl_4]^j$	$4[Ph]^{oct} - 4[Cl]^{oct}$	3.80	(234)	3.32
$M_2[RSnCl_5]^f$	$2[R]^{oct} - 2[Cl]^{oct}$	1.90^f	f	1.88
$(pyH)_2[PhSnCl_5]$	$2[Ph]^{oct} - 2[Cl]^{oct}$	1.92	(234)	1.66
$(R_3SnF)_n{}^g$	$4[F]^{tba} - 3[R]^{tbe}$	3.88^g	g	h
$(R_3SnCl)_n{}^g$	$4[Cl]^{tba} - 3[R]^{tbe}$	3.57^g	g	h
$(R_3SnBr)_n{}^g$	$4[Br]^{tba} - 3[R]^{tbe}$	3.35^g	g	h
$(R_3SnI)_n{}^g$	$4[I]^{tba} - 3[R]^{tbe}$	3.08^g	g	h
Neo_3SnF^j	$2[F]^{tet} - 2[R]^{tet}$	2.79	(336)	2.08
Neo_3SnCl	$2[Cl]^{tet} - 2[R]^{tet}$	2.65	(336)	1.88
Neo_3SnBr	$2[Br]^{tet} - 2[R]^{tet}$	2.65	(336)	1.82
Neo_3SnI	$2[I]^{tet} - 2[R]^{tet}$	2.40	(336)	1.68
Ph_3SnCl	$2[Cl]^{tet} - 2[Ph]^{tet}$	2.54^i	i	1.66
Ph_3SnBr	$2[Br]^{tet} - 2[Ph]^{tet}$	2.50^i	i	1.60
Ph_3SnI	$2[I]^{tet} - 2[Ph]^{tet}$	2.15^i	i	1.46

[a] Taken from Table IV.
[b] Data given in mm/sec.
[c] Calculated using the point charge parameters from Ref. (442).
[d] Average of R = Me, Et.
[e] Average of R = Me, Et, M = Cs and R = Me, M = pyH.
[f] Average of R = Et, M = Me_4N (442) and R = Bu, M = Et_4N (171).
[g] Average of R = Me, Et; data taken from Table XX.
[h] Data used in calculation of point charge parameters.
[i] Data taken from Table XX.
[j] PyH = pyridinium, Neo = Me_2CPhCH_2.

to assess the expected variation of quadrupole splitting with structure (234, 440). From Table XXI it can be seen that the relative quadrupole splittings calculated for organotin compounds fall into two main groups. Compounds with tetrahedral R_nSnX_{4-n} ($n = 1–3$) and octahedral $RSnX_5$ and cis-R_2SnX_4 structures should all have comparable quadrupole splittings, whereas compounds with $trans$-R_2SnX_4 and trigonal-bipyramidal R_3SnX_2 (X axial) coordination should have quadrupole splittings which are roughly twice as large. The data available at the time (Table XIX) for compounds whose structures are reasonably well established appeared to be in general agreement with this conclusion as noted by Fitzsimmons *et al.* (*234*) for octahedral compounds and by Parish and Platt (*440*) for a more general range of compounds.

One important conclusion from these original additivity model calculations concerns the effect of intermolecular association on quadrupole splitting. Thus, the formation of associated structures for tri- and dialkyltin species involving trigonal-pyramidal R_3SnX_2 (X axial) and octahedral *trans*-R_2SnX_4 coordination of the tin atom, respectively, should result in a significant increase in quadrupole splitting compared with the unassociated forms. As center shifts of tri- and dialkyltin species generally fall into a rather narrow range, this increase in quadrupole splitting forms the basis of the correlation of ρ with structure. In contrast, an associated six-coordinate $RSnX_5$ structure for a monoalkyltin compound would be expected to show approximately the same quadrupole splitting as a monomeric tetrahedral species.

Although the data in Table XIX show that, at least in a general way, the additivity model provides a realistic description of quadrupole splitting trends, it has recently been shown (*122*) (see Section II,D) that strict adherence to the simple additivity model would not be expected for compounds with differing coordination numbers. However, before more detailed consideration can be given to the degree of precision with which the additivity model allows a prediction of quadrupole splitting values, it is necessary to describe some structural conclusions for trialkyltin halides, which have been reached with the aid of Mössbauer data.

Organotin fluorides are generally involatile insoluble solids in keeping with polymeric structures, observed crystallographically for Me_3SnF (*118*) and Me_2SnF_2 (*507*). In contrast, other organotin halides are low-melting solids or liquids and the observation (*51, 123, 428*) of two Sn–C stretching frequencies in both solid and solution has often been taken as evidence for monomeric tetrahedral structures. However, IR evidence is not unambiguous in that changes in the Sn–X stretching frequency for Me_3SnX (X = Cl, Br) in the solid melt and carbon disulfide solution indicate some degree of intermolecular association (*379*). By the same criterion, the triphenyltin halides Ph_3SnX (X = Cl, Br, I) do not seem to be associated (*378*). Mössbauer data for some trialkyl- and triaryltin halides and some halide complexes are summarized in Table XX. Infrared data suggests that the complexes have trigonal-bipyramidal R_3SnX_2 structures (*442*).

The quadrupole splitting observed for Me_3SnF is large as expected from the trigonal-bipyramidal coordination of the tin atom. The quadrupole splittings of other trimethyltin halides are also large and show a regular decrease in quadrupole splitting with increasing size of the halogen atom. Further, there is a close similarity in the quadrupole splittings of Me_3SnX (X = Cl, Br) and the corresponding halide anions $[Me_3SnX_2]^-$ (X = Cl, Br). As monomeric tetrahedral structures would be anticipated to give much lower quadrupole splittings, the data provide strong evidence that, like Me_3SnF, the compounds Me_3SnX (X = Cl, Br, I) are strongly associated with five-coordinate tin atoms (442). This has been confirmed in the case of Me_3SnCl, by an X-ray diffraction study (164). Quadrupole splittings for trialkyltin halides R_3SnX (R = Et, n-Pr, n-Bu, i-Bu; X = F, Cl, Br, I) are similar to those of their trimethyltin analogs indicative of similar polymeric structures. There is, however, a trend (128) to lower quadrupole splitting for R_3SnBr and R_3SnI species as the size of the alkyl group is increased from Et to i-Bu. This may be reflection of a weakening of the intermolecular association due to steric hindrance of the alkyl group.

The quadrupole splitting of triphenyltin fluoride is slightly lower than that of trimethyltin fluoride, consistent with a fully associated polymeric structure (442). The lowering of quadrupole splitting is probably due to the greater polarity of the tin–phenyl bond as illustrated by the relative quadrupole splittings of the anions $[R_3SnCl_2]^-$ (R = Me, Ph). In contrast, the quadrupole splittings of the triphenyltin halides Ph_3SnX (X = Cl, Br, I) are considerably lower than those of the fluoride and also less than that of the $[Ph_3SnCl_2]^-$ ion, indicating a reduction in the degree of intermolecular association (442). The trineophyltin halides also show much lower quadrupole splittings than the trimethyltin halides. In order to assign structures unambiguously to the compounds Ph_3SnX (X = Cl, Br, I) and Neo_3SnX (X = F, Cl, Br, I) it is necessary to compare the quadrupole splittings with that of a known tetrahedral organotin halide.

The crystal structure (131) of the compound $Ph_2Sn(I) \cdot [CH_2]_4$-$Sn(I)Ph_2$ shows a slightly distorted tetrahedral environment of the tin atoms with no close intermolecular Sn–I distances, and this compound can be regarded as a model tetrahedral compound. The quadrupole splitting of 2.37 mm/sec (400) observed for $Ph_2Sn(I) \cdot [CH_2]_4 \cdot Sn(I)Ph_2$ is closely similar to those found for Neo_3SnI and Ph_3SnI, and the relative quadrupole splittings of the three compounds are consistent with the differing number of tin–phenyl bonds. This observation, together with the good correlation observed between quadrupole splitting and Taft σ^* constant of the halogen atom, strongly suggests that the compounds R_3SnX (R = Neo, X = F, Cl, Br, I; R = Ph, X = Cl, Br, I) are isostructural with tetrahedral coordination of the tin atom. The lack of intermolecular association is probably due to the steric hindrance of the bulky neophyl and phenyl groups.

Ensling et al. (207) have recently suggested that the large difference in quadrupole splitting between Ph_3SnX (X = Cl, Br) and the four-coordinate species $Ph_3Sn(S_2CNEt_2)$ [$\Delta = 1.85$ mm/sec (207)] argues against a monomeric structure for the triphenyltin halides. However, this difference in quadrupole splitting is more probably due to the difference in bond polarity between the tin–halide and tin–sulfur bonds, as illustrated by the large difference in quadrupole splitting between the trans-octahedral species $[Me_2SnX_4]^{2-}$ [X = Cl, Br, Δ = ca. 4.1–4.2 mm/sec (Table XXIV)] and Me_2Sn $(S_2CNEt_2)_2$ ($\Delta = 3.14$ mm/sec) (Table XXIV). Indeed the relative quadrupole splittings of Ph_3SnX (X = Cl, Br) and $Ph_3Sn(S_2CNEt)$ are in excellent agreement with the additivity model calculations (see Table XXIII), and a recent crystal structure determination of triphenyltin chloride has shown monomeric tetrahedral coordination of the tin atom (70).

The assignment of structures to the trialkyl- and triphenyltin halides makes possible a more rigorous investigation of the additivity description of quadrupole splitting. The compounds contained in Table XXI form a series of four-, five-, and six-coordinate tin species with closely similar ligands, and the appropriate additivity expressions for V_{ZZ}, written in terms of the $L^{superscript}$ parameters defined in Eqs. (29.1)–(29.4) are also included in Table XXI.

In the point charge formulation of the additivity model (440, 442), it is assumed that the parameters [L]^{oct}, [L]^{tet}, [L]^{tba}, and [L]^{tbe} are identical. On this basis a series of partial quadrupole splitting parameters (p.q.s.)$_L$, where (p.q.s.)$_L = \frac{1}{2}e^2|Q|$[L], has been calculated (442).* The quadrupole splittings of the anions [SnCl$_5$]$^-$ (Δ in the range 0.46–0.77 mm/sec (316, 442)) were used to obtain a value of (p.q.s.)$_{Cl}$, while (p.q.s.)$_R$ was calculated from the quadrupole splittings of the five-coordinate $(R_3SnCl_2)_n$ and [R_3SnCl_2]$^-$ (R = Me, Et) species. Values of (p.q.s.)$_L$ for other ligands were obtained in a similar manner. Calculated quadrupole splittings obtained using these (p.q.s.)$_L$ values are included in Table XXI. Clearly, although the overall trends of the quadrupole splitting patterns are reproduced, detailed numerical agreement is not obtained.

In Section II, D, it was shown using a simple molecular orbital model (122), that it is not a good approximation to describe octahedral, tetrahedral, trigonal-bipyramidal axial, and trigonal-bipyramidal equatorial bonds with the same [L] values. It is probably the inadequacy of this approximation which gives rise to the discrepancies between the observed and calculated quadrupole splittings in Table XXI.

A more realistic application of the additivity model involves the

* In Parish and Platt (442), the (p.q.s.)$_L$ of a ligand L is given the symbol [L].

calculation of separate partial quadrupole splitting parameters for each type of bond. Calculations of partial quadrupole splitting for octahedral and tetrahedral compounds have been described and these will be discussed below. Some discussion of the quadrupole splittings of five-coordinate compounds will also be given.

Before describing the derivation of $(p.q.s.)_L^{oct}$ and $(p.q.s.)_L^{tet}$ values, it is necessary to pay some attention to the signs of the quadrupole splittings. Using the additivity model, assuming a dominance of the valence contribution to the EFG, the sign of V_{ZZ} expected for any particular structure may be predicted. For example, consider a $trans$-R_2SnX_4 compound. Substitution of $[R]^{oct}$ and $[L]^{oct}$ into the appropriate equation in Table IV yields the expression

$$V_{ZZ} = \{4[R]^{oct} - 4[X]^{oct}\}e \tag{39}$$

Clearly, if X is more electron-withdrawing than R, then $\sigma_R^{oct} > \sigma_L^{oct}$, $-[R]^{oct} > -[X]^{oct}$ and, hence, V_{ZZ} is negative. In pictorial terms, the greater electron-withdrawing power of the X ligand produces a deficiency of negative charge in the XY plane and, hence, a negative value of V_{ZZ}. It is probable that the sign of the quadrupole moment for ^{119}Sn is negative (see Section IV,E) and, hence, the sign of $\tfrac{1}{2}e^2qQ$ ($\tfrac{1}{2}eQV_{ZZ}$) for a $trans$-R_2SnX_4 species is expected to be positive. In a similar manner, positive values of $\tfrac{1}{2}e^2qQ$ are predicted for $RSnX_5$ and $RSnX_3$ species, while cis-R_2SnX_4, R_3SnX_2 (X axial), and R_3SnX compounds would be expected to have negative signs (*439*).

The sign of $\tfrac{1}{2}e^2qQ$ may be determined by application of an external magnetic field, and the analysis of this type of spectra has been discussed by Gibb (*265*). The signs which have been reported (*64, 217, 232, 288–291, 402, 439*) are summarized in Table XXII. With the exception of cis-R_2SnX_4 species, which will be discussed in detail later, all the signs are in agreement with those expected from the additivity model, as noted by Parish and Johnson (*439*) for a general range of structures and by other groups (*217, 232, 288, 290, 291*) for more limited data. Goodman and Greenwood (*290*) have suggested that the negative sign observed for Ph_3SnCl indicates a polymeric five-coordinate structure rather than a monomeric tetrahedral configuration (*400*). In contrast, additivity calculations do not predict a change of sign of $\tfrac{1}{2}e^2qQ$ for R_3SnX and R_3SnX_2 (X axial) species, and this is confirmed by the experimental data in Table XXII and the recent crystal structure (*70*).

The positive signs of $\tfrac{1}{2}e^2qQ$ observed (*288*) for the compounds [π-cpFe(CO)$_2$]$_2$SnX$_2$ (X = Cl, NCS) are interesting as an idealized SnA_2B_2 system would be expected to have $\eta = 1$. However, η is very sensitive to small distortions (*122*) and the crystal structure of [π-cpFe(CO)$_2$]$_2$SnCl$_2$

TABLE XXII
Sign $\frac{1}{2}e^2qQ$ Observed for Some Tin(IV) Compounds

Compound	Q.S. = $\frac{1}{2}e^2qQ^a$	Ref.
trans-R_2SnX_4		
Me_2SnF_2	+4.65	b
$Me_2SnCl_2(pyo)_2$	+4.10	(232)
$Cs_2[Me_2SnCl_4]$	+4.28	(439)
$K_2[Me_2SnF_4]$	+4.12	(439)
Bu_2Sn maleate	+3.74	(439)
Me_2SnMoO_4	+4.20[c]	(289)
$Me_2Sn(acac)_2$	+3.93[d]	(217)
n-Pr_2SnCl_2 2β-pic[e]	+3.99	(291)
n-Pr_2SnCl_2 4-pic[e]	+3.42	(291)
cis-R_2SnX_4		
$Me_2Sn(oxin)_2$	+2.06	(439)
$Ph_2Sn(oxin)_2$	+1.67	(439)
$Ph_2Sn(S_2CNEt_2)_2$	+1.72	(439)
$Ph_2Sn(NCS)_2phen$	+2.36	(439)
$Ph_2SnCl_2(morph)_4$[f]	+1.92	(291)
n-$Pr_2SnCl_2(morph)_2$[f]	+2.41	(291)
$RSnX_5$		
n-$PrSnCl_3(pip)_2$[g]	+1.99	(291)
$(Me_4N)_2[EtSnCl_5]$	+1.94	(439)
R_3SnX		
$Me_3SnC_6F_5$	−1.39	(439)
$Ph_3SnC_6F_5$	−0.97	(439)
Ph_3SnCl	−2.51[h]	(290)
$Ph_2Sn(I)\cdot(CH_2)_4$-$(I)SnPh_2$	−2.37[i]	(400)
$Bu_3Sn(\pi\text{-}cp)Fe(CO)_2$	−0.59	(288)
R_2SnX_2[j]		
$[Mn(CO)_5]MeSnCl_2$	+2.56	(288)
$(\pi\text{-}cp)Fe(CO)_2SnCl_2$	+2.35	(288)
$(\pi\text{-}cp)Fe(CO)_2Sn(NCS)_2$	+2.57	(288)
$RSnX_3$[j]		
$Mn(CO)_5SnCl_3$	+1.58	(288)
$(\pi\text{-}cp)Fe(CO)_2SnCl_3$	+1.77, 1.82	(288, 364)
R_3SnX_2(X axial)[k]		
$(Me_4N)[Me_3SnCl_2]$	−3.31	(439)
$(Ph_3PCH_2Ph)[Et_3SnCl_2]$	−3.49	(439)
$(Me_4N)[Ph_3SnCl_2]$	−3.02	(439)

TABLE XXII—continued

Compound	Q.S. = $\frac{1}{2}e^2qQ^a$	Ref.
R_3SnX_2(X axial)[k] (continued)		
$(Et_3SnCN)_n$	−3.17	(439)
$(Me_3SnNCS)_n$	−3.77[h]	(290)
Me_3SnOH	−2.91[l]	(290)
Ph_3SnF	−3.62	(290)
$Ph_3SnCl \cdot pip^g$	−2.95	(291)
$Ph_3SnCl \cdot 2\beta\text{-pic}^e$	−2.97	(291)
R_2SnX_3(R equatorial)[h]		
$Et_4N[Me_2SnBr_3]$	+3.39	(439)
Me_2SnCl_2	+3.4[c]	(289)
Bu_2SnO	+2.13	(439)
Me_2SnO	+2.09[m]	(290)

[a] Data given in mm/sec.
[b] References (217, 290); magnitude of Q.S. from Ref. (327).
[c] Original data in Ref. (289); reinterpretation as described in Refs. (217, 438).
[d] Magnitude of Q.S. taken from Ref. (323).
[e] β-pic = β-picoline.
[f] morph = Morpholine.
[g] pip = Piperidine.
[h] Magnitude of Q.S. taken from Ref. (439).
[i] Magnitude of Q.S. taken from Ref. (400).
[j] Definition of R extended to include $Mn(CO)_5$ and $\pi\text{-cpFe}(CO)_2$.
[k] Trigonal-bipyramidal structure.
[l] Magnitude of Q.S. taken from Ref. (141).
[m] Magnitude of Q.S. taken from Ref. (440).

(426) shows a distorted tetrahedral coordination with an enlarged Fe–Sn–Fe angle and reduced X–Sn–X angle. Both the point charge (288) and molecular orbital (122) treatments of this type of distortion predict a positive sign for $\frac{1}{2}e^2qQ$.

The crystal structure of $Me_2Sn(ox)_2$ (505) shows a distorted cis-R_2SnX_4 configuration and the magnitudes of the quadrupole splittings of this compound and the species $Ph_2Sn(ox)_2$, $Ph_2Sn(S_2CNEt)_2$, $Ph_2Sn(NCS)_2phen$, $Ph_2SnCl_2 \cdot 4morph$ and $N\text{-}Pr_2SnCl_2 \cdot 2morph$ are those expected for cis-octahedral structures (vide infra). It is rather surprising therefore that the predicted negative values of $\frac{1}{2}e^2qQ$ are not observed. The origins of this inconsistency probably lie in deviations of the structure from a regular geometry (291, 439). For example, Parish and Johnson (439) have calculated, using the point charge model, the contribution of one pair of ligands to the EFG as a function of L–Sn–L bond angle (α) (Fig. 9). Only when α lies between 70.5° and 109.5° does V_{ZZ} lie perpendicular to the L–Sn–L plane. For other values of α, V_{ZZ} lies in the L–Sn–L plane and, hence, has opposite sign to that expected for a regular cis geometry with α = 90°. The positive sign of $\frac{1}{2}e^2qQ$ observed for cis compounds may, therefore, be associated with an enlargement of the R–Sn–R angle as observed for $Me_2Sn(ox)_2$, in which the Me–Sn–Me angle is 111°. Such distortions probably have only a marginal effect on the magnitude of the quadrupole splitting (439). The positive

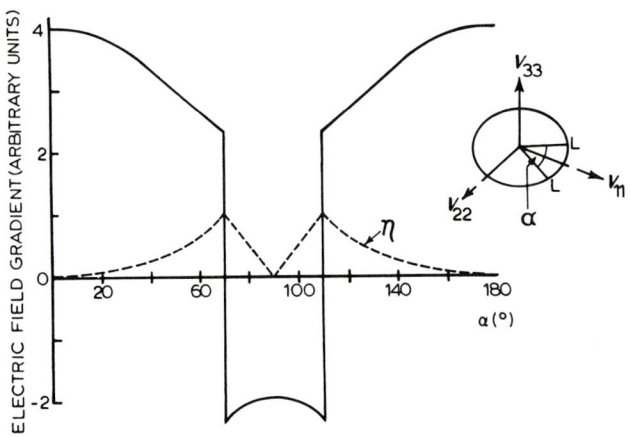

FIG. 9. Relative magnitude of the electric field gradient for an L–Sn–L system as a function of the bond angle (α). The principal component (V_{ZZ}) coincides with V_{11} for 0° < α < 70.5°, with V_{33} for 70.5° < α < 109.5°, and with V_{22} for 109.5° < α < 180° (439).

signs observed for Me_2SnCl_2 and $(Et_4N)(Me_2SnBr_3)$ are also consistent (439) with distorted structures as observed for Me_2SnCl_2 (167) and suggested for $(Et_4N)(Me_2SnBr_3)$ (442).

The information used in determining partial quadrupole splittings is collected in Table XXIII for tetrahedral compounds and Table XXIV for octahedral species. The first twenty-four compounds in Table XXIII and the first thirty-six compounds in Table XXIV have been used to derive partial quadrupole splitting values, while the remaining compounds serve as cross-checks. Most of the information is considered in Clark et al. (122), although several additions have been made.

The structures of the compounds used in calculating partial quadrupole splitting values are reasonably well established. Crystal structures of a wide range of tin–transition metal species have been reported [see references contained in Goodman et al. (288)] and all are tetrahedral, although severe distortions are sometimes found. Structural assignments for the triphenyl- and trineophyltin halides have been discussed above, and a tetrahedral structure of Neo_3SnO_2CMe has been deduced by Herber et al. (336) and Ford and Sams (241). Crystal structures of $Me_2SnCl_2L_2$ [L = dimethyl sulfoxide (353) and pyridine N-oxide (69)] show trans-octahedral structures and the large quadrupole splittings found for the compounds in Table XXIV (compounds 1–14 and 16–35) also indicate trans structures as noted by the authors referred to in Table XXIV. Poller et al. (463) have assigned structures on the basis of quadrupole splitting data, to a range of $Sn(S_2R)_2L_2$ species (where S_2R is 1,2-ethane dithiol, 1,3-propane dithiol, or 1,2-propane dithiol, and L is a nitrogen or oxygen donor) and the ethane dithiol species are included as representative examples. The sign of $\frac{1}{2}e^2qQ$ has been included in Tables XXIII and XXIV when available. In some other cases a sign has been assigned by consideration of structural type, and these are given in parentheses.

A survey of the compounds in Tables XXIII and XXIV reveals that phenyl, alkyl, and halogen are relatively more common than most others. Further, it is found that the differences between different alkyl ligands are not significant in relation to the overall accuracy of the additivity model. All alkyl ligands may, therefore, be assigned a single octahedral or tetrahedral partial quadrupole splitting value. By the same criterion, F, Cl, or Br may also be assigned a single parameter, and we arbitrarily assign $(p.q.s.)_X = 0$.* In view of these observations it was decided to concentrate (122) on careful determinations of $(p.q.s.)_R^{tet}-(p.q.s.)_X^{tet}$, $(p.q.s.)_{Ph}^{tet}-(p.q.s.)_X^{tet}$, and $(p.q.s.)_R^{oct}-(p.q.s.)_X^{oct}$ where X = F, Cl, Br and $(p.q.s.)_L$ is the partial quadrupole splitting for ligand L. Values of these parameters were calculated by taking unweighted averages of statistically independent estimates from several different sets of compounds, using the appropriate additivity expressions from Table IV. The calculations are summarized in Table XXV.

For the remaining ligands the partial quadrupole splitting values were obtained by the rather more subjective procedure of calculating values from data on compounds believed to be relatively close to idealized geometry. The values obtained are summarized in Table XXVI. For trans-$R_2SnCl_2L_2$ systems, in which L is monodentate, the calculated values are independent of the arrangement of the Cl and L ligands as $(p.q.s.)_L^{oct}-(p.q.s.)_{Cl}^{oct}$ is small in all cases. For the bidentate ligands pic

* It should be emphasized that we do not consider this to be an accurate value; $(p.q.s.)_X$ almost certainly is *not* zero. Relative p.q.s. values and calculated Q.S. values are *independent* of what we choose as a reference value.

TABLE XXIII
Observed and Calculated Quadrupole Splittings for Some Tetrahedral Compounds

			Quadrupole splitting[c]	
Code No.[a]	Compound[b]	Ref.	Obs.	Calc.
1	Neo_3SnF	(336)	(−)2.79	—
2	Neo_3SnCl	(336)	(−)2.65	—
3	Neo_3SnBr	(336)	(−)2.65	—
4	Neo_3SnI	(336)	(−)2.40	—
5	$Neo_3Sn(O_2C \cdot Me)$	(336)	(−)2.45	—
6	$Me_3SnC_6F_5$	(439, 440)	−1.35	—
7	$Me_3SnC_6Cl_5$	(440)	(−)1.09	—
8	Me_3SnCF_3	(150, 440)	(−)1.48	—
9	$Me_3Sn(o\text{-}CF_3 \cdot C_6H_4)$	(115)	(−)0.66	—
10	$(p\text{-}F \cdot C_6H_4)_3SnI$	(336)	(−)1.91	—
11	$Ph_3SnCo(CO)_4$	(364)	(−)1.00	—
12	$Cl_3SnMn(CO)_5$	(29, 288, 364, 431, 569)	+1.59	—
13	$ClSn[Re(CO)_5]_3$	(364)	(−)1.60	—
14	Ph_3SnCl	(258, 290, 442)	−2.54	—
15	Ph_3SnBr	(258, 442)	(−)2.50	—
16	$MeSn(C_6F_5)_3$	(528)	(+)1.14	—
17	$Ph_3SnC_6F_5$	(439, 440, 528)	−0.95	—
18	$[cpFe(CO)_2]SnCl_3$	(29, 64, 288, 331)	+1.81	—
19	$(C_6F_5)_3SnCl$	(138)	(−)1.55	—
20	$(C_6F_5)_3SnBr$	(138)	(−)1.60	—
21	$Cl_3SnRh(PPh_3)_3$	(227)	+1.73	—
22	$Cl_3SnIr(C_8H_{12})_2$	(227)	(+)1.64	—
23	$[cpFe(CO)_2]Sn(NCS)_3$	(64)	(+)2.24	—
24	$[cpFe(CO)_2]Sn(HCO_2)_3$	(64)	(+)1.45	—
25	Ph_3SnI	(258, 442)	2.15	−2.18
26	$Ph_2SnI(CH_2)_4SnIPh_2$	(400, 402)	−2.37	−2.26
27	$Me_2Sn(C_6F_5)_2$	(114, 440, 528)	1.51	1.55[d]
28	$Ph_3SnC_6Cl_5$	(440)	0.84	−0.86
29	$Ph_2Sn(C_6Cl_5)_2$	(140, 440)	1.14	0.99[d]
30	$PhSn(C_6Cl_5)_3$	(140)	0.80	+0.86
31	$Ph_2Sn(C_6F_5)_2$	(528)	1.11	1.29[d]
32	$PhSn(C_6F_5)_3$	(528)	0.92	+1.12
33	$(4\text{-}MePh)Sn(C_6F_5)_3$	(528)	1.02	+1.12[e]
34	$(4\text{-}MePh)_2Sn(C_6F_5)_2$	(528)	1.18	1.29[d,e]
35	$(m\text{-}CF_3 \cdot C_6H_4)_3SnBr$	(336)	1.94	−2.08[f]
36	$(PhCH_2)_3SnCl$	(367)	2.80	−2.74[g]
37	$PhSn[Co(CO)_4]_3$	(226)	1.28	+1.00
38	$Ph_2Sn[Co(CO)_4]_2$	(226, 364)	1.29	1.15[d]
39	$Br_3SnMn(CO)_5$	(364, 431, 569)	1.54	+1.60

TABLE XXIII—continued

Code No.[a]	Compound[b]	Ref.	Quadrupole splitting[c]	
			Obs.	Calc.
40	BrSn[Re(CO)$_5$]$_3$	(364)	1.60	−1.60
41	ClSn[Mn(CO)$_5$]$_3$	(364)	1.55	−1.60
42	Me$_3$SnMn(CO)$_5$	(29, 431, 569)	0.75	−1.14
43	Me$_2$Sn[Mn(CO)$_5$]$_2$	(569)	0.92	1.32[d]
44	MeSn[Mn(CO)$_5$]$_3$	(569)	0.95	+1.14
45	Me$_2$ClSnMn(CO)$_5$	(29, 431)	2.63	−2.59[h]
46	MeCl$_2$SnMn(CO)$_5$	(234, 288, 431)	+2.59	+2.67[h]
47	Me$_2$BrSnMn(CO)$_5$	(431)	2.54	−2.59
48	MeBr$_2$SnMn(CO)$_5$	(431)	2.51	+2.67
49	Ph$_2$ClSnMn(CO)$_5$	(364, 431)	2.55	−2.38
50	PhCl$_2$SnMn(CO)$_5$	(431)	2.36	+2.50
51	Ph$_2$BrSnMn(CO)$_5$	(431)	2.28	−2.38
52	PhBr$_2$SnMn(CO)$_5$	(431)	2.63	+2.50
53	Ph$_2$ClSn[cpFe(CO)$_2$]	(29)	2.54	−2.38[h]
54	PhCl$_2$Sn[cpFe(CO)$_2$]	(29)	2.84	+2.57[h]
55	Ph$_2$Sn[Mn(CO)$_5$][Co(CO)$_4$]	(364)	1.15	−1.11
56	Ph$_2$SnCl[Co(CO)$_4$]	(364)	2.22	−2.38
57	ClSn[Mn(CO)$_5$][cpFe(CO)$_2$]$_2$	(275)	2.02	−1.76
58	[cpFe(CO)$_2$]$_2$SnCl$_2$	(64, 288, 331)	+2.38	2.10[d]
59	[cpFe(CO)$_2$]$_2$SnBr$_2$	(64)	2.42	2.10[d]
60	[cpFe(CO)$_2$]$_2$SnI$_2$	(64)	2.25	1.71[d]
61	[cpFe(CO)$_2$]$_2$Sn(NCS)$_2$	(64, 288)	+2.56	2.59[d]
62	[cpFe(CO)$_2$]$_2$Sn(HCO$_2$)$_2$	(64)	2.19	1.69[d]
63	[cpFe(CO)$_2$]SnBr$_3$	(64)	1.60	+1.82
64	[cpFe(CO)$_2$]SnI$_3$	(64)	1.50	+1.48
65	[cpFe(CO)$_2$]Sn(O$_2$CMe)$_3$	(64)	1.87	+1.52
66	[cpFe(CO)$_2$]$_2$Sn(O$_2$CMe)$_2$	(64)	2.60	1.75[d]
67	[Re(CO)$_5$][Mn(CO)$_5$]SnCl$_2$	(364)	2.48	1.85[d]
68	Ph$_3$Sn(S·CS·NEt$_2$)	(207)	1.85	−1.86[i]

[a] Code number will be preceded by table number in text.
[b] Neo = 2-methyl-2-phenylpropyl, cp = π-cyclopentadienyl.
[c] Data given in mm/sec. Observed values are unweighted averages; all measurements at liquid nitrogen or below.
[d] Denotes $\eta = 1$.
[e] Calculated using (p.q.s.)$_{Ph}^{tet}$.
[f] Assuming (p.q.s.)$_{o-CF_3Ph}^{tet}$ = (p.q.s.)$_{m-CF_3Ph}^{tet}$.
[g] Assuming (p.q.s.)$_R^{tet}$ = (p.q.s.)$_{PhCH_2}^{tet}$.
[h] Values obtained in Ref. (29); compound 45, $\Delta = -2.28$; compound 46, $\Delta = +2.42$; compound 53, $\Delta = -2.06$; compound 54, $\Delta = +2.30$ (2.79 with distortion).
[i] Calculated assuming (p.q.s.)$_L^{tet}$: (p.q.s.)$_L^{oct}$ is 1 : 0.75 for the ligand S$_2$CNMe$_2$.

TABLE XXIV
Observed and Calculated Quadrupole Splittings for some Octahedral Compounds

Code No.[a]	Compound[b]	Ref.	Quadrupole splitting[c] Obs.	Calc.
1	$K_2[Me_2SnF_4]$	(439, 442)	+4.12	—
1a	$K_2[Et_2SnF_4]$	(442)	(+)4.44	—
2	$Cs_2[Me_2SnCl_4]$	(439, 442)	+4.30	—
3	$(pyH)_2[Me_2SnCl_4]$	(234)	(+)4.32	—
4	$(Me_4N)_2[Et_2SnCl_4]$	(442)	(+)3.99	—
5	$Cs_2[Me_2SnBr_4]$	(442)	(+)4.22	—
6	$(pyH)_2[Ph_2SnCl_4]$	(442)	(+)3.80	—
7	Bu_2SnCl_2phen	(417)	(+)4.07	—
8	Me_2SnCl_2phen	(323)	(+)4.03	—
9	Me_2SnCl_2bipy	(323)	(+)4.09	—
10	Bu_2SnCl_2bipy	(417)	(+)3.83	—
11	Bu_2SnI_2phen	(417)	(+)3.75	—
12	$Me_2SnCl_2(dmso)_2$	(168, 464)	(+)4.13	—
13	$Me_2SnCl_2(pyO)_2$	(168, 232)	+4.03	—
14	$Me_2SnCl_2(py)_2$	(10, 141)	(+)3.92	—
15	$(edt)_2Snphen$	(210, 463)	(+)1.03	—
16	$(Me_4N)_2[(CH_2{=}CH)_2SnCl_4]$	(400)	(+)3.84	—
17	$Bu_2Sn(NCS)_2phen$	(418)	(+)4.18	—
18	$Et_2SnCl_2dipyam$	(465)	(+)3.78	—
19	$(Me_4N)_2[EtSnCl_5]$	(439, 442)	+1.94	—
20	$(Et_4N)_2[BuSnCl_5]$	(168, 171)	(+)1.86	—
21	$Bu_2Sn(2\text{-}SpyO)_2$	(455)	(+)3.20	—
22	$Bu_2Sn(pic)_2$	(421)	(+)4.35	—
23	$n\text{-}Pr_2SnCl_22pip$	(291)	(+)4.10	—
24	$n\text{-}Pr_2SnCl_22\beta\text{-pic}$	(291)	+3.99	—
25	$Bu_2Sn(trop)_2$	(420)	(+)3.68	—
26	$Bu_2Sn(koj)_2$	(420)	(+)3.60	—
27	$Bu_2SnCl_2(Bu_3PO)_2$	(416)	(+)4.13	—
28	$Bu_2SnCl_2(Ph_3AsO)_2$	(416)	(+)4.04	—
29	$Bu_2SnCl_2(Ph_3PO)_2$	(416)	(+)4.11	—
30	$Me_2Sn(S_2CNEt_2)_2$	(416)	(+)3.14	—
31	$Me_2Sn(S_2CNPh_2)_2$	(233)	(+)3.20	—
32	$Bu_2Sn(S_2CNPh_2)_2$	(233)	(+)3.21	—
33	$Me_2Sn[S_2CN(CH_2)_4]_2$	(233)	(+)2.85	—
34	$Bu_2Sn[S_2CN(CH_2)_4]_2$	(233)	(+)3.06	—
35	$Bu_2Sn[S_2CN(CH_2Ph)_2]_2$	(233)	(+)3.38	—
36	Bu_2SnBr_2phen	(417)	3.94	+4.04[m]
37	Bu_2SnBr_2bipy	(417)	3.95	+3.96[m]
38	$Bu_2Sn(NCS)_2bipy$	(418)	4.04	+4.10[m]
39	$Ph_2Sn(NCS)_2phen$	(418, 439)	+2.35	−2.26[d]

TABLE XXIV—continued

Code No.[a]	Compound[b]	Ref.	Quadrupole splitting[c]	
			Obs.	Calc.
40	Ph$_2$Sn(NCS)$_2$bipy	(418)	2.13	See text
41	Ph$_2$SnCl$_2$bipy	(418, 465)	3.45	+3.64[e]
42	Ph$_2$SnCl$_2$phen	(418)	3.37	+3.72[e]
43	Me$_2$SnF$_2$	(168, 217, 327)	+4.38	+4.12[f]
44	Ph$_2$SnCl$_2$(py)$_2$	(465)	3.39	+3.60[e]
45	Ph$_2$SnCl$_2$dipyam	(465)	3.58	+3.46[e]
46	Ph$_2$SnCl$_2$(dmso)$_2$	(465)	3.54	+3.82[e]
47	Ph$_2$SnCl$_2$4β-pic	(291)	+3.42	+3.66[e]
48	Ph$_2$SnCl$_2$4pip	(291)	3.49	+3.78[e]
49	Ph$_2$SnBr$_2$bipy	(465)	3.52	+3.64[e]
50	Ph$_2$SnBr$_2$(py)$_2$	(465)	3.49	+3.60[e]
51	Ph$_2$SnI$_2$bipy	(465)	3.35	+3.36[e]
52	Ph$_2$SnCl$_2$dipyam	(465)	3.58	+3.46[e]
53	Ph$_2$SnBr$_2$dipyam	(465)	3.45	+3.46[e]
54	Et$_2$SnBr$_2$dipyam	(465)	3.64	+3.78[e]
55	(edt)$_2$Sn(py)$_2$	(210, 463)	1.85	−1.84[g]
56	(edt)$_2$Snbipy	(336)	1.17	+0.96
57	Ph$_2$Sn(2-SpyO)$_2$	(455)	1.45	−1.44[h]
58	BuSnCl(2-SpyO)$_2$	(415)	1.72	+1.87[i]
59	PhSnCl(2-SpyO)$_2$	(415)	1.52	+1.71[i]
60	Bu$_2$SnI$_2$bipy	(417)	3.82	+3.68[m]
61	Ph$_2$Sn(pic)$_2$	(421)	1.94	−2.02[h]
62	(pyH)$_2$[PhSnCl$_5$]	(234)	1.92	+1.90
63	(CH$_2$=CH)$_2$SnCl$_2$(dmso)$_2$	(420)	3.80	+3.86[j]
64	(CH$_2$=CH)$_2$SnCl$_2$(py)$_2$	(420)	3.63	+3.64[j]
65	(CH$_2$=CH)$_2$SnCl$_2$(pyO)$_2$	(420)	3.74	+3.82[j]
66	(CH$_2$=CH)$_2$SnCl$_2$phen	(420)	3.71	+3.76[j]
67	(CH$_2$=CH)$_2$SnCl$_2$bipy	(420)	3.73	+3.68[j]
68	(CH$_2$=CH)$_2$Sn(pic)$_2$	(420)	4.02	+4.08[j]
69	Ph$_2$Sn(trop)$_2$	(420)	1.88	−1.68[h]
70	(CH$_2$=CH)$_2$Sn(trop)$_2$	(420)	1.92	−1.70[k]
71	(CH$_2$=CH)$_2$Sn(NCS)$_2$	(420)	4.28	+4.12[j]
72	(CH$_2$=CH)$_2$Sn(NCS)$_2$(py)$_2$	(420)	3.81	+3.78[j]
73	(CH$_2$=CH)$_2$Sn(NCS)$_2$(pyO)$_2$	(420)	3.88	+3.88[j]
74	(CH$_2$=CH)$_2$Sn(NCS)$_2$phen	(420)	2.62	+2.28[l]
75	(CH$_2$=CH)$_2$Sn(NCS)$_2$bipy	(420)	2.27	See text
76	(Et$_4$N)$_2$[(CH$_2$=CH)$_2$Sn(NCS)$_4$]	(420)	3.94	+4.12[j]
77	(Et$_4$N)$_2$[Bu$_2$Sn(NCS)$_4$]	(420)	4.35	+4.40[m]
78	(Et$_4$N)$_2$[Ph$_2$Sn(NCS)$_4$]	(420)	3.82	+4.08[e]
79	(CH$_2$=CH)$_2$Sn(2-SpyO)$_2$	(420)	1.76	−1.46[k]
80	Ph$_2$Sn(koj)$_2$	(420)	1.98	−1.64[h]
81	(CH$_2$=CH)$_2$Sn(S$_2$CNEt$_2$)$_2$	(420)	2.67	+2.84[j]
82	Ph$_2$SnCl$_2$(Bu$_3$PO)$_2$	(416)	3.81	+3.80[e]

continued

TABLE XXIV—continued
OBSERVED AND CALCULATED QUADRUPOLE SPLITTINGS FOR SOME OCTAHEDRAL COMPOUNDS

Code No.[a]	Compound[b]	Ref.	Quadrupole splitting[c]	
			Obs.	Calc.
83	$Ph_2Sn[S_2CNPh_2]_2$	(233)	1.69	-1.44^e
84	$Ph_2Sn[S_2CNEt_2]_2$	(233)	1.76	-1.40^h
85	$Ph_2Sn[S_2CN(CH_2Ph)_2]_2$	(233)	1.66	-1.52^h
86	$Ph_2Sn[S_2CN(CH_2)_4]_2$	(233)	1.68	-1.32^h

[a] Code number will be preceded by table number in text.

[b] phen = 1,10-Phenanthroline; bipy = 2,2'-bipyridyl; dmso = dimethyl-sulfoxide; pyO = pyridine oxide; py = pyridine; edtH = ethanedithiol; dipyam = di-(2-pyridylamine); 2-HSpyO = 2-pyridinethiol 1-oxide; picH = picolinic acid; pip = piperidine; β-pic = β-picoline; Htrop = tropolone; koj = kojate anion.

[c] When appropriate an unweighted average has been taken, mm/sec at liquid N_2 or below. Bracketed signs are assumed.

[d] cis-Ph$_2$, trans-(NCS)$_2$.
[e] trans-Ph$_2$.
[f] trans-Me$_2$.
[g] trans-py$_2$.
[h] cis-Ph$_2$.
[i] See text.
[j] trans-(CH$_2$=CH)$_2$.
[k] cis-(CH$_2$=CH).
[l] cis-(CH$_2$=CH), trans-(NCS)$_2$.
[m] trans-Bu$_2$.

and SpyO, in which the donor atoms are different, it is assumed that, at least in this semiempirical treatment, one $(p.q.s.)_L^{oct}$ value will suffice to represent the average effect of the ligands.

It should be noted that the partial quadrupole splittings are quoted in units of magnitude of quadrupole splitting; in other words, the tabulated quantities are $\frac{1}{2}e^2|Q|([L]-[X])$. Thus, when using these values to calculate quadrupole splittings, the negative sign of the quadrupole moment of ^{119}Sn must be included. It is also important to note that the partial quadrupole splittings have only a *relative* significance and no importance should be placed on the absolute magnitudes.

In the remainder of the section, we will use the term $(p.q.s.)_L$ to describe the partial quadrupole splitting of a ligand L with the implication that

$$(p.q.s.)_L = \tfrac{1}{2}e^2|Q|([L]-[X]) \qquad (X = F, Cl, \text{or } Br) \qquad (40)$$

Calculated quadrupole splittings are compared with experimental

TABLE XXV

CALCULATION OF PARTIAL QUADRUPOLE SPLITTING DIFFERENCES FOR KEY LIGANDS (122)

Parameter[a]	Estimator	Estimate[d]	Mean value[d]
$[R]^{tet} - [X]^{tet}$	$+\frac{1}{2}QS(1)^b$	−1.40	−1.37 ± 0.06
	$+\frac{1}{2}QS(2)^b$	−1.33	
	$+\frac{1}{2}QS(3)^b$	−1.32	
	$+\frac{1}{2}[QS(6) + QS(19)]^b$	−1.45	
	$+\frac{1}{2}[-QS(16) + QS(20)]^b$	−1.37	
$[Ph]^{tet} - [X]^{tet}$	$+\frac{1}{2}QS(14)^b$	−1.27	−1.26 ± 0.01
	$+\frac{1}{2}QS(15)^b$	−1.25	
	$+\frac{1}{4}[2QS(17) + QS(19) + QS(20)]^b$	−1.26	
$[R]^{oct} - [X]^{oct}$	$-\frac{1}{4}QS(1)^c$	−1.03	−1.03 ± 0.06
	$-\frac{1}{4}QS(1a)^c$	−1.11	
	$-\frac{1}{4}QS(2)^c$	−1.08	
	$-\frac{1}{4}QS(3)^c$	−1.08	
	$-\frac{1}{4}QS(4)^c$	−1.00	
	$-\frac{1}{4}QS(5)^c$	−1.05	
	$-\frac{1}{4}QS(19)^c$	−0.97	
	$-\frac{1}{4}QS(20)^c$	−0.93	

[a] R = alkyl; X = F, Cl, or Br.
[b] Code numbers refer to Table XXIII.
[c] Code numbers refer to Table XXIV.
[d] Quantity tabulated is $\frac{1}{2}e^2|Q|$ ([L] − [X]) (X = F, Cl, or Br) in units of mm/sec, i.e., (p.q.s.)$_L$.

values in Tables XXIII and XXIV. Once again the additivity expressions contained in Table IV have been employed. Calculations for compounds 45, 46, 53, and 54 in Table XXIII were first reported by Bancroft et al. (29) using different (p.q.s.)$_L$ values, and these calculations are included as a footnote to Table XXIII.

The additivity approximation is expected to be satisfactory only if terms arising from nonadditivity or distortions contribute no more than 10–20% of the total EFG. It is suggested, therefore (122), that a discrepancy between observed and calculated quadrupole splittings should be considered exceptional if it exceeds approximately 0.4 mm/sec. The vast majority of cross-checks in Tables XXIII and XXIV lie well within this limit, lending greater confidence to the additivity description of quadrupole splittings. It is also interesting to note that the largest discrepancies are observed for compounds of type X_2SnM_2 (M = transition metal, X = electronegative ligand), and in each case the predicted quadrupole splitting is too low. These discrepancies probably arise from deviations from a regular geometry (29, 64, 288). For example, the crystal structure (426) of $[cpFe(CO)_2]_2SnCl_2$ shows a very distorted tetrahedral structure with an Fe–Sn–Fe bond angle of 128.6°. Both the point charge

TABLE XXVI

VALUES OF PARTIAL QUADRUPOLE SPLITTING PARAMETERS (122)

Tetrahedral structures			Octahedral structures		
Ligand[a]	Data used[b]	Value[c]	Ligand[a]	Data used[d]	Value[e]
Alkyl	Table XXV	−1.37	Alkyl	Table XXV	−1.03
Ph	Table XXV	−1.26	Ph	6	−0.95
I	4	−0.17	I	11	−0.14
NCS	23	+0.21	NCS	17	+0.07
$MeCO_2$	5	−0.15	$\frac{1}{2}$(phen)	7, 8	−0.04
C_6F_5	6	−0.70	$\frac{1}{2}$(bipy)	9, 10	−0.08
C_6Cl_5	7	−0.83	dmso	12	+0.01
CF_3	8	−0.63	pyO	13	−0.05[f]
o-CF_3-C_6H_4	9	−1.04	py	14	−0.10
p-F-C_6H_4	10	−1.12	$\frac{1}{2}$(edt)	15	−0.56
$Co(CO)_4$	11	−0.76	$CH_2=CH$	16	−0.96
$Mn(CO)_5$	12	−0.80[f]	$\frac{1}{2}$(dipyam)	18	−0.17
$Re(CO)_5$	13	−0.80	$\frac{1}{2}$(pic)	22	+0.06
$cpFe(CO)_2$	18	−0.91	$\frac{1}{2}$(2-SpyO)	21	−0.23
HCO_2	24	−0.18	pip	23	−0.01
$Rh(PPh_3)_3$	21	−0.87	β-pic	24	−0.07
$Ir(C_8H_{12})_2$	22	−0.82	$\frac{1}{2}$(trop)	25	−0.11
			$\frac{1}{2}$(koj)	26	−0.13
			Bu_3PO	27	0
			Ph_3PO	29	0
			Ph_3AsO	28	−0.04
			$\frac{1}{2}(S_2CNEt_2)$	30	−0.25
			$\frac{1}{2}(S_2CNPh_2)$	31, 32	−0.23
			$\frac{1}{2}[S_2CN(CH_2)_4]$	33, 34	−0.29
			$\frac{1}{2}[S_2CH(CH_2Ph)_2]$	35	−0.19

[a] See Tables XXIII and XXIV for abbreviations.
[b] Code numbers refer to Table XXIII.
[c] Quantity tabulated is $(p.q.s.)_L^{tet} = \frac{1}{2}e^2|Q|q([L]^{tet} - [X]^{tet})$, where X = F, Cl, or Br.
[d] Code numbers refer to Table XXIV.
[e] Quantity tabulated is $(p.q.s.)_L^{oct} = \frac{1}{2}e^2q|Q|([L]^{oct} - [X]^{oct})$, where X = F, Cl, or Br.
[f] Values differ slightly from Ref. (122), owing to the inclusion of additional data.

(*64, 288*) and molecular orbital (*122*) treatments of this type of distortion predict an increase in quadrupole splitting. Bancroft *et al.* (*29*) have also improved the agreement between observed and calculated quadrupole splittings of $PhCl_2Sn$ $[cpFe(CO)_2]$ (Table XXIII, compound 54) by consideration of distortions.

For octahedral compounds which have more than one geometric isomer, the calculated splitting listed is that which gives the best agreement. The structures predicted in this manner are indicated in the footnotes to Table XXIV. For *trans*-$R_2SnX_2L_2$ species (L is a monodentate ligand), the calculated values vary very little with the arrangement of the X and L ligands and the values given in the table are for a cis arrangement of L_2 and X_2. In most cases the structural assignments are those put forward in the references cited in Table XXIV. For example, Curran *et al.* (*417, 418, 420, 421, 455*) have assigned structures with *trans*-alkyl, phenyl, or vinyl groups in the following species in Table XXIV, compounds 36–38, 60, 63–68, 71–73, 76–78, and 81. They assigned structures with *cis*-phenyl or vinyl groups to these species in Table XXIV, compounds 39, 40, 57, 61, 69, 79, and 80. In many cases, the above assignments were confirmed by dipole moment data. Similarly, the structures assigned to compounds 55 and 56 are those of Poller *et al.* (*463*), while the structures of compounds 47 and 48, Table XXIV were first deduced by Goodman *et al.* (*291*). Fitzsimmons *et al.* (*233*) have argued that the species in Table XXIV, compounds 83–86 have *cis*-phenyl groups.

In some cases, the calculations allow a more detailed appraisal of the structure. For example, the calculated quadrupole splittings of $RSnCl$ $(2-SpyO)_2$ (R = Ph, Bu) for Cl cis to R (Table XXIV, compounds 58 and 59) are much closer to the observed values than those obtained for Cl trans to R (−1.14 mm/sec, R = Bu and −0.98 mm/sec, R = Ph). A structure with Cl cis to R is also indicated by dipole moment data (*415*). Similarly, Poller *et al.* (*465*) have suggested that the relatively low quadrupole splittings of the compounds $Ph_2SnX_2L_2$ (X = Cl, Br, or I; L = py, bipy, dipyam, or tripyam) indicate deviations from a regular geometry, whereas the calculations in Table XXIV show that the lowering is more probably due to the polarity of the tin–phenyl bond.

The additivity model does not always resolve structural ambiguities. For example, Mullins and Curran (*418*) deduced from dipole moment data that, at least in solution, the NCS groups in $Ph_2Sn(NCS)_2L_2$ (L = bipy or phen) are trans to each other. For $Ph_2Sn(NCS)_2phen$ (Δ = 2.34 mm/sec) the calculated quadrupole splittings for a *cis*-Ph_2-*trans*-$(NCS)_2$ isomer (Table XXIV, compound 39) (−2.26 mm/sec) is in much closer numerical agreement with the experimental value than that calculated for the cis-cis isomer (−1.94 mm/sec), confirming the assignment of Mullins and Curran (418). In contrast, the calculated quadrupole splittings for the cis-trans and cis-cis isomers of $Ph_2Sn(NCS)_2bipy$ (Table XXIV, compound 40) are −2.34 and −1.89 mm/sec. A similar situation is encountered for the species $(CH_2=CH)_2Sn(NCS)_2L$ [L = phen (Table XXIV, compound 74); L = bipy (Table XXIV, compound 75)], in which cases the calculated quadrupole splittings indicate a cis-$(CH_2=CH)_2$-trans-$(NCS)_2$ structure for L = phen, whereas the observed value for L = bipy falls between the calculated values for cis-$(CH_2=CH)_2$-cis-$(NCS)_2$ ($\Delta = 1.91$ mm/sec) and cis-$(CH_2=CH)_2$-trans-$(NCS)_2$ ($\Delta = 2.36$ mm/sec).

A further example is provided by the complexes of type $RSnCl_3L_2$ (168, 415), data for which are summarized in Table XXVII, together with the calculated quadrupole splitting values for the various structural isomers. In most cases it is not possible to assign structures confidently in view of the lack of variation in the calculated quadrupole splittings and the relatively poor agreement between observed and calculated values. However, the results for 2L=phen or bipy (Table XXVII, compounds 6–13) do show an indication that a structure with all cis-X groups is the most probable, although such a structure is at variance with dipole moment data (415).

The $(p.q.s.)_L$ values listed in Table XXVI, may be used to check the prediction made in Section II,D, that $[L]^{oct}$ should be approximately 70% of $[L]^{tet}$. From Eqs. (29.1)–(29.4),

$$\frac{(p.q.s.)_L^{oct} - (p.q.s.)_X^{oct}}{(p.q.s.)_L^{tet} - (p.q.s.)_X^{tet}} = \frac{2}{3}\left\{\frac{\sigma_L^{oct} - \sigma_X^{oct}}{\sigma_L^{tet} - \sigma_X^{tet}}\right\} = r[L]$$

Values of $r[L]$ for L = alkyl, phenyl, iodine, and NCS are 0.75 ± 0.06, 0.75 ± 0.07, 0.82 ± 0.51, and 0.33 ± 0.33, respectively, where the errors quoted are standard deviations calculated on the basis that an effective standard deviation of 0.067 mm/sec can be assigned to all parameters listed in Table XXVI. These values of $r[L]$ support the prediction, although $r[NCS]$ seems rather low.

From Eqs. (29.1)–(29.4), the $(p.q.s.)_L$ values may be written in the form

$$(p.q.s.)_L^{tet} = \tfrac{3}{8}\Delta_0 \sigma_L^{tet} \tag{41}$$

$$(p.q.s.)_L^{oct} = \tfrac{1}{4}\Delta_0 \sigma_L^{oct} \tag{42}$$

where Δ_0 is the quadrupole splitting due to one $5p_z$ electron. As both σ_L^{tet} and σ_L^{oct}, in the limit of the additivity approximation, are equal to $2c_1^2$ [Eq. (25)], $(p.q.s.)_L$ values should be a reflection of the relative

TABLE XXVII
Observed and Calculated Quadrupole Splittings for Compounds of Type RSnX₃L₂

Code No.[a]	Compound[a]	Ref.	Obs.	Calc.[b]	Calc.[c]	Calc.[d]
				Quadrupole splitting		
1	BuSnCl₃(py)₂	(168)	1.86	+2.17	+1.86	+1.89
2	BuSnCl₃(Ph₃PO₄)₂	e	2.34	+2.06	+2.06	+2.06
3	PhSnCl₃(Ph₃PO)₂	(416)	2.01	+1.90	+1.90	+1.90
4	BuSnCl₃(Ph₃AsO)₂	(416)	1.81	+2.10	+1.98	+2.00
5	BuSnCl₃(dmso)₂	(168)	1.73	+2.05	+2.08	+2.08
6	BuSnCl₃phen	(168, 415)	1.64	+2.10	+1.98	—
7	BuSnCl₃bipy	(168, 415)	1.64	+2.14	+1.90	—
8	PhSnCl₃phen	(415)	1.48	+1.94	+1.82	—
9	PhSnCl₃bipy	(415)	1.50	+1.98	+1.74	—
10	BuSn(NCS)₃phen	(415)	1.80	+2.32	+1.98	—
11	BuSn(NCS)₃bipy	(415)	1.75	+2.37	+1.90	—
12	PhSn(NCS)₃phen	(415)	1.56	+2.16	+1.82	—
13	PhSn(NCS)₃bipy	(415)	1.55	+2.21	+1.74	—
14	n-PrSnCl₃(piperidine)₂	(291)	+1.99	+2.07	+2.04	+2.04
15	n-PrSnCl₃(β-picoline)₂	(291)	1.87	+2.13	+1.92	+1.94
16	PhSnCl₃(piperidine)₂	(291)	1.70	+1.91	+1.88	+1.88
17	PhSnCl₃(β-picoline)₂	(291)	1.40	+1.97	+1.76	+1.78

[a] Code number is preceded by table number in text, py = pyridine; dmso = dimethyl sulfoxide; bipy = 2,2′-bipyridyl; phen = 1,10-phenanthroline.

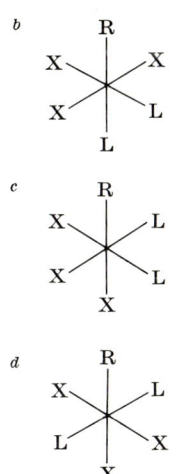

[e] Average of Refs. (168, 416).

populations of the tin–ligand bonds. Williams and Kocher (566) have recently reported a method of calculating orbital populations for tetrahedral compounds using Mössbauer data and it is of interest to compare these calculations with the $(p.q.s.)_L$ values.

In their original paper, Williams and Kocher (566) derived a value of $\Delta_0 = +5.8$ mm/sec by consideration of Br NQR and ^{119}Sn Mössbauer data for the compound Et$_3$SnBr. However, in view of the probable trigonal-bipyramidal structure of this compound (442), this value of Δ_0

TABLE XXVIII

CALCULATED ORBITAL POPULATIONS FOR Sn^{IV}
TETRAHEDRAL COMPOUNDS[a]

Compound	$\alpha_R{}^b$	$\alpha_X{}^b$	$\alpha_X(NQR)^c$
Neo$_3$SnF[d,e]	1.128	0.263	—
Neo$_3$SnCl[d,e]	1.127	0.305	—
Neo$_3$SnBr[d,e]	1.131	0.311	—
Neo$_3$SnI[d,e]	1.111	0.366	—
Ph$_3$SnCl[f]	1.107	0.313	0.307
Ph$_3$SnBr[f]	1.110	0.341	0.365
Ph$_3$SnI[f]	1.065	0.367	0.447

[a] From Ref. (401).
[b] Orbital populations derived from Mössbauer data, $\Delta_0 = 4.3$ mm/sec.
[c] Orbital populations derived from halogen NQR data (566).
[d] Neo = CH$_3$–CH(Ph) CH$_3$–CH$_2$.
[e] Mössbauer data from Ref. (336).
[f] Mössbauer data from Ref. (442).

is probably inflated (401). An alternative value of $\Delta_0 = 4.3$ mm/sec has been reported (401) based on ^{119}Sn Mössbauer and Cl NQR data for Ph$_3$SnCl. Calculated orbital populations (401) for Neo$_3$SnX (X = F, Cl, Br, or I) and Ph$_3$SnX (X = Cl, Br, or I) are given in Table XXVIII. Recently Williams and Kocher (567) have amended their equations and, using a value of $\Delta_0 = 5.5$ mm/sec, obtained a second set of orbital populations for such compounds as Ph$_3$SnX and Ph$_4$Sn. It is interesting to note that from William's and Kocher's data, with $\Delta_0 = 4.3$, values of $(p.q.s.)_R^{tet} - (p.q.s.)_X^{tet} = -1.35$ mm/sec and $(p.q.s.)_{Ph}^{tet} - (p.q.s.)_X^{tet} = -1.26$ mm/sec can be calculated and these differences are in excellent agreement with those in Table XXVI obtained directly from Q.S. data.

Perhaps the most interesting (p.q.s.)$_L$ values are those found for the transition metal groups π-cpFe(CO)$_2$ (−0.91 mm/sec), Rh(PPh$_3$)$_3$ (−0.87 mm/sec), Ir(C$_8$H$_{12}$)$_2$ (−0.82 mm/sec), Re(CO)$_5$ (−0.80 mm/sec), Mn(CO)$_5$ (−0.80 mm/sec), and Co(CO)$_4$ (−0.76 mm/sec). These values indicate that these groups are poorer σ donors than alkyl or phenyl. In semiquantitative terms, using $\Delta_o = 4.3$ mm/sec, $\sigma_R^{tet} - \sigma_M^{tet}$ values for these compounds are in the approximate range 0.29–0.37 and have the relative ordering Co(CO)$_4$ > Mn(CO)$_5$ = Re(CO)$_5$ > Ir(C$_8$H$_{12}$)$_2$ > Rh(PPh$_3$)$_3$ > π-cpFe(CO)$_2$. These conclusions are supported by the quadrupole splitting observed for Me$_3$SnMn(CO)$_5$ ($\Delta = 0.62$–0.82 mm/sec) (*29, 431, 569*) and Me$_3$Sn[π-cpFe(CO)$_2$] ($\Delta = 0.46$ mm/sec) (*149*) compared with the absence of or small quadrupole splittings in their triphenyltin analogs (*29, 149, 431*), and by the negative sign observed for $\tfrac{1}{2}e^2qQ$ of the compound Bu$_3$Sn[πcpFe(CO)$_2$] (*288*). In many ways these (p.q.s.)$_L$ values are rather surprising as Fenton and Zuckerman (*226*) and Onaka *et al.* (*431*) have shown that NMR, X-ray diffraction, and center shift data (see Section IV,A,2) all indicate that these transition metal groups are all better σ donors than alkyl or phenyl. In this context the quadrupole splitting trends found for the compounds R$_3$SN[π-cpFe(CO)L] (R = Me, Ph; L = Ph$_3$P, Ph$_3$As, Ph$_3$Sb, Ph$_2$CF$_3$, Ph$_2$AsCF$_3$, Ph$_2$PMe, PhPMe$_2$, or Ph$_2$AsCF$_3$) reported by Sams *et al.* (*149*) are of interest. Thus, the triphenyltin species all have small quadrupole splittings (Δ in the range 0.44–0.78 mm/sec), while the trimethyltin analogs show single-line spectra or, in the case of L = Ph$_3$P, a very small quadrupole splitting. Although these data strongly suggest that the groups π-cpFe(CO)L, in contrast to π-cpFe(CO)$_2$, are better σ donors than methyl or phenyl, this is not reflected in the magnitude of the quadrupole splitting of Cl$_3$Sn[π-cpFe(CO)PPh$_3$], which is closely similar to that of Cl$_3$Sn[π-cpFe(CO)$_2$].

An additivity model treatment of five-coordinate compounds is rather more complicated than that for tetrahedral or octahedral compounds. A literal point charge model interpretation of the relative quadrupole splittings of [R$_n$SnCl$_{5-n}$]$^-$ ($n = 0$–3) ions has been reported (*442*), and this type of treatment has been extended to other systems (*152, 207*). In the point charge model, differences between equatorial and axial bonds were ignored and the quadrupole splittings of the [SnCl$_5$]$^-$ ions were taken as an absolute value of (p.q.s.)$_{Cl}$. It is probable that both these assumptions are incorrect.

An SnA$_5$ system will have an axially symmetric EFG with principal component:

$$V_{ZZ} = (4[A]^{tba} - 3[A]^{tbe})e \tag{43}$$

substitution for $[A]^{tba}$ and $[A]^{tbe}$ from Eqs. (29.1)–(29.4) yields

$$V_{ZZ} = \tfrac{4}{5}\langle r^{-3}\rangle_p e\,(\sigma_A^{tbe} - \sigma_A^{tba}) \qquad (44)$$

and the corresponding quadrupole splitting is given by

$$\Delta = \tfrac{1}{2}eQV_{ZZ} = \Delta_0(\sigma_A^{tbe} - \sigma_A^{tba}) \qquad (45)$$

From Eq. (45), if axial and equatorial bonds are identical in $[SnCl_5]^-$, i.e., $\sigma_{Cl}^{tbe} = \sigma_{Cl}^{tba}$, then zero quadrupole splitting is expected. The small quadrupole splittings which are observed show that σ_{Cl}^{tbe} does not equal σ_{Cl}^{tba}. Further, from Eq. (45), it can be seen that the quadrupole splittings of $[SnCl_5]^-$ ions give a measure of $(p.q.s.)_{Cl}^{tba} - \tfrac{3}{4}(p.q.s.)_{Cl}^{tbe}$ and not the absolute value of $(p.q.s.)_{Cl}$. Indeed, as explained in Section II,D, it is not possible to calculate absolute $(p.q.s.)_L$ values direct from quadrupole splitting data (*122*).

At this stage, it does not seem worthwhile to pursue the detailed calculations of five-coordinate additivity parameters for two reasons. First, although the quadrupole splittings of $[SnCl_5]^-$ ions can be used to calculate a value of $(p.q.s.)_{Cl}^{tba} - \tfrac{3}{4}(p.q.s.)_{Cl}^{tbe}$, the sign of $\tfrac{1}{2}e^2qQ$ for these species has not yet been reported. Second, as most five-coordinate tin compounds, for which both Mössbauer and structural data are available, are of type R_3SnX_2 (X axial), calculation of additivity parameters would amount merely to a parameterization of the data. The best course of action, at present, seems to be to use quadrupole splitting data for five-coordinate compounds in a purely empirical manner.

The relationship of the quadrupole splittings of five-coordinate R_3SnX_2 (X axial) species to those of tetrahedral R_3SnX compounds is of interest in the study of intermolecular association. From Eqs. (29.1)–(29.4),

$$\Delta(R_3SnX) = -\tfrac{3}{5}\langle r^{-3}\rangle_p (\sigma_X^{tet} - \sigma_R^{tet})\frac{e^2Q}{2}$$

$$\Delta(R_3SnX_2) = -\tfrac{4}{5}\langle r^{-3}\rangle_p (\sigma_X^{tba} - \sigma_R^{tbe})\frac{e^2Q}{2}$$

and, hence,

$$\frac{\Delta(R_3SnX)}{\Delta(R_3SnX_2)} = \tfrac{3}{4}\frac{(\sigma_X^{tet} - \sigma_R^{tet})}{(\sigma_X^{tba} - \sigma_R^{tbe})} \qquad (46)$$

It may be anticipated, therefore, that for a compound of stoichiometry R_3SnX, the formation of an associated structure would result in a quadrupole splitting approximately 1.33 times that expected for a monomeric tetrahedral structure. The quadrupole splittings observed for the five-coordinate species Me_3SnX (X = F, Cl, Br, or I) and the four coordinate species Neo_3SnX (X = F, Cl, Br, or I) show an average ratio of 1.28, in good agreement with the predicted value.

c. *Organometallic Compounds and Halogen Complexes.* The variation of quadrupole splitting with structure discussed in the preceding section makes ^{119}Sn Mössbauer spectroscopy a very powerful means of studying the structures of organotin compounds. In particular, it has greatly helped in the study of intermolecular association.

One of the best examples of this approach is provided by the trialkyltin carboxylates, data for which are summarized in Table XXIX. Infrared evidence (*355*) strongly suggests that the compounds R_3SnO_2CMe (R = Me, Et, Bu) have polymeric structures with trigonal-bipyramidal R_3SnX_2 (X axial) coordination of the tin atom. The quadrupole splittings observed for these compounds are high (Table XXIX, compounds 2, 12, 14) as expected for R_3SnX_2 (X axial) species (*440*). In contrast, monomeric tetrahedral structures would be expected to result in much smaller quadrupole splittings (*440*). Using these considerations Sams *et al.* (*240, 241, 459*) have assigned polymeric five-coordinate structures to the following species in Table XXIX, compounds 3–5, 7, 9–11, and 23–30, and monomeric four-coordinate structures to Table XXIX, compounds 31–33. In a similar manner Debye *et al.* (*170*) have assigned polymeric structures to compounds 6, 8, and 17–22 in Table XXIX, and this assignment may be extended to the remaining compounds in Table XXIX. In some cases, infrared evidence provides further confirmation of the assignments in that the OCO bands are shifted further from the normal ester frequencies in the five-coordinate compounds as compared with the four-coordinate species (*240, 241, 459*).

The most probable reason for the lack of association in the species 31–33 in Table XXIX are steric effects. Sams *et al.* (*240, 241*) have suggested that the major factor in producing monomeric structures is the presence of bulky groups bonded to the α-carbon atom of the carboxylate group. However, the steric effects of the phenyl groups are also important (*400*), as evidenced by the associated structures found for $R_3SnO_2CCMe=CH_2$ (R = Et, Bu). Tricyclohexyltin acetate provides an interesting example as the crystal structure (*4*) shows essentially monomeric coordination with an intermolecular Sn–O distance of 3.84 Å. The observed quadrupole splitting [$\Delta = 3.27$ mm/sec (*400*)], however, is between the values expected for tetrahedral and five-coordinate structures. At present, no obvious explanation of this anomaly is apparent, although intramolecular interaction with the carbonyl oxygen is possible, as this Sn–O distance is only 2.95 Å.

TABLE XXIX

Mössbauer Parameters for Some Compounds of Type $R_3SnO_2CR^1$

Code No.[a]	R^1	C.S.[b,c]	Q.S.[b]	Ref.
		R = Me		
1	H	1.31	3.55	(170, 323, 516)
2	Me	1.34	3.53	(240, 323, 336, 516)
3	CH_2I	1.37	3.83	(459)
4	CH_2Br	1.34	3.90	(459)
5	CH_2Cl	1.41	3.89	(459)
6	CH_2F	1.37	3.86	(170)
7	$CHCl_2$	1.37	4.08	(459)
8	CHF_2	1.40	4.02	(459)
9	CBr_3	1.43	4.13	(459)
10	CCl_3	1.44	4.15	(459)
11	CF_3	1.40	4.20	(170, 459)
		R = Et		
12	Me	1.49	3.35	(84, 367)
13	$C(Me)=CH_2$	1.35	3.00	(83)
		R = Bu		
14	Me	1.40	3.61	(170, 240, 323, 367)
15	$C(Me)=CH_2$	1.45	3.70	(8)
16	$(CH_2)_{11}Me$	1.40	3.62	(516)
17	CH_2Cl	1.40	3.94	(170)
18	$CHCl_2$	1.47	4.00	(170)
19	CCl_3	1.57	3.96	(170)
20	CH_2F	1.42	3.96	(170)
21	CHF_2	1.59	3.92	(170)
22	CF_3	1.62	4.04	(170)
		R = Ph		
23	H	1.37	3.58	(240)
24	Me	1.24	3.30	(240, 323, 464)
25	$(CH_2)_nMe$[d]	1.24–1.33	3.31–3.46	(240, 464)
26	$(CH_2)_nCHMe_2$[e]	1.25–1.27	3.21–3.39	(84, 240, 367)
27	$(CH_2)_7CH=CH$-$(CH_2)_7$–Me	1.27	3.38	(240)
28	$CH_2CHMeEt$	1.29	3.39	(241)
29	$CHMePr$	1.26	3.34	(241)
30	$CH=CH_2$	1.28	3.41	(241)
31	$CMe=CH_2$	1.15	2.18	(84, 240, 367)
32	$CHEtBu$	1.21	2.26	(241)
33	CMe_3	1.22	2.35	(240, 464)
34	CH_2Cl	1.30	3.46	(464)
35	$CHCl_2$	1.35	3.72	(516)

[a] Code number will be preceded by table number in text.
[b] Units are mm/sec at liquid nitrogen temperature; when appropriate an unweighted average has been taken.
[c] Relative to SnO_2, assuming center shift of $BaSnO_3$ is zero.
[d] Range of data for n = 1, 2, 4, 6, 7, 8, 10, 14, or 16.
[e] Range of data for n = 0, 1, 2.

The quadrupole splitting and center shift patterns observed for the halogeno-substituted trimethyl- and tributyltin carboxylates are of interest. Poder and Sams (*459*) could detect little variation in center shift for the compounds 2–5, 7, and 9–11 in Table XXIX, whereas the quadrupole splitting increases uniformly with increasing Taft σ^* of the substituent and decreasing pK_a of the acid. These data were interpreted in terms of a progressive weakening and, hence, lengthening of the intermolecular Sn–O bond with increasing inductive power of the halogen substituent. Infrared data (*459*) are also in agreement with this interpretation, as increasing quadrupole splitting is reflected in a shift of the carbonyl and carboxyl bands to higher and lower frequency, respectively. Substantially similar quadrupole splitting trends were found by Debye *et al.* (*170*) for the compounds 6, 8, 9, and 22 in Table XXIX, although there is a marked saturation effect in the variation of quadrupole splitting with halogen substitution for the tributyltin species.

Debye *et al.* (*170*) also observed a regular increase in center shift with inductive power of the halogen group, which was attributed to shielding effects. Both Debye *et al.* (*170*) and Poder and Sams (*459*) use the $J_{^{119}Sn-CH_3}$ coupling constants observed in NMR spectra of the trimethyltin derivatives in discussing their Mössbauer data. Poder and Sams (*459*) take the data as evidence for a constant electron density in the tin–methyl bonds, while Debye *et al.* (*170*) find an increase in $J_{^{119}Sn-CH_3}$ with both center shift and quadrupole splitting. It should, however, be remembered that as NMR measurements are made on dilute solutions, the correlation with Mössbauer results obtained on the solid compounds may be rather tenuous.

Mössbauer data for compounds of type R_3SnX (X = NCS, NCO, N_3, CN, OH, and ON=C_6H_{10}) are collected in Table XXX. All the compounds of type R_3SnX (X = OH, CN, NCS) show relatively large quadrupole splittings allowing the assignment (*258*) of polymeric five-coordinate structures as found crystallographically for Me_3SnCN (*506*), Me_3SnOH (*366*), and Me_3SnNCS (*243*). Cheng and Herber (*112*) have shown that both quadrupole splitting data and the temperature dependence of the recoil free fraction (see Section IV,A,2) of Me_3SnN_3 are consistent with an associated structure and similar structures seem probable for the higher alkyl analogs. The relatively small quadrupole splittings found for $(PhCH_2)_3SnNCO$ and Ph_3SnNCO argue against any appreciable degree of association (*383*). However, some form of intermolecular interaction seems probable for the species R_3SnNCO (R = Me, Et, Pr, and Bu), confirmed in the case of Me_3SnNCO by the temperature dependence of the recoil free fraction. Harrison and Zuckerman (*318*) have demonstrated that the quadrupole splittings of

TABLE XXX

MÖSSBAUER DATA FOR SOME R_3SnX SPECIES*

X =	CN			NCS			OH			N_3			NCO			ON = C_6H_{10}		
R	C.S.§	Q.S.	Ref.	C.S.§	Q.S.	Ref.	C.S.§	Q.S.	Ref.	C.S.§	Q.S.	Ref.	C.S.§	Q.S.	Ref.	C.S.§	Q.S.	Ref.
Me	1.35	3.10	a	1.40	−3.77	b	1.11	−2.86	c	1.34	3.45	d	1.36	3.31	e	1.43‡	2.96‡	f
Et	1.37	−3.11	g	1.57	3.80	h	1.35	3.00	h	1.24	3.04	i	1.46	3.29	e	1.58	1.96	f
n-Pr	—	—		—	—		—	—		1.21	2.96	i	1.48	3.33	e	1.42	2.03	f
n-Bu	1.37	3.27	h	1.60	3.69	h	1.46	3.21	h	1.26	3.17	i	1.36	3.19	e	1.48	1.76	f
n-Ph	—	—		1.35	3.50	h	1.23	2.73	j	1.40	3.19	k	1.30	2.47	e	1.38	1.44	f
Cyclo-C_6H_{11}	—	—		1.68	3.82	l	1.40	2.99	l	—	—		—	—		—	—	
$PhCH_2$	—	—		—	—		—	—		—	—		1.51	2.85		—	—	
Neo	—	—		—	—		1.13·	1.08	k	1.33	2.48	k	—	—		—	—	

* When appropriate data is an average of the references quoted. Data in mm/sec at liquid nitrogen temperature.
§ Relative to SnO_2 assuming that the center shifts of SnO_2 and $BaSnO_3$ are identical.
‡ X=ON=CMe₂, C.S. = 1.40 mm/sec, Q.S. = 2.93 mm/sec.

Key to references:
a(258, 367, 525) b(258, 290) c(141, 258, 290, 516) d(131, 336) e(383) f(318) g(258, 367, 439) h(258) i(131) j(258, 279, 336, 525) k(336) l(400).

TABLE XXXI

MÖSSBAUER DATA FOR COMPOUNDS OF TYPE $(R_3Sn)_2XO_4$

Compound	Temp.(°K)	C.S.[a,b]	Q.S.[a]	Ref.
$(Me_3Sn)_2SO_4$	80	1.37	4.06	(242)
	295	No effect		
$(Me_3Sn)_2SeO_4$	80	1.39	4.09	(242)
	295	1.33	4.14	
$(Me_3Sn)_2CrO_4$	80	1.36	3.77	(242)
	295	1.35	3.73	
$(Bu_3Sn)_2SO_4$	80	1.56	4.01	(520)
	295	No effect		
Me_3SnNO_3	80	1.44	4.14	(141)

[a] Data given in mm/sec.
[b] Relative to SnO_2, assuming a shift of 2.1 mm/sec for α-tin.

$R_3SnON\!=\!C_6H_{10}$ (R = Me, Et, Pr, Bu, and Ph) and $Me_3SnON\!=\!CMe_2$, together with infrared and mass spectra, indicate that only the trimethyltin derivatives are appreciably polymeric.

Data for compounds of type $(R_3Sn)_2XO_4$ (X = S, Se, Cr) are given in Table XXXI. Ford et al. (242) have shown that both Mössbauer and infrared data for $(Me_3Sn)_2XO_4$ (X = S, Se, and Cr) are consistent with a polymeric structure as illustrated in Fig. 10b. The structure of $(Bu_3Sn)_2SO_4$, however, has been the subject of some controversy. Stapfer et al. (326, 335, 520) have suggested a monomeric structure (Fig. 10a) based on consideration of frozen solution data, IR spectra, and the temperature dependence of the recoil free fraction (326), whereas Garrod et al. (257) have pointed out that the similarity of the spectra of $(Me_3Sn)_2SO_4$ and $(Bu_3Sn)_2SO_4$ argues against a change in structure.

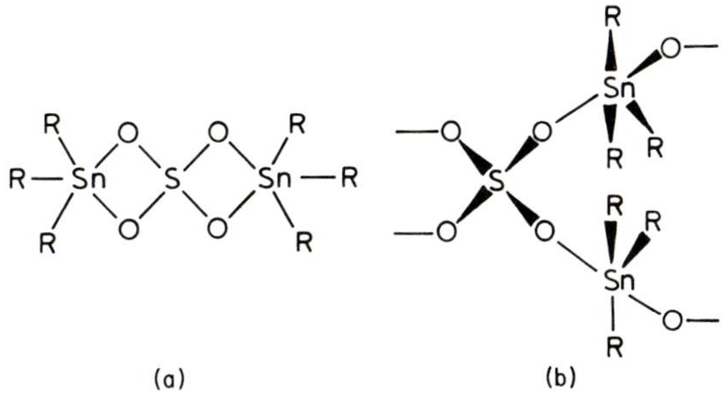

Fig. 10. Two possible structures for $[R_3Sn]_2SO_4$ compounds (257).

It is also interesting to note that adducts formed between $(Me_3Sn)_2(XO_4)$ and water, methanol, dimethylformamide, or pyridine show only slight changes compared with the parent compound (*242*).

In Table XXXII are collected the Mössbauer parameters for the compounds R_2SnX_2 (X = F, Cl, Br, I, NCS, N_3, or NCO), while in Table XXXIII are given data for dialkyltin dicarboxylates and related species. The high quadrupole splittings observed for the compounds R_2SnF_2 (R = Me, Et, Pr, Bu, Ph, Oct), $R_2Sn(NCS)_2$ (R = Me, Et, Bu, Ph), Me_2SnXO_4 (X = S, Se, Mo, W), Bu_2SnSO_4, $Me_2Sn(NO_3)_2$, $Me_2Sn(acac)_2$, and Me_2SnSO_3X (X = F, Cl, CF_3, Me, Et) indicate (*167, 234, 242, 258, 300, 418, 572*) polymeric structures with trans-octahedral coordination of the tin atom as found crystallographically for Me_2SnF_2 (*507*) and $Me_2Sn(NCS)_2$ (*116, 244*). The crystal structure of $Me_2Sn(NCS)_2$ does, in fact, show a considerable distortion with a Me–Sn–Me angle of 148.9°, and this may be reflected in the quadrupole splittings of the $R_2Sn(NCS)_2$ (R = Me, Et, Bu) species, which are considerably lower than the calculated value for a *trans*-$R_2Sn(NCS)_4$ species ($\Delta = 4.40$ mm/sec). Ford *et al.* (*242, 572*) have shown that infrared data support the quadrupole splittings in indicating trans-octahedral structures for Me_2SnXO_4 (X = S, Se) and $Me_2Sn(SO_3X)_2$ (X = F, Cl, CF_3, Me, Et). It is also interesting to note that, unlike $(Me_3Sn)_2XO_4$ (X = S, Se, Cr), adduct formation of pyridine, dimethyl sulfoxide or dimethylformamide with Me_2SnXO_4 (X = S, Se) produces a marked lowering of quadrupole splitting (*242*). The low quadrupole splitting observed for Me_2SnWO_4 is rather curious, and ^{182}W Mössbauer shows a marked asymmetry around the tungsten atom in this compound (*254*).

Using the rather crude assumption that the (p.q.s.)^{oct} value for half of a bidentate carboxylate group is approximately 75% of (p.q.s.)$^{tet}_{O_2CMe}$, an estimate of ca. 3.7 mm/sec can be obtained for the quadrupole splittings expected for a trans-octahedral structure for the dialkyltin dicarboxylates. Such structures could arise from either inter- or intramolecular bidentate coordination of the carboxylate groups. The data for the dialkyltin dicarboxylates in Table XXXIII are in reasonable agreement with this prediction, although the quadrupole splittings for the dimethyltin species appear to be rather high.

The quadrupole splittings of the dialkyltin dihalides R_2SnX_2 (X = Cl, Br, I) fall between the values calculated for monomeric tetrahedral structures (X = Cl, Br, $\Delta = 3.16$ mm/sec; X = I, $\Delta = 2.77$ mm/sec) and associated trans-octahedral structures (X = Cl, Br, $\Delta = 4.12$ mm/sec; X = I, $\Delta = 3.56$ mm/sec), consistent with highly distorted octahedral structure as observed for Me_2SnCl_2 (*167*). In contrast, the quadrupole splittings of Ph_2SnX_2 (X = Cl, Br, I) are those expected for

TABLE XXXII

Mössbauer Data for Some R_2SnX_2 Species[a,b]

X = R	F* C.S.[c]	F* Q.S.	F* Ref.	Cl† C.S.[c]	Cl† Q.S.	Cl† Ref.	Br C.S.[c]	Br Q.S.	Br Ref.	I C.S.[c]	I Q.S.	I Ref.	NCS C.S.[c]	NCS Q.S.	NCS Ref.	NCO§ C.S.[c]	NCO§ Q.S.	NCO§ Ref.	N_3 C.S.[c]	N_3 Q.S.	N_3 Ref.
Me	1.33	+4.38	d	1.56	+3.55	e	1.60	3.36	f	—	—	—	1.48	3.87	g	1.29	2.84	h	1.06	2.61	i
Et	1.42	4.27	j	1.64	3.64	k	1.70	3.27	l	1.75	3.09	m	1.56	3.96	g	—	—	—	1.23	2.94	i
n-Pr	1.45	4.36	j	1.70	3.60	n	—	—	—	—	—	—	—	—	—	—	—	—	1.15	2.74	i
n-Bu	1.42	4.07	o	1.62	3.40	p	1.65	3.28	q	1.80	2.65	r	1.56	3.90	s	1.31	3.15	h	1.29	2.99	i
Ph	1.28	3.43	j	1.38	2.82	t	1.43	2.54	l	1.51	2.38	l	1.45	3.96	u	—	—	—	—	—	—

*R = Oct
1.45 4.31 j

†R = CH₂=CH
1.45 3.34 l

R = Cyclo-C_6H_{11}
1.68 3.44 v

§R = i-Bu
1.45 3.51 h

R = Ph·CH₂
1.12 2.21 h

[a] Where appropriate an average of the data has been taken.
[b] Data given in mm/sec at liquid nitrogen temperature.
[c] Relative to SnO_2 assuming center shifts of SnO_2 and $BaSnO_3$ are identical.

Key to references:
[d](167, 217, 290, 327) [e](138, 141, 167, 171, 289, 336, 367, 442, 528) [f](442, 528) [g](258) [h](383) [i](112) [j](167) [k](167, 442) [l](442) [m](81, 442) [n](84) [o](6, 167, 367) [p](6, 84, 167, 516) [q](6, 7, 367, 516) [r](6, 367) [s](258, 418, 516) [t](84, 141, 167, 171, 338, 418, 442, 531) [u](418) [v](336, 400)

TABLE XXXIII

Mössbauer Data for Some Further R₂SnX₂ Species

Compound	C.S.[a,b]	Q.S.[a]	Ref.
Me₂Sn(acac)₂	1.18	3.93	(323)
Me₂Sn(O₂CH)₂	1.30	4.60	(323, 531)
Me₂Sn(O₂CPH)₂	1.40	3.96	(323)
Me₂Sn(O₂CC₅H₄N)₂	1.28	4.43	(323)
Me₂SnC₂O₄H₂O	1.55	4.65	(551)
Me₂SnSO₄	1.61	5.00	(242)
Me₂SnSeO₄	1.52	4.82	(242)
Me₂SnMoO₄	1.42	4.10	(551)
Me₂SnWO₄	1.39	3.53	(551)
Me₂Sn(NO₃)₂	1.62	4.13	(516)
Me₂Sn(SO₃F)₂	1.82	5.54	(572)
Me₂Sn(SO₃CF₃)₂	1.79	5.51	(572)
Me₂Sn(SO₃Cl)₂	1.75	5.20	(572)
Me₂Sn(SO₃Me)₂	1.52	5.05	(572)
Me₂Sn(SO₃Et)₂	1.52	4.91	(572)
Bu₂Sn[O₂C(CH₂)ₙMe]₂[c]	1.34–1.49[c]	3.23–3.70[c]	(5, 279, 380)
Bu₂Sn[O₂C(CH₂)ₙCH₂Cl]₂[d]	1.30–1.60[d]	2.89–3.65[d]	(5, 367, 380)
Bu₂Sn(O₂CHCl₂)₂	1.54	3.73	(367)
Bu₂Sn(O₂CCl₃)₂	1.58	3.93	(5, 7, 367)
Bu₂Sn(O₂CMe=CH₂)₂	1.43	3.70	(5, 9)
Bu₂Sn maleate	1.44	3.58	(5, 367, 400)
Bu₂Sn(O₂CPh)₂	1.60	3.50	(323, 367)
Bu₂SnSO₄	1.66	4.78	(6, 367)
Ph₂Sn(acac)₂	0.74	2.14	(234)

[a] Data given in mm/sec at liquid nitrogen temperature.
[b] Relative to SnO_2 assuming that the center shifts of SnO_2 and $BaSnO_3$ are identical.
[c] Range of data observed for $n = 0, 1, 3, 5, 6, 8, 10,$ or 16.
[d] Range of data observed for $n = 0, 3, 5,$ or 13; when appropriate an average of the data has been taken.

tetrahedral structures (400).* Using a similar criterion, Herber et al. (112, 383) have assigned tetrahedral structures to the compounds R_2SnX_2 (X = N_3, R = Me, Et, Pr, or Bu; X = NCO, R = Me, Bu, i-Bu, or PhCH₂).

In Table XXXIV are given quadrupole splitting data for some organotin trihalides and $BuSn(SCN)_3$, together with calculated quadrupole splittings for monomeric and associated structures. These data are

* A very recent crystal structure of Ph_2SnCl_2 (296) confirms the tetrahedral structure.

clearly consistent with some degree of association, although the quadrupole splitting of BuSn(SCN)$_3$ appears rather anomalous.

The organotin oxinates (Table XXXV) form an interesting series of compounds. The quadrupole splittings of all dialkyltin dioxinates fall in a very narrow range, while, as expected, that of diphenyltin dioxinate is rather lower, indicative (10, 234, 418, 462) of common cis-octahedral structures as found crystallographically for Me$_2$Sn(ox)$_2$ (505). From the

TABLE XXXIV

OBSERVED AND CALCULATED QUADRUPOLE SPLITTINGS
FOR SOME RSnX$_3$ SPECIES

	Quadrupole splitting^a			
Compound	Obs.	Ref.	Calculated^b octahedral	Calculated^b tetrahedral
MeSnBr$_3$	1.91	(528)	+2.06	+2.74
EtSnCl$_3$	1.97	(442)	+2.06	+2.74
EtSnBr$_3$	1.85	(442)	+2.06	+2.74
(CH$_2$=CH)SnCl$_3$	1.86	(442)	+1.92	+2.56
BuSnCl$_3$	1.88	c	+2.06	+2.74
PhSnCl$_3$	1.79	d	+1.90	+2.52
PhSnBr$_3$	1.62	(442)	+1.90	+2.52
BuSn(NCS)$_3$	1.46	(258)	+2.20	+3.16

^{a} Data given in mm/sec at liquid nitrogen temperature.
^{b} Using (p.q.s.)$_L$ values from Table XXVI, assuming (p.q.s.)$_{CH_2=CH}^{oct}$ = 0.75 (p.q.s.)$_{CH_2=CH}^{tet}$.
^{c} Average of Refs. (168, 171, 415).
^{d} Average of Refs. (84, 275, 415, 442, 528).

magnitude of the quadrupole splittings of R$_2$Sn(ox)$_2$ species, it is apparent that ½ox has a similar (p.q.s.)^{oct} value to that of Cl.

Monomeric five-coordinate structures seem probable for the species R$_2$Sn(ox)X (Table XXV, compounds 4–10) as the observed quadrupole splittings are midway between the values expected for associated structures with either cis or trans R groups (418, 462). Unfortunately, the additivity description of the quadrupole splittings for the five-coordinate compounds is not sufficiently advanced to speculate on the geometric isomerism of these species or of the compound Ph$_3$Sn(ox). Mullins and Curran (418) have concluded from dipole moment data that the most probable structure for Ph$_2$Sn(ox)NCS is one with a phenyl group and nitrogen atom in the axial position. These authors (418) also find some evidence for association in $Bu_2Sn(ox)NCS$, but this does not seem to be reflected in the quadrupole splitting.

Six-coordinate structures have been assigned (415, 462) to the species $RSn(ox)_2X$ (Table XXXV, compounds 11–14), although the

TABLE XXXV

Mössbauer Data for Some Organotin Oxinates

Code No.[a]	Compound[b]	C.S.[c,d]	Q.S.[c]	Ref.
1	$Ph_3Sn(ox)$	1.07	1.75	(462)
2	$R_2Sn(ox)_2$[e]	0.77–1.13	1.81–2.21	(10, 141, 234, 300, 323, 367, 418, 439, 462)
3	$Ph_2Sn(ox)_2$	0.73	1.65	(10, 234, 323, 418, 439, 462)
4	$R_2Sn(ox)Cl$[f]	1.26–1.56	2.78–3.36	(462)
5	$Et_2Sn(ox)Br$	1.39	3.08	(462)
6	$Et_2Sn(ox)I$	1.43	2.85	(462)
7	$Et_2Sn(ox)NCS$	1.31	3.07	(462)
8	$Bu_2Sn(ox)NCS$	1.33	3.25	(418)
9	$Ph_2Sn(ox)Cl$	1.10	2.40	(418, 462)
10	$Ph_2Sn(ox)NCS$	0.98	2.48	(418)
11	$BuSn(ox)_2Cl$	0.82	1.68	(168, 415, 462)
12	$BuSn(ox)_2NCS$	0.76	1.73	(415)
13	$PhSn(ox)_2Cl$	0.67	1.48	(415, 462)
14	$PhSn(ox)_2NCS$	0.58	1.57	(415)
15	$BuSn(ox)_3$	0.69	1.76	(168, 462)

[a] Code number will be preceded by table number in text.
[b] ox = 8-hydroxyquinoline anion.
[c] Data given in mm/sec at liquid nitrogen temperature; when appropriate an average has been taken.
[d] Relative to SnO_2, assuming identical center shift for SnO_2 and $BaSnO_3$.
[e] Range of data for R = Me, Et, Pr, Bu, i-Bu, or Oct.
[f] Range of data for R = Me, Et, Pr, Bu, or Oct.

quadrupole splittings are rather lower than expected. This seems to be a general phenomenon of six-coordinate monoalkyl- or monophenyltin compounds as shown by the data in Table XXVII. For the compound $BuSn(ox)_3$ a seven-coordinate structure seems the most probable (168, 462).

Mössbauer data for a wide range of organotin oxygen and sulfur derivatives of general type R_3SnXR', $(R_2SnX)_n$, and $R_2Sn(XR')_2$ (X = O, S) have been reported. Most of the compounds give quadrupole splittings in the general range 1–3 mm/sec and convenient tabulations of the data are available in Refs. (*167, 577*). At the present time, the relationship between quadrupole splitting and structure, and especially the degree of association, for these species has not been fully elucidated (*167*). As mentioned in Section II,E, Goldanskii and his co-workers (*424*) have exploited frozen solution data to study intermolecular association in the species of type $R_3Sn–S–R'$ and $Et_3Sn–O–C_6H_4·X$-p, while Davies *et al.* (*169*) have concluded that compounds of the type $XBu_2SnOSnBu_2X$ [X = F, Cl, Br, NCS, OSiMe$_3$, O$_2$CMe, OC$_6$H$_4$·X'-4 (X' = H, Me, OMe, Cl)], which show quadrupole splittings in the range 2.74–3.36 mm/sec, have a ladder-type dimeric structure with five-coordinate tin atoms (*169*). Ford *et al.*(*240*) from correlation of Mössbauer, IR, and chromatographic studies of compounds Ph(OCOR')O [R' = (CH$_2$)$_8$CH=CH$_2$, (CH$_2$)$_{16}$Me] have assigned trimeric structures, whereas for the analogous compounds with R' = CMe$_3$, CCl$_3$, and CF$_3$, Poller *et al.* (*464*) prefer a polymeric structure.

Organotin nitrogen derivatives of type $(R_3Sn)_{3-n}NR_n$ ($n = 0$–2) show relatively small quadrupole splittings (Δ in the range 0–1.84 mm/sec), in keeping with tetrahedral coordination (*159*). It is interesting that there is a trend to decreasing quadrupole splitting as n varies from 2→0. In contrast, trialkyl and triphenyl derivatives of the bidentate ligands imadazole, 1,2,4-triazole, benzimidazole, and 1,2,3-benzitriazole show large splittings [Δ in the range 2.59–3.18 mm/sec (*336, 440*)], consistent with five-coordinate associated structures (*336*).

Data for a series of five-coordinate complexes of type R_3SnXL are summarized in Table XXXVI. As expected, the coordination of a further ligand to Me$_3$SnCl produces relatively little variation in quadrupole splitting reflecting the common five-coordinate structure. In contrast, the coordination of a ligand to Ph$_3$SnX (X = Cl, Br) gives rise to a marked increase in quadrupole splitting, probably arising mainly from a change in structure from tetrahedral to trigonal-bipyramidal (*556*).

Mössbauer data for a great many halide complexes have been reported and a selection of the available data is given in Table XXXVII. The majority of the center shifts show a decrease compared with SnX$_4$ (*10, 103, 457, 574*) and show some variation with the nature of the donor atom (*103, 457*). For example, phosphine derivatives tend to give the most negative shifts, and Carty *et al.* (*103*) have interpreted this trend in terms of a concentration of 5s density in the Sn–P bonds. A more detailed discussion of center shifts is given in Section IV,A,2.

The small or zero quadrupole splittings observed are expected in view of the very similar (p.q.s.)^{oct} values of ligands such as dmso, Ph$_3$XO (X = As, P), Bu$_3$PO, py and $\frac{1}{2}$(bipy) to those of the halogens

(Table XXVI). Yeats et al. (574) have suggested that the observation of quadrupole splittings for SnX_4L_2 species in which L is an oxygen donor reflects a weaker donor interaction for the L→Sn as opposed to the X→Sn bond. This suggestion is supported by the following observations.

(1) For ligands of type R_2SO, R_2SO_2, and R_3PO, the shift of M–O (M = S, P) stretching frequency on complexation, which gives some guide to the strength of the donor bond, tends to be largest for compounds which give single-line spectra.

TABLE XXXVI

MÖSSBAUER DATA FOR SOME FIVE-COORDINATE COMPLEXES[a]

Compound[b]	C.S.[c,d]	Q.S.[c]	Ref.
Me_3SnCl	1.39	3.30	e
$(Et_4N)Me_3SnCl_2$	1.33	3.28	f
$Me_3SnClpy$	1.43	3.44	e
$Me_3SnClPh_3PO$	1.45	3.49	(338)
$Me_3SnClMeCONMe_2$	1.50	3.69	(338)
$Me_3SnCl(C_5H_4NOMe\text{-}4)$	1.44	3.45	(338)
$Me_3SnClHMPA$	1.44	3.52	(338)
Ph_3SnCl	1.35	2.49	g
$(Me_4N)[Ph_3SnCl_2]$	1.32	3.00	(442)
$(Ph_3PC_{10}H_{21})[Ph_3SnCl_2]$	1.23	2.87	(207)
Ph_3SnClR_2SO	1.26–1.30	3.08–3.25	h
$Ph_3SnCl(RO)_3PO$	1.30–1.31	3.07–3.18	i
$Ph_3SnClPh_3PO$	1.29	3.19	j
$Ph_3SnClPh_3AsO$	1.29	3.09	(556)
$Ph_3SnClpyO$	1.30	3.03	(556)
$Ph_3SnCl(Me_2NCHO)$	1.31	2.84	(556)
Ph_3SnBr	1.32	2.48	g
$(Ph_3PC_{10}H_{21})[Ph_3SnBr_2]$	1.29	2.87	(207)
$Ph_3SnBrpyO$	1.28	3.03	(207)
$Ph_3SnBrPh_3PO$	1.20	3.20	(207)
$Ph_3SnBrMe_2SO$	1.31	3.22	(207)

[a] Only a selection of the available data is given; note that Ph_3SnCl and Ph_3SnBr are four-coordinate.
[b] HMPA = hexamethylphosphoramide, pyO = pyridine N-oxide.
[c] Data given in mm/sec at liquid nitrogen temperature.
[d] relative to SnO_2, assuming zero shift for $BaSnO_3$.
[e] Average of data quoted in Ref. (516).
[f] Average of data from Refs. (171, 442).
[g] Average of data from Refs. (207, 516).
[h] Range of data for R = Me, Pr, Bu, $\frac{1}{2}(CH_2)_4$ (207, 556).
[i] Range of data for R = Me, Et, Ph (556).
[j] Average of data from Refs. (207, 556).

(2) Larger quadrupole splittings are observed for derivatives of the weaker acceptor $SnBr_4$ (457).
(3) The Sn–O bond distances in $SnCl_4 \cdot 2POCl_3$ [2.30 and 2.25 Å (77)], which shows a quadrupole splitting of 1.12 mm/sec (574), are larger than those of $SnCl_4 \cdot 2MeSO$ [2.17 and 2.10 Å (315)] or $SnCl_4 \cdot 2SeOCl_2$ [2.12 Å (337)] for which no quadrupole splittings are observed.
(4) Most complexes with nitrogen donor ligands, which are thought to be better donors than analogous oxygen derivatives, do not give quadrupole splittings. Of particular interest in this context are the complexes of the bidentate ligand pyrazine. Thus, the 1 : 2 derivative, $SnCl_4 \cdot 2pyz$ does not show quadrupole splitting, whereas the 1 : 1 derivatives, $SnX_4 \cdot pyz$ (X = Cl, Br, I) (Table XXXVII, compounds 65–67), which probably have bridging pyz groups and, hence, a weaker interaction, show quadrupole splitting which increase in the order Cl > Br > I (286). Similar quadrupole splittings are observed for RCN adducts (Table XXXVII, compounds 25, 26) in which the sp hybridization of the nitrogen atom will also be expected to result in a weaker interaction.

Yeats et al. (574) also note that the trend of quadrupole splitting for the species $SnCl_4 \cdot R_2SO$ (Table XXXVII, compounds 1–4) suggest that steric effects may be important.

In contrast to oxygen and nitrogen donor adducts, the data for phosphine and arsine complexes indicate that the quadrupole splittings arise from an increased donor interaction in the Sn–P as compared with the Sn–Cl bond. Thus, Cunningham et al. (152) and Carty et al. (103) have

TABLE XXXVII

Mössbauer Parameters for Some Halide Complexes of Sn^{IV}[a]

Code No.	Compound[b]	C.S.[c,d]	Q.S.[c]	Ref.
	$SnCl_4 \cdot 2L$			
	(L)			
1	Me_2SO	0.38	—	e
2	Et_2SO	0.36	Small	(574)
3	$n\text{-}Pr_2SO$	0.38	0.59	(574)
4	$n\text{-}Bu_2SO$	0.37	0.67	(574)
5	$(Me_2O)_2SO$	0.34	—	(574)
6	$\frac{1}{2}(R_2SO_2)$[f]	0.38–0.51	0.83–1.57	(574)
7	$Ph_nCCl_{3-n}PO$[g]	0.27–0.51	0.50–1.61	e
8	$n\text{-}Bu_3PO$	0.24	—	(574)
9	$Cl_n(PhO)_{3-n}PO$[h]	0.34–0.42	0.71–1.13	(574)
10	Ph_3AsO	0.44	0.70	(574)

continued

TABLE XXXVII—continued

MÖSSBAUER PARAMETERS FOR SOME HALIDE COMPLEXES OF Sn^{IV} [a]

Code No.[a]	Compound[b]	C.S.[c]		Ref.
11	pyO	0.42	—	(574)
12	Cl_2SeO	0.37	—	(574)
13	$(Me_2N)_3PO$	0.31	0.70	(457)
14	$(Me_2N)_2CO$	0.35	0.75	(457)
15	$\frac{1}{2}[MeO(CH_2)_3OMe]$	0.51	0.80	(457)
16	oxH	0.45	—	i
17	SalH	0.43	1.10	(10)
18	ROH[j]	0.33–0.43	0.50–0.70	(349)
19	$(Me_2N)_2CS$	0.70	—	(457)
20	$\frac{1}{2}[MeS(CH_2)_2SMe]$	0.70	—	(457)
21	$\frac{1}{2}$(bipy)	0.45	—	k
22	py	0.51	—	(457)
23	$\frac{1}{2}$(en)	0.61	—	(457)
24	NMe_2	0.59	—	(152)
25	MeCN	0.41	0.81	(152, 349)
26	PhCN	0.41	0.77	(152)
27	F^-, Cl^-, Br^-, I^-[l]	0.29–0.78	—	l
28	Ph_3P	0.77	—	(103, 152)
29	RPh_2P[m]	0.63–0.81	0.45–0.58	(103, 152)
30	R_2PhP[n]	0.85–0.88	0.97–1.04	(103, 152)
31	R_3P[o]	0.85–0.89	0.95–1.15	(103, 152, 457)
32	$\frac{1}{2}(Ph_2P \cdot CH_2 \cdot PPh_2)$	0.69	—	(152)
33	$\frac{1}{2}[Ph_2P(CH_2)_2PPh_2]$	0.72	—	(103)
34	$AsPh_3$	0.81	—	(152)
35	$AsEt_3$	0.87	0.90	(152)
	$SnBr_4 \cdot 2L$ (L)			
36	Me_2SO	0.66	—	(457)
37	$(Me_2N)_2CO$	0.70	—	(457)
38	$(Me_2N)_3PO$	0.56	0.76	(457)
39	Ph_3PO	0.63	0.61	(457)
40	$\frac{1}{2}[MeO(CH_2)_2OMe]$	0.81	—	(457)
41	oxH	0.65	—	(10)
42	SalH	0.73	1.22	(10)
43	C_4H_8S	0.99	—	(457)
44	$(Me_2N)_2CS$	0.94	—	(457)
45	$MeS(CH_2)_2SMe$	0.97	—	(457)
46	NH_2CS	0.80	—	(278)
47	$\frac{1}{2}$(bipy)	0.70	—	k
48	py	0.74	—	(457)
49	NMe_2	0.90	—	(152)
50	F^-, Cl^-, Br^-, I^-	0.53–1.01	—	l
51	Ph_3P	0.63	0.66	(457)
52	n-Bu_3P	1.03	1.01	(152)

TABLE XXXVII—continued

Code No.	Compound[b]	C.S.[c,d]	Q.S.[c]	Ref.
53	AsPh$_3$	1.06	—	(152)
54	AsEt$_3$	1.02	0.83	(152)
	SnI$_4$·2L			
	(L)			
55	oxH	0.91	—	(10)
56	½(bipy)	0.95	—	(300)
57	NMe$_3$	1.30	—	(152)
58	Cl$^-$, Br$^-$, I$^-$	0.98–1.60	—	l
59	Sn(SpyO)$_2$X$_2$[p]	0.32–0.90	0.60–0.82	(455)
60	PcSnX$_2$[q]	0.03–0.45	0 –1.13	(433, 530)
61	Sn(porph)X$_2$[r]	(–)0.06–0.24	r	(433)
62	Sn(ox)$_2$X$_2$[s]	0.30–0.61	—	s
63	Sn(Sal)$_2$X$_2$[t]	0.23–0.41	—	t
64	SnCl$_4$(pyz)$_2$	0.43	—	(286)
65	SnCl$_4$pyz	0.38	0.60	(286)
66	SnBr$_4$pyz	0.85	0.92	(286)
67	SnI$_4$pyz	1.48	0.96	(286)

[a] Only a selection of data is given. Further data may be found in the references quoted in Table XXXVIII and in Ref. (577). In text code number will be preceded by table number.
[b] oxH = 8 oxyquinoline; SalH = salicylaldehyde; en = Me$_2$N-(CH$_2$)$_2$NMe$_2$; bipy = 2,2'-bipyridyl; py = pyridine; HSpyO = 2-pyridinethiol-1-oxide; Pc = phthalocyanine; pyz = pyrazine.
[c] Data given in mm/sec.
[d] Relative to SnO$_2$, assuming center shift of BaSnO$_3$ is zero and Pd(Sn) is 1.52 mm/sec.
[e] Average of data from Refs. (457, 574).
[f] Range of data for R = Me, Et, n-Pr, n-Bu, Ph, and ½(CH$_2$)$_4$.
[g] Range of data for n = 0, 1, 3.
[h] Range of data for n = 0, 1, 2.
[i] Average of data from Refs. (10, 300).
[j] Range of data for R = Me, Et, n-Pr, i-Pr.
[k] Average of data from Refs. (300, 457).
[l] Range of data from Table XXXVIII.
[m] Range of data for R = Me, Et, MeO.
[n] Range of data for R = Et, Me.
[o] Range of data for R = Et, n-Pr, n-Bu.
[p] Range of data for X = F, Cl, Br, I.
[q] Range of data for X = F, Cl, Br, I, OH.
[r] X = F, Cl, OH; porph = tetra(4-X'-C$_6$H$_4$)porphine (X' = MeOH, Me, Cl, H); only X = OH, X' = Cl show a quadrupole splitting (0.76 mm/sec).
[s] Range of data for X = Cl, Br, I from Table XXXVIII.
[t] Range of data for X = Cl, Br, I from Table XXXVIII.

noted that while the complexes $SnCl_4 \cdot 2LPh_3$ (L = As, P) do not show quadrupole splittings, the substitution of alkyl for phenyl, which will increase the donor ability of the LR_3 group, gives rise to quadrupole splitting. The overall quadrupole splitting trends are consistent with the order of donor ability of the ligands $R_3P \sim R_3As > Ph_3P \sim Ph_3As \geqslant N(sp^2) \sim N(sp^3) \sim Cl > N(sp)$.

The quadrupole splittings observed for the species $Sn(SpyO)_2X_2$ (455) are of interest as from (p.q.s.)$_L^{oct}$ values, these may be attributed to the greater donor power of SpyO as opposed to halogen. The $PcSnX_2$ derivatives (433) form the only series of sp^2 nitrogen donors to give quadrupole splittings and it is curious that analogous tetraarylporphin derivatives [with the exception of dihydroxytetra(p-tolyl)porphinotin-(IV)] give single-line spectra (433).

Very little correlation between quadrupole splitting and ligand geometry has been found for the halide complexes. One example is provided by species $SnCl_4(Ph_2P(CH_2)_nPPh_2)$ ($n = 1,2$), for which the lack of quadrupole splitting may be associated with cis geometry, as opposed to $SnCl_4 \cdot 2PRPh_2$ species, which are probably trans and which give quadrupole splittings in the range 0.45–0.58 mm/sec (103, 152). However, the quadrupole splittings of the species $Sn(SpyO)_2X_2$ (X = F, Cl, Br) fall between the values calculated for trans (ca. 0.92 mm/sec) and cis (ca. 0.46 mm/sec) geometries. Clausen and Good (127) have attributed the large line-width found for the species SnX_4F_2 (X = Cl, Br) ($\Gamma = 1.09$, 1.43 mm/sec) compared with the other halide anions ($\Gamma = 0.94$–1.19 mm/sec) to trans structures, whereas Yeats et al. (574) have isolated two forms of $SnCl_2 \cdot 2(n-Pr_2SO_2)$ with quadrupole splittings approximating to a 2 : 1 ratio as expected for a cis-trans pair.

2. Center Shifts

The center shifts of tin(IV) compounds are smaller than those of gray tin, reflecting the transition from the $5sp^3$ configuration of gray tin to the $4d^{10}$ configuration anticipated for a "perfect" Sn^{4+} ion. Increasing polarity of the tin–ligand bonds would be expected to result in a decrease in center shift with a limiting value corresponding to that of a perfect Sn^{4+} ion.

Some evidence for such trends has been found for compounds in which all the tin–ligand bonds are reasonably polar; a selection of data for these types of compounds is given in Table XXXVIII. The center shifts of the series of halide anions (Table XXXVIII, compounds 9–20) correlate (127, 168, 330) well with the average electronegativity of the ligands, and similar correlations have been found (273) for the tetrahalides [excluding SnF_4 which has a different structure (374) to the other halides]. These trends are illustrated in Fig. 11. The center shifts of these compounds have also been correlated with S.C.F.M.O. calculations (299). Further evidence of a general trend to lower center shifts with decreasing size of the halogen atom is provided by the series $PcSnX_2$ (Table XXXVIII, compounds 23–26) (433); $Sn(SpyO)_2X_2$ (compounds 28–31) (455); $SnX_4 \cdot 2oxH$ (compounds 38–40) (10); $Sn(sal)_2X_2$ (compounds 41–43) (10); $Sn(ox)_2X_2$ (compounds 35–37) (10); $SnX_4 \cdot 2NR_3$ (R = Me or Et) (compounds 49–52) (152), and $(SnCl_4 \cdot pyz)_n$ (compounds 53–55) (286), while there is also a gradual decrease in center shift as chlorine is replaced by ligand in the series $SnCl_6^{2-}$, $SnCl_4$ (2oxH),

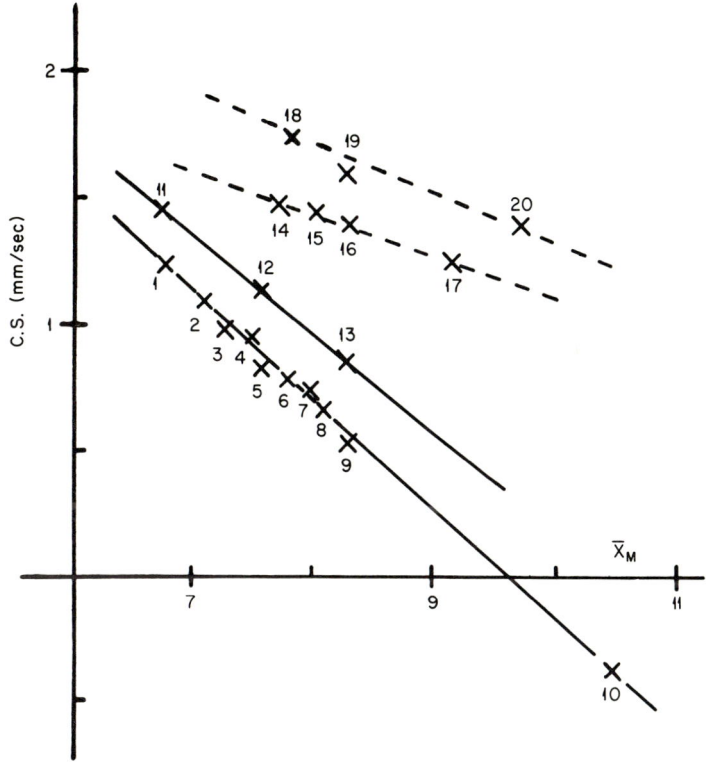

FIG. 11. Center shift versus average electronegativity (441). Data from (330) (points 1–9), Pauling scale, 2.2 assumed for methyl group (14–20). Data from (127) (points 1–10), Mulliken scale, 8.3 assumed for methyl group (points 14–20). Key: (1) SnI_6^{2-}, (2) $SnBr_2I_4^{2-}$, (3) $SnCl_2I_4^{2-}$, (4) $SnBr_4I_2^{2-}$, (5) $SnBr_6^{2-}$, (6) $SnCl_4I_2^{2-}$, (7) $SnCl_2Br_4^{2-}$, (8) $SnCl_4Br_2^{2-}$, (9) $SnCl_6^{2-}$, (10) SnF_6^{2-}, (11) SnI_4, (12) $SnBr_4$, (13) $SnCl_4$, (14) Me_3SnI, (15) Me_3SnBr, (16) Me_3SnCl, (17) Me_3SnF, (18) $Me_2SnBr_4^{2-}$, (19) $Me_2SnCl_4^{2-}$, (20) $Me_2SnF_4^{2-}$.

$SnCl_3ox \cdot oxH$, and $SnCl_2ox_2$ (*10*). These types of correlations have been used by Herber et al. (*112, 383*) to assign Mulliken group electronegativities to N_3^- [8.54 (*112*)] and NCO^- [9.68 (*383*)]. Unexpectedly, the compounds $SnOX_2$ (Table XXXVIII, compounds 56–58) show a gradual decrease in center shift with increasing size of the halogen atom (*113*), and this trend has been explained in terms of the probable bonding (*113*).

Correlations have also been found between center shift and coordination number (*10, 103, 316, 440, 457, 513, 574*). Thus, the series $SnCl_4$, $[SnCl_5]^-$, and $[SnCl_6]^{2-}$ show a decrease in center shift, and a further decrease is found for the seven-coordinate compounds $(SnX(pic)_3$ (X = Cl, Br) (Table XXXVIII, compounds 60, 61) (*421*) and the eight-coordinate species $Sn(ox)_4$ and Pc_2Sn (Table XXXVIII, compounds 48, 59) (*10, 421*). It is, in fact, a general observation, exemplified by the data in Table XXXVIII, that coordination of donor ligands to $SnCl_4$ results in a decrease in center shift.

The effect of coordination number on center shift has been attributed (*440*) to a lengthening of the tin–ligand bonds with increasing coordination number, as illustrated by the bond lengths of $SnCl_4$ (2.32 Å), $SnCl_5^-$ (2.37 Å), and $SnCl_6^{2-}$ (2.42 Å) (*79, 111*). An alternative explanation (*513*) is that the participation of 5d orbitals in the bonding produces a greater shielding of the 5s electrons. Both effects are, in fact, complementary, as increased d orbital participation will produce bond lengthening owing to the larger radial functions.

The dependence of center shift on coordination number for these types of compounds has enabled structural assignments to be made. For example, Ali et al. (*10*) have used the similarity of the center shifts of $SnCl_3ox$ (Table XXXVIII, compound 44), $SnCl_2ox_2$ (compound 35), and $SnCl_3ox \cdot oxH$ (compound 45) to assign a polymeric structure to $SnCl_3ox$. In a similar manner, the center shifts of the pairs of compounds $SnClox_3$, $Snox_4$ (Table XXXVIII, compounds 47, 48) and $SnCl_4 \cdot oxH$, $SnCl_4 \cdot 2oxH$ (Table XXXVIII, compounds 46, 47) indicate polymeric structures for $SnClox_3$ and $SnCl_4 \cdot oxH$ containing eight- and six-coordinate tin atoms, respectively (*10*). These assignments have been confirmed (*10*) by the observation of room temperature Mössbauer effects for $SnCl_3ox$, $SnClox_3$, and $SnCl_4 \cdot oxH$.

Attempts have been made to estimate the center shift of a "perfect" Sn^{4+} ion. Some authors have suggested that the $[SnF_6]^{2-}$ ion, which gives the lowest observed center shift, is essentially ionic and that the center shifts of these compounds may be taken as representative of that of an Sn^{4+} ion (*381*). An alternative approach has been suggested by Goldanskii et al., who compared the relative center shifts of the tin tetrahalides with estimates of the ionic characters of the tin–halogen bonds from electro-negativity, NQR, and dielectric permativity data (*273, 276*). Extrapolation to 100% ionic character yielded a center shift of 2.9 mm/ sec relative to SnO_2 for the Sn^{4+} ion. This value is much lower than the center shift of $[SnF_6]^{2-}$ ions and, if correct, indicates that the $[SnF_6]^{2-}$ ion has a substantial degree of covalent character. Such a conclusion is supported by S.C.F.M.O. calculations (*299*), which show a significant 5s and 5p population for $[SnF_6]^{2-}$. It is also interesting that the calculated center shifts of Sn^{4+} (−5.0 mm/sec) and Sn^{2+} (5.6 mm/sec; see Section IV, E) ions from gray tin, which correspond to a loss and a gain of 5s electrons, respectively, are remarkably similar in magnitude.

The straightforward correlation of center shift with bond polarity and coordination number does not extend to organotin compounds. From the trends outlined above, it might be anticipated that the substitution of an electronegative ligand, X, into R_4Sn would produce a reduction in center shift, which is proportional to the electron-withdrawing ability of X. Further, it would be expected that an increase in coordination number would result in a further lowering of center shift and that for isostructural compounds the center shift would increase with the number of Sn–R bonds and vary with the nature of the ligands. An appraisal of the available center shift data reveals that many of these expected trends are absent.

Let us first consider the variation of center shift with bond polarity. The center shift data for the tetrahedral R_3SnX species in Table XVIII, the tetrahedral halides Neo_3SnX (X = F, Cl, Br, I) and Ph_3SnX (X = Cl, Br, I) in Table XX, and the tetrahedral acetates Ph_3SnO_2CR' (R' = $CMe=CH_2$, CHEtBu, CMe_3) in Table XXIX show a remarkable consistency and give no evidence of trend to decreasing center shift with increasing bond polarity (*114, 115, 440*). On the other hand, some evidence for such a trend may be found for five- and six-coordinate compounds as illustrated by the series of compounds $Et_2Sn(ox)X$ (X = Cl, Br, I), Ph_2SnX_2phen (X = Cl Br, I), and R_2SnX_2bipy (R = Bu, Ph, X = Cl, Br, I) (Table XXXIX), which show a decrease in center shift with decreasing halogen size (*417, 462, 465*). Similarly, linear relationships have been found (Fig. 11) between center shift and Mulliken electronegativity of the X ligand for the series Me_3SnX (X = F, Cl, Br, I, OH) (*383*) and $[Me_2SnX_4]^{2-}$ (X = Cl, Br, I) (*441*). However, these types of correlations are not general for five- and six-coordinate compounds as illustrated by the trimethyl- and triethyltin haloacetates (Table XXIX).

The effects of varying the number of R groups are also rather unexpected; for example, consider the series of compounds in Table XL. The expected increase in center shift with alkyl or aryl substitution is

TABLE XXXVIII
Some Center Shift Data for Sn^{IV} Compounds

Code No.[a]	Compound[b]	C.S.[c]	Ref.
1	SnF_4	−0.25	(139)
2	$SnCl_4$	0.85	(139)
3	$SnBr_4$	1.14	(139)
4	SnI_4	1.45	(139)
5	SnS_2	1.20	(139)
6	SnO_2	0.00	(139)
7	$SnCl_5^-$	0.47–0.63[d]	d
8	$SnBr_5^-$	0.93–0.99[e]	(316)
9	SnF_6^{2-}	(−)0.26–(−)0.50[f]	f
10	$SnCl_4F_2^{2-}$	0.29[i]	(127)
11	$SnCl_6^{2-}$	0.48–0.52[g]	g
12	$SnCl_4Br_2^{2-}$	0.62–0.67[h]	h
13	$SnCl_4I_2^{2-}$	0.53–0.78[j]	j
14	$SnBr_4F_2^{2-}$	0.53[i]	(127)
15	$SnBr_4Cl_2^{2-}$	0.65–0.77[j]	j
16	$SnBr_6^{2-}$	0.84–0.90[k]	k
17	$SnBr_4I_2^{2-}$	0.89–1.01[p]	p
18	$SnI_4Cl_2^{2-}$	0.98–1.17[j]	j
19	$SnI_4Br_2^{2-}$	1.09–1.35[j]	j
20	SnI_6^{2-}	1.23–1.60[l]	l
21	$Sn(NCO)_6^{2-}$	(−)0.05–(−)0.10[m]	(383)
22	$Sn(N_3)_6^{2-}$	0.48[n]	(112)
23	$PcSnF_2$	0.03[o]	(433)
24	$PcSnCl_2$	0.28[o]	(433)
25	$PcSnBr_2$	0.34[o]	(433)
26	$PcSnI_2$	0.45[o]	(433)
27	$PcSn(OH)_2$	0.09[o]	(433)
28	$Sn(SpyO)_2F_2$	0.32	(455)
29	$Sn(SpyO)_2Cl_2$	0.59	(455)
30	$Sn(SpyO)_2Br_2$	0.69	(455)
31	$Sn(SpyO)_2I_2$	0.90	(455)
32	$Sn(pic)_2Cl_2$	0.31	(421)
33	$Sn(pic)_2Br_2$	0.44	(421)
34	$Sn(pic)_2I_2$	0.64	(421)
35	$Sn(ox)_2Cl_2$	0.30, 0.32	(10, 300)
36	$Sn(ox)_2Br_2$	0.44	(10)
37	$Sn(ox)_2I_2$	0.61	(10)
38	$SnCl_4(oxH)_2$	0.45	(10, 300)
39	$SnBr_4(oxH)_2$	0.65	(10)
40	$SnI_4(oxH)_2$	0.91	(10)
41	$Sn(sal)_2Cl_2$	0.23	(10)
42	$Sn(sal)_2Br_2$	0.28	(10)
43	$Sn(sal)_2I_2$	0.41	(10)

MÖSSBAUER SPECTRA OF INORGANIC COMPOUNDS 155

TABLE XXXVIII—continued

Code No.[a]	Compound[b]	C.S.[c]	Ref.
44	Sn(ox)Cl$_3$	0.34	(10)
45	Sn(ox)Cl$_3$oxH	0.37	(10)
46	SnCl$_4$oxH	0.42, 0.43	(10, 300)
47	Sn(ox)$_3$Cl	0.11	(10)
48	Sn(ox)$_4$	0.03	(10)
49	SnF$_4$(NEt$_3$)$_2$	−0.22	(152)
50	SnCl$_4$(NMe$_3$)$_2$	0.59	(152)
51	SnBr$_4$(NMe$_3$)$_2$	0.90	(152)
52	SnI$_4$(NMe$_3$)$_2$	1.30	(152)
53	(SnCl$_4$pyz)$_n$	0.38	(286)
54	(SnBr$_4$pyz)$_n$	0.85	(286)
55	(SnI$_4$pyz)$_n$	1.48	(286)
56	SnOCl$_2$	0.25	(113)
57	SnOBr$_2$	0.22	(113)
58	SnOI$_2$	0.16	(113)
59	Pc$_2$Sn	0.11	(433)
60	Sn(pic)$_3$Cl$_2$	0.15	(421)
61	Sn(pic)$_3$Br$_2$	0.18	(421)

[a] In the text, the code number will be preceded by the table number.
[b] oxH = 8-hydroxyquinoline; Pc = phthalocyanine; HSpyO = 2-pyridinethiol 1-oxide; picH = picolinic acid; salH = salicylaldehyde.
[c] Data given in mm/sec relative to SnO$_2$, assuming identical center shifts for SnO$_2$ and BaSnO$_3$; only a selection of the available data has been included.
[d] Range of values for the cations Et$_4$N$^+$ (442), Ph$_3$C$^+$ (316), Ph$_2$(4-Me–C$_6$H$_4$)C$^+$ (316), Ph(4-Me–C$_6$H$_4$)$_2$C$^+$ (316), and (4-Me–C$_6$H$_4$)$_3$C$^+$ (316).
[e] Range of values for the cations Ph$_3$C$^+$ (316), Ph$_2$(4-Me–C$_6$H$_4$)C$^+$ (316), Ph(4-Me–C$_6$H$_4$)C$^+$ (316).
[f] Range of values for the cations Li$^+$ (534), K$^+$ (127, 168, 278, 534), Rb$^+$ (534), Cu^{2+} (534), Sr^{2+} (534), Be^{2+} (534), and Cs$^+$ (137, 278).
[g] Range of values for the cations K$^+$ (316, 330), NH$_4^+$ (316), Me$_4$N$^+$ (316, 330), Et$_4$N$^+$ (127, 168), MeNH$_3^+$ (300, 457), tropenylium (316), Ph(4-Me–C$_6$H$_4$)$_2$C$^+$ (316), and (4-Me–C$_6$H$_4$)$_3$C$^+$ (316).
[h] Range of data for the cations Me$_4$N$^+$ (330), and Et$_4$N$^+$ (127, 168).
[i] Cation = Et$_4$N$^+$.
[j] Range of data for the cations Me$_4$N$^+$ (330), and Et$_4$N$^+$ (127, 168).
[k] Range of data for the cations Et$_4$N$^+$ (127, 168, 457), Me$_4$N$^+$ (316, 330), K$^+$ (330), NH$_4^+$ (300),tropenylium(316), Ph(4-Me–C$_6$H$_4$)$_2$C$^+$(316), and(4-Me–C$_6$H$_4$)$_3$C$^+$ (316).
[l] Range of data for the cations Et$_4$N$^+$ (127, 168, 330), Me$_4$N$^+$ (300, 330), and K$^+$ (330).
[m] Values for the cations Me$_4$N$^+$ and Et$_4$N$^+$, respectively.
[n] Cation is Me$_4$N$^+$.
[o] For an alternative set of data, see Ref. (530).
[p] Range of data for the cations Me$_4$N$^+$ (330), Et$_4$N$^+$ (127,168), and NH$_4^+$ (330).

TABLE XXXIX

Some Center Shift Data for Five- and Six-Coordinate Sn^{IV} Compounds

Compound[a]	C.S.[b]	Ref.
Bu_2SnCl_2phen	1.59	(417)
Bu_2SnBr_2phen	1.63	(417)
Bu_2SnI_2phen	1.69	(417)
Bu_2SnCl_2bipy	1.56	(417)
Bu_2SnBr_2bipy	1.62	(417)
Bu_2SnI_2bipy	1.70	(417)
Ph_2SnCl_2bipy	1.26, 1.22	(418, 465)
Ph_2SnBr_2bipy	1.33	(465)
Ph_2SnI_2bipy	1.41	(465)
$Et_2Sn(ox)Cl$	1.34	(462)
$Et_2Sn(ox)Br$	1.39	(462)
$Et_2Sn(ox)I$	1.43	(462)
$K_2[Me_2SnF_4]$	1.38	(442)
$[Me_2SnCl_4]^{2-}$	1.59[c], 1.63[d]	(234, 442)
$Cs_2[Me_2SnBr_4]$	1.76	(442)

[a] phen = 1,10-Phenanthroline; bipy = 2,2'-bipyridyl; ox = 8-oxyquinoline.
[b] Data given in mm/sec assuming identical center shifts for SnO_2 and $BaSnO_3$; data from Refs. (417), 418) have been converted assuming a center shift of 1.52 for Pd(Sn).
[c] Cation is pyridinium.
[d] Cation is Cs^+.

TABLE XL

Center Shift[a] as a Function of the Number of Sn–R Bonds

Compound	n = 0	1	2	3	4
$Et_nSnCl_{6-n}^{2-}$	0.52[b]	1.10[b]	1.64[b]	—	—
$Et_nSnCl_{5-n}^{-}$	0.59[b]	1.18[b]	1.54[b]	1.50[b]	—
$Me_nSn(C_6F_5)_{4-n}$	1.04[c]	1.19[c]	1.25[c]	1.27[d]	1.21[c]
$Ph_nSn(C_6F_5)_{4-n}$	1.04[c]	1.16[c]	1.22[c]	1.25[c]	1.22[c]

[a] Data given in mm/sec relative to SnO_2 at liquid nitrogen temperature.
[b] Ref. (442).
[c] Ref. (528).
[d] Ref. (440).

observed in early stages of the series, i.e., $[SnCl_6]^{2-} < [EtSnCl_5]^{2-} < [Et_2SnCl_4]^{2-}$; $[SnCl_5]^- < [EtSnCl_4]^- < [Et_2SnCl_3]^-$, and $Sn(C_6F_5)_4 < RSn(C_6F_5)_3 < R_2Sn(C_6F_5)_2$ (R = Me, Ph). Further substitution may arrest the decrease $[R_nSn(C_6F_5)_{4-n}(n = 2, 3, 4; R = Me, Ph)]$ or even reverse it $[EtSnX_{5-n}^-$ $(n = 2, 3)]$. There is also strong evidence that octahedral cis-R_2SnX_4 species have lower center shifts than octahedral $trans$-R_2SnX_4 compounds (421, 441). Thus, Table XLI contains center shifts for pairs of R_2SnX_4 compounds with similar X ligands but with

TABLE XLI

CENTER SHIFT DATA FOR SOME COMPOUNDS OF TYPE cis- AND $trans$-R_2SnX_4

Code No.[a]	Compound (cis)[c]	C.S.[b]	Ref.	Compound (trans)	C.S.[b]	Ref.
1	Ph₂Sn(S₂CNPh₂)₂	1.19	(233)	R₂Sn(S₂CNPh₂)₂[d]	1.54–1.72	(233)
2	Ph₂Sn(S₂CNEt₂)₂	1.17	(233)	Me₂Sn(S₂CNEt₂)₂	1.57	(233)
3	Ph₂Sn[S₂CN(CH₂)₄]₂	1.17	(233)	R₂Sn[S₂CN(CH₂)₄]₂[d]	1.53–1.59	(233)
4	Ph₂Sn(S₂CNCH₂Ph)₂	1.08	(233)	Bu₂Sn(S₂CNCH₂Ph)₂	1.69	(233)
5	Bu₂Sn(ox)₂	0.92	(421)	Bu₂Sn(pic)₂	1.45	(421)
6	Ph₂Sn(acac)₂	0.74	(234)	Me₂Sn(acac)₂	1.18	(323)
7	Ph₂Sn(NCS)₂bipy	0.82	(418)	Bu₂Sn(NCS)₂bipy	1.43	(418)
8	Ph₂Sn(NCS)₂phen	0.81	(418)	Bu₂Sn(NCS)₂phen	1.42	(418)
9	Ph₂SnCl₂·4morph	0.94	(291)	Ph₂SnCl₂·4pip	1.33	(291)
10	n-PrSnCl₂·2morph	0.98	(291)	n-Pr₂SnCl₂·2pip	1.63	(291)

[a] In the text code number will be preceded by table number.
[b] Data given in mm/sec relative to SnO₂ at liquid nitrogen temperature, assuming identical center shifts for SnO₂ and BaSnO₃. Data from Ref. (418) have been converted assuming a center shift of 1.52 for Pd(Sn).
[c] bipy = 2,2'-Bipyridyl; oxH = 8-hydroxyquinoline; acac = acetylacetonate; phen = 1,10-phenanthroline; picH = picolinic acid; pip = piperidine; morph = morpholine.
[d] Range of values for R = Me, Bu.

cis and trans arrangement of the R groups. Clearly, there is a marked trend to lower shifts for the cis species. For the pairs of compounds 1–4 and 6–8 in Table XLI, some of the difference may be due to the differing polarities of the tin–phenyl and tin–alkyl bonds, but this effect is not large enough to account for the whole of the change.

Unlike the complexes in Table XXXVIII, organotin compounds do not seem to show a dependence of center shift on coordination number, except perhaps for the small differences in center shift between the species $[EtSnCl_4]^-$ and $[EtSnCl_5]^{2-}$ and the low center shifts found for the seven-coordinate compound BuSn(ox)₃ (168, 462). Thus, for example, the data in Tables XX and XXIX show that the center shifts of trialkyltin halides and carboxylates with associated five-coordinate structures are very similar to those for analogous monomeric tetrahedral species. Similarly, the coordination of an extra ligand to Ph_3SnCl to form a five-coordinate complex, Ph_3SnClX, produces no significant change in center shift either for $X = Cl^-$ (207, 442) or an oxygen donor such as $R_2SO[R = Me, Et, Bu, \frac{1}{2}(CH_2)_4]$ or R_3PO ($R = MeO, EtO, PhO$) (207, 556) (Table XXXVI). Six coordinate trans-R_2SnX_4 species have some of the highest center shifts, as exemplified by the data for the $[Me_2SnX_4]^{2-}$ ions (Table XXXIX), despite the high coordination number and the presence of four Sn–X bonds.

At present, no complete interpretation of the center shifts found for organotin compounds has appeared. Two explanations have been proposed to account for the constant center shifts of tetrahedral R_3SnX species. Chivers and Sams (114, 115) have suggested that the substitution of an electronegative ligand causes a rehybridization of the bonding orbitals of the tin atom, which results in an asymmetric distribution of $5p$ electrons (and, hence an EFG), a reduction in s character of the Sn–X bond, but no net loss of charge. Although the distortions of bond angles observed (131) for the compound $Ph_2Sn(I) \cdot (CH_2)_4 \cdot (I)SnPh_2$ indicate some degree of rehybridization, halogen NQR data provide strong evidence (566) for loss of charge from the tin orbital of the Sn–X bond. Alternatively, it was suggested (441) that an electronegative substituent removes charge from the valence shell, but that the resultant residual positive charge provides a deshielding and contraction of the $5s$ orbitals, i.e., an increase in the effective nuclear charge. Such an effect would compensate for the loss of $5s$ electron density to the ligand. Any rehybridization which does occur can then be seen as a complementary effect which increases the s character of the R–Sn bond and, hence, accentuates the deshielding effect. Similar explanations (459) have been proposed to account for the relative insensitivity of the center shifts of five- and six-coordinate compounds.

It has been argued (421, 441) that the difference in the center shifts of cis- and trans-R_2SnX_4 isomers reflects a variation in the $5s$ character of the Sn–R bonds, and this suggestion is supported by the higher $J_{^{119}Sn-CH_3}$ NMR coupling constant observed for $Me_2Sn(pic)_2$ [$J_{^{119}Sn-CH_3} = $ 77.6 Hz (395)] compared with $Me_2Sn(ox)_2$ [$J_{^{119}Sn-CH_3} = 71.2$ Hz (395)] and by simple molecular orbital considerations (441). In fact, there may be a general correlation between the stereochemistry of Sn–R bonds and center shifts (441). Gassenheimer and Herber (258) have proposed an electron delocalization effect (i.e., deshielding) associated with a change in hybridization to explain the high center shifts of five-coordinate species relative to four-coordinate compounds.

In summary, it is evident from the above discussion that the factors controlling the center shifts of organotin compounds are poorly understood. In contrast, complexes without Sn–R bonds seem to conform reasonably accurately to an interpretation based on bond polarity charges and it may, in fact, be possible to calculate partial center shift values for these species in an analogous manner to low-spin Fe^{II} (*44*).

The center shifts of tin–transition metal species are of interest, and some of the available data are summarized in Table XLII. These data show a consistent general trend to increased center shift with increasing number of Sn–M bonds (*64, 226, 275, 431, 569*), as illustrated, for example, by the series $Me_{4-n}Sn[cpFe(CO_2)_2]_n$ ($n = 0$–2, 4) (Table XLII, compounds 1, 3, 20, 40), $Ph_{4-n}Sn[cpFe(CO)_2]_n$ ($n = 0$–4) (Table XLII, compounds 2, 5, 22, 32, 40), $Me_{4-n}Sn[Mn(CO)_5]_n$ ($n = 0$–3) (Table XLII, compounds 1, 6, 26, 35), $Cl_{4-n}Sn[cpFe(CO)_2]_n$ ($n = 0$–2, 4) (Table XLII, compounds 40, 42, 48, 62), and $Br_{4-n}Sn[cpFe(CO)_2]_n$ ($n = 0$–2, 4) (Table XLII, compounds 40, 43, 50, 63). This trend has been interpreted (*64, 226, 431, 569*) in terms of an increased 5s character of the Sn–M bond as opposed to alkyl–, phenyl–, or halide–tin bonds. Support for this interpretation is derived from proton NMR data for Me_nSnM_{4-n} ($n = 1$–4) species which show a decrease in $J_{119Sn-CH_3}$ (*226, 431*) and, hence, in 5s character of the Me–Sn bond with decreasing n. Further, X-ray diffraction data reveal a tendency for larger than tetrahedral M–Sn–M bond angles and smaller than tetrahedral C–Sn–C and X–Sn–X bond angles, and also for relatively short Sn–M and long Sn–C and Sn–X bond lengths [(*288*) and references quoted therein]. All these trends are those expected for a high 5s character in the Sn–M bond.

The series of compounds $R_{3-n}X_nSnMn(CO)_5$ (R = Me, Ph; X = Cl, Br; $n = 0$–3) (Table XLII) show an increase in center shift as R is replaced by X (*431*). This suggests (*431*) that $Mn(CO)_5$ is a stronger electron donor than methyl, phenyl, or halogen and the more halogen atoms that are attached to the tin atom, the more σ electrons are transferred from manganese to tin. Support for this interpretation is provided by proton and ^{55}Mn NMR data, which show a decrease both in $J_{119Sn-CH}$ and in the ^{55}Mn chemical shift with increasing center shift (*431*).

Although the center shift data discussed above provide convincing evidence that the metal moieties are better σ donors than alkyl or phenyl groups, such a conclusion is at variance with quadrupole splitting data, which indicates that the metal moieties are poorer σ donors than phenyl or alkyl (Section IV, A, b). At present, the most satisfactory explanation for this apparent discrepancy involves a difference in degree of utilization of the 5s and 5p orbitals (*437*). The quadrupole splitting measures the p donor capacity, whereas the center shift is most sensitive to the s

TABLE XLII

Some Center Shift Data for Tin–Transition Metal Species

Code No.[a]	Compound	C.S.[b]	Ref.
1	Me_4Sn	1.31	(516)
2	Ph_4Sn	1.22	(516)
3	$Me_3SncpFe(CO)_2$[c]	1.38	(65, 149)
4	$BuSncpFe(CO)_2$	1.47	(288)
5	$Ph_3SncpFe(CO)_2$	1.44	(149, 275, 324, 331)
6	$Me_3SnMn(CO)_5$	1.40	(431, 569)
7	$Ph_3SnMn(CO)_5$	1.46	(364, 431, 569)
8	$Ph_3SnCo(CO)_4$	1.50	(364)
9	$Ph_3SnRe(CO)_5$	1.45	(364)
10	$Ph_3SnRe(CO)_4PPh_3$	1.50	(364)
11	$Me_3SncpCr(CO)_3$	1.41	(65)
12	$Me_3SncpMo(CO)_3$	1.43	(65)
13	$Me_3SncpW(CO)_3$	1.36	(65)
14	$Me_3SnIrHCl(CO)(PPh_3)_2$	1.84	(65)
15	$Me_3SnIrDCl(CO)(PPh_3)_2$	1.84	(65)
16	$Ph_3SnIrHCl(CO)(PPh_3)_2$	1.42	(65)
17	$Ph_3SnIrHCl(CO)(PPh_2Me)_2$	1.46	(65)
18	$R_3SncpFe(CO)L$[d]	1.39–1.50	(149)
19	$R_3SncpFeL_2$[e]	1.47–1.71	(149)
20	$Me_2Sn[cpFe(CO)_2]_2$	1.68	(324)
21	$Et_2Sn[cpFe(CO)_2]_2$	1.74	(324)
22	$Ph_2Sn[cpFe(CO)_2]_2$	1.74	(275)
23	$Ph_2Sn[Co(CO)_4]_2$	1.68	(364)
24	$Ph_2Sn[Mn(CO)_5]Co(CO)_4$	1.65	(364)
25	$Ph_2Sn[Re(CO)_5]_2$	1.70	(364)
26	$Me_2Sn[Mn(CO)_5]_2$	1.68	(569)
27	$Me_2ClSnMn(CO)_5$	1.54	(431)
28	$Me_2BrSnMn(CO)_5$	1.54	(431)
29	$Ph_2ClSnMn(CO)_5$	1.60	(364, 431)
30	$Ph_2BrSnMn(CO)_5$	1.59	(431)
31	$Ph_2ClSnCo(CO)_4$	1.56	(364)
32	$PhSn[cpFe(CO)_2]_3$	2.00	(275)
33	$PhSn[Co(CO)_4]_3$	1.54	(226)
34	$PhSn[Re(CO)_5]_3$	1.75	(364)
35	$MeSn[Mn(CO)_5]_3$	1.83	(569)
36	$MeCl_2SnMn(CO)_5$	1.67	(288, 431)
37	$MeBr_2SnMn(CO)_5$	1.69	(431)
38	$PhCl_2SnMn(CO)_5$	1.68	(431)
39	$PhBr_2SnMn(CO)_5$	1.80	(431)
40	$Sn[cpFe(CO)_2]_4$	2.14	(275)
41	$Sn[Co(CO)_4]_4$	1.96	(227)
42	$SnCl_4$	0.85	(139)
43	$SnBr_4$	1.14	(139)
44	SnI_4	1.45	(139)
45	$(NH_4)[SnCl_3]$	3.71	(64)

TABLE XLII—continued

Code No.[a]	Compound	C.S.[b]	Ref.
46	$(NH_4)[SnBr_3]$	3.79	(64)
47	$(NH_4)[SnI_3]$	4.03	(64)
48	$Cl_3SncpFe(CO)_2$	1.70	(64, 149, 275, 331)
49	$Cl_3SncpFe(CO)PPh_3$	1.88	(149)
50	$Br_3SncpFe(CO)_2$	1.75	(64)
51	$I_3SncpFe(CO)_2$	1.88	(64)
52	$Cl_3SnMn(CO)_5$	1.66	(288, 364, 431, 569)
53	$Br_3SnMn(CO)_5$	1.79	(364, 431, 569)
54	$Cl_3SnRh(PPh_3)_3$	1.78	(227)
55	$Cl_3SnIr(C_8H_{12})_2$	1.80	(227)
56	$[(Cl_3Sn)_2RuCl_2]^{2-}$	1.94	f
57	$[(Cl_3Sn)_2PtCl_2]^{2-}$	1.75	f,g
58	$[(Cl_3Sn)_5Pt]^{3-}$	1.65	h
59	$[(Cl_3Sn)_2Pt_3(C_8H_{12})_3MeNO_2]$	1.50	(227)
60	$[(Cl_3Sn)_2PdCl_2]^{2-}$-$(Me_4N^+)_2$	1.52	(46)
61	$[(Cl_3Sn)_4Rh_2Cl_2]^{4-}$-$(Me_4N^+)_4$	1.90	(227)
62	$Cl_2Sn[cpFe(CO)_2]_2$	1.99	(64, 275, 288, 324, 331)
63	$Br_2Sn[cpFe(CO)_2]_2$	1.99	(64)
64	$I_2Sn[cpFe(CO)_2]_2$	2.00	(64)
65	$Cl_2Sn[Mn(CO)_5]cpMo(CO)_3$	1.98	(364)
66	$Cl_2Sn[Mn(CO)_5]Re(CO)_5$	1.96	(364)
67	$ClSn[Mn(CO)_5]_3$	1.92	(275, 364)
68	$ClSn[Re(CO)_5]_3$	1.82	(275)
69	$BrSn[Re(CO)_5]_3$	1.82	(364)
70	$ClSn[Mn(CO)_5][cpFe(CO)_2]_2$	2.10	(275)
71	$(NCS)_2Sn[cpFe(CO)_2]_2$	1.85	(64, 288)
72	$(HCO_2)_2Sn[cpFe(CO)_2]_2$	1.61	(64)
73	$(MeCO_2)_2Sn[cpFe(CO_2)]_2$	1.63	(64)
74	$(NCS)_3SncpFe(CO)_2$	1.65	(64)
75	$(HCO_2)_3SncpFe(CO)_2$	1.09	(64)
76	$(MeCO_3)_3SncpFe(CO)_2$	1.17	(64)
77	$Me_4Sn_3Fe_4(CO)_{16}$	2.20, 1.45	(169)
78	$Bu_2SnFe(CO)_8SnBu_2$	1.70	(169)

[a] Code number will be preceded by Table number in text.
[b] Data given in mm/sec relative to SnO_2 at liquid nitrogen temperature, assuming that center shift of $BaSnO_3$ is zero and that of α-tin is 2.10 mm/sec. When appropriate, data are an unweighted average.
[c] cp = π-Cyclopentadienyl.
[d] Range of data for R = Me or Ph, L = Ph_3M (M = P, As, Sb), Ph_2PCF_3, Ph_2PMe, $PhPMe_2$, f_6fos [$Ph_2P \cdot C{=}C(PPh_2)(CF_2)_2CF_2$], Ph_2AsCF_3, or $(C_6H_5O)_3P$.
[e] Range of data for R = Me, L = $SbPh_3$, $\frac{1}{2}Ph_2P(CH_2)_2PPh_2$; R = Ph, L = Ph_2PMe, $PhPMe_2$.
[f] Cation = Me_4N^+ (46, 227).
[g] Cation = Et_4N^+, δ = 1.56 mm/sec (227).
[h] Average of cations = Me_4N^+ (46) and Et_4N^+ (227).

donor capacity. As indicated by the crystal structure data, bonds to the metal moiety involve considerably more 5s character than 5p character relative to an alkyl or phenyl or halide group. If the bonds between Sn and the alkyl, phenyl, and halide groups involve mainly 5p electrons on Sn in these compounds, a strong donor such as alkyl would then increase the 5p electron density and *decrease* $[\Psi(0)_s]^2$ and the center shift relative to a halide as is observed *(28)*. Onaka *et al.* *(431)* suggest π-bonding effects may be important, but this seems unlikely. Thus, π bonding will have only a secondary effect on quadrupole splitting due to the large 5d radial function. Further, as noted by Fenton and Zuckerman *(227)*, π donation would be expected to give a decrease in center shift (from shielding effects); and from the data in Table XLII, it is evident that if π-bonding effects are present at all, they are not the dominant factor in the center shift trends.

Fenton and Zuckerman *(226)* have concluded that as the center shift for $SnCl_3$ transition metal derivatives (Table XLII, compounds 48–61) are smaller than that of α-tin (C.S. = 2.1 mm/sec), they should be regarded as derivatives of Sn^{IV}. However, Bird *et al.* *(65)* have argued that this more properly indicates the *valence state* rather than the formal oxidation state of the tin atom. Thus, when $SnCl_3^-$ acts as a ligand, the involvement of the lone pair in bonding means that the tin atom cannot be regarded as bivalent. It is probable that the concepts of formal oxidation state have little meaning in these systems.

Finally, perhaps one of the most elegant experiments recently reported is the correlation found by Barber and Swift *(47)* between center shift and the 4d binding energy of tin (as measured by high-energy photoelectron spectroscopy) for the series $Y_2Sn(ox)_2$ (Y = Et, Ph, Cl, Br, or I). This correlation is illustrated in Fig. 12, and indicates the potential of combining these two techniques.

3. *Temperature Dependence of the Mössbauer Effect and the Goldanskii–Karyagin Effect*

In early studies of ^{119}Sn Mössbauer spectra, it was realized that certain compounds give rise to a room temperature Mössbauer effect, while such an effect is absent from other species. It was suggested *(323, 531)* that the observation of a room temperature effect can be taken as evidence for a polymeric structure. This has been confirmed by Stöckler and Sano *(529)*, who have demonstrated that both the Debye temperature and characteristic temperatures are higher for polymeric materials than monomeric species. Stöckler *et al.* *(531)* have also concluded that the recoil free fraction (*f*) has no simple dependence on; (*a*) the nearest neighbor atom mass, (*b*) the nearest neighbor ligand mass, (*c*) the molecular weight, (d) the coordination number, (e) the macro properties of the material, or (f) its center shift and quadrupole splitting.

Further studies on the correlation between a room temperature Mössbauer effect and molecular structure have been reported by Poller et al. (464). These authors measured the ratio (R) of the room temperature effect to that at liquid nitrogen for a series of organotin compounds and the R values obtained are summarized in Table XLIII. For compounds 1–11 in Table XLIII, which are believed to have polymeric structure, finite R values were observed; whereas the compounds 17–26,

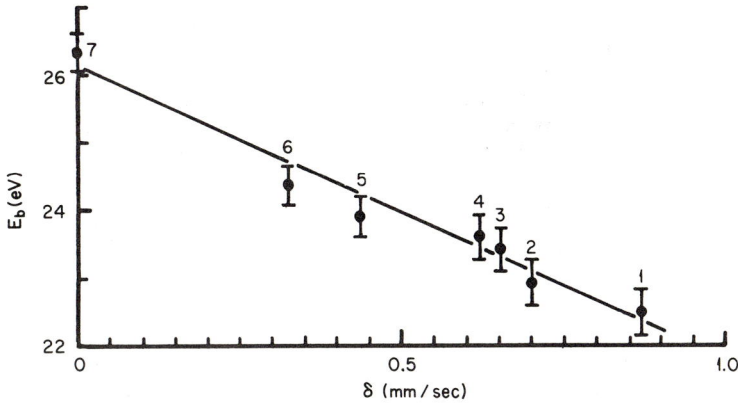

Fig. 12. The 4d binding energies of tin, E_b(EV) against Mössbauer center shifts for compounds 1–7. (1) $Et_2Sn(OX)_2$; (2) $Ph_2Sn(OX)_2$; (3) $SnBr_4 \cdot 2oxH$; (4) $SnI_2(OX)_2$; (5) $SnBr_2(OX)_2$; (6) $SnCl_2(OX)_2$; and (7) SnO_2 (47).

which are probably monomeric, and compound 16, which is thought to be trimeric, did not give a significant R value. The compound 12 in Table XLIII is of particular interest as the observed R value is in conflict with infrared evidence (467), which suggests a monomeric five-coordinate structure.

The work of Poller et al. (464) and Stöckler et al. (529, 531) strongly suggests that the observation of a room temperature Mössbauer effect may be taken as firm evidence for a polymeric structure, and this criterion has been exploited by, for example, Ford et al. (242, 572, 573) in assigning polymeric structures to the species $(Me_3Sn)_2XO_4$ (X = Se, Cr), Me_2SnXO_4 (X = S, Se), $Me_2Sn(SO_3X)$ (X = F, Cl, CF_3, Me, Et), $Sn(SO_3F)_4$, and $SnCl_2(SO_3F)_2$. It should, however, be emphasized that the absence of a room temperature effect for a compound in no way allows the elimination of an associated structure. For example, there are excellent reasons to assign associated structures to the compounds $(Me_3Sn)_2SO_4$ (206), Bu_3SnO_2CMe (240), and Me_3SnNCO (383), none of which shows a detectable room temperature Mössbauer effect. Another example is provided by the compounds (464) $PhSn(OCOCX_3)O$

TABLE XLIII

R Values for Some Organotin Compounds[a]

Code No.[b]	Compound[e]	R^c	Ref.[d]
1	$Me_3SnL \cdot H_2O$	0.035	(468)
2	$Bu_3SnL \cdot H_2O$	0.044	(468)
3	$Ph_3SnL \cdot H_2O$	0.045	(468)
4	$Bu_2SnCl_2 \cdot 4,4'$-bipy	0.057	(466)
5	$Ph_2SnCl_2 \cdot 4,4'$-bipy	0.040	(466)
6	$Ph_2SnCl_2 \cdot$ pyrazine	0.035	(466)
7	$Ph_2SnCl_2 \cdot t$-DTDO	0.159	(467)
8	Ph_3SnO_2CMe	0.078	(240)
9	Ph_3SnO_2CEt	0.110	(240)
10	$Ph_3SnO_2C \cdot CHMe_2$	0.080	(241)
11	$Ph_3SnO_2C \cdot CH_2Cl$	0.183	(464)
12	$Ph_2SnCl_2 \cdot c$-DTDO	0.091	(467)
13	$PhSn[O_2C \cdot CMe_3]O$	0.385	(464)
14	$PhSn[O_2C \cdot CCl_3]O$	0.057	(464)
15	$PhSn[O_2C \cdot CF_3]O$	0	(464)
16	Oct_2SnL	0	(468)
17	$Me_2Sn(ox)_2$	0	(462, 505)
18	$Bu_2Sn(ox)_2$	0	(462)
19	$Oct_2Sn(ox)_2$	0	(462)
20	$Ph_2Sn(ox)_2$	0	(462)
21	$Bu_2SnCl_2 \cdot 2(4$-phepy)	0	(466)
22	$Ph_2SnCl_2 \cdot 2,2'$-bipy	0	(465)
23	$Ph_2SnCl_2 \cdot 2DTO$	0	(467)
24	$Ph_2SnCl_2 \cdot 2Me_2SO$	0	(464)
25	$Me_2SnCl_2 \cdot 2Me_2SO$	0	(353)
26	$Ph_3SnO_2C \cdot CMe_3$	0	(240)

[a] Data from (464), in which reference will be found values of center shift and quadrupole splitting.
[b] In text code number will be preceded by table number.
[c] See text for definition.
[d] References are to structural studies.
[e] LH_2 = Bis(8-hydroxy-5-quinolyl)methane; oxH = 8-hydroxyquinoline; 4,4'-bipy = 4,4'-bipyridine; 4-phepy = 4-phenylpyridine; 2,2'-bipy = 2,2'-bipyridine; t-DTDO = $trans$-1,4-dithiane 1,4-dioxide; c-DTDO = cis-1,4-dithiane 1,4-dioxide; DTO = 1,4-dithiane 1-oxide.

(X = H, Cl and F). All three of these compounds are very high melting solids (m.p. > 360°) indicative of polymeric structures, but the R factor falls dramatically from R = H to R = Cl and is equal to zero when R = F. Studies of the variation of the magnitude of the Mössbauer effect over a wide range of temperatures have been reported by Stöckler and Sano (526) for the compounds $Me_3SnCl \cdot py$, Me_3SnF, Ph_3SnF, and Me_3SnO_2CH. As might be anticipated, the Mössbauer effect varied much more markedly with temperature for the monomeric compound $Me_3SnClpy$ (348) than for the polymeric species Me_3SnF and $(C_6H_5)_3SnF$, while the compound Me_3SnO_2CH showed a temperature dependence between the two extremes. This type of study has been extended by Herber (326) to the compounds $(Bu_3Sn)_2SO_4$, $Bu_3SnO_2CH_3$, Bu_3SnF, and Bu_2SnF_2. In these cases the ratios of the temperature dependence of the Mössbauer effect (as measured by the area under the resonance curve) are $(Bu_3Sn)_2SO_4 : Bu_3SnO_2CMe : Bu_3SnF : Bu_2SnF_2 = 1 : 0.918 : 0.576 : 0.302$. These ratios clearly indicate that the tributyl- and dibutyltin fluorides have stronger polymeric lattices than either Bu_3SnO_2CMe or $(Bu_3Sn)_2SO_4$, and Herber (326) has taken these data as further evidence in favor of a monomeric structure for $(Bu_3Sn)_2SO_4$. However, infrared (355) and Mössbauer (240) evidence clearly show the presence of intermolecular association in the compound Bu_3SnO_2CMe, and the similarity of the temperature dependence of the Mössbauer effect in Bu_3SnO_2CMe and $(Bu_3Sn)_2SO_4$ makes it difficult to eliminate the possibility of association in $(Bu_3Sn)_2SO_4$. Herber et al. (112, 383) have also used the temperature dependence of the Mössbauer effect to assign polymeric structures to the compounds Me_3SnN_3 and Me_3SnNCO.

Another phenomenon, which is of interest in studying polymeric materials, is the Goldanskii–Karyagin effect (365). This effect, which arises from a lattice dynamic anisotropy in the recoil free fraction, is manifested in an asymmetry in the intensities of the components of a quadrupole doublet. The presence of this effect was detected by Stöckler and Sano (527) for the polymeric materials $(Me_3SnF)_n$ and $(Me_3SnOH)_n$, whereas it was found (527) to be absent for the monomeric species Ph_3SnCl. However, the absence of a significant effect in Me_3SnO_2CH demonstrates that polymeric structure is not a sufficient criterion to produce a significant Goldanskii–Karyagin effect (526). Stöckler and Sano (526, 527) and Herber et al. (326–329) have exploited the temperature dependence of this effect in the study of lattice dynamics for the species Me_3SnOH (527), Me_3SnF (328, 527), Ph_3SnF (526), Bu_3SnF (326), Bu_2SnF_2 (326), Me_2SnF_2 (327), and Me_3SnCN (329, 525), and Herber (326) has noted its absence in the Mössbauer spectrum of $(Bu_3Sn)_2SO_4$.

B. Fe^{II} Low-Spin Compounds

In the absence of π bonding, an octahedral low-spin Fe^{II} compound can be represented by the simple molecular orbital picture illustrated in Fig. 13. The metal $3d$ electrons form a t_{2g}^6 configuration, while ligand donation from appropriate σ orbitals populates the molecular orbitals formed by overlap with the iron $3d_{z^2}$, $3d_{x^2-y^2}$, $4s$, and $4p$ orbitals. In hybridization terms, the σ bonding can be represented by overlap between d^2sp^3 hybrids of the iron atom with the σ orbitals of the ligands.

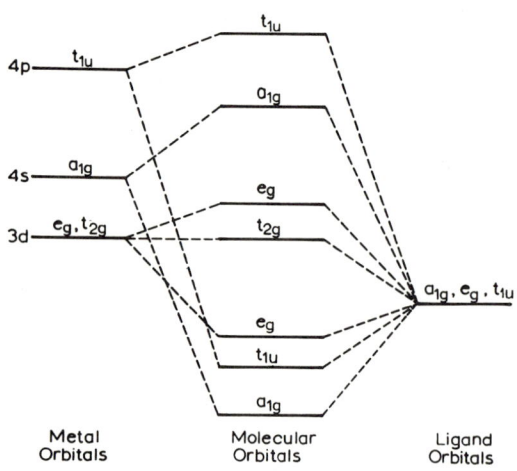

Fig. 13. Molecular orbital diagram for an octahedral MX_6 molecule considering only σ bonding.

In many cases the ligands also have empty antibonding π^* orbitals, which have suitable symmetry for overlap with the t_{2g}^6 electrons of the metal atom. Such overlap results in a donation of charge from the metal to the ligand.

As the t_{2g}^6 shell has cubic symmetry and is diamagnetic, any variations in center shift and quadrupole splitting found for octahedral compounds must arise from variations in the nature of the ligands, and Mössbauer spectroscopy has proved a very powerful means of studying the variations of metal–ligand interactions in low-spin Fe^{II} systems. This is in contrast to many other oxidation states of iron (*vide infra*) for which the asymmetry of the free ion electrons obscures the dependence of the C.S. and Q.S. on ligand properties.

1. Quadrupole Splitting

As for Sn^{IV} compounds, the quadrupole splittings for Fe^{II} low-spin compounds are due to q_{lat} and/or $q_{m.o.}$, and the quadrupole splitting will be sensitive to changes in both the nature and geometric distribution of the ligands. In Section II, D, it was shown that for contributions to the EFG arising from σ-bonding asymmetry, the additivity model should provide a realistic guide to the variation of quadrupole splitting with structure. However, for contributions from π-bonding asymmetries, some deviation from additivity might be anticipated. Table XLIV lists a wide range of quadrupole splittings for low-spin Fe^{II} compounds. We will first discuss the relationship between quadrupole splitting and structure, derive empirical partial quadrupole splittings, and comment on the nature of the metal–ligand bonds, particularly the $Fe-N_2$ bond as deduced from p.q.s. values.

a. Derivation of p.q.s. Values and Structural Determination. Berrett and Fitzsimmons (55) first showed that a $trans$-FeA_2B_4 isomer has twice the Q.S. of the corresponding cis-FeA_2B_4 isomer. They studied compounds of the type $trans$- and cis-$Fe(CN)_2(CNR)_4$ and $[Fe(CN)(CNR)_5^+]$-ClO_4^- (R = Et, Me, CH_2Ph) (Table XLIV, compounds 12, 39, 40) and used the point charge model to show that the observed magnitudes correspond reasonably well to the predicted 2:1:1 ratio. Bancroft et al. (44) extended the additivity model to a wide range of compounds and calculated empirical p.q.s. values for a wide range of ligands (p.q.s. = $\frac{1}{2}e^2Q[L]$ and Table IV). To calculate p.q.s. values it was necessary to assign structures to key compounds. For example, the exact 2:1 ratio for the isomers of $FeCl_2(ArNC)_4$ (Table XLIV, compounds 1, 30) and $Fe(SnCl_3)_2(ArNC)_4$ (Table XLIV, compounds 3, 32) allows the immediate assignment of structures to these four compounds, and the similarity of the quadrupole splittings of $FeX_2(depe)_2$ (X = Cl, Br, I) and $FeCl_2(dmpe)_2$ (Table XLIV, compounds 2, 4–6) to that of $trans$-$FeCl_2(ArNC)_4$ suggests trans-octahedral structures for these compounds. Although the quadrupole splitting of $FeCl_2(depb)_2$ (Table XLIV, compound 7) is too low to make an unambiguous assignment of structure, Chatt and Hayter (109) have used NMR and dipole moment data to assign trans structures to $FeHCl(depe)_2$, $FeHI(depe)_2$, $FeCl_2(depe)_2$, and $FeCl_2(depb)_2$ in solution, and it is a reasonable assumption that these compounds have trans structures in the solid state. All the quadrupole splittings of the pseudohalides $FeY_2(depe)_2$ (Y = NCO, NCS, N_3) compounds (Table XLIV, compounds 9–11) are rather smaller, but it appears likely that they are also trans like the other bisdiphosphine complexes.

TABLE XLIV

QUADRUPOLE SPLITTINGS FOR Fe^{II} LOW-SPIN COMPOUNDS[a]

Code No.[b]	Compound[d]	Quadrupole splitting[c]		Ref.
		Obs.	Calc.	
1	trans-$FeCl_2(ArNC)_4$	+1.55	—	(34, 44)
2	trans-$FeCl_2(depe)_2$	+1.29	—	(34, 44)
3	trans-$Fe(SnCl_3)_2(ArNC)_4$	(+)1.05	—	(34, 44)
4	trans-$FeCl_2(dmpe)_2$	(+)1.51	—	(34, 44)
5	trans-$FeBr_2(depe)_2$	(+)1.37	—	(34, 44)
6	trans-$FeI_2(depe)_2$	(+)1.33	—	(34, 44)
7	trans-$FeCl_2(depb)_2$	(+)1.13	—	(34, 44)
8	trans-$FeH_2(depb)_2$	(−)1.84	—	(34, 44)
9	trans-$Fe(NCO)_2(depe)_2$	(+)0.49	—	(34, 44)
10	trans-$Fe(NCS)_2(depe)_2$	(+)0.53	—	(34, 44)
11	trans-$Fe(N_3)_2(depe)_2$	(+)0.98	—	(34, 44)
12	trans-$Fe(EtNC)_4(CN)_2$	−0.60	—	(34, 55)
13	$Na_2[Fe(CN)_5NO]\cdot 2H_2O$	+1.73	—	(34, 161, 238)
14	$Na_3[Fe(CN)_5NH_3]\cdot H_2O$	+0.67	—	(34, 44, 76, 238)
15	$K_3[Fe(CN)_5H_2O]\cdot 7H_2O$	+0.80	—	(34, 142)
16	$Na_5[Fe(CN)_5SO_3]\cdot 9H_2O$	(+)0.76	—	(142, 238)
17	$(Na,K)_4[Fe(CN)_5NO_2]$	(+)0.85	—	(76, 142, 238)
18	$Fe(CNH)_4(CN)_2$	~0.0	—	(312)
19	$Fe(CNH)_4(CNBF_3)_2$	~0.0	—	(312)
20	$Na_3[Fe(CN)_5P(C_6H_5)_3]$	(+)0.62	—	(239)
21	$Na_3[Fe(CN)_5As(C_6H_5)_3]$	(+)0.92	—	(239)
22	$Na_3[Fe(CN)_5Sb(C_6H_5)_3]$	(+)0.94	—	(239)
23	cis-$Fe(CO)_4Cl_2$	(−)0.25	—	(39)
24	$K_2PcFe(CN)_2$	(+)0.56	—	(156)
25[e]	$PcFe(py)_2$	+2.04	—	(156, 347)
26	$PcFe(Im)_2$	(+)1.77	—	(156, 347)
27	$PcFe(but)_2$	(+)1.94	—	(156)
28	$PcFe(pip)_2$	(+)2.21	—	(156)
29	$Fe(niox)_2(NH_3)_2$	(+)1.75 (η = large)	—	(156, 347)
30	cis-$FeCl_2(ArNC)_4$	−0.78	−0.78	(34, 44)
31	$[FeCl(ArNC)_5]ClO_4$	0.73	+0.78	(44)
32	cis-$Fe(SnCl_3)_2(ArNC)_4$	0.50	−0.52	(44)
33	cis-$FeCl(SnCl_3)(ArNC)_4$	0.61	−0.65	(44)
34	$[Fe(SnCl_3)(ArNC)_5]ClO_4$	0.32	+0.52	(44)
35	trans-$FeHCl(depe)_2$	<0.12	−0.20	(44)
36	trans-$FeHI(depe)_2$	<0.19	−0.18	(44)
37	trans-$FeClSnCl_3(depe)_2$	1.28	+1.02	(44)
38	trans-$FeBr_2(depb)_2$	1.22	+1.20	(44)
39	cis-$Fe(CN)_2(EtNC)_4$	0.29	+0.30	(55)
40	$[Fe(CN)(EtNC)_5]ClO_4$	0.17	−0.30	(55)

TABLE XLIV—continued

Code No.[b]	Compound[d]	Quadrupole splitting[c]		Ref.
		Obs.	Calc.	
41	Fe(niox)$_2$(Im)$_2$	1.38	+1.64	(156)
42	Fe(niox)$_2$(py)$_2$	1.75	+1.92	(156)
43	Fe(niox)$_2$(but)$_2$	1.83	+1.84	(156)
44	K$_2$Fe(niox)$_2$(CN)$_2$	0.80	+0.44	(156)
45	KFe(niox)$_2$Im·CN	0.93	+1.04	(156)
46	Fe(niox)$_2$Im·CO	0.77	+1.86	(156)
47	[FeH(ArNC)(depe)$_2$]$^+$BPh$_4^-$	−1.14	−0.98	(34, 37)
48	[FeH(CO)(depe)$_2$]$^+$BPh$_4^-$	(−)1.00	−0.46	(37, 39)
49	cis-FeH$_2$(CO)$_4$	0.55	+1.22	(39, 44)
50	[FeH(P(OMe)$_3$)(depe)$_2$]$^+$BPh$_4^-$	(−)0.90	—	(37)
51	[FeH(P(OPh)$_3$)(depe)$_2$]$^+$BPh$_4^-$	(−)0.72	—	(37)
52	[FeH(PhCN)(depe)$_2$]$^+$BPh$_4^-$	(−)0.58	—	(37)
53	[FeH(MeCN)(depe)$_2$]$^+$BPh$_4^-$	(−)0.46	—	(37)
54	[FeH(N$_2$)(depe)$_2$]$^+$BPh$_4^-$	(−)0.33	—	(37)
55	Fe(DMG)$_2$·2py	1.84	—	(3, 236)
56	Fe(DMG)$_2$·2β-pic	1.62	—	(3, 236)
57	Fe(DMG)$_2$·2α-pic	0.7	—	(3, 236)
58	Fe(DMG)$_2$·22,4-lut	0.5	—	(3, 236)
59	[Fe(DTOH$_2$)$_2$]Cl$_2$	0.66	—	(1, 236)
60	[Fe(DTOH)$_2$]	2.02	—	(1, 236)
61	[Fe(DTOCH$_3$)$_2$]	0.88	—	(1, 236)
62	Fe(NCS)$_2$(qp)	0.86	—	(237)
63	Fe(CN)$_2$(qp)	1.29	—	(237)

[a] Data given in mm/sec at room temperature. Many "octahedral" complexes such as K$_4$Fe(CN)$_6$, Fe(phen)$_3$(ClO$_4$)$_2$ and others usually have zero or very small splittings. They are not given here. See Refs. (135, 195, 209, 215, 397, 404)

[b] Code Number will be preceded by table number in the text.

[c] Signs without brackets have been determined (see Table XLV), those with brackets have not been determined, but deduced in the references by analogy with the compounds of known sign. Calculated values use the p.q.s. values in Table XLVI. Many of these are reported for the first time.

[d] ArNC = p-Methoxyphenylisocyanide; depe = 1,2-bis(diethylphosphino)ethane; dmpe = 1,2-bis(dimethylphosphino)ethane; depb = o-phenylenebisdiethylphosphine; Pc = phthalocyanine; py = pyridine; Im = imidazole; but = n-butylamine; pip = piperidine; niox = 1,2-cyclohexanedione dioxime; DMG = dimethylglyoxime; 2β-pic = 2β-picoline; 2α-pic = 2α-picoline; 22,4-lut = 22,4-lutidine; DTOH$_2$ = diacetylthiosemicarbozoneoxime; DTOCH$_3$ = O-methyldiacetylthiosemicarbazoneoxime; qp = tris-(O-diphenylphosphinophenyl)phosphine.

[e] Similar quadrupole splittings are given for other PcFeX$_2$ compounds (347). (X = β-picoline, γ-picoline, or α-picoline.)

As the original calculations of p.q.s. values were undertaken before any measurement of the signs of the Q.S., it was necessary to make assumptions for certain key compounds. For example, consider the compounds trans-FeX_2B_4 (X = Cl, Br, I; L = RNC, depe/2, etc.). It is a reasonable assumption, supported by center shift data (vide infra) and general chemistry, that halide ligands carry a higher negative charge and are poorer σ donor and π acceptors than the neutral B ligands. It would therefore be expected (Section I,C) that the q_{lat} and π-bonding inequalities would give negative contributions to the Q.S., whereas the σ bonding inequalities would give a positive contribution to the Q.S. Clearly, both the sign and the magnitude of these Q.S. values depends on the relative importance of these contributions. Originally, Bancroft et al. (44) assumed that the q_{lat} contribution was most important after calculating the q_{lat} contribution to the Q.S. from a trans-$FeCl_2$ linkage to be -1.2 mm/sec, in reasonable agreement with the Q.S. values for most of the trans-$FeCl_2B_4$ compounds. The reference p.q.s. value for Cl^- was thus taken as -0.30 mm/sec, the signs of trans-FeX_2B_4 compounds taken as negative, and p.q.s. values for other ligands derived. Predicted values were compared with observed, and good agreement obtained in most cases (Table XLIV, compounds 30–40) demonstrating that the additivity model provides a reasonable guide to variations of the Q.S. with structure. The predicted values calculated originally (44) are identical with those calculated in Table XLIV.

Measurement of the signs of the Q.S. of a number of Fe^{II} low-spin compounds using the magnetic field technique (36) showed that the wrong signs had been assumed. For example, Fig. 14 illustrates the spectra of cis- and trans-$FeCl_2(ArNC)_4$ in an applied magnetic field. These spectra indicate that the trans and cis isomers have positive and negative Q.S. values, respectively. These measurements provide an elegant confirmation of the opposite signs of the trans and cis isomers predicted by the additivity model, but show that the initially assumed signs of the Q.S. were incorrect.

Other signs have been measured (Table XLV), and have been invaluable in calculating p.q.s. values for a wide variety of ligands using the first twenty-nine compounds in Table XLIV. Using these p.q.s. values (Table XLVI), the quadrupole splitting values for compounds 30–49 have been calculated. In deriving the p.q.s. values, it was assumed that all the trans-FeX_2B_4 compounds have positive quadrupole splittings. Some of the signs of other compounds (such as compounds 20–22 in Table XLIV) were deduced by analogy with similar compounds to give reasonable p.q.s. values for such ligands as $P(C_6H_5)_3$. In addition, when some doubt exists about the sign of the quadrupole splitting, the ambiguity can be eliminated by comparing quadrupole splittings for similar compounds. For example, the sign of the quadrupole splitting in trans-Fe(SnCl$_3$)$_2$(ArNC)$_4$ is not known, and p.q.s. values of SnCl$_3^-$ of -0.43 and -0.95 mm/sec can be calculated from this compound assuming a positive and negative sign, respectively. However, the calculated quadrupole splitting for the compound trans-FeClSnCl$_3$(depe)$_2$ (observed

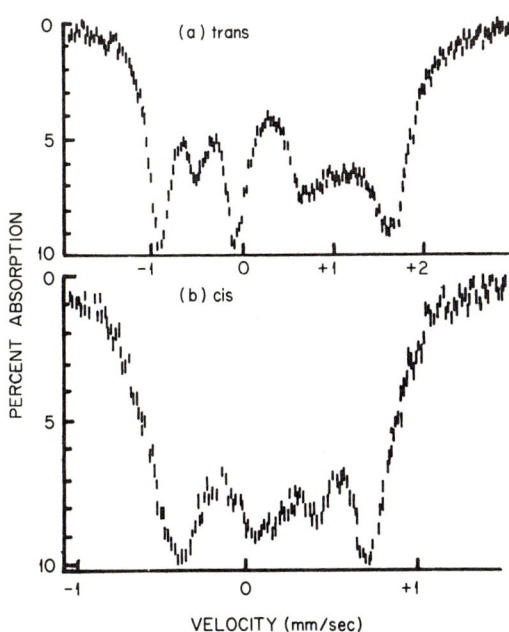

Fig. 14. Mössbauer spectra of (a) trans-FeCl$_2$(p-MeO·C$_6$H$_4$·NC)$_4$ at 4°K in a longitudinal magnetic field of 36kG; (b) cis-FeCl$_2$(p-MeO·C$_6$H$_4$·NC)$_4$ at 4°K in a longitudinal field of 28 kGauss (36).

1.28 mm/sec) is $+1.02$ mm/sec with (p.q.s.)$_{SnCl_3^-}$ = -0.43, and -0.02 mm/sec with (p.q.s.)$_{SnCl_3^-}$ = -0.95 mm/sec, clearly indicating a positive sign for the Q.S. of trans-Fe(SnCl$_3$)$_2$(ArNC)$_4$. Similarly, the positive sign of trans-FeCl$_2$(depe)$_2$, and the very small Q.S. for trans-FeHCl(depe)$_2$ allows the assignment of a negative sign to the Q.S. of trans-FeH$_2$(depb)$_2$. This sign is confirmed by the measured negative sign observed for trans-[FeH(ArNC)(depe)$_2$]$^+$BPh$_4^-$ (Table XLV), and the predicted Q.S. for the above compound is in good agreement with the observed value (Table XLIV, compound 47). The measured negative sign for trans-[FeH(ArNC)(depe)$_2$]$^+$BPh$_4^-$ makes it highly probable that all the other trans-[FeHL(depe)$_2$]$^+$ species [L = CO, Me$_3$CNC, P(OMe)$_3$, P(OPh)$_3$,

TABLE XLV

Signs of the EFG for Fe^{II} Low-Spin Complexes

Compound No.	Compound	Q.S.	Ref.
1	PcFe(Py)$_2$	+1.96	(157)
2	Fe^{II}(niox)$_2$(im)$_2$	1.30 (η = large)	(157)
3	Fe^{II}(niox)$_2$(NH$_3$)$_2$	1.72 (η = large)	(157)
4	trans-[FeH(ArNC)(depe)$_2$$^+$]BPh$_4$$^-$	−1.14	(34)
5	trans-FeCl$_2$(ArNC)$_4$	+1.55	(34, 36)
6	cis-FeCl$_2$(ArNC)$_4$	−0.83	(34, 36)
7	trans-FeCl$_2$(depe)$_2$	+1.29	(34)
8	trans-Fe(CN)$_2$(EtNC)$_4$	−0.60	(34)
9	Na$_3$[Fe(CN)$_5$NH$_3$]·H$_2$O	+0.67	(34)
10	K$_3$[Fe(CN)$_5$H$_2$O]·7H$_2$O	+0.80	(34)
11	Na$_2$[Fe(CN)$_5$NO]·2H$_2$O	+1.73	(34, 161)

N$_2$, PhCN, and MeCN] (Table XLIV, compounds 48, 50–54) also have negative signs.

It is apparent from the general agreement between predicted and observed Q.S. values in Table XLIV that the additivity model holds rather well. We take satisfactory agreement to mean that predicted and observed values are within ±0.2 mm/sec. These p.q.s. values should now be extremely useful for rationalizing Q.S. values for other Fe^{II} compounds containing these ligands in different combinations (vide infra; the carbonyl compounds), as well as rationalizing signs and magnitudes of Q.S. values in Co^{III}, Ru^{II}, and Ir^{III} compounds (vide infra).

However, there are at least three types of compounds where the agreement between predicted and observed values might be expected to be unsatisfactory. First, for cationic or anionic compounds, a q_{lat} contribution from the anion or cation, respectively, or a significant lowering or raising of the Fe d orbitals relative to the ligand orbitals might lead to discrepancies. The agreement for such cationic compounds as trans-[FeH(ArNC)(depe)$_2$]BPh$_4$, [Fe(CN)(EtNC)$_5$]ClO$_4$, and [Fe(SnCl$_3$)-(ArNC)$_5$]ClO$_4$ is generally not as good as that for most of the analogous neutral compounds (Table XLIV, compounds 30–40).

Second, satisfactory agreement for such compounds as Fe^{II}(niox)$_2$L$_2$ (Table XLIV, compounds 41–46) is not always observed, but this might be due to large distortions from regular symmetry. The large values of η for these compounds (157) are consistent with this interpretation.

TABLE XLVI
PARTIAL QUADRUPOLE SPLITTINGS FOR Fe^{II} LIGANDS[a]

$NO^+ = +0.02$	$NCO^- = 0.50$
$Br^- = -0.28$	$NH_3 = -0.51$
$I^- = -0.29$	$PPh_3 = -0.53$
$Cl^- = -0.30$	$im = -0.54$
$SbPh_3 = -0.37$	$depb/2 = -0.58$
$AsPh_3 = -0.38$	$depe/2 = -0.62$
$N_3^- = -0.38$	$dmpe/2 = -0.67$
$NO_2^- = -0.41$	$RNC = -0.69$
$SnCl_3^- = -0.43$	$CNH = -\sim0.8$
$pip = -0.43$	$CNBF_3 = -\sim0.8$
$CO = -0.43$	$CN^- = -0.84$
$H_2O = -0.44$	$niox/2 = -0.95$
$SO_3^{2-} = -0.46$	$Pc/4 = -0.98$
$py = -0.47$	$H^- = -1.04$
$NCS^- = -0.49$	
$but = -0.49$	

[a] These are derived from compounds 1–28 in Table XLV. Data given in mm/sec. Some of these values are given in Ref. (34); the others are newly calculated values.

Chelating ligands, in general, would be expected to give more variable p.q.s. values owing to steric effects. One interesting anomaly to the additivity model predictions concerns the compounds cis- and trans-Fe(phen)$_2$(CN)$_2$ and K$_2$Fe(phen)(CN)$_4$. The existence of the two isomers is based on infrared evidence (135, 504). Preliminary Mössbauer data show that all the Q.S. values are positive with values of ~0.6 mm/sec (35). Since the nitrogens in K$_2$Fe(phen)(CN)$_4$ must certainly be cis-, the predicted Q.S. ratios for the species trans-Fe(phen)$_2$(CN)$_2$, cis-Fe(phen)$_2$(CN)$_2$, and K$_2$Fe(phen)(CN)$_4$ is 2: −1:1. Clearly these prepredicted ratios are incompatible with the near equality in the sign and magnitude of the observed quadrupole splittings. Even if it is assumed that the cis and trans isomers of Fe(phen)$_2$(CN)$_2$ are, in fact, the same compound, it is still difficult to explain the data. Thus, if Fe(phen)$_2$-(CN)$_2$ is cis, the magnitude of the quadrupole splitting is as expected, but the sign is anomalous, while a trans structure gives a correct prediction of sign, but an anomalous magnitude. Analogous bipyridyl complexes (76) show similar quadrupole splittings, although measurements of the signs are not available. At present no clear explanation of these data is apparent, although one possible cause might be distortions from a regular geometry owing to the steric properties of the phenanthroline.

A similar explanation has been used to rationalize the anomalous signs for Sn^{IV} complexes (439).

Finally, for strong π-accepting groups such as CO or NO^+, we might expect considerable variations in p.q.s. values, leading to marked discrepancies between predicted and observed values (Table XLIV, compounds 46, 48, 49). For example, taking the sign of the Q.S. of cis-$Fe(CO)_4X_2$ compounds (X = Cl, Br, I) to be negative, the p.q.s. for CO is -0.43 mm/sec. Using this value to predict the Q.S. for trans-$[FeH(CO)(depe)_2]^+BPh_4^-$ (compound 48), we obtain -0.46 mm/sec which is not in satisfactory agreement with the observed value. This compound perhaps represents one of the most extreme cases, since the very low CO

TABLE XLVII

Q.S. VALUES FOR CARBONYL COMPOUNDS AT $80°K^a$

Compound No.	Compound	Q.S.
1	cis-$Fe(CO)_4I_2$	0.31
2	cis-$Fe(CO)_4Br_2$	0.31
3	cis-$Fe(CO)_4Cl_2$	0.25
4	$Fe(CO)_3I_2(PPh_2Et)$	0.43
5	$Fe(CO)_2I_2(PPh_2Me)_2$	0.39
6	$Fe(CO)_2Br_2(PPh_2Me)_2$	0.56
7	$Fe(CO)_2Cl_2(PPh_2Me)_2$	0.60

^{a} From Ref. (39). Data given in mm/sec.

stretching frequency (45) indicates that the bonding properties (and, thus, the p.q.s. value) of CO have changed markedly. To examine the effect of a strong π-accepting ligand such as CO, the spectra of such compounds as $Fe(CO)_3X_2P$ and $Fe(CO)_2X_2P_2$ (X = Cl, Br, I; P = PPh_2Me, PPh_2Et, $P(OPh)_3$, etc.) are now being recorded (39). The latter compounds have five possible geometric isomers, some of which are extremely difficult to distinguish by other spectroscopic methods, and the p.q.s. treatment could be extremely useful. Results for a few compounds are given in Table XLVII and the components of the EFG are given in terms of p.f.g. values in Table IV. The two phosphine ligands give identical values within experimental error. The structure of $Fe(CO)_3I_2(PPh_2Et)$ is taken as in Table IV (structure 18), and a p.q.s. for PPh_2Et is calculated to be -0.51 mm/sec. Then the predicted Q.S. values for structures 20 and 22 (Table IV) (both possible from CO IR) for the iodide derivative (compound 5, Table XLVII) are -0.60 and

-0.39 mm/sec, respectively. The Q.S. for compound 5 suggests then that it has the all cis structure, while the Q.S. for compounds 6 and 7 suggest that they have the *trans*-PPh$_2$Me structure. However, the solution NMR data for all three compounds suggests *trans*-P configurations, and the Q.S. for the two compounds are not different enough to be conclusive about structural assignments, especially because we would expect the p.q.s. of CO to vary somewhat. The few results given above certainly indicate the potential power of the additivity model in predicting such structures once we can be more confident about variations in p.q.s. values for such ligands as CO.

b. Bonding Properties of Ligands. As we have mentioned previously, three possible factors may contribute to the quadrupole splitting in Fe^{II} compounds; inequalities in σ or π bonding ($q_{M.O.}$) and inequalities in ligand charge (q_{lat}). The q_{lat} term will be expected to make a significant contribution only for relatively ionic ligands such as the halogens. The measured positive signs for the compounds *trans*-FeCl$_2$(ArNC)$_4$ *(36)* and *trans*-FeCl$_2$(depe)$_2$ *(34)* clearly show that the q_{lat} contribution is not the dominant contribution to the EFG, and indicate that σ-bonding inequalities are more important than π-bonding inequalities. A dominance of the σ-bonding contribution may also be assumed for the species FeX$_2$B$_4$ [X = Cl, Br, I, B = depe/$_2$, dmpe/$_2$, depb/$_2$ *(34)*], Fe(niox)$_2$L$_2$ (L = py, im, NH$_3$, but, CN) *(156)*, and PcFeL$_2$ (L = im, but, pip, CN) *(156)*. However, the strong π-accepting characteristics of NO$^+$ and CO, for example, are obviously important in determining the positive Q.S. of Na$_2$Fe(CN)$_5$NO·2H$_2$O, the probably positive Q.S. of Fe(niox)$_2$im·CO, and the large differences in Q.S. between corresponding Fe(CO)$_4$X$_2$ and Fe(ArNC)$_4$X$_2$ (X = Cl, Br, I) compounds (e.g., Table XLIV, compounds 23, 30). For example, the much smaller Q.S. of the CO compounds can be attributed to the better π-acceptor properties of CO relative to ArNC (*vide infra*). In addition, the fairly small differences in Q.S. between *trans*-FeCl$_2$B$_4$ compounds (B = ArNC, depe, dmpe, etc.) could be due to the differences in π-accepting ability of the neutral ligands. Similarly, a π-bonding interpretation has been suggested to explain the Q.S. variation in PcFeL$_2$ (L = py, im, etc.) species *(347)*. However, as mentioned above, it seems more likely that σ-bonding inequalities play the dominant role in determining the Q.S. in these latter compounds.

The above discussion can be considered more quantitatively by considering the p.q.s. values. It is apparent from Section II,C that the p.q.s. values will become more negative as q_{lat} and σ_L for a ligand increases, but as π_L decreases *(44)*, i.e.,

$$\text{p.q.s.} \propto [-q_{lat} + (\pi_L - \sigma_L)] \qquad (47)$$

Thus, considering the p.q.s. values in Table XLVI, it is apparent that H^-, the best σ donor, gives the most negative p.q.s., whereas NO^+, the best π acceptor gives the most positive p.q.s. value. Comparing other neutral ligands such as ArNC, CO, and phosphines, the p.q.s. values indicate that $\pi_L - \sigma_L$ increases in the order ArNC < phosphines < CO. Thus, CO, a good π acceptor, gives a low p.q.s. value for a neutral ligand. Combined with center shift data (*vide infra*), which gives a measure of $\sigma_L + \pi_L$, we can qualitatively separate the σ_L and π_L contributions for a wide range of ligands.

To illustrate the above concepts, we now discuss the quadrupole splitting trend in two series of compounds: the $[Fe(CN)_5X^{n-}]^{(3-n)-}$ (Table XLIV, compounds 13–17) species and the *trans*-$[FeHL(depe)_2]^+$-BPh_4^- compounds (Table XLIV, compounds 48, 50–54). The quadrupole splittings for the $[Fe(CN)_5X^{n-}]^{(3-n)-}$ series have been the subject of much discussion (*76, 160, 238*). The order of center shifts, $L = NH_3 > H_2O > SO_3^{2-} > NO_2^- > CN^- > NO^+$ parallelled the expected π-bonding ability of the ligands, and the magnitude of the Q.S. values observed $L = NO^+ > NO_2^- > SO_3^{2-} > H_2O > NH_3$ has also been attributed to the differing π-acceptor abilities of the above ligands. The sign of the Q.S. for $Na_2Fe(CN)_5NO \cdot 2H_2O$ has been shown to be positive (Table XLV), as expected if NO^+ is a better π acceptor (and/or weaker σ donor) than CN^-, and Oosterhius and Lang (*432*) have found that the magnitude of the Q.S. is consistent with the calculated t_{2g} orbital populations as expected by the π-bonding argument. However, for the other compounds, the above order of C.S. indicated that CN^- is a better π acceptor than the other ligands, and on a π-bonding argument, a negative Q.S. sign would be expected. In contrast, positive signs have been observed for $Na_3[Fe(CN)_5NH_3]$ and $K_3[Fe(CN)_5 \cdot H_2O]$ (*34*), indicating that the greater σ-donor ability of CN^- largely determines the Q.S. The Q.S. data for this series of compounds strongly indicate that both σ *and* π bonding can be important in determining the magnitude and signs of the Q.S., as was indicated previously when discussing the variations of p.q.s. values. Further, although the present results indicate that σ-bonding inequalities are usually dominant, it is apparent that the sign of the Q.S. for a series of compounds should be known before the results are interpreted. For example, the observed Q.S. values in the series $[Fe(CN)_5MPh_3]^{3-}$ (M = P, As, Sb) (Table XLIV, compounds 20–22) (*239*) may not be a reflection of the decreasing π-bonding ability $PPh_3 > AsPh_3 > SbPh_3$ (*239*). The Q.S. for all these compounds is most likely positive since the p.q.s. values for phosphines are probably substantially less negative than for CN^-. A decrease in π bonding in the above order would give the PPh_3 compound the most positive of the Q.S. values, in opposition to the observed values (taking the Q.S. values to be positive). More likely, the trend in Q.S. could be attributed to a decrease in σ-donor ability in the order $PPh_3 > AsPh_3 > SbPh_3$.

The cationic compounds trans-$[FeHL(depe)_2]^+BPh_4^-$ (Table XLIV, compounds 47, 48, 50–54) provide another interesting series from which the bonding of N_2 can be compared with a large number of other neutral ligands. The sign of the Q.S. for $L = p\text{-MeO}\cdot C_6H_4\cdot NC$ has been determined to be negative (37), but unfortunately the sign of the Q.S. for the N_2 compound could not be determined unambiguously because of its small magnitude. However, it is reasonable to assume that all signs are negative since the range of p.q.s. values observed for a large number of neutral ligands (−0.40 to −0.70) (Table XLVI) gives an expected range of Q.S. values for any trans-$FeHL(depe)_2^+$ species of −0.40 to −1.00 mm/sec.

The N_2 complex has the most positive Q.S. indicating that it is the best $\pi_L-\sigma_L$ ligand, although in contrast, the center shift data (vide infra) indicate that it is one of the poorest $\sigma_L + \pi_L$ ligands; π acceptance relative to σ donation is more important in N_2 than in the other ligands. As discussed in more detail in the next section, these data suggest that N_2 is a moderate π acceptor, but a weak σ donor.

The compounds $Fe(DMG)_2L_2$ where L is pyridine or a methyl-substituted pyridine (Table XLIC, compounds 55–58) show variations in both center shift and quadrupole splitting (3, 236), which have been discussed in terms of a quantitative molecular orbital description. Quantitative predictions (11, 12) of the electronic structure of these compounds have also been reported, and reasonable agreement between observed and calculated quadrupole splittings obtained. Goldanskii et al. (1) have considered the relative quadrupole splittings of the species $[Fe(DTOH_2)_2]Cl_2$, $Fe(DTOH)_2$, and $Fe(DTOMe)_2$, where $DTOH_2$ is diacetylthiosemicarbazonoxime and DTOH and DTOMe are ionic derivatives formed by removal of a proton from the oximide (=NOH) or thiosemicarbazene (=N·NH·CS·NH$_2$) group, respectively. A simple molecular orbital description (2) was used to rationalize the quadrupole splittings, but this is of doubtful validity as a dominant contribution from the diffuse $4p_z$ electrons was assumed.

2. *Center Shifts*

Variations of both the σ- and π-bonding abilities of the ligands will be expected to affect the center shifts of low-spin Fe^{II} compounds (44, 76, 142, 156, 238). An increase in the σ-donating power will result in an increased population of iron d^2sp^3 hybrids and, as the center shift is more sensitive to 4s augmentation than 3d or 4p augmentation, a net decrease in center shift will result. Greater π-acceptor power of the ligand will produce a greater delocalization of the t_{2g} electrons, with a consequent reduction in the shielding of the s electrons and this will also lead to a decrease in center shift.

The first observation (76, 142, 238) of a systematic variation of center shift was for the ions $[\text{Fe(CN)}_5 X^{n-}]^{(3-n)-}$ (Table XLVIII, compounds 15, 17–20, 56), for which it was observed initially (142)* that the center shift order $\text{NH}_3 > \text{H}_2\text{O} > \text{SO}_3^{2-} > \text{NO}_2^- > \text{CN}^- > \text{NO}^+$ inversely parallels the π-accepting abilities of these ligands. Other indications of a dependence of center shift on ligand–iron bonds were found (156), for the compounds $\text{Fe(niox)}_2\text{L}_2$ and FePcL_2 (Table XLVIII, compounds 26–32, 66–70), and generally the center shift decreased with increasing σ-donor ability of the ligands. In contrast, ring substituents in the compound Fe(L)_3^{2+} (L = 1, 10-phenanthroline or dipyridyl) have only a small effect upon the isomer shift (135, 209).

a. Derivation of p.c.s. Values. A comprehensive survey of center shifts for Fe^{II} low-spin compounds has been reported by Bancroft et al. (44), and much of the data in this paper, as well as other Fe^{II} center shifts, are given in Table XLVIII. These values are all relative to nitroprusside. It was postulated (44) that the observed center shifts could be represented as an algebraic sum of contributions from each ligand. The contribution of each ligand to the center shift (C.S.) was termed the partial center shift (p.c.s.). Relative to nitroprusside, we obtain the equation:

$$\text{C.S.} = \sum_{i=1}^{6} (\text{p.c.s.}) + 0.16 \qquad (48)$$

The use of p.c.s. values involves certain assumptions similar to those used for the quadrupole splitting additivity model. The center shifts must not be sensitive to small distortions in geometry, and variations in center shift must be chiefly dependent on variations in isomer shift, i.e., the S.O.D. shift must remain relatively constant (319). There is some evidence for this last assumption in the observation of an almost constant temperature dependence of center shift. It is also necessary to assume that the p.c.s. of a particular ligand is not dependent on the other ligands, i.e., that p.c.s. values are additive.

Center shifts relative to stainless steel at 295°K were used in the calculation of p.c.s. values assuming that the p.c.s. value of ArNC is zero (44); p.c.s. values for thirty-nine ligands (Table XLIX) are derived (from $trans$-FeA_2B_4 compounds where possible), and these p.c.s. values are used to predict the C.S. values for 46 other compounds. Except for

---

* Examination of the more recent data in Table XLVIII shows that the order is slightly different.

ten of these compounds, agreement between predicted and observed values is excellent (within ±0.05 mm/sec), and the data lend credence to an additivity model for the C.S. However, the significantly smaller C.S. of cis-$FeCl_2(ArNC)_4$ compared to trans-$FeCl_2(ArNC)_4$, indicates that the p.c.s. values of π-acceptor ligands such as ArNC do vary from compound to compound. The above discrepancy may be attributed to the increased π-acceptor properties of ArNC when it is trans to a chloride, as opposed to another ArNC ligand. More serious π-bonding discrepancies will be noted shortly. Variations in p.c.s. values of σ-bonding ligands from compound to compound seem to be small as illustrated by the good agreement of the calculated and observed center shifts for trans-FeHCl-(depe)$_2$ (Table XLVIII, compound 45).

The availability of p.c.s. data allowed a more careful study of the relationship between center shift and bonding properties of the ligands first observed for the $Fe(CN)_5X$ derivatives (76, 142, 238). As stated earlier, an increase in both the σ- donor and π- acceptor abilities of the ligands will be expected to produce a decrease in the p.c.s. values. In agreement with this expectation, the most positive p.c.s. values (I$^-$, Br$^-$, and Cl$^-$) are associated with the most ionic ligands, while H$^-$ (very strong σ donor) and NO$^+$ (very strong π acceptor) have the most negative values. In fact, using the partial ligand field strengths of the ligands (δ) calculated from the optical spectra of Co^{III} compounds (524), a good general correlation was observed between p.c.s. value and the ranking of the ligand in the spectrochemical series (44). Although an exact relationship is not expected, the correlation demonstrates that p.c.s. values give a good guide to relative δ values. For example, the p.c.s. of H$^-$ provides strong evidence that H$^-$ occupies a position close to CN$^-$

TABLE XLVIII

CENTER SHIFT VALUES FOR Fe^{II} LOW-SPIN COMPOUNDS[a]

Code No.	Compound	Center shift		Ref.
		Obs.	Pred.	
1	trans-$FeCl_2(ArNC)_4$	+0.36	—	(44)
2	trans-$Fe(SnCl_3)_2(ArNC)_4$	+0.24	—	(44)
3	trans-$FeCl_2(depe)_2$	+0.59	—	(44)
4	trans-$FeBr_2(depe)_2$	+0.66	—	(44)
5	trans-$FeI_2(depe)_2$	+0.65	—	(44)
6	trans-$FeCl_2(depb)_2$	+0.59	—	(44)
7	trans-$FeH_2(depb)_2$	+0.23	—	(44)

continued

TABLE XLVIII—continued
CENTER SHIFT VALUES FOR Fe^{II} LOW-SPIN COMPOUNDS[a]

Code No.	Compound	Center shift Obs.	Pred.	Ref.
8	trans-$FeCl_2(dmpe)_2$	+0.54	—	(44)
9	cis-$Fe(CO)_4Cl_2$	+0.24	—	(39)
10	trans-$Fe(N_3)_2(depe)_2$	+0.56	—	(44)
11	trans-$Fe(NCO)_2(depe)_2$	+0.51	—	(44)
12	trans-$Fe(NCS)_2(depe)_2$	+0.49	—	(44)
13	$[Fe(MeNC)_6](HSO_4)_2$	+0.14	—	(55)
14	$[Fe(EtNC)_6](ClO_4)_2$	+0.16	—	(55)
15	$K_4Fe(CN)_6$	+0.21	—	(215)
16	$[Fe(PhCH_2NC)_6](ClO_4)_2$	+0.12	—	(55)
17	$Na_2[Fe(CN)_5NO] \cdot 2H_2O$	0.00	—	(238)
18	$Na_4[Fe(CN)_5NO_2]$	+0.26	—	(238)
19	$Na_5[Fe(CN)_5SO_3]$	+0.22	—	(238)
20	$Na_3[Fe(CN)_5H_2O]$	+0.31	—	(142)
21	$[Fe(bipy)_3](ClO_4)_2$	+0.52	—	(215)
22	$[Fe(phen)_3](ClO_4)_2$	+0.58	—	(215)
23	$Na_3[Fe(CN)_5P(C_6H_5)_3]$	+0.23	—	(239)
24	$Na_3[Fe(CN)_5As(C_6H_5)_3]$	+0.29	—	(239)
25	$Na_3[Fe(CN)_5Sb(C_6H_5)_3]$	+0.26	—	(239)
26	$Fe(niox)_2(NH_3)_2$	+0.46	—	(156)
27	$Fe(niox)_2(py)_2$	+0.46	—	(156)
28	$K_2[Fe(noix)_2(CN)_2]$	+0.34	—	(156)
29[b]	$PcFe(py)_2$	+0.51	—	(156)
30	$Fe(niox)_2(im)_2$	+0.49	—	(156)
31	$Fe(niox)_2(but)_2$	+0.47	—	(156)
32	$PcFe(pip)_2$	+0.51	—	(156)
33	$Fe(NCS)_2(qp)$	+0.47	—	(237)
34	$[Fe(pyim)_3]X_2$	+0.62	—	(195)
35	$[Fe(tripyam)_2](ClO_4)_2$	+0.63	—	(406)
36[c]	$Fe(phen-derivatives)_3(X)_2$	+0.52-+0.56	—	(135, 209, 236)
37[b]	$Fe(DMG)_2(py)_2$	+0.39	—	(3, 236)
38	$K_4[Fe(CNBF_3)_6]$	+0.15	—	(312)
39	$Fe(CNH)_4(CN)_2$	+0.16	—	(312)
40	cis-$FeCl_2(ArNC)_4$	+0.28	+0.36	(44)
41	cis-$Fe(SnCl_3)_2(ArNC)_4$	+0.27	+0.24	(44)
42	cis-$FeClSnCl_3(ArNC)_4$	+0.23	+0.30	(44)
43	$[FeCl(ArNC)_5]ClO_4$	+0.22	+0.26	(44)
44	$[Fe(SnCl_3)(ArNC)_5]ClO_4$	+0.18	+0.20	(44)
45	trans-$FeHCl(depe)_2$	+0.39	+0.42	(44)
46	trans-$FeHI(depe)_2$	+0.39	+0.45	(44)
47	trans-$FeCl(SnCl_3)(depe)_2$	+0.55	+0.54	(44)
48	trans-$FeBr_2(depb)_2$	+0.61	+0.66	(44)
49	trans-$Fe(CN)_2(MeNC)_4$	+0.16	+0.18	(55)
50	cis-$Fe(CN)_2(MeNC)_4$	+0.16	+0.18	(55)

TABLE XLVIII—continued

Code No.	Compound	Center shift Obs.	Pred.	Ref.
51	trans-Fe(CN)$_2$(EtNC)$_4$	+0.21	+0.18	(55)
52	cis-Fe(CN)$_2$(EtNC)$_4$	+0.21	+0.18	(55)
53	[Fe(CN)(EtNC)$_5$]ClO$_4$	+0.20	+0.17	(55)
54	trans-Fe(CN)$_2$(PhCH$_2$NC)$_4$	+0.15	+0.14	(55)
55	[Fe(CN)(PhCH$_2$NC)$_5$]ClO$_4$	+0.14	+0.12	(55)
56	Na$_3$[Fe(CN)$_5$NH$_3$]H$_2$O	+0.26	+0.28	(238)
57	Fe(phen)$_2$(NO$_2$)$_2$	+0.53	+0.54	(372)
58	K$_2$[CaFe(NO$_2$)$_6$]	+0.55	+0.46	(501)
59	[Fe(bipy)$_2$(NCS)py]NCS	+0.61	+0.52	(372)
60	K$_3$[Fe(CN)$_5$CO]	+0.15	+0.18	(215)
61	K$_2$[Fe(bipy)(CN)$_4$]	+0.32	+0.32	(215)
62	K$_2$[Fe(phen)(CN)$_4$]	+0.33	+0.34	(215)
63	cis-Fe(bipy)$_2$(CN)$_2$	+0.43	+0.42	(215)
64	cis-Fe(phen)$_2$(CN)$_2$	+0.42	+0.46	(215)
65	trans-Fe(phen)$_2$(CN)$_2$	+0.48	+0.46	(215)
66	K$_2$[PcFe(CN)$_2$]	+0.37	+0.38	(156)
67	KFe[(niox)$_2$Im·CN]	+0.34	+0.41	(156)
68	Fe(niox)$_2$Im·CO	+0.26	+0.37	(156)
69	PcFe(Im)$_2$	+0.47	+0.52	(156)
70	PcFe(but)$_2$	+0.52	+0.50	(156)
71	Fe(CN)$_2$(qp)	+0.31	+0.36	(237)
72	Fe(CNH)$_4$(CNBF$_3$)$_2$	+0.14	+0.16	(312)
73	trans-Fe(CNMe)$_4$(CNBF$_3$)$_2$	+0.18	+0.16	(312)
74	trans-Fe(CNEt)$_4$(CNBF$_3$)$_2$	+0.16	+0.16	(312)
75	[FeH(ArNC)(depe)$_2$]$^+$BPh$_4^-$	+0.19	+0.32	(45)
76	[FeH(CO)(depe)$_2$]$^+$BPh$_4$	+0.12	+0.29	(45)
77	cis-FeH$_2$(CO)$_4$	+0.01	−0.12	(44)
78	[Fe(DTOH$_2$)$_2$]Cl$_2$	+0.50	—	(1, 236)
79	[Fe(DTOH)$_2$]	+0.30	—	(1, 236)
80	[Fe(DTOCH$_3$)$_2$]	+0.53	—	(1, 236)
81	[FeHP(OMe)$_3$(depe)$_2$]$^+$BPh$_4^-$	+0.25	—	(45)
82	[FeHP(OPh)$_3$(depe)$_2$]$^+$BPh$_4^-$	+0.26	—	(45)
83	[FeH(PhCN)(depe)$_2$]$^+$BPh$_4^-$	+0.33	—	(45)
84	[FeH(MeCN)(depe)$_2$]$^+$BPh$_4^-$	+0.35	—	(45)
85	[FeHN$_2$(depe)$_2$]$^+$BPh$_4^-$	+0.32	—	(45)

[a] Data given in mm/sec at room temperature relative to sodium nitroprusside. This table is largely taken from Table 5, Ref. (44). All C.S. values have been converted from stainless steel to nitroprusside. C.S. = 0.16 + $\sum_i$(p.c.s.)$_i$.

[b] C.S. values for analogous compounds with py replaced by other N bases give very similar C.S. values (3, 236, 347).

[c] Many substituted phenanthroline compounds with different X groups have been run with very similar results (135, 209, 236).

in the spectrochemical series, and not a position between H_2O and NH_3 as previously suggested (*434, 537*). The relationship between p.c.s. and δ also suggests (*44*) that there may be a limiting value of C.S. for low-spin Fe^{II} compounds (~0.5 relative to stainless steel; ~0.7 relative to nitroprusside), above which there is a transition from low-spin to high-spin Fe^{II}.

As with the p.q.s. treatment, there are situations where the p.c.s. values may vary widely, thus leading to large discrepancies between observed and predicted values. For strong π-accepting ligands such as

TABLE XLIX

Partial Center Shift Values for Fe^{II} Low-Spin Compounds[a]

NO^+ = −0.20	$SnCl_3^-$ = 0.04	py = 0.07
H^- = −0.08	qp/4 = 0.05	NH_3 = 0.07
CO = −0.03	niox/2 = 0.04	but = 0.07
$PhCH_2NC$ = −0.01	$SbPh_3$ = 0.05	$AsPh_3$ = 0.08
$CNBF_3$ = 0.00	NO_2^- = 0.05	pip = 0.08
CNH = 0.00	Pc/4 = 0.05	pyim/2 = 0.08
ArNC = 0.00	dmpe/2 = 0.05	tripyam/3 = 0.08
MeNC = 0.00	NCS^- = 0.05	Im = 0.08
EtNC = 0.00	NCO^- = 0.06	N_3^- = 0.08
CN^- = 0.01	depe/2 = 0.06	H_2O = 0.10
SO_3^{2-} = 0.01	depb/2 = 0.06	Cl^- = 0.10
$P(C_6H_5)_3$ = 0.02	bipy/2 = 0.06	Br^- = 0.13
DMG/2 = 0.03	phen/2 = 0.07	I^- = 0.13

[a] Data given in mm/sec. Most of these values are taken from Ref. (*44*).

CO, it has become apparent that the C.S. values are not additive. For example, consider $[FeHCO(depe)_2]^+$ (Table XLVIII, compound 76). Although a *trans*-CO compound was not available to derive (p.c.s.)$_{CO}$ initially, the predicted C.S. value using the p.c.s. in Table XLIX is 0.29 mm/sec, while the observed value is 0.12 mm/sec. Similarly, for compounds such as $Fe(CO)_2X_2P_2$ and $Fe(CO)_3X_2P$ (X = Cl, Br, I; P = PPh_2Me, etc.) (*39*), the C.S. values are not additive. The CO group tends to "draw off" any excess charge on the Fe atom, and this effect, of course, results in a lowering of the CO stretching frequency.

In anionic compounds, there is also some evidence that the cation has an effect on the C.S. For example, a series of $M_4Fe^{II}(CN)_6$ and $M_2Fe^{II}(CN)_5NO$ (M = H, Li, Na, K, etc.) complexes (*255, 404*) shows a systematic trend in C.S. This trend has been attributed to a change in S.O.D. shift (*404*), although the changing polarizing power of the cation may also be a factor. A small difference in C.S. has also been observed between $K_4Fe(CN)_6 \cdot 3H_2O$ and $H_4Fe(CN)_6$ *(500)*.

Despite these discrepancies, it appears that for the majority of six-coordinate Fe^{II} low-spin compounds which do not contain strong π acceptors such as CO or NO^+, the p.c.s. treatment should provide a reasonably accurate guide to C.S. values. For series of compounds such as the *trans*-[FeHL(depe)$_2$]$^+$BPh$_4^-$ series, where some of the L are π acceptors, it is probably more informative to discuss variations in C.S. from compound to compound rather than derive p.c.s. for ligands.

b. Bonding Properties from C.S. and Q.S. Data. As discussed previously, p.c.s. values decrease with increasing σ bonding and π back-bonding, while the p.q.s. values become more positive with increasing π back-bonding, but more negative with increasing q_{lat}. Thus *(44)*,

$$\text{p.c.s.} = k(\sigma_L + \pi_L) \tag{49}$$

$$\text{p.q.s.} = q_{lat} + c(\pi_L' - \sigma_L') \tag{50}$$

These two equations provide a very useful method for at least qualitatively characterizing both the σ-donor and π-acceptor properties of ligands. As expected from the above equations, H^- and NO^+, the best σ-donor and π-acceptor ligands, respectively, give the most negative p.c.s. values of all ligands (Table XLIX), but have the most negative and positive p.q.s. values, respectively (Table XLVI). Considering some of the neutral ligands, the p.c.s. values show that $\sigma + \pi$ increases in the order $H_2O < NH_3 < \text{depe}/2 \sim \text{depb}/2 < \text{dmpe}/2 < \text{ArNC} < \text{CO}$, whereas the p.q.s. values show that $\pi-\sigma$ increases in the order $\text{RNC} < \text{dmpe}/2 < \text{depe}/2 < \text{depb}/2 < NH_3 < H_2O \sim \text{CO}$ (Fig. 15). For the first five ligands, the σ-donor properties probably dominate both the p.c.s. and p.q.s. values, and a correlation of p.c.s. *vs.* p.q.s. is observed (Fig. 15). However, RNC and CO (and NO^+) lie to the left of the line, because the π-acceptor properties are becoming more important in the order $\text{RNC} < \text{CO} < NO^+$. In the extreme case where the π-acceptor ability dominates both the p.c.s. and p.q.s. values, a line of opposite slope to that observed for the "σ-bonding" ligands would be observed.

For charged ligands, it is more difficult to evaluate σ and π because of possible q_{lat} effects.

Similarly, for the *trans*-[FeHL(depe)$_2$]$^+$BPh$_4^-$ series of compounds, a plot of C.S. *vs.* Q.S. (Fig. 15) gives a reasonable correlation for all ligands but CO and N_2. The slope of the line indicates that, except for CO and N_2, the σ-donor properties are dominant in determining both the C.S. and Q.S. The C.S. data indicate that $\sigma + \pi$ increases in the order: $\text{MeCN} < \text{PhCN} < N_2 < P(OPh)_3 < P(OMe)_3 < Me_3CNC < p\text{-MeO} \cdot C_6H_4 \cdot$

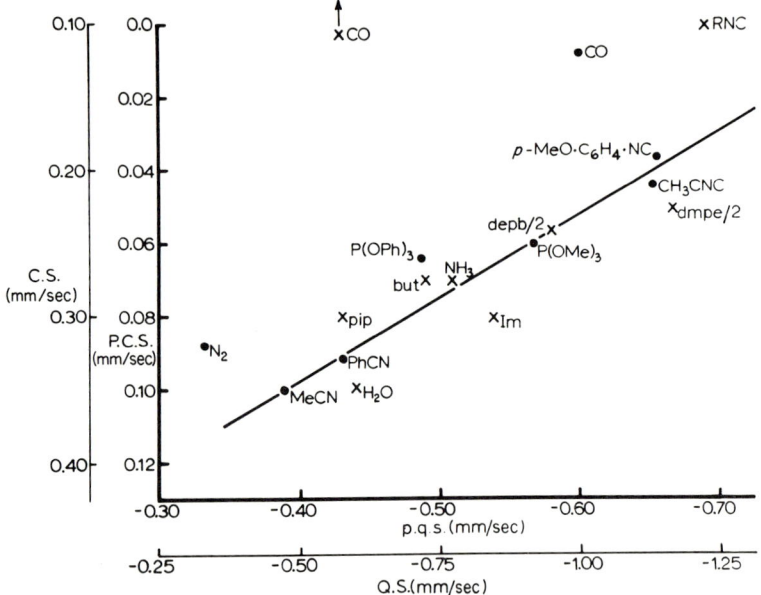

Fig. 15. Plots of (a) C.S. vs. Q.S. for [FeHL(depe)₂] BPh₄ compounds (denoted ●); (b) p.q.s., vs. p.q.s. for a number of neutral ligands (Tables XLIV, XLVI, XLVIII, and XLIX) (denoted ×).

NC < CO, while π–σ (from the Q.S.) increases in the order p-MeO·C_6H_4·NC < Me_3CNC < CO < P(OMe)$_3$ < P(OPh)$_3$ < PhCn < MeCN < N_2. The correlation for all ligands but CO and N_2 again indicates that N_2 is a very poor (relatively) σ donor and a moderate π acceptor, whereas CO is a good π acceptor and moderately good σ donor (37).

C. Ru^{II} and Ir^{III} Compounds

It would seem likely that the p.c.s. and/or p.q.s. treatments might be applicable to other analogous t_{2g}^6 systems such as Co^{III}, Ru^{II}, and Ir^{III}. Although Co does not have a Mössbauer nuclide, NQR data on Co^{III} compounds can be compared with Mössbauer data for Fe^{II} low-spin compounds discussed previously to obtain the sign of the Q.S. for Co^{III} compounds (22). The sign of the Co Q.S. cannot be obtained from NQR. Using the p.q.s. values derived from Fe^{II} low-spin compounds, the expected Q.S. for a Co^{III} species such as [Co(NH$_3$)$_4$Cl$_2$]$^+$ is given by

$$(e^2qQ)_{\text{Co cpd}} = \frac{e^2q_{3d(\text{Co})}Q_{\text{Co}}}{e^2q_{3d(\text{Fe})}Q_{\text{Fe}}} (e^2qQ)_{\text{Fe cpd}} \qquad (51)$$

assuming that the bonding in the analogous Co and Fe compounds is the same. For example, the expected Q.S. for a low-spin $trans$-Fe(NH$_3$)$_4$Cl$_2$ species is 0.84 mm/sec, and the Q.S. for $trans$-[Co(NH$_3$)$_4$Cl$_2$]$^+$ should also be positive, instead of negative as calculated previously (517). Q.S. values for Fe compounds should be about one-third that of the Co

TABLE L

MÖSSBAUER PARAMETERS FOR Ru^{II} COMPOUNDS AT 4.2°K[a]

Code No.[b]	Compound	C.S.[c]	Q.S.	Ref.
1	Ru(NH$_3$)$_6$Cl$_2$	−0.82	—	(470, 472)
2	[Ru(NH$_3$)$_5$CH$_3$CN](ClO$_4$)$_2$	−0.71	—	(472)
3	[Ru(NH$_3$)$_5$N$_2$]Cl$_2$	−0.62	0.20	(472)
4	[Ru(NH$_3$)$_5$N$_2$]X$_2$[d]	−0.81	0.26	(470)
5	[Ru(NH$_3$)$_5$SO$_2$]Cl$_2$	−0.61	0.30	(472)
6	[Ru(NH$_3$)$_5$CO]Br$_2$	−0.54	—	(472)
7	[Ru(NH$_3$)$_5$NO]Cl$_3$·nH$_2$O	−0.18	0.36	(294, 470, 472)
8	[Ru(NH$_3$)$_4$(SO$_2$)Cl]Cl	−0.50	0.56	(470)
9	[Ru(NH$_3$)$_4$(NO)(OH)]X$_2$[d]	−0.16	0.27	(470)
10	K$_4$[Ru(CN)$_6$]	−0.24	—	(129, 362)
11	K$_2$[Ru(CN)$_5$NO]·2H$_2$O	−0.06	0.42	(129, 294, 470)
12	K$_4$[Ru(CN)$_5$NO$_2$]	−0.37	0.28	(129, 470)
13	M$_2$[RuCl$_5$(NO)][d]	−0.37	0.21	(294, 470)
14	Cs$_2$[RuBr$_5$(NO)]	−0.47	0.08	(294)
15	Rb$_2$[Ru(NCS)$_5$(NO)]	−0.30	0.24	(294)

[a] Data given in mm/sec. C.S. and Q.S. values quoted are averages of available data.
[b] Code number will be preceded by table number in text.
[c] Relative to Ru metal at 4.2°K.
[d] X = Cl, Br and/or I; M = K, Rb.

compounds (mainly due to $Q_{Co} \sim 0.4$ barns, and $Q_{Fe} \sim 0.2$ barns), and it should be possible to use the above equation to calculate approximate e^2qQ values for either Fe or Co compounds, if the e^2qQ value for the corresponding Co or Fe compound is known, or can be predicted from the p.q.s. values in Table XLVI (22).

For Ru^{II} and Ir^{III}, the bonding characteristics of ligands cannot be assumed identical in the analogous Ru^{II}, Ir^{III}, and Fe^{II} compounds, but it should still be possible to use the relative p.q.s. values in Table XLVI to predict the signs and relative magnitudes of Q.S. values in Ru^{II} and Ir^{III} compounds containing the ligands in Table XLVI. Recent data for Ru^{II} compounds (Table L) provides an excellent framework to discuss the possibilities of this idea. It should be recognized, however, that the Ru^{II} quadrupole splittings are rather small and have relatively large errors.

For example, the averaged Q.S. values for $K_2[Ru(CN)_5NO] \cdot 2H_2O$ (*129, 294, 470*) and $K_4[Ru(CN)_5NO_2]$ (*129, 470*) are 0.42 and 0.28 mm/sec, respectively, while the corresponding Fe compounds have quadrupole splittings of +1.72 and 0.85 mm/sec (most likely positive), respectively. The ratios of the splittings are similar, and it would seem most likely that the signs of the Ru Q.S. values are positive. From the Fe^{II} p.q.s. values for NO^+, Br^-, Cl^-, NCS^-, and CN^- (Table XLVI), we would predict that

TABLE LI

Q.S. VALUES FOR
$XYIr^{III}Cl(CO)(PPh_3)_2$ COMPOUNDS[a]

X	Y	Q.S.[b]
H	H	4.76
H	Cl	1.44
Cl	Cl	3.10

[a] From Ref. (*562*).
[b] Data given in mm/sec.

the signs of the Q.S. in the series $[Ru(X_5)NO]^{2-}$ (X = Br^-, Cl^-, NCS, CN^-) (Table L, compounds 11, 13–15) are positive, and that the magnitudes would vary in the observed order X = $CN^- > NCS^- > Cl^- > Br^-$. Similarly, with the ammonia derivatives (Table L, compounds 1–9), we would expect a large positive Q.S. for $[Ru(NH_3)_5NO]^{3+}$, and a very small positive Q.S. for $[Ru(NH_3)_5CO]^{2+}$, because of the very similar p.q.s. values for NH_3 and CO (Table XLVI). In addition, it seems likely that the sign of the Q.S. in $[Ru(NH_3)_5N_2]Cl_2$ is positive, since N_2 should have a more positive p.q.s. than NH_3 from the Fe^{II} low-spin work. This positive sign would be expected since N_2 is likely a poorer σ donor and a better π acceptor than NH_3.

Q.S. values reported recently for Ir^{III} compounds (Table LI) (*562*) should obey the additivity model reasonably well, and from the Fe^{II} p.q.s. values (Table XLVI), it would seem possible that the H,H and Cl,Cl derivatives have negative and positive V_{zz}, respectively, with the H,Cl derivative probably having a negative V_{zz} (assuming trans X and Y groups).

Little systematic C.S. work on series of Ru and Ir compounds has yet been carried out, but recent interesting Ru Mössbauer studies indicate that the order of C.S. values is similar to that expected from the Fe^{II} low-spin work. The C.S. values for the series $[Ru(NH_3)_5X]$ (Table L, compounds 1–7) decrease in the order $NO^+ > CO > SO_2 > N_2 > CH_3CN > NH_3$, Although attributed to a decrease in π-acceptor properties of X (470, 472, 473), the above order probably reflects a decrease in $(\sigma + \pi)$ of these ligands. Thus, as concluded from the $[Fe^{II}HL(depe)_2]^+BPh_4$ series, N_2 is a poorer $(\sigma + \pi)$ ligand than CO, but slightly better than nitriles. Combined with the Q.S. data, N_2 appears to be a better π acceptor, but poorer σ donor than nitriles. This result is also consistent with that deduced from the Fe^{II} low-spin data (37). As with the Fe^{II} low-spin compounds, there is a correlation of the C.S. values for the $[RuX_5NO]^{2-}$ compounds (Table L, compounds 11, 13–15) with the spectrochemical ranking of the X ligands (294).

D. Iodine Compounds

1. General Principles

Mössbauer spectra of iodine compounds may be obtained by utilizing either the 57.6 keV $\frac{7}{2} + \rightarrow \frac{5}{2} +$ transition of ^{127}I or the 27.7 keV $\frac{5}{2} + \rightarrow \frac{7}{2} +$ transition of ^{129}I. Unfortunately, both isotopes present problems. In the case of ^{127}I, the resolution of the spectra is often poor, and detailed computer analysis is needed to extract meaningful parameters; although the resolution of ^{129}I spectra is excellent, the necessity of synthesizing all absorbers from active ^{129}I presents a formidable experimental inconvenience. The majority of work published so far deals with ^{129}I spectra, although considerable effort has also been given to ^{127}I. For both isotopes the spins of the ground and excited states allow the determination of the sign and magnitude of e^2qQ, as well as η and the center shift. The values of e^2qQ and η are, of course, complimentary to values available from ^{127}I NQR spectroscopy, but the extra parameters (i.e., the center shift and the sign of e^2qQ) available from Mössbauer spectra often enable a more meaningful analysis of the data.

In this section we will discuss the general principles used in the interpretation of the iodine Mössbauer spectra and in the following section we will survey the experimental data.

The quadrupole coupling parameter e^2qQ and η are usually analyzed using the Townes–Dailey theory (163, 540) developed for the interpretation of NQR spectra. The basic assumption of this theory is that the major contributions of the EFG arise from an aspherical distribution of the electronic charge of the valence shell. Neglecting contributions from 5d electrons and cross-terms, q for an iodine atom is given by Eq. (9), and e^2qQ is given by:

$$e^2qQ = e^2QK_p\,[-N_{p_z} + \tfrac{1}{2}(N_{p_x} + N_{p_y})] \tag{52}$$

where Q is the quadrupole moment of ^{129}I and is negative. The term $[-N_{p_z} + \tfrac{1}{2}(N_{p_x} + N_{p_y})]$ is often referred to as U_p and, hence,

$$e^2qQ = e^2QK_p U_p \tag{53}$$

The asymmetry parameter is given by the expression

$$\eta = \tfrac{3}{2}(N_{p_x} - N_{p_y})/-U_p \tag{54}$$

For an iodine atom with a single I–A bond, e^2qQ may be related to bonding parameters. Thus, the I–A bond can be described by a molecular orbital (ψ) formed by a linear combination of atomic orbitals of I(ϕ_I) and A(ϕ_A). This molecular orbital may be written

$$\psi = \frac{a\phi_I + b\phi_A}{(a^2 + b^2 + 2abS)^{1/2}} \tag{55}$$

where $S = \int \phi_I \phi_A\, \delta\Gamma$. The parameters a and b determine the relative importance of ϕ_I and ϕ_A in the molecular orbital and are often expressed in terms of the ionic character I.

$$I = \frac{1 - b^2/a^2}{1 + b^2/a^2} = \frac{a^2 - b^2}{a^2 + b^2} \tag{56}$$

Thus, if $a = b = 1$, the bond is completely covalent, whereas if $a = 0$, $I = -1$ and the bond is completely ionic in the sense I$^+$A$^-$. For $b = 0$, the bond is ionic in the sense I$^-$A$^+$.

The atomic orbital of ϕ_I may be written as a linear combination of the $5s$, $5p_z$, and $5d_{z^2}$ orbitals.

$$\phi_I = (1 - s^2 - d^2)^{1/2}\phi_{p_z} + s\phi_s + d\phi_{d_{z^2}} \tag{57}$$

while the involvement of $5s$ electrons in ϕ_I means that the $5s$ lone pair will be contained in an orbital of type

$$\phi_t = (1 - s^2)^{1/2}\phi_s + s\phi_{p_z} \tag{58}$$

The population of ϕ_I is given to a first approximation [i.e., neglecting $2abS$ (292)] by $2a^2/(a^2 + b^2)$ and, hence, N_{p_z} may be written as

$$N_{p_z} = \frac{2a^2}{a^2 + b^2}(1 - s^2 - d^2) + 2s^2 \tag{59}$$

From the definition of I (Eq. 56),

$$I + 1 = \frac{2a^2}{a^2 + b^2} \tag{60}$$

and, hence,

$$N_{p_z} = 1 + I + s^2 - d^2 - I(s^2 + d^2) \tag{61}$$

The term $I(s^2 + d^2)$ is normally neglected and, hence,

$$N_{p_z} = 1 + I + s^2 - d^2 \tag{62}$$

In the absence of π bonding, N_{p_x} and N_{p_y} will be equal to 2, but if π bonding removes π electrons from the p_x and p_y orbitals, then

$$N_{p_y} = N_{p_x} = 2 - \pi \tag{63}$$

Substitution from Eqs. (62) and (63) into Eq. (52) yields

$$\frac{e^2qQ}{e^2QK_p} = 1 - I - \pi - s^2 + d^2 \tag{64}$$

From Eq. (64) the relationship of e^2qQ to the iodine bonding may be assessed. Thus, e^2qQ will have a maximum negative value ($q = +ve$) for an I^+ ion ($a^2 = 0$, $I = -1$) and a minimum value for an I^- ion ($b^2 = 0$, $I = 1$). In the case of I_2 ($a^2 = b^2$, $I = 0$), V_{ZZ} will be equal to $K_p(1 - s^2 + d^2)$ and, hence, in the limiting case of pure p bonding with $\pi = 0$, e^2qQ will be equal to e^2QK_p. The quantity e^2QK_p has been found to be equal to -2293 Mc/sec (*354*) and this value allows $1 - I - \pi - s^2 + d^2$ (U_p) to be expressed in number of electrons.

The type of argument outlined above may be extended to more complex iodine bonding and these situations will be discussed as they arise.

The center shifts of iodine compounds will depend directly on the number of $5s$ electrons and indirectly (via shielding effects) on the number of $5p$ electrons. If h_s and h_p are the magnitudes of the s and p electron holes in the closed I^- valence shell, then the center shift with respect to an arbitrary source may be written (*310, 452*)

$$\text{C.S.} = K[-h_s + \gamma(h_p + h_s)(2 - h_s)] + S \tag{65}$$

where the term h_s represents the direct contribution from the loss of $5s$ electrons and the second term in the square brackets represents the increased s density at the nucleus resulting from the change in h_p and h_s. S is the center shift of the source from I^- and the constant K will depend on the value of $\Delta R/R$. Assuming $h_s \ll 1$ and, hence, neglecting terms in h_s^2 and $h_s h_p$, Eq. (65) reduces to

$$\text{C.S.} = Ah_s + Bh_p + C \tag{66}$$

where $A = K - 2K\gamma$ and $B = 2K\gamma$.

For compounds in which the iodine atoms have single or two colinear bonds, a relationship between center shift and e^2qQ may be derived (203). Thus, neglecting π bonding,

$$h_p = 6 - (N_{p_z} + N_{p_x} + N_{p_y}) = |U_p| = 2 - N_{p_z} \tag{67}$$

and as

$$N_{p_z} = (1 + s^2 - d^2 + I) \quad [\text{Eq. (62)}]$$
$$h_p = (1 - s^2 + d^2 - I) \tag{68}$$

If the effect of d-orbital participation is also neglected, expressions for h_p and h_s may be derived from the orbitals ϕ_l [Eq. (57)] and ϕ_t [Eq. (58)] which will have populations of $2a^2$ and 2 electrons, respectively. Thus,

$$h_p = 2 - 2a^2(1 - s^2) - 2s^2 = s^2(2a^2 - 2) + 2 - 2a^2 \tag{69}$$

and

$$h_s = 2 - 2a^2 s^2 - 2(1 - s^2) = s^2(2 - 2a^2) \tag{70}$$

and, hence,

$$h_p + h_s = 2 - 2a^2 \tag{71}$$

and

$$\frac{h_s}{h_p + h_s} = \frac{s^2(2 - 2a^2)}{(2 - 2a^2)} = s^2 \tag{72}$$

The equations derived above allow, at least in principle, the determination of the parameters h_p, h_s, s^2, and I from a combination of quadrupole coupling and center shift data and, thus, provide a framework for the discussion of the iodine Mössbauer spectra.

2. Experimental Data

The majority of the available ^{127}I and ^{129}I Mössbauer data are summarized in Table LII. All values of e^2qQ have been expressed relative to the quadrupole moment of ^{127}I, and values of U_p have been calculated assuming $e^2QK_p = -2293$ Mc/sec. Also included in Table LII are some NQR data; in general, the agreement with the Mössbauer results is excellent, especially as the Mössbauer and NQR measurements are often taken at different temperatures.

One of the most important tasks in the interpretation of the data in Table LII is the calibration of the center shifts in terms of h_p and h_s. This was first attempted by Hafemeister et al. (310) for ^{129}I using the data for the alkali iodides (Table LII, compounds 1–5). Independent estimates of h_p for these species are available from NMR (68) and dynamic quadrupole coupling (408) measurements, and the variations of h_p reflect those of the center shifts. From these data, Hafemeister et al. (310) concluded that $\Delta R/R$ for ^{129}I is 3×10^{-5}. Pasternak et al. (443), assuming pure p bonding, took the center shifts of crystalline I_2 (Table LII, compound 9) to represent $h_p = 1$. Combination of the data for I_2 with those of the alkali iodides, gives the equation

$$\text{C.S.}(\text{ZnTe})^{129} = 0.136 h_p - 0.054 \, (\text{cm/sec}) \qquad (73)$$

More recent measurements (89) of frozen solutions of iodine in hexane, carbon tetrachloride, and argon (Table LII, compounds 10–12) have shown increased center shifts compared with crystalline iodine. This increase in center shift is probably due to intermolecular association in the solid and, hence, the frozen solution data provide a better guide to the center shift for $h_p = 1$. Interpretation of frozen solution data modifies Eq. (73) to give

$$\text{C.S.}(\text{ZnTe})^{129} = 0.15 h_p - 0.054 \, (\text{cm/sec}) \qquad (74)$$

The effect of variations in h_s has been assessed (445) by consideration of the center shifts of the ions IO_3^-, IO_4^-, and IO_6^{5-} (Table LII, compounds 21–29). The iodine atoms in the ions IO_4^- and IO_6^{5-} have a formal oxidation of VII and are situated in regular tetrahedral and octahedral environments. In contrast, IO_3^- contains an I^V atom situated on top of a pyramid with an O–I–O angle close to 90°. Assuming pure p bonding in IO_3^-, Pasternak and Sonnino (445) deduced from Eq. (73) that $h_p = 1.54$, which can be interpreted in terms of the removal of 0.51 electrons from the I^- configuration per I–O bond. The value of A in Eq. (66) was then deduced from the center shifts of IO_4^- and IO_6^{5-} assuming that $s^2 = \frac{1}{4}$ and $\frac{1}{6}$, respectively, and that each I–O bond withdraws 0.51 electrons. A mean value of $A = -0.82$ was obtained and Eq. (74) may be extended to give

$$\text{C.S.}(\text{ZnTe})^{129} = -0.92 h_s + 0.15 h_p - 0.054 \, (\text{cm/sec}) \qquad (75)$$

where the original formula of Pasternak and Sonnino (445) has been modified as suggested by Bukshpan et al. (89). Equation (75) shows that the effect on the center shift of removing an s electron is of opposite sign and six times as large as that of removing a p electron.

The first calibration of the center shift scale for ^{127}I was by Perlow and Perlow (452). These authors, assuming pure p bonding in the interhalogen compounds, used the quadrupole coupling data of HI, ICl, $KICl_4 \cdot H_2O$, and $KICl_2 \cdot H_2O$ (Table LII, compounds 13, 14, 18, 20)

TABLE LII
Mössbauer Data for Some Iodine Compounds

Code No.[a]	Compound[b]	Isotope	Mössbauer data					NQR data		
			C.S.[c]	e^2qQ^d	η	$U_p{}^e$	Ref.[f]	e^2qQ^d	η	Ref.[g]
1	LiI	129	−0.38[h]	—	—	—	(310)	—	—	—
2	NaI	129	−0.46[h]	—	—	—	(310)	—	—	—
3	KI	127	+0.14[i]	—	—	—	(452)	—	—	—
		129	−0.51[h]	—	—	—	(310)	—	—	—
		127	+0.14[i]	—	—	—	(452)	—	—	—
4	RbI	129	−0.43[h]	—	—	—	(310)	—	—	—
5	CsI	129	−0.37[h]	—	—	—	(310)	—	—	—
		127	+0.12[i]	—	—	—	(452)	—	—	—
6	AgI	129	−0.22[h]	—	—	—	(419)	—	—	—
7	CuI	129	−0.38[h]	—	—	—	(153)	—	—	—
8	HI(aq.)	127	+0.16[i]	—	—	—	(452)	—	—	—
9	I$_2$	127	−0.58[i]	−2238[i]	0.12	0.98	(452)	2153[k]	0.15[k]	(173)
		129	+0.83[h]	−2156[h]	0.16	0.94	(445)	—	—	—
10	I$_2$/hexane	129	+0.98[l]	−2263[l]	—	0.99	(89)	—	—	—
11	I$_2$/CCl$_4$	129	+0.91[l]	−2273[l]	—	0.99	(89)	—	—	—
12	I$_2$/argon	129	+0.93[m]	−2231[m]	—	0.97	(89)	—	—	—
13	HI(anhydrous)	127	−0.62[i,j]	−1640[i]	—	0.72	(452)	−1831[n]	—	(144)
14	ICl	127	−0.62[i]	−2868[i]	0.06	1.25	(452)	3008[o]	0.02[o]	(371)
		129	+1.73[h]	−3131[h]	0.06	1.36	(445)	−3046[h]	0.03[h]	(571)
15	IBr	129	+1.23[h]	−2892[h]	—	1.26	(445)	—	—	—
16	ICN	129	+1.19[p]	−2640[p]	0.06	1.15	(446)	−2549[k]	—	(105)
17	I$_2$Cl$_6$	129	+3.50[h]	+3060[h]	—	−1.33	(445)	3035[h]	0.08[h]	(571)
18	KICl$_4$·H$_2$O	127	−1.39[i]	+3094[i]	—	−1.35	(452)	3059[q]	—	(571)
19	I$_2$Cl$_4$Br$_2{}^r$	129	+3.48	+3040[h]	0.06	−1.33	(445)	—	—	—
			+2.82	+2916[h]	0.06	−1.27	(445)	—	—	—

#	Compound							
20	$KICl_2 \cdot H_2O$	127	-0.58^i					
21	KIO_3	129	$+1.56^h$			1.39		
22	NH_4IO_3	129	$+1.31^h$					
23	$Ba(IO_3)_2$	129	$+1.11^h$	$+1030^h$	(452)	-0.45		
24	$NaIO_3$	127	-0.44^i	$+1089^{i,s}$	(310)	-0.47		
25	$NaIO_3 \cdot H_2O$	127	-0.41^i	$+1107^{i,s}$	(310)	-0.48		
26	$Na_2Mn(IO_3)_6$	127	-0.55^i	$+1080^{i,s}$	(310)	-0.47		
27	KIO_4	129	-2.34^h		(361)			
28	$NaIO_4$	127	$+0.70^i$		(361)			
29	$Na_3H_2IO_6$	127	$+0.85^i$		(361)			
30	GeI_4	127	$+1.02^i$		(310)			
31	SiI_4	129	$+0.48^i$	-1500^i	(452)	0.65	$1492^{h,u}$	(493)
32	SnI_4	129	$+0.26^i$	-1335^i	(361)	0.58	$1329^{h,u}$	(493)
33	CI_4	129	$+0.43^i$	-1364^i	(361)	0.59	$1389^{h,u}$	(493)
34	CHI_3	129	$+0.20^i$	-2102^i	(88)	0.92	2130^h	(493)
		127	-0.35^i	-2160^i	(88)	0.94		
		129	$+0.53^i$	-2029^i	(91)	0.88		
		129	$+0.52^i$	-2065^i	(93)	0.90		
35	CH_2I_2	127	-0.21^i	-2060^i	(293)	0.90	$2056^{h,v}$	(493)
36	CH_3I	129	-0.14^i	-1920^i	(93)	0.84		
		129	$+0.20^i$	-1739^i	(203)	0.76		
		127	-0.01^i	-1775^i	(203)	0.77		
37	IF_5	129	$+3.00^w$	$+1073^w$	(93)	-0.47	1897^h	(493)
38	IF_7	129	-4.56^w	-148^w	(203)	0.06	1766^h	(493)
39	$IF_6^+ AsF_6^-$	129	-4.68^i		(90)			
40	$CsIF_6$	129	$+2.45^i$	-1414^i	(90)	0.62		
41	I_2/benzene	129	$+0.76^i$	-2412^i	(92)	1.05		
42	$I_2 \cdot PH$	129	$+0.93^i$	-2223^i	(92)	0.97		
43	$I_2 \cdot AC^r$	129	$+1.64^i$	-2840^i	(89)	1.24		
			$+0.29^i$	-1308^i	(350)	0.57		
44	I_2HMTA^r	129	$+1.51^i$	-2582^i	(350)	1.13		(571)
			$+0.28^i$	-1272^i	(350)	0.55		(392)

continued

TABLE LII—continued

MÖSSBAUER DATA FOR SOME IODINE COMPOUNDS

Code No.[a]	Compound[b]	Isotope	Mössbauer data					NQR data		
			C.S.[c]	e^2qQ[d]	η	U_p[e]	Ref.[f]	e^2qQ[d]	η	Ref.[g]
45	2(IBr)2,2′-bipy	129	+1.35[i]	−2910[i]	0.15	1.27	(570)	—	—	—
46	2(IBr)4,4′-bipy	129	+1.48[i]	−3110[i]	—	1.36	(570)	—	—	—
47	IBr·py	129	+1.61[i]	−3224[i]	—	1.41	(570)	—	—	—
48	2(ICl)2,2′-bipy	129	+1.90[i]	−3424[i]	—	1.49	(570)	—	—	—
49	2(ICl)4,4′-bipy	129	+1.43[i]	−3428[i]	—	1.49	(570)	—	—	—
50	ICl·py	129	+1.74[i]	−3310[i]	—	1.44	(570)	—	—	—
51	ICl·PMT	129	+1.90[i]	−3395[i]	—	1.48	(570)	—	—	—
42	PhI	127	−0.27[i]	−1840[i]	—	0.80	(205)	—	—	—
53	PhICl$_2$	127	−0.78[i]	+2525[i]	0.70	−1.10	(205)	—	—	—
54	PhIO	127	−0.73[i]	+2345[i]	0.60	−1.02	(205)	—	—	—
55	PhIO$_2$	127	−0.75[i]	+1050[i]	0.18	−0.46	(205)	1046[x]	0.17[x]	(392)
56		127	−0.83[i]	+2585[i]	0.92[i]	−1.13	(205)	—	—	—
57	CsI$_3$[y]	129	+1.40[i]	−2500[i]	—	1.09	(204)	2477[h]	—	(502)
				−1460[i]	—	0.64	—	1436[h]	0.04	—
				−830[i]	—	0.36	—	819[h]	0.02	—
58	Benz·HI$_3$[r]	129	+1.33[i]	−2460[i]	—	1.07	(204)	—	—	—
				−1180[i]	—	0.51	—	—	—	—
59	Am·HI$_3$[r]	129	+1.31[i]	−2450[i]	—	1.07	(204)	—	—	—
				−930[i]	—	0.41	—	—	—	—

60	CrI$_3$	129	+0.23^h				(285)
61	SaI$_3$	129	−0.25^h				(153)
62	GdI$_3$	129	−0.38^h				(153)
63	ErI$_3$	129	−0.29^h	+662^h	0.35	−0.29	(153)

^{a} Code number will be prefixed by Table number in the text.
^{b} PH = Phenazine; AC = acridine; HMTA = hexamethyltetramine; 2,2′bipy = 2,2′bipyridyl; 4,4′-bipy = 4,4′-bipyridyl; py = pyridine; PMT = pentamentylene tetrazole; Benz = benzamide; Am = amylose.
^{c} Data given in mm/sec relative to ZnTe.
^{d} Mc/sec for ^{127}I ground state, values quoted relative to ^{129}I have been converted assuming $Q(129)/Q(127) = 0.701$.
^{e} Calculated assuming $eQKp = −2292.7$ (354).
^{f} Reference to Mössbauer data.
^{g} Reference to NQR data.
^{h} At 77°K.
^{i} At 4°K.
^{j} Converted from H$_6$TeO$_6$ source by addition of +0.95 mm/sec. This conversion is an average of the data for compounds, which were measured relative to both sources in Ref. (458).
^{k} At 300°K.
^{l} At 88°K.
^{m} At 22°K.
^{n} Data for gaseous compound obtained by microwave spectroscopy.
^{o} At 295°K.
^{p} At 100°K.
^{q} At 298°K.
^{r} Spectrum shows two nonequivalent iodine atoms.
^{s} Converted from mm/sec.
^{t} At 85°K.
^{u} Average of two frequencies.
^{v} Spectrum of CHI$_3$ · 3S$_8$.
^{w} At 90°K.
^{x} At 274°K.
^{y} Spectrum shows three inequivalent iodine atoms.

to obtain values of h_p. Thus, for the species $KICl_2 \cdot H_2O$, ICl, and HI, assuming no π bonding, Eq. (67) gives

$$h_p = |U_p| \qquad (76)$$

whereas for $KICl_4 \cdot H_2O$ which is square-planar

$$h_p = |2U_p| \qquad (77)$$

For I_2, h_p was taken as 1.11 to allow for an 11% admixture of p^4d character (493) in the bonding. These values of h_p gave a good linear relationship with center shift, from which it was deduced that $B = 2K\gamma = -0.56$ mm/sec/h_p. Thus,

$$C.S.(I^-)^{127} = -0.56 h_p \, (\text{mm/sec}) \qquad (78)$$

Then, taking $\gamma = 0.1$, $B = -K + 2K\gamma = 2.2$ mm/sec/h_s, and Eq. (78) becomes

$$C.S.(I^-)^{127} = 2.2 h_s - 0.56 h_p \, (\text{mm/sec}) \qquad (79)$$

From Eq. (79) it will be noted that $\Delta R/R$ for ^{127}I is negative and, thus, has an opposite sign to that for ^{129}I. Ruby and Shenoy (498) have calculated the ratio of $\Delta R/R(^{127}I)/\Delta R/R(^{129}I)$ to be -0.78, while a combination of the results of Pasternak et al. (443) and Perlow and Perlow (452) yields a ratio of -0.85 (498).

The combination of center shift and quadrupole coupling data for an iodine molecule provides a good means of determining the degree of s hybridization in the iodine bonds. Thus, for linear or square-planar arrangements of the bonds (neglecting π bonding), values of h_p calculated from U_p data [Eqs. (76) and (77), together with Eqs. (74) and (78)] should give a reasonable prediction of center shift. However, as the center shift caused by an s-electron hole is six times the shift due to a p-electron hole, even a small degree of s hybridization will lead to large inaccuracies in using Eqs. (74) and (78) to calculate center shifts. In Table LIII are given calculated center shifts for the ^{129}I spectra of IBr, I_2, ICN, ICl, and I_2Cl_4, together with the observed values. In all cases, save ICl, good agreement is obtained confirming pure p bonding (445, 446). The parameters for ICl may indicate some degree of π bonding. However, the extent of the π bonding required ($\pi = 10\%$) seems far too large, and Pasternak and Sonnino (445) have concluded that the value of U_p for ICl is not a good measure of the p-electron density.

Ehrlich and Kaplan (205) have made a detailed study of the bonding in the species in Table LII, compounds 52–56. The crystal structure (16) of $PhICl_2$ shows a linear ICl_2 group at an angle of $85°$ to the phenyl plane. The bonding of this compound may be described either by

dsp^3 hybridization with two equatorial lone pairs or by a p-bonding scheme utilizing a three-center two-electron molecular orbital between the iodine and two chlorine atoms. The former bonding scheme, which would involve a minimum value of $h_s = 0.37$ (205) is not compatible with either the center shift or the values of e^2qQ and η found for $PhICl_2$ and must, therefore, be rejected. Assuming, as required for a p-bonding

TABLE LIII

CALCULATED AND OBSERVED CENTER SHIFTS FOR SOME IODINE COMPOUNDS

Compound[a]	Isotope	Obs. C.S.[b]	h_p[c]	Calc. C.S.	π
IBr	129	+1.23	1.26	+1.35[d]	—
ICl	129	+1.73	1.36	+1.50[d]	10[g]
ICN	129	+1.19	1.15	+1.19[d]	—
I_2Cl_6	129	+3.50	2.66	+3.45[d]	—
PhI	127	−0.27	0.80	−0.29[e], −0.24[f]	—
$PhICl_2$	127	−0.78	1.91	−0.91[e], −0.80[f]	—
PhIO	127	−0.73	1.78	−0.84[e], −0.73[f]	—
[benzene-I(CO-O)₂ structure]	127	−0.83	1.98	−0.95[e], −0.83[f]	—
$PhIO_2$	127	−0.75	2.74	−1.37[e], −1.21[f]	—
I_2/benzene	129	+0.76	1.05	+1.04[d]	—
GeI_4	129	+0.48	0.65	+0.44[d]	—
SiI_4	129	+0.26	0.58	+0.33[d]	—
SnI_4	129	+0.43	0.59	+0.35[d]	—
2(IBr)2,2'-bipy	129	+1.35	1.27	+1.37[d]	4[g]
2(IBr)4,4'-bipy	129	+1.48	1.36	+1.50[d]	4[g]
IBrpy	129	+1.61	1.41	+1.58[d]	6[g]
2(ICl)2,2'-bipy	129	+1.90	1.49	+1.70[d]	7[g]
2(ICl)4,4'-bipy	129	+1.43	1.49	+1.70[d]	0[g]
IClpy	129	+1.74	1.44	+1.62[d]	4[g]
IClPMT	129	+1.90	1.48	+1.68[d]	10[g]

[a] 2,2'-bipy = 2,2'-Bipyridyl; 4,4'-bipy = 4,4'-Bipyridyl; py = pyridine; PMT = pentamentylene tetrazole.
[b] Data from Table LII, mm/sec relative to ZnTe.
[c] h_p derived from e^2qQ and Table LII as described in text.
[d] Using Eq. (74).
[e] Using Eq. (78), assuming center shift of I^- is +0.16.
[f] Assuming $2K\gamma = 0.50$ and center shift of I^- is +0.16.
[g] Reference (570).

scheme, a filled p_z orbital perpendicular to the molecular plane, e^2qQ and η are given (205) by the expressions

$$\frac{e^2qQ \,(\text{Mc/sec})}{2293 \times 1.15} = \frac{2525}{2637} = 2 - \tfrac{1}{2}(N_{p_x} + N_{p_y}) \tag{80}$$

$$\eta = 0.70 = \frac{\tfrac{3}{2}(N_{p_x} - N_{p_y})}{2 - \tfrac{1}{2}(N_{p_x} + N_{p_y})} \tag{81}$$

where the correction term 1.15 has been applied to e^2QK_p to allow for a change in $\langle r^{-3} \rangle$ resulting from the positive charge on the iodine atom. Solution of Eqs. (80) and (81) yields

$$N_{p_z} = 2.00, \qquad N_{p_x} = 1.27, \qquad N_{p_y} = 0.82 \tag{82}$$

from which it follows that $h_p = 1.91$. Similar analysis may be applied to the e^2qQ and η values of PhIO, PhIO$_2$, and iodosodilactone assuming a C–I–O bond angle of 90° in PhIO and T-shaped IO$_2$ groupings in both PhIO$_2$ and iodosodilactone; for iodobenzene $U_p = h_p$. Values of h_p and calculated center shifts using Eq. (78) are contained in Table LIII. In all cases except PhIO$_2$, good agreement with experimental center shift is obtained confirming pure p bonding. The discrepancy between the observed and calculated center shifts for PhIO$_2$ is probably due to s hybridization and a value of $h_s = 0.16$ (205) would be sufficient to account for the observed center shift.

The probability of pure p bonding in Table LII, compounds 52–54 and 56 provides a further means of calibrating the ^{127}I center shift scale. A plot of C.S. against h_p gives (205) a value of $2K\gamma = -0.50$ mm/sec/ h_p compared with -0.56 mm/sec/h_p from Eq. (78) and -0.47 mm/sec/h_p from Eq. (73) when converted to ^{127}I units.

No π bonding would be expected in the series CH$_n$I$_{4-n}$ ($n = 0$–3) (Table LII, compound 33–36) and, hence, it would be anticipated that $|U_p| = h_p$. The calculated center shifts for the ^{127}I spectra using Eq. (78) show an increasing deviation from the observed values as n increases, indicative of an increasing degree of s hybridization from $n = 0$ to $n = 3$. Erhlich and Kaplan (203) have calculated, using Eqs. (68)–(72), values of h_p, h_s, s^2, and I for these compounds, and the results are given in Table LIV. Calculations using both available values of $2K\gamma$ (205, 452) have been made as well as calculations using the ^{129}I data of Bukshpan and Sonnino (93). For CI$_4$ and CHI$_3$, values of s^2 from ^{129}I and ^{127}I agree with the uncertainty of the calibration constant, but for CH$_3$I a large discrepancy of experimental origin is observed. The data in Table LIV also show (as noted above) a trend to increasing s^2 and I as n increases from 0→3.

The calculated center shifts, using Eq. (74) assuming $U_p = h_p$, for the species MI_4 (M = Si, Ge, or Sn) (Table LII, compounds 30–32) are summarized in Table LIII. For GeI_4 the agreement with the experimental value is good, indicating pure p bonding; while for SiI_4, the data suggest a small degree of s hybridization. In the case of SnI_4, the calculated center shift is rather low, perhaps indicating a small degree of π bonding (91).

Wynter et al. (570) have reported ^{129}I Mössbauer spectra of the addition complexes of IBr and ICl with nitrogen bases (Table LII,

TABLE LIV

BONDING PARAMETERS FOR SOME IODOMETHANES[a]

Compound	2Kγ = −0.56 mm/sec				2Kγ = −0.50 mm/sec			
	h_p	h_s	$s^2\%$	$I\%$	h_s	$s^2\%$	$I\%$	$s^2\%$[b]
CI_4	0.94	0.004	0.4	6	−0.012	−1.3	7	0.6
CHI_3	0.90	0.034	3.6	6	0.022	2.4	8	1.8
CH_2I_2	0.84	0.046	5.2	11	0.036	4.1	12	—
CH_3I	0.77	0.071	8.5	14	0.066	7.9	15	4.4

[a] Taken from Ref. (203), see text for definition of parameters.
[b] Data from Ref. (93), recalculated in Ref. (203).

compounds 45–51). For the IBr complexes the p-orbital imbalance (U_p) increases linearly with the basicity of the amine, suggesting a linear N–I–Br linkage (570). The center shifts also increase with basicity, and calculated center shifts using Eq. (74) are in good agreement with the observed values confirming pure p bonding (Table LIII). A systematic relationship between Mössbauer parameters and basicity is not observed for the ICl derivatives, and there is also substantially worse agreement between observed and calculated center shifts (Table LIII). This is not altogether surprising as difficulty has been encountered in interpreting the Mössbauer data for ICl itself. Wynter et al. (570) have also calculated the percent π-bonding character (π) for the iodine bonds in these complexes. Thus, deriving U_p from e^2qQ and h_p from Eq. (74), a value of π may be calculated from the relationship

$$\pi = \tfrac{1}{3}(h_p - U_p) \qquad (83)$$

The values of π obtained are listed in Table LIII, but in view of the uncertainties in the calibration of the center shift, it is doubtful if they have more than qualitative significance.

An increase in e^2qQ and, hence, U_p is observed for an I_2 solution in benzene (Table LII, compounds 35–41) as compared with crystalline iodine or solutions of iodine in inert matrices (Table LII, compounds 9–12) (*89*, *445*). This is rather surprising as a transfer of electrons from the benzene ring to the iodine bond would be expected. Further, the calculated center shift, using Eqs. (74) and $h_p = U_p$ (Table LIII) is much higher than the experimental value and corresponds to $h_s =$ ca. 0.03. Bukshpan *et al.* (*89*) have calculated a value of $h_p = 0.87$ by substituting experimental values of center shift into Eq. (74) and have used this result to infer the transfer of 0.26 electrons from the ring to the iodine atoms. This conclusion is, however, of doubtful validity. Thus, use of Eq. (74) implies that $h_s = 0$ and, hence, $h_p = U_p$. This is clearly not the case for the I_2/benzene solution in which h_p from Eq. (74) is equal to 0.87, while $U_p = 1.05$.

The Mössbauer spectrum of $I_2 \cdot$ phenazine (Table LII, compound 42) (*350*) indicates only one type of iodine atom in agreement with the crystal structure (*544*), which shows a straight-chain configuration of alternating iodine and phenazine molecules resulting in equivalent N–I bonds. In contrast, I_2-acridine and I_2-hexamethylenetetramine (Table LII, compounds 43, 44) show two chemically inequivalent iodine atoms in the Mössbauer spectrum (*350*), indicative of only one N–I bond per iodine molecule. Ichiba *et al.* (*350*) have derived orbital populations for these complexes assuming pure p bonding.

In agreement with NQR data (*502*), Erhlich and Kaplan found evidence for three nonequivalent iodine atoms in the Mössbauer spectrum of CsI_3 (Table LII, compounds 35–37). Assuming a linear structure, the largest value of e^2qQ may be assigned to the terminal iodine atoms. A sum of U_p for all three types of iodine is remarkably close to 2, indicating pure p bonding. Inequivalent iodine atoms are also indicated in the species benzamide-HI_3 (Table LII, compound 58) and amylose-HI_3 (Table LII, compound 59), although it was not possible to resolve separate spectra for the two terminal atoms. These data are in agreement with the structural studies of Robin *et al.* (*475*, *492*).

The Mössbauer parameters of the fluoride species IF_7, IF_5, IF_6^-, and IF_6^+ (Table LII, compounds 37–40) (*90*, *92*) are of interest because of the very high oxidation states and coordination numbers. The structures of IF_5 (*521*) and IF_7 (*521*) have been determined by use of X-ray diffraction and are illustrated in Fig. 16, while a regular octahedral structure may be inferred for IF_6^+ (*117*).

Bukshpan *et al.* (*90*, *92*) have reported a detailed analysis of the

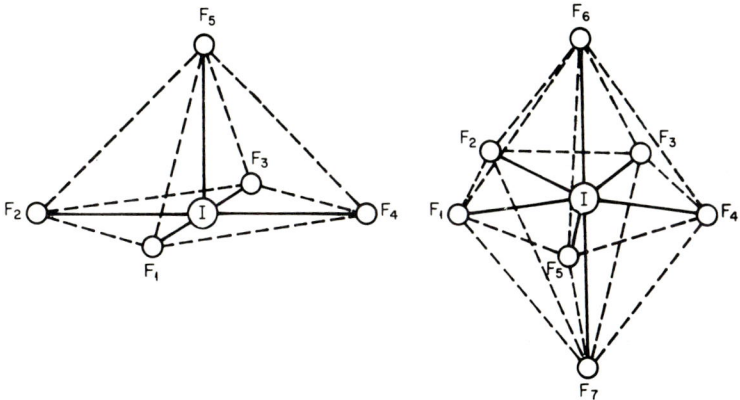

FIG. 16. Molecular structures of IF_5 and IF_7 (90).

Mössbauer parameters of these species. Thus, assuming that for IF_7 $N_{p_z} = 0$, it follows that $N_{p_x} + N_{p_y} = 2U_p = 0.13$. These populations give a value of $h_p = 5.87$ and from Eq. (75), $h_s = 1.56$.* The total number of electrons transferred from the iodine to fluorines is, therefore, 6.43, and each fluorine atom can be considered to have removed 0.92 electrons. If it is now assumed that the fluorine atoms also remove 0.92 electrons from the iodine atom in IF_5, the experimental center shift and U_p may be used to derive the parameters $h_p = 5.08$, $h_s = 0.5$, $N_{p_z} = 0.62$, $N_{p_x} = N_{p_y} = 0.3$. These results seem realistic as they indicate a lone pair directed along the C_4 axes as would be expected from the stereochemistry (90). The Mössbauer spectrum of IF_6^+ is a single line, as expected for an octahedral structure. From the experimental center shift, assuming $h_p = 6$, h_s equals 1.60, which corresponds to a loss of 5.60 electrons to the fluorine atoms, i.e., 0.93 electrons per fluorine atom. The spectrum of IF_6^- shows a positive value of U_p suggesting a similar structure to IF_7 with one of the equatorial fluorine atoms replaced by a lone pair. Assuming $N_{p_z} = 0$, $h_p = 4.76$, $h_s = 0.50$ and, hence, 0.88 electrons are removed per I–F bond (92).

E. Sn^{II} COMPOUNDS

The crystal structures of many tin(II) compounds [e.g., SnS (341), SnF_2 orthorhombic (188), $Na_2Sn_2F_5$ (394), K_2SnCl_4 (363), SnSe (429), $SnCl_2$ (499), and $SnSO_4$ (488)] show a distorted octahedral environment

* In this, and other calculations of h_s, the authors (90, 92) have used C.S. $(ZnTe)^{129} = -0.82h_s + 0.15 h_p - 0.054$ (cm/sec) instead of Eq. (75).

with three short tin–ligand bonds, resulting in pyramidal coordination of the tin atom, and three longer tin–ligand distances completing the octahedron. This situation is often considered (*182*) in terms of sp^3 hybridization in which three orbitals form covalent bonds to the nearest neighbor atoms and the fourth orbital contains the lone pair and prevents the close approach of the atoms along the direction in which it points. The pyramidal bond angles would be expected to be reduced by lone pair–bond pair repulsions and in the limiting case of pure p bonding, bond angles of ca. 90° are anticipated. It is, however, difficult to deduce any information about bonding characteristics from the experimental bond angles owing to both the partial ionic character of the bonding and the unknown effect of lattice forces (*182*). The crystal structure of tetragonal SnO (*409*) shows a square-pyramidal arrangement of the oxygen atoms suggesting $sp^3d_{z^2}$ hybridization resulting in four covalent bonds and a lone pair perpendicular to the plane.

As a result of the $5s^2$ electrons, a "perfect" Sn^{2+} ion would be expected to have a much greater center shift than gray α-tin. For Sn^{II} compounds, although the involvement of the 5s electrons in bonding will produce a reduction in center shift compared with an Sn^{2+} ion, the presence of a lone pair of electrons in an orbital of reasonably high 5s character should always ensure center shifts in excess of that of gray tin.

Table LV contains a collection of data for Sn^{II} compounds and, as anticipated, all center shifts are much larger than that of gray tin. From these data, attempts have been made to assess the relationship between bond character and center shift and, hence, obtain an estimate of the center shift of an Sn^{2+} ion ($\delta_{Sn^{2+}}$).

Early workers (*74, 137*) regarded $SnCl_2$ as ionic and took the center shift of this compound as a value of $\delta_{Sn^{2+}}$. However, as noted by Donaldson (*190*), the structure, volatility, and any estimate of the bonding of $SnCl_2$ are inconsistent with an ionic formalism. As a result, this estimate of $\delta_{Sn^{2+}}$ is probably too low.

An alternative approach by Lees and Flinn (*381*) uses a plot of quadrupole splitting against center shift. Most of the compounds considered lie on one of two straight lines, the gradients of which differ by a factor of two. It was assumed that the steeper of these two lines [$SnSO_4$, $SnCl_2 \cdot 2H_2O$, SnC_2O_4, SnF_2, $Sn_2P_2O_7$, $Sn(MeCO_2)_2$, $Sn(C_4H_4O_6)_2$, $Sn_3(PO_4)_2$, and $Sn_3(AsO_4)_2$], which shows an approximately linear increase in quadrupole splitting with decreasing center shift, refers to compounds in which sp_z hybridization is important, while the other slope ($SnBr_2$ and SnS) refers to compounds in which equal amounts of p_x and p_y character are used in hybridization. Extrapolation of both lines to zero quadrupole splitting gave a value of 4.76 mm/sec for $\delta_{Sn^{2+}}$. Recent work (182, 271), however, has shown that the EFG of compounds lying on both lines [SnF_2, $Sn_3(PO_4)_2$, SnC_2O_4, $Sn(O_2CMe)_2$, SnC_2O_4, SnS, and $SnSO_4$] have the same sign, thus rendering Lees and Flinn's interpretation untenable.

A third approach to the problem has been described by Donaldson and Senior (190). These authors plotted the center shift of the Sn^{II} halides and chalcogenides, corrected for shielding effects, against the percentage sp^3 character estimated from the electronegatives of Hinze

TABLE LV

MÖSSBAUER PARAMETERS FOR SOME TIN COMPOUNDS

Compound	C.S^{a,b,e}	Q.S.^{a}	Ref.^{c}	Ref.^{d}
SnF_2(monoclinic)	3.67	+1.73	(74, 137, 189, 190, 381)	(271)
SnF_2(orthorhombic)	3.30	2.18	(189)	—
$NaSnF_3$	3.17	+1.83	(191)	(183)
$NaSn_2F_5$	3.37	+1.86	(191)	(183)
$SnCl_2$	4.08	—	f	—
$SnCl_2 \cdot 2H_2O$	3.72	1.15	(137, 381)	—
$SnBr_2$	3.92	—	f	—
SnI_2	4.02	—	(182, 190, 303)	—
SnO(tetragonal)	2.80	+1.48	(74, 137, 190, 381)	(271)
SnO(orthorhombic)	2.70	2.20	(74)	—
SnS	3.41	+0.89	(74, 137, 190, 381)	(271)
SnSe	3.40	0.60	(190)	(271)
SnTe	3.33	0.50	(82, 190)	—
$Sn_3(PO_4)_2$	3.18	+1.75	(137, 381)	(271)
$Sn_2P_2O_7$	3.35	1.61	(381)	—
Sn_3AsO_4	2.97	2.03	(381)	—
$Sn(O_2CMe)_2$	3.33	+1.72	(184, 381)	(183)
SnC_2O_4	3.75	+1.46	(137, 381)	(271)
$SnC_4H_4O_6$	3.11	1.97	(381)	—
$Sn(HCO_2)_2$	3.15	+1.56	(184)	—
$K_2Sn(C_2O_4)_2 \cdot H_2O$	—	+1.97	(183)	(183)
$SnSO_4$	3.99	+1.21	(137, 183, 187, 381)	(183)

^{a} Data given in mm/sec.
^{b} Relative to SnO_2 assuming center shift of α-tin is 2.1 mm/sec and that of PdSn is 1.52; data from Ref. (381) have converted from Mg_2Sn (77°K) using the value of the center shift SnO_2 quoted in this reference.
^{c} Reference to measurement of C.S., Q.S.
^{d} Reference to measurement of sign of $\tfrac{1}{2}e^2qQ$.
^{e} When appropriate an unweighted average has been taken.
^{f} Only recent data included. Refs. (187, 194, 303, 185).

and Jaffe (*339, 340*). A straight-line relationship was obtained for all compounds save SnO and SnF_2, and extrapolation to zero sp^3 character yielded a center shift of 7.7 mm/sec for an Sn^{2+} ion. As mentioned previously (Section IV,A,2), this value is close to that expected from consideration of the probable center shift of an Sn^{4+} ion. Donaldson and Senior have suggested that the anomalously low center shifts of SnO and SnF_2 are due to extra sp mixing arising from crystal field effects. As such effects are very sensitive to the tin–anion bond length, they are only expected to be important for the oxide and fluoride.

The origins of the quadrupole splittings found for Sn^{II} compounds has been the subject of some discussion. The crystal structure of tetragonal SnO leaves little doubt that the lone-pair electrons are contained in a dsp^3 hybrid orbital perpendicular to the plane of the oxygen atoms. Because of the axial symmetry, the p contribution to the lone pair will be entirely p_z and, thus, a negative value of V_{ZZ} is anticipated. Boyle *et al.* (*74*) deduced from the asymmetry in the intensities of the spectrum of an oriented sample that the sign of $\frac{1}{2}e^2qQ$ for SnO was positive, thus allowing a negative sign to be assigned to Q for the excited state of ^{119}Sn.

For compounds with pyramidal structures the presence of a lone pair in an approximately sp^3-hybridized orbital would also be expected to give a positive sign for $\frac{1}{2}e^2qQ$, and this has been confirmed by application of a large magnetic field for the compounds SnF_2 (*271*), $NaSnF_3$ (*183*), $NaSn_2F_5$ (*183*), SnS (*271*), $Sn_3(PO_4)_2$ (*271*), $Sn(O_2CMe)_2$ (*183*), SnC_2O_4 (*271*), $Sn(HCO_2)_2$ (*183*), $K_2Sn(C_2O_4)_2 \cdot H_2O$ (*183*), and $SnSO_4$ (*183*).

An alternative interpretation of the origins of the quadrupole splittings in pyramidal Sn^{II} compounds has been suggested by Donaldson and Senior (*192*), who noted that for compounds of known structure (SnF_2, $NaSn_2F_5$, SnS, $K_2SnCl_4 \cdot H_2O$, $SnSO_4$, and $SnCl_2$) the largest quadrupole splittings seem to be found for compounds (SnF_2, $NaSn_2F_5$) with the greater degree of asymmetry in the bond lengths. In contrast, there does not seem to be any correlation between quadrupole splitting and center shift, which might be expected if quadrupole splitting is chiefly dependent on lone-pair electrons. From these observations, Donaldson and Senior suggested that the EFG is produced primarily from inequalities in the tin–ligand bonds. Using the assumption that the lone-pair electrons make no contribution to the EFG, these authors calculated values of the p-orbital imbalance for the compounds SnF_2, $NaSn_2F_5$, SnS, $K_2SnCl_4 \cdot H_2O$, $SnSO_4$, and $SnCl_2$, which are a reasonable reflection of quadrupole splitting trends. However, the calculations predict opposite signs of $\frac{1}{2}e^2qQ$ for SnS and SnF_2 and such a change is not observed experimentally.

Although the available evidence seems to favor an interpretation of quadrupole splitting in terms of a dominant contribution from the lone pair, the lack of a general relationship between center shift and quadrupole splitting is surprising (*183*). Further, no rationalization of the relative magnitudes of the quadrupole splittings has been possible. For example, the point charge model (*183*) does not account for the widely different quadrupole splittings of $SnSO_4$ and $Na_2Sn_2F_5$. Clearly it is not yet possible to gain a complete understanding or the origins of the EFG's found for Sn^{II} compounds.

Mössbauer center shifts of a wide range of Sn^{II} complexes have been reported and a selection of the data is given in Table LVI. These compounds show a consistent trend to decreasing center shift compared with the parent compounds, and this shift has been associated (*106, 185–187, 194, 303*) with an increase in sp^3 character caused by the replacement of a bridging tin–ligand bond in the parent compound with a terminal bond in the complex. Another example of this effect (*191*) is found for the $[Sn_2F_5]^{2-}$ ions, which have consistently higher shifts than $[SnF_3]^-$ species. Coordination of a second ligand results in a further decrease in center shift—indicative of further participation of the $5s$ orbital in bonding. Two five-coordinate complexes have been reported ($SnCl_2 \cdot$ terpyridyl and $SnCl_2 \cdot PMAQ$), and these species have lower center shifts than four-coordinate analogs (*194*).

The amount by which the center shift is decreased on complex formation can be taken as a guide to the Sn–ligand bond strengths. For example, the sequence (*186, 303*) found for the amine complexes with $SnCl_2$—N-methylmorpholine > morpholine ~ piperidine ~ piperazine > pyridine ~ α-picoline > γ-picoline > 1,10-phenanthroline ~ 2,2'-bipyridyl—roughly parallels the basic strength of the ligands as measured by pK values. Donaldson and Nicholson (*185*) have used the center shifts of 1:1 complexes with $SnCl_2$ to deduce the order of bond strengths: triphenylarsine oxide > triphenylphosphine oxide > pyridine > thiourea > pyridine 1-oxide > diglyme > urea > water.

The complexes in Table LVI and related species (*106, 185–187 194, 303*) show quadrupole splittings in the range 0–2.25 mm/sec. However, very little correlation between quadrupole splitting and the nature of the complexes has been possible. This is probably due to the influence of the lone-pair electrons, as a result of which the quadrupole splitting is no longer an additive function of the ligands.

Mössbauer data for the species $Sn(O_2CR)_2$ [R = H, Me, Et, n-Pr, i-Pr, i-Bu, t-Bu, CH_nCl_{3-n} ($n = 0$–2), CH_nF_{3-n} ($n = 0$–2), $CH_3 \cdot CHCl$, and $ClCH_2 \cdot CH_2$] allow the assignment (*184*) of polymeric structures with pyramidally coordinated Sn^{II} units, and the Mössbauer parameters for the ions $Sn(O_2CR')_3^-$ (R' = H, Me, $ClCH_2$, and FCH_2) (184) and $[Sn(HPO_3)_3]^{4-}$ (166) are also consistent with pyramidal coordination. Mössbauer data has also proved useful in the study of the products of the tin(II)–thiourea–anion system (106) and the decomposition products of trihydroxystannates (165).

The ternary halides of type SnXF (X = Cl, Br, I) comprise an interesting series of compounds (193). The Mössbauer spectra show only one type of tin environment. Structures based on the parent halides are, therefore, indicated. The center shifts of the SnXF (X = Cl, Br, I) species (3.64–3.78 mm/sec) are much closer to the value for SnF_2 (C.S. = 3.67 mm/sec) than $SnCl_2$ (C.S. = 4.17 mm/sec), $SnBr_2$ (C.S. = 4.03 mm/sec) or SnI_2 (C.S. = 3.95 mm/sec) suggesting a pyramidal structure with bridging fluorine atoms and terminal X atoms. Similarly, the center shifts of SnIX (X = Cl, Br) (C.S. = 3.76 mm/sec) are close to that of SnI_2, indicating a structure with bridging iodine atoms. These types of arguments have been extended to assign structures to the more complex species Sn_2XF_3 (X = Cl, I), Sn_2BrCl_3, Sn_3BrF_5, and Sn_2NCSX_3 (X = F, Cl, Br, I).

TABLE LVI

SELECTION OF CENTER SHIFT DATA FOR TIN(II) COMPLEXES[a]

Compound[b]	C.S.[c]	Ref.
SnF_2	3.67	(187)
$MSnF_3$[d]	3.03–3.29	(191)
MSn_2F_5	3.17–3.44	(191)
$SnF_2 \cdot py$	3.24	(187)
$SnCl_2$	4.08	(185, 187, 194, 303)
$(pyH)SnCl_3$	3.12	(187)
$SnCl_2 \cdot py$	3.29	(187)
$SnCl_2 \cdot 2py$	3.12	(187)
$SnCl_2 \cdot Ph_3PO$	3.21	(185)
$SnCl_2 \cdot 2Ph_3PO$	3.34	(185)
$SnCl_2 \cdot pyO$	3.50	(185)
$SnCl_2 \cdot 2pyO$	3.30	(185)
$SnCl_2 \cdot Ph_3AsO$	3.00	(185)
$SnCl_2 \cdot CO(NH_2)_2$	3.70	(185)
$SnCl_2 \cdot CS(NH_2)_2$	3.41	(106)
$SnCl_2 \cdot H_2O$	3.87	(185)
$2SnCl_2 \cdot dg$	3.58	(185)
$SnCl_2 \cdot ppz$	2.92	(186)
$SnCl_2 \cdot 2morph$	2.91	(186, 303)
$SnCl_2 \cdot Me\text{-morph}$	2.45	(303)

TABLE LVI—continued

Compound[b]	C.S.[c]	Ref.
$SnCl_2 \cdot 2pip$	2.86	(186)
$SnCl_2 \cdot bipy$	3.53	(194, 303)
$SnCl_2 \cdot phen$	3.59	(194, 303)
$SnCl_2 \cdot 2p\text{-tol}$	3.47	(194)
$SnCl_2 \cdot 2(\gamma\text{-pic})_2$	3.40	(303)
$SnCl_2 \cdot 2(\alpha\text{-pic})_2$	3.27	(303)
$SnCl_2 \cdot terp$	3.17	(194)
$SnCl_2 \cdot PMAQ$	3.33	(194)
$SnBr_2$	3.92	(185, 187, 194, 303)
$(pyH)SnBr_3$	3.65	(187)
$SnBr_2 \cdot py$	3.45	(187)
$SnBr_2 \cdot 2py$	3.36	(187)
$Sn(NCS)_2$	3.52	(187)
$(pyH)Sn(NCS)_3$	3.28	(187)
$Sn(NCS)_2 \cdot 2py$	3.29	(187)
$Sn(HCO_2)_2$	3.15	(187)
$MSn(HCO_2)_3$[e]	2.90–3.13	(184)
$Sn(HCO_2)_2 \cdot py$	3.05	(187)
$Sn(HCO_2)_2 \cdot 2py$	3.04	(187)
$SnSO_4$	4.00	(187)
$SnSO_4 \cdot py$	3.52	(187)

[a] Further data for similar complexes may be found in Refs. (106, 184–187, 194, 303).
[b] dg = Diglyme; ppz = piperazine; morph = morpholine; Me-morph = N-methylmorpholine; p-tol = p-toluidine; pic = picoline; terp = terpyridyl; PMAQ = 8-(2-pyridylmethyleneamine)quinoline.
[c] Data given in mm/sec at liquid nitrogen temperature relative to SnO_2, assuming the center shifts of Pd(Sn) and α-tin are +1.52 and +2.10 mm/sec, respectively; when appropriate an unweighted average has been taken.
[d] Range of data for M = NH_4, Na, K, Rb, Cs, Sr, and Ba.
[e] Range of data for M = Rb, Cs, NH_4.

Clark et al. (124) have reported the Mössbauer spectra of the ions SnX_3^- (X = F, Cl, Br, or I) and SnX_2Y^- (X, Y = Cl, Br, or I). Although the SnX_3^- species do show some correlation between center shift and ligand electronegativity, a linear relationship is not observed and the correlation does not extend to the mixed salts. This is in marked contrast to the octahedral Sn^{IV} halide complexes (Section IV, A, 2), and probably illustrates the ability of the lone pair to mask the effect of the ligands.

F. OTHER OXIDATION STATES OF IRON

An enormous number of Mössbauer spectra of Fe compounds have been recorded, and these spectra have been extremely useful in describing the structure and bonding in many of these compounds. For Fe^{-II} and Fe^{III} high spin, $q_{C.F.} = 0$, and any quadrupole splitting will be due to q_{lat} and/or $q_{M.O.}$ [Eqs. (6)–(11)]. In contrast, for Fe^{II} high-spin and Fe^{III} low-spin compounds, the major part of the Q.S. is due to $q_{C.F.}$ [Eq. (11)]. The magnitudes of these Q.S. values over a wide range of temperature have been interpreted with varying degrees of success using models developed by Ingalls (*351*) and Golding [*281–283*]. Bonding information is usually more difficult to extract for these two species than for other Mössbauer compounds containing Fe^{II} low spin, Sn^{IV}, or Fe^{-II} for which $q_{C.F.} = 0$. For other oxidation states of iron, i.e., Fe^0, Fe^{-I}, and intermediate spin states of Fe^{II} and Fe^{III}, the separation of q into lattice and valence contributions [Eq. (6)], and the separation of q_{val} into $q_{M.O.}$ and $q_{C.F.}$ [Eq. (11)] no longer appears to be useful. Both $q_{M.O.}$ and $q_{C.F.}$ are of similar magnitudes. As a result, rationalization of Q.S. values—except on the purely empirical level—has usually not been possible.

In this section, we will review very briefly what we consider to be some interesting Mössbauer applications for iron compounds. Enough recent data is presented to enable the interested reader to find easily most of the papers in this area. A recent review of magnetic data (*479*) has recently been published, and as noted in the Preface, the very interesting magnetic properties of many of these iron compounds will not be discussed here. Many iron spectra have also been discussed recently (*110*).

1. Fe^{-II} Compounds

Mössbauer spectra of a number of Fe^{-II} (formally) compounds have been obtained, and a selection of data is given in Table LVII. Fe^{-II} has the spherosymmetric electronic configuration d^{10}, and the partial quadrupole splitting treatment outlined in Section II should be applicable to Fe^{-II} quadrupole splittings. Indeed, Mazak and Collins (*402*) were among the first to apply a point charge treatment. All the compounds in Table LVII contain strong π-accepting ligands such as CO or NO^+, and as noted previously for Fe^{II} low-spin compounds (Section IV,B), the partial quadrupole splitting treatment may not be as useful for such ligands. However, the p.q.s. treatment still should be useful for rationalizing the signs of the Q.S. values, and their approximate magnitudes.

The bonding in tetrahedral Fe^{-II} compounds can be considered to be a combination of σ donation to the sp^3 Fe hybrids and π acceptance from the t_{2g} $3d$ orbitals. Because of the smaller $\langle r^{-3} \rangle$ for $4p$ electrons than $3d$ electrons, we would expect that σ donation in Fe^{-II} compounds would not affect the Q.S. values as much as in Fe^{II} low-spin compounds.

Mazak and Collins (406) determined the signs of the Q.S. for $KFe(CO)_3NO(+ve)$, $Fe(CO)_2(NO)_2(-ve)$, and $Fe(Ph_3P)_2(NO)_2(-ve)$. The first sign is expected from the p.q.s. values for Fe^{II} low-spin compounds (Table XLVI). NO^+ has a much more positive p.q.s. than CO, mainly due to the very strong π-acceptor properties of NO^+. The magnitude of the Q.S. for $Fe(CO)_2(NO)_2$ is similar to that of $KFe(CO)_3NO$. η should be 1 for this compound (Table IV, compound 3), but the negative Q.S.

TABLE LVII

Mössbauer Parameters for Fe^{-II} Compounds[a]

Compound[b]	C.S.[c]	Q.S.	Ref.
$Hg\{Fe(CO)_2NO[P(C_6H_5)_3]\}_2$	0.26	1.42	(333)
$Hg\{Fe(CO)_3NO\}_2$	0.26	1.26	(333)
	0.28	1.32	(406)
$Fe(CO)_2(NO)_2$	0.32	-0.33 ($\eta \sim 0.85$)	(406)
	0.32	0.34	(146)
$NaFe(CO)_3NO$	0.16	0.37	(406)
$KFe(CO)_3NO$	0.18	$+0.36$	(406)
$Fe(Ph_3P)(CO)(NO)_2$	0.20	0.55	(406)
$Fe(Ph_3P)_2(NO)_2$	0.35	0.69	(406)
	0.33	0.70	(146)
$Fe(Ph_2MeP)_2(NO)_2$	0.30	0.60	(146)
$Fe[(PhO)_3P]_2(NO)_2$	0.29	0.50	(146)
$Fe(Ph_3As)_2(NO)_2$	0.42	0.59	(146)
$Fe(Ph_3As)(CO)(NO)_2$	0.34	0.64	(146)
$Fe(diphos)(NO)_2$	0.29	0.54	(146)
$Fe(arphos)(NO)_2$	0.36	0.58	(146)
$Fe(arphos)(CO)(NO)_2$	0.28	0.47	(146)
$Fe(f_4fos)(NO)_2$	0.35	0.29	(146)
$Fe(f_6fos)(NO)_2$	0.33	0.29	(146)

[a] Data given in mm/sec at 80°K.
[b] diphos = Ph_2 P CH_2CH_2 PPh_2; arphos = $Ph_2PCH_2CH_2AsPh_2$; f_4fos = Ph_2P $\overline{C = C\ P(Ph)_2CF_2CF_2}$; f_6fos = Ph_2P $\overline{C = C\ P(Ph)_2(CF_2)_2CF_2}$.
[c] With respect to sodium nitroprusside; unfortunately, Mazak and Collins (406) do not give their C.S. standard. To bring their values into line with the others, it appears as if the values in (406) were quoted relative to Fe metal. 0.26 mm/sec has been added to these values.

implies that the C–Fe–C angle is greater than 109°. The substantially larger and more negative Q.S. for $Fe(Ph_3P)_2(NO)_2$ implies that the p.q.s. value for Ph_3P is more negative than that for CO, and that the P–Fe–P bond angle is greater than 109° These results suggest that the order of p.q.s. values in tetrahedral Fe^{II} compounds is (p.q.s.)$_{NO^+}$ > (more positive than) (p.q.s.)$_{CO}$ > (p.q.s.)$_{(PhO)_3P}$ > (p.q.s.)$_{Ph_3As}$ ~ (p.q.s.)$_{Ph_2MeP}$ > (p.q.s.)$_{Ph_3P}$. For the ligands NO^+, CO, and Ph_3P, the above order is the same as that found for these ligands in Fe^{II} low-spin compounds (Table XLVI). Since the Q.S. is probably most sensitive to π-bonding inequalities, the above order probably reflects a decrease in π-acceptor properties from NO^+ to Ph_3P (406).

The C.S. values for the compounds in Table LVII become more positive in the order NO^+ < CO < PR_3 < $AsPh_3$. This order reflects a decrease in σ donor and/or π acceptor properties from NO^+ to $AsPh_3$.

For the compounds containing the chelating ligands such as diphos (146), the Q.S. values are substantially smaller than for the phosphine–NO compounds such as $Fe(Ph_3P)_2(NO)_2$. These smaller values have been attributed to distortions (146), although it is quite possible that the bonding properties of NO vary appreciably from one compound to another and give rise to these significant differences and/or that the bonding properties of diphos are appreciably different than Ph_3P.

2. Fe^{III} High-Spin Compounds

Fe^{III} high spin has the electronic configuration $(t_{2g})^3(e_g)^2$. Any Q.S. is thus due to q_{lat} or $q_{M.O.}$ and, in general, Q.S. values should be small (Table LVIII). The octahedral and tetrahedral species (Table LVIII, compounds 1, 2, 5–7) should all have small or zero quadrupole splittings as is observed (41, 125, 201). However, the quite large splitting for $(NEt_4)Fe(NCO)_4$ (201) shows that considerable distortion from tetrahedral symmetry is present. Mössbauer spectroscopy appears to be able to detect such distortions more easily than infrared in this case (245).

The C.S. values for six-coordinate complexes (Table LVIII, compounds 1, 2) are about 0.20 mm/sec larger than for four-coordinate complexes containing the same ligand (Table LVIII, compounds 5, 6) and this difference can be used to distinguish six-coordinate from four-coordinate species. For example, the very similar C.S. values for $M(acac)_3FeCl_4$ (M = Si, Ge) (compounds 3 and 4) to those of the $FeCl_4^-$ species strongly indicates (41) that the $FeCl_4^-$ anion is present in the acac compounds, and no bridging acac groups are present. Similarly, the C.S. values in frozen solution for ferric bromide and thiocyanate species (Table LVIII, compounds 8, 10) allowed the assignment of a tetrahedral structure to the bromide species (HFeBr$_4$), but an octahedral structure to the thiocyanate species (399). These differences in C.S. probably reflect the different bond lengths in tetrahedral and octahedral coordination. For example, r_{Fe-Cl} in NaFeCl$_4$ and FeCl$_3$ are 2.19 (484) and 2.48 Å (304), respectively.

Clausen and Good (125, 126) have studied extensively a number of tetrahedral ferric systems. The Q.S. in the (R$_n$NH$_{4-n}$)FeX$_4$ species (125) (Table LVIII, compounds 13–23) has been attributed to hydrogen bonding. The identical solid and solution spectra strongly support this suggestion and tend to rule out lattice effects. Spectra of [FeCl$_{4-n}$Br$_n$]$^-$ species (126) show that the C.S. increases from 0.55 mm/sec for FeCl$_4^-$ to 0.62 for FeBr$_4^-$, but very little, if any, line broadening is observed. This indicates that the q_{lat} contributions from Cl$^-$ and Br$^-$ are very similar.

A combination of Mössbauer and other spectroscopic data has provided good evidence in EDTA and DTPA complexes (Table LVIII, compounds 27–31) that the coordination number changes (313, 519). It was concluded that compounds 27 and 29 contain six-coordinate Fe; compounds 28 and 30, seven-coordinate Fe; and compound 31, eight-coordinate iron.

A number of substituted ferric acetylacetonates (Table LVIII, compounds 39–43) have been prepared, and their Mössbauer spectra recorded (32, 42, 256, 405). Very broad lines are usually obtained, and these have been attributed to relaxation effects. Surprisingly, γ irradiation (32) narrows the line width for Fe(acac)$_3$ markedly. The most likely reason for the small and variable Q.S. in these compounds is π acceptance from the Fe d orbitals to the acac ring.

Several groups have recorded spectra of Fe^{III} compounds containing the N,N'-ethylenebis(salicylaldiminato)anion (salen) (43, 54, 56, 86, 87, 480, 481). The crystal structure of [Fe(salen)Cl]$_2$ (261) proves that this is a dimer; and the magnetic susceptibilities and molecular weights suggest that [Fe(salen)]$_2$O and related compounds are dimers. However, the structure of a nitromethane adduct of Fe(salen)Cl (260) showed that this was monomeric. The magnetic susceptibilities have been fit rather successfully to a simple binuclear model involving exchange between $S = \frac{5}{2}$ iron ions in the dimeric species (54, 259, 387). However, for Fe(salen)Cl·½MeNO$_2$ (86), the magnetic and Mössbauer data strongly suggest that this is a dimer instead of the monomeric nitromethane adduct reported earlier (260). For example, no enhancement of the external magnetic field was observed at low temperatures in large magnetic fields (86). This appears to be conclusive proof that there is a spin zero ground state originating from antiferromagnetic coupling

TABLE LVIII
Mössbauer Parameters for Fe^{III} High-Spin Compounds[a]

Code No.	Compound[b]	C.S.[c]	Q.S.	Ref.
1	FeCl$_3$(anhydrous)	+0.76	—	(41)
2	[Co(NH$_3$)$_6$][FeCl$_6$]	+0.75	—	(41)
3	Si(acac)$_3$FeCl$_4$	+0.55	>0	(41)
4	Ge(acac)$_3$FeCl$_4$	+0.55	>0	(41)
5	(NMe$_4$)FeCl$_4$	+0.55	0.00	(125, 201)
6	(NEt$_4$)FeCl$_4$	+0.55	0.00	(125, 201)
7	(NEt$_4$)FeBr$_4$	+0.62	0.00	(201)
8	HFeBr$_4$[d]	+0.61	0.00	(399)
9	(NEt$_4$)Fe(NCO)$_4$	+0.60	0.86	(201)
10	Fe(SCN)$_4$X$_2$$^{-e}$	+0.76	0.56	(399)
11	FeCl$_3$·TPA	+0.54	0.21	(58)
12	FeCl$_3$·TPP	+0.57	0.23	(58)
13	MeNH$_3$FeCl$_4$	+0.54	0	(125)
14	Me$_2$NH$_2$FeCl$_4$	+0.53	0.33	(125)
15	Me$_3$NHFeCl$_4$	+0.52	0	(125)
16	BuNH$_3$FeCl$_4$[f]	+0.55	0.41	(125)
17	Bu$_2$NH$_2$FeCl$_4$[f]	+0.55	0.37	(125)
18	Bu$_3$NHFeCl$_4$[f]	+0.55	0.28	(125)
19	Bu$_4$NFeCl$_4$[f]	+0.52	0	(125)
20	BuNH$_3$FeBr$_4$[d]	+0.61	0.32	(125)
21	Bu$_2$NH$_2$FeBr$_4$[d]	+0.61	0.30	(125)
22	Bu$_3$NHFeBr$_4$[d]	+0.62	0.22	(125)
23	Bu$_4$NFeBr$_4$	+0.56	0	(125)
24	α-FephenCl$_3$	+0.68	0.85	(56)
25	β-FephenCl$_3$	+0.70	0.80	(56)
26	[Fe(phen)$_2$Cl$_2$]ClO$_4$	+0.65	0.05	(56)
27	HFe(OH$_2$)Y[b]	+0.71	0.42	(519)
28	KFe(OH$_2$)Y·H$_2$O[b]	+0.83	0.81	(519)
29	H$_2$FeDTPA·2H$_2$O[i]	+0.63	0.84	(313)
30	NH$_4$HFeDTPA·H$_2$O[i]	+0.62	1.10	(313)
31	K$_2$FeDTPA·2H$_2$O[i]	+0.66	0.99	(313)
32	[Fe$_3$(RCOO)$_6$(OH)$_2$]X'·H$_2$O[l]	+0.66	0.63	(403)
33	[Fe$_3$(RCOO)$_6$](CH$_3$COO)$_3$	+0.63	0.62	(403)
34	Fe(sal)$_3$·3H$_2$O	+0.79	1.06	(211)
35	Fe(ac)$_3$	+0.79	1.06	(211)
36	Fe(ol)$_3$	+0.74	0.81	(211)
37	Fe(nic)$_3$·H$_2$O	+0.77	1.05	(211)
38	Fe(AcNH)$_3$	+0.32[g]	0.90	(256)
39	Fe(acac)$_3$	+0.78	0.67[h]	(32, 42, 405)
40	Fe(BzAc)$_3$	+0.86	0.33	(42)

MÖSSBAUER SPECTRA OF INORGANIC COMPOUNDS 213

TABLE LVIII—continued

Code No.	Compound[b]	C.S.[c]	Q.S.	Ref.
41	Fe(TfAc)$_3$	+0.78	0.67	(42)
42	Fe(dbm)$_3$	+0.67	0.60	(401)
		+0.81	—	(42)
43	Fe(tta)$_3$	+0.78	0.39	(401)
44	Fe(salen)$_2$O	+0.66	0.83	(43, 56, 480)
45	[Fe(salen)Br]$_2$	+0.74	1.64	(43)
46	[Fe(salen)Cl]$_2$	+0.67	1.40	(43, 54, 87, 405)
47	Fe(salen)Cl·2MeNO$_2$	+0.65	1.33	(43, 481)
48	Fe(salen)Cl·2Y[j]	+0.70	1.42	(43)
49	Fe(salen)Cl·MeCN	+0.63	0.85	(43)
50	Fe(salen)Br·2Z[j]	+0.71	1.62	(43)
51	Fe(salen)Br·MeNO$_2$	+0.72	1.12	(43)
52	Fe(salen)Br·MeCN	+0.74	0.54	(43)
53	Fe(salen)Br·2MeOH	+0.76	0.86	(43)
		+0.72	1.65	
54	Fe(salen)Cl·½MeNO$_2$	+0.64	1.40	(86)
55	Fesal-N-(2 hydroxyphenyl)Cl[k]	+0.65	1.42	(54)
		+0.60	0.95	
56	Fe[N(SiMe$_3$)$_2$]$_3$	+0.43	5.12	(13)

[a] Data given in mm/sec at 80°K except where noted.
[b] acac = acetylacetonate; TPA = triphenylarsine; TPP = triphenylphosphine; Y = anion of ethylenediaminetetraacetic acid; DTPA = anion of diethylenetriaminepentaacetic acid; sal = salicylhydroxamate; ac = acetylhydroxamate; ol = oleylhydroxamate; nic = nicotinylhydroxamate; AcNH = acetoacetanilide; BzAc = benzoylacetylacetonate; TfAc = trifluoroacetylacetonate; dbm = dibenzoylmethane; tta = thenocyltrifluoroacetonate; salen = N,N'-ethylenebis-(salicylaldiminate)anion.
[c] With respect to sodium nitroprusside.
[d] In frozen solutions.
[e] In frozen solution; X = nitrobenzene or H$_2$O.
[f] These spectra were run in frozen benzene solution with very similar results (125).
[g] This shift appears to be erroneous.
[h] After γ irradiation this splitting was resolved (32). Other spectra did not resolve the Q.S.
[i] At room temperature.
[j] Y = MeOH, CHCl$_3$ and C$_5$H$_5$N; Z = CHCl$_3$, C$_5$H$_5$N, MeNO$_2$. An average of values is given.
[k] Other similar compounds give very similar four-line spectra.
[l] R = CH$_3$, CH$_2$Cl, CHCl$_2$, CCl$_3$: an average of the very similar results has been taken.

between two iron ions. Apparently, yet another $MeNO_2$ adduct was prepared (43), and Mössbauer and infrared evidence was consistent with a monomeric species. Indeed, a very recent communication (478) shows that *rapid* crystallization of $[Fe(salen)Cl]_2$ in $MeNO_2$ or C_5H_5N yields products which are almost certainly monomeric. Application of small magnetic fields results in large internal fields in contrast to the dimeric compounds $[Fe(salen)Cl]_2$ and $[Fe(salen)]_2O$ (476). On the basis of Mössbauer and infrared evidence, monomeric structures were assigned to compounds 47, 49, 51, and 52 in Table LVIII (43), while from this evidence, most of the others appear to be dimeric. However, some compounds (Table LVIII, compounds 53, 55) gave at least two doublets. These two doublets could be due to a mixture of monomeric and dimeric species, or two iron atoms in a dimeric unit which are not in identical environments with respect to the neighboring units in the crystal (54). It is apparent from the above results that there is a delicate balance between the monomeric and dimeric species, and much more work needs to be done to obtain pure products of one form in all cases.

Finally, the very unusual spectrum of $Fe[N(SiMe_3)_2]_3$ should be noted. This compound contains three-coordinate iron [(13) and references], V_{ZZ} is positive and $\mu = 5.91$ B.M. The quadrupole splitting (5.12 mm/sec) is by far the largest observed for any iron compound to date and the C.S. is the smallest yet observed for Fe^{III} high-spin compounds. Interesting magnetic properties have also been observed at low temperatures (13).

3. Fe^{II} *High-Spin and* Fe^{III} *Low-Spin Compounds*

As noted in Section II of this review, a large q_{val} contribution [specifically $q_{C.F.}$ in Eq. (11)] becomes dominant when considering Q.S. values in Fe^{II} high-spin ($t_{2g}^4 \, e_g^2$) and Fe^{III} low-spin (t_{2g}^5) compounds. Normally, a large temperature-dependent Q.S. is observed. The $q_{C.F.}$ contribution often masks changes in Q.S. owing to differences in covalency of the iron–ligand bonds, but at least for some Fe^{II} high-spin compounds, the Q.S. values appear to be useful for estimating bonding properties of ligands (19, 320, 321). The theoretical bases for interpreting Fe^{II} high-spin and Fe^{III} low-spin quadrupole splittings are similar, and a brief outline of the method (351) for Fe^{II} high-spin compounds will now be made.

The energy level diagram for Fe^{II} high spin is given in Fig. 17. The primary effect of the crystal field is to remove the degeneracy of the d orbitals. If the sixth d electron exclusively occupied the low-energy d_{xy} orbital, then $q_{val} = \frac{4}{7} \langle r^{-3} \rangle_{3d}$ which should give rise to a Q.S. of over 4 mm/sec. However, the Q.S. is reduced from this value by (a) thermal population of the other t_{2g} levels, (b) spin orbit coupling, (c) covalency

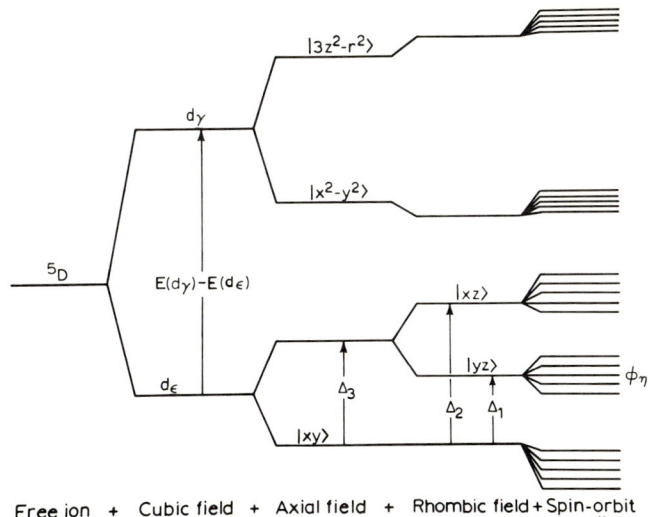

Fig. 17. Energy level diagram for Fe^{II} high spin (351).

effects, and (d) contributions from q_{lat}. q_{val} can be written as (351)

$$q_{val} = (1 - R)\tfrac{4}{7}\langle r^{-3}\rangle_{3d} F(\varDelta_1, \varDelta_2, \alpha^2\lambda_0, T)\alpha^2 \quad (84)$$

where the function F expresses the decrease in q_{val} due to thermal population and spin-orbit coupling; λ_0 is the spin orbit coupling constant; and α^2 is the covalency parameter and takes the values 0.6 to 0.9. Neglecting spin-orbit coupling, and taking $\varDelta_1 = \varDelta_2 \equiv \varDelta_3$, then

$$F(\varDelta_3, T) = \frac{1 - \exp(-\varDelta_3/kT)}{1 + 2\exp(-\varDelta_3/kT)} \quad (85)$$

If the orbital doublet is lowest in energy, then

$$F(\varDelta_3, T) = -\left[\frac{1 - \exp(-\varDelta_3/kT)}{2 + \exp(-\varDelta_3/kT)}\right] \quad (86)$$

A similar expression (390) is easily derived for the Boltzmann population for tetrahedral Fe^{II} species.

The spin-orbit coupling lifts the fivefold degeneracy of each orbital state (Fig. 17). This decreases F by an amount depending on the ratios $\varDelta_1/\lambda$ and $\varDelta_2/\lambda$. Ingalls has computed the decrease of F as a function of these ratios (351). The Q.S. is also decreased directly and indirectly by the expansion of the radial part of the $3d$ wave function on bonding. The covalency parameter α^2 takes this into account.

The lattice contribution for the axial compression and elongation considered in Eqs. (85) and (86) is given as (351):

$$q_{\text{lat}} = \pm (1 - \gamma) \frac{14\Delta_3}{3e^2 \langle r^2 \rangle} \tag{87}$$

The lattice contribution is always of opposite sign to q_{val}, and has often been neglected. However, for square-planar complexes of Fe^{II}, (121) the lattice contribution can be larger than q_{val}.

There are several difficulties in applying this method. The Q.S. for one d electron is still not precisely known, and it is not always obvious whether the decrease in Q.S. from the large value of 4 mm/sec is due to covalency, q_{lattice}, or spin-orbit coupling.

Some Q.S. and C.S. values for Fe^{II} high-spin compounds are given in Table LIX, and the crystal field splittings calculated using Ingall's

TABLE LIX

MÖSSBAUER PARAMETERS FOR Fe^{II} HIGH-SPIN COMPOUNDS[a]

Code No.	Compound[b]	C.S.[c]	Q.S.	Ref.
1	$(NMe_4)_2FeCl_4$	1.26	2.61	(202)
2	$(NEt_4)_2FeCl_4$	1.27	2.59	(202, 268)
3	$(PQ)FeCl_4$	1.25	2.99	(202)
4	$(Cat)Fe(NCSe)_4$	1.24	2.69	(202)
5	$(NMe_4)_2Fe(NCS)_4$	1.22	2.10	(202)
6[d]	$FeLCl_2$	1.08	2.71	(94)
7[d]	$FeLBr_2$	1.08	2.42	(94)
8[d]	$FeLI_2$	1.07	1.83	(94)
9	$Fe(quin)_2Cl_2$	1.23	3.06	(390)
10[d]	$Fe(quin)_2Cl_2$	1.12	2.72	(95)
11[d]	$Fe(quin)_2Br_2$	1.08	2.71	(95)
12[d]	$(Et_4N)Fe(quin)Cl_3$	1.13	2.08	(95)
13[d]	$Fe(Ph_3PO)_2Cl_2$	1.22	0.80	(95)
14[d]	$Fe(Ph_3PO)_2Br_2$	1.26	2.21	(95)
15[d]	$Fe(IQ)_4Cl_2$	1.29	3.18	(97)
16[d]	$Fe(IQ)_4Br_2$	1.26	2.21	(97)
17[d]	$Fe(IQ)_4I_2$	1.23	0.40	(97)
18[d]	$Fe(IQ)_4(NCS)_2$	1.37	1.50	(97)
19[d]	$Fe(IQ)_4(NCO)_2$	1.29	2.47	(96)
20[d]	$Fe(IQ)_4(NCSe)_2$	1.30	1.41	(96)
21[d]	$Fe(\gamma\text{-pic})_4Cl_2$	1.31	2.98	(97)
22[d]	$Fe(\gamma\text{-pic})_4Br_2$	1.25	1.28	(97)
23[d]	$Fe(\gamma\text{-pic})_4I_2$	1.19	0.19	(97)
24[d]	$Fe(\gamma\text{-pic})_4(NCS)_2$	1.33	1.67	(97)
25[d]	$Fe(py)_4Cl_2$	1.30	3.11	(97, 284)

TABLE LIX—continued

Code No.	Compound[b]	C.S.[c]	Q.S.	Ref.
26[d]	Fe(py)$_4$(NCO)$_2$	1.35	2.49	(96, 284)
27[d]	Fe(py)$_4$(NCSe)$_2$	1.30	0.71	(96)
28[d]	Fe(py)$_4$(NCS)$_2$	1.35	1.54	(97, 220, 284)
29[d]	FeL$_3$(ClO$_4$)$_2$	1.29	2.37	(94)
30[d]	FeL$_2$Cl$_2$	1.28	2.23	(94)
31[d]	FeL$_2$Br$_2$	1.27	2.12	(94)
32	FeCl$_2$·2AM	1.55	2.80	(58)
33	FeCl$_2$·2FA	1.59	3.00	(58)
34	FeCl$_2$·BM	1.54	2.87	(58)
35	FeCl$_2$·AN	1.49	2.18	(58)
36	Fe(phen)$_2$(N$_3$)$_2$	1.20	3.02	(135, 372)
37	Fe(phen)$_2$Cl$_2$	1.27	3.23	(135, 372, 390)
38	Fe(phen)$_2$I$_2$	1.30	2.80	(135)
39	Fe(phen)$_2$(SCN)$_2$	1.28	2.96	(135, 284)
40	Fe(phen)$_2$Br$_2$	1.29	3.14	(135, 284, 372)
41	FeCl$_2$·4H$_2$O	1.49	3.05	(135)
42	Fe(phen)$_2$(SeCN)$_2$	1.26	2.61	(135)
43	Fe(phen)$_2$(OCN)$_2$	1.20	3.15	(284, 372)
44	Fe(phen)$_2$(HCOO)$_2$	1.21	3.03	(284, 372)
45	Fe(phen)$_2$(CH$_3$COO)$_2$	1.12	3.17	(372)
46	Fe(Htcaz)$_2$(NCS)$_2$)[d]	1.25	2.98	(98)

[a] Data in mm/sec at 80°K unless otherwise noted.
[b] PQ = N,N'-Dimethyl-4,4'-dipyridyl; Cat = α,α-(bistriphenylphosphonium)-p-xylene; L = di-2-pyridylamine; quin = quinoline; IQ = isoquiniline; γ-pic = γ-picoline; AM = acetamide; FA = formamide; BM = benzamide; AN = aniline; Htcaz = thiocarbohydrazide.
[c] Relative to sodium nitroprusside.
[d] Room temperature results.

method are given in Table LX along with the ground state orbital. The Q.S. values are very sensitive to small distortions from tetrahedral or octahedral symmetry, and the Mössbauer spectra of the FeX$_4^{2-}$ species (X = Cl$^-$, Br$^-$, NCS$^-$, NCSe$^-$) all indicate small distortions from tetrahedral symmetry which had gone undetected (202, 268). As outlined in the introduction (Section II), a perfect tetrahedral or octahedral species would give no splitting. Neglecting spin-orbit coupling and q_{lat}, the Q.S. as a function of T for the tetrahedral compounds gave a reasonable fit for most compounds to Ingalls treatment (202). Many of the crystal field splittings for the octahedral species were calculated from the Q.S. at only one temperature, and it has been shown recently (390) that the Q.S. does not fit the theory well over a large temperature range. The more rigorous calculations given by Ingalls for such compounds as $FeSiF_6 \cdot 6H_2O$, $FeSO_4 \cdot 7H_2O$, and others gave more acceptable agreement. Golding et al. (199, 284) have correlated Q.S. and magnetic moments for $Fe(py)_4X_2$ compounds (X = Cl⁻, Br⁻, I⁻, NCO⁻, NCS⁻, etc.). The quadrupole splitting is very sensitive to small distortions from cubic symmetry, whereas the average magnetic moment is not sensitive to small distortions.

TABLE LX

Crystal Field Splittings in Fe^{II} High-Spin Compounds

Compound	Δ_3 (cm⁻¹) (or Δ_1 and Δ_2)		Ground state orbital	Ref.
$PQFeCl_4$	470		d_z^2	(202)
$(Cat)Fe(NCSe)_4$	292		d_z^2	(202)
$(NEt_4)_2FeCl_4$	135		d_z^2	(202)
	185		d_z^2	(268)
$(NMe_4)_2FeCl_4$	125		d_z^2	(202)
$(NEt_4)_2FeBr_4$	96		d_z^2	(202)
$(NMe_4)_2Fe(NCS)_4$	101		d_z^2	(202)
$Fe(quin)_2Cl_2$	~600		d_{xy}	(390)
$Fe(IQ)_4Cl_2$	600		d_{xy}	(97)
$Fe(IQ)_4Br_2$	360		d_{xy}	(97)
$Fe(IQ)_4I_2$	80		d_{xy}	(97)
$Fe(\gamma\text{-pic})_4Cl_2$	520		d_{xy}	(97)
$Fe(\gamma\text{-pic})_4Br_2$	200		d_{xy}	(97)
$Fe(\gamma\text{-pic})_4I_2$	40		d_{xy}	(97)
$Fe(\gamma\text{-pic})_4(NCS)_2$	~500		d_{xz}, d_{yz}	(97)
$FeSiF_6 \cdot 6H_2O$	760		d_z^2	(351)
$FeSO_4 \cdot 7H_2O$	480	1300	d_{xy}	(351)
$Fe(NH_4SO_4)_2 \cdot 6H_2O$	240	320	d_{xy}	(351)
$FeC_2O_4 \cdot 2H_2O$	100	960	d_{xy}	(351)
$FeSO_4$	360	1680	$d_{x^2-y^2} + \delta d_z^2$	(351)
FeF_2	1000	2200	$d_{x^2-y^2} + \delta d_z^2$	(351)

From linear correlations of Q.S. and isomer shift (corrected for S.O.D. shift) for Fe^{II} high-spin compounds (19, 320, 321), Hazony et al. have suggested that the main cause of variation in the Q.S. in a series of compounds such as FeF_2, $FeCl_2$, $FeBr_2$, and FeI_2 is due to a variation in α^2 in Eq. (84). Using the value of $\alpha^2 = 0.60$ for FeF_2 from ESR data (539), they obtained (19) $\alpha^2 = 0.34$, 0.39, and 0.42 for the iodide, bromide, and chloride, respectively, in contrast to the much higher values given previously. This order parallels the nephelauxetic series. Also, the variation in C.S. values shows an excellent correlation with the nephelauxetic series, in contrast to the correlation with the spectrochemical series for Fe^{II} low-spin (44) and Au compounds (49).

Similarly (321), for $[FeCl_{6-n} \cdot nH_2O]^{(n-4)}$ compounds, plots of Q.S. and isomer shift versus n are reasonably linear, and the correlation of Q.S. with I.S. has been attributed to central field covalency—the expansion of the radial portion of the $3d$ wave function due to the reduction of the metal ion's effective charge via σ and π bonding. Hazony et al. have shown that it should be possible to estimate both σ- and π-bonding properties of ligands from these correlations. They have also recently extended these ideas to other iron species (320). These papers should be very important for further interpretation of Fe^{II} high-spin data.

The degeneracy of the $^2T_2$ ground state in Fe^{III} low spin is lifted by the combined effects of spin-orbit coupling and the ligand field. Unfortunately, application of Golding's method (281, 282) for explaining the temperature dependence of Fe^{III} low-spin Q.S. values has not proven to be very successful [for example, see (480) and (489)]. The most serious disagreement apparently arises from the neglect of lattice contributions and from inadequate recognition of covalency. Qualitatively, however, Fe^{III} low-spin quadrupole splittings should be larger than Fe^{III} high-spin and Fe^{II} low-spin quadrupole splittings, and examination of Table LXI shows that this is indeed true. The very large splittings for compounds 8, 22, and 23, however, cannot be reconciled with Golding's estimate of a maximum Q.S. of 2.54 mm/sec in Fe^{III} low-spin compounds.

A very small selection of the Fe^{III} low-spin data is given in Table LXI, mainly because the C.S. is remarkably insensitive to variations in ligand properties within similar series of compounds (e.g., Table LXI, compounds 1-5, 19-23), and because the Q.S. values have not been readily rationalized because of the difficulty in estimating the relative magnitudes of such contributions as covalency and q_{lat}. Recent determinations of signs of the Q.S. for such compounds as $Fe(bipy)_3(ClO_4)_3$ and $Fe(ethylenediamine)_3Cl_3$ (477) should enable a more rigorous interpretation of these Q.S. values. For example, the positive sign of V_{ZZ} in the above two compounds indicates an orbital doublet ground state, with the magnitude of the Q.S. being reduced by covalency.

The magnetic properties of, and the relaxation effects in, these compounds have perhaps been of much greater interest (145, 432, 487-489), but they are beyond the scope of this article.

The dithiocarbamate compounds, $Fe(RR'NCS_2)_3$, are of considerable interest because the strength of the ligand field is close to the value of the mean pairing energy of the d electrons. The magnetic susceptibilities

(*485* and references) indicate that compound 15 is pure high spin, while the others exhibit high spin–low spin equilibria. However, all compounds show a single spectrum at all temperatures. The time of change from one spin state to another must be less than 1.5×10^{-7} sec, and the C.S. and Q.S. are, therefore, averages which depend on the proportion of high-spin to low-spin species (*246, 485*). This thermal admixture makes it even more difficult to interpret variations in Q.S. The sign of the EFG in these compounds is negative (*485*), and the very large C.S. values result from electron donation from ligand σ orbitals into metal d orbitals (*485*).

TABLE LXI

MÖSSBAUER PARAMETERS FOR Fe^{III} LOW-SPIN COMPOUNDS[a]

Code No.[b]	Compound[c]	C.S.[d]	Q.S.	Ref.
1	$K_3Fe(CN)_6$[e]	+0.14	0.28	(*135, 142*)
2	$Cu_3[Fe(CN)_6]_2$[e]	+0.10	0.48	(*142*)
3	$Ag_3Fe(CN)_6$[e]	+0.12	0.76	(*142*)
4	$Na_2[Fe(CN)_5H_2O]$[e]	+0.12	1.82	(*238*)
5	$Na_2[Fe(CN)_5NH_3] \cdot H_2O$[e]	+0.12	1.78	(*238*)
6	$[Fe(phen)_3](ClO_4)_3$	+0.35	1.67	(*135*)
		+0.31[e]	1.62[e]	(*56*)
7	$[Fe(dipy)_3](ClO_4)_3$	+0.32	1.76	(*135*)
8	$[Fe(terpy)_2](ClO_4)_3$	+0.32	3.43	(*480*)
9	$Na_2[Fe(CN)_5PPh_3]$[e]	+0.14	1.04	(*239*)
10	$Na_2[Fe(CN)_5AsPh_3]$[e]	+0.20	1.00	(*239*)
11	$Na_2[Fe(CN)_5SbPh_3]$[e]	+0.26	0.94	(*239*)
12	$[Fe(bipyr)_2(CN)_2]ClO_4$[e]	+0.24	1.63	(*56*)
13	$Fe(SacSac)_3$	+0.55	1.90	(*53*)
14	$Fe(Sacac)_3$	+0.57	1.47	(*145*)
15	$Fe[(CH_2)_4NCS_2]_3$	+0.76	0.42	(*485*)
16	$Fe[(CH_3)_2NCS_2]_3$	+0.71	0.73	(*213, 485*)
17	$Fe[(i\text{-}C_4H_9)_2NCS_2]_3$	+0.73	0.64	(*213, 485*)
18	$(Ph_4P)_3[Fe\{S_2C_2(CN)_2\}_3]$	+0.66	1.85	(*489*)
19	$(Et_4N)_2[Fe\{S_2C_2Ph_2\}_2]_2$	+0.61	2.37	(*61*)
20	$(Et_4N)_2[Fe\{S_2C_2(CF_3)_2\}_2]_2$	+0.59	2.50	(*61*)
21	$(Et_4N)_2[Fe\{S_2C_2(CN)_2\}_2]_2$	+0.59	2.76	(*61*)
22	$(Bu_4N)_2[Fe\{S_2C_6H_3Me\}_2]_2$	+0.60	2.95	(*61*)
23	$(Bu_4N)_2[Fe\{S_2C_6Cl_4\}_2]_2$	+0.58	3.02	(*61*)

[a] Data given in mm/sec at 80°K unless otherwise noted.
[b] Code number will be preceded by Table number in text.
[c] SacSac = Dithioacetylacetonate; Sacac = monothioacetylacetonate.
[d] Relative to sodium nitroprusside.
[e] At room temperature.

In contrast, Fe(Sacac)$_3$ did show two resolvable patterns (*145*), indicating a longer relaxation time in this compound compared with that in the dithiocarbamate complexes. The high-spin form of Fe(Sacac)$_3$ gave a very similar Q.S. (0.56 mm/sec) to Fe(acac)$_3$ (Table LVIII), whereas the low spin isomer gave the very much larger splitting due to the q_{val} contribution. Similarly, Fe(SacSac)$_3$ gave a larger splitting (*53*) indicating a more substantial splitting of the ground $^2T_2$ state than in Fe(Sacac)$_3$.

A larger number of spectra of dithiolene complexes have been reported (*59, 61, 489*), but because of the complications outlined earlier, even a qualitative description of trends in Q.S. has been hampered. For (Ph$_4$P)$_3$[Fe{S$_2$C$_2$(CN)$_2$}$_3$], the sign of the Q.S. is −ve (*489*), and this has been taken to be a d_{xy} hole well separated by higher states. Application of Golding's theory to the Q.S. values over a wide range of temperatures did not give good agreement. One series of these compounds (Table LXI, compounds 19–23) shows a very large variation in Q.S., but no change in C.S. (*59*). No adequate explanation for this phenomenon can be given at the present time.

4. Fe^0, Fe^{-I}, and π-$cpFe$ Complexes

An enormous number of Mössbauer spectra of Fe^0, Fe^{-I}, and π-cpFe compounds have been reported. A very small selection of data is given in Table LXII. Much of the earlier work is reviewed in reference (*110*). Generally, variations in C.S. and Q.S. within isostructural and isoelectronic series of compounds are comparatively small. Variations in C.S. have been successfully related to qualitative differences in bonding properties of ligands, and in some cases, useful empirical correlations have been noted between the Q.S. and coordination number of the iron. However, the variations in Q.S. for fairly similar groups of compounds have not proved to be amenable to any consistent qualitative or semiquantitative treatment (with the exception of some π-cp complexes noted later). These difficulties are probably largely due to the large and variable q_{val} term for Fe^0 (d^8) and Fe^{-1} (d^9). The relative energies of the Fe atomic orbitals and ligand orbitals probably vary greatly in fairly similar compounds. This situation is in direct contrast to another d^8 system of considerable Mössbauer interest, Au^{III}, whose energy levels have greater stability, giving rise to large and predictable changes in Q.S. with variations in ligand (see Section IV, G, 1).

Herber, King, and Wertheim (*333*) first proposed that center shifts could be expressed as a sum of partial center shifts [Eq. (48)]. They derived p.c.s. values for a large number of ligands, and calculated C.S. values for Fe compounds of varying coordination number and formal

TABLE LXII

^{57}Fe Mössbauer Parameters for Fe⁰, Fe⁻¹, and π-cpFe Complexes[a]

Code No.[b]	Compound[c]	C.S.[d]	Q.S.	Ref.
1	Fe(CO)$_5$	0.17	2.57	(134, 224, 333)
2	(Et$_4$N)$_2$[Fe$_2$(CO)$_8$]	0.18	2.22	(224)
3	(Et$_4$N)[Fe(CO)$_4$H]	0.09	1.36	(224)
4	Na$_2${Fe(CO)$_4$}	0.08	<0.18	(224)
		0.01[e]	~0[e]	(218)
5	Fe$_2$(CO)$_9$	0.39	0.48	(224, 333)
6	Fe$_3$(CO)$_{12}$	1. 0.37	1.13	(224)
		2. 0.30	0.13	—
7	(Et$_4$N)[Fe$_2$(CO)$_8$H]	0.33	0.50	(224)
8	[Fe(CO)$_3$(PMe$_2$)]$_2$	0.21	0.69	(267)
9	[Fe(CO)$_3$(SMe)]$_2$	0.30	0.89	(333)
10	[Fe(CO)$_3$(SPh)]$_2$	0.32	1.07	(267)
11	[Fe(CO)$_3$I(PMe$_2$)]$_2$	0.26	0.99	(267)
12	[Fe(CO)$_4$(PMe$_2$)]$_2$	0.23	2.58	(267)
13	[Fe(CO)$_3$SC(Me)$_3$]$_2$	0.29	0.93	(333)
14	cis-[Fe(CO)cp(PMe$_2$)]$_2$	0.40	1.61	(267)
15	trans-[Fe(CO)cp(PMe$_2$)]$_2$	0.42	1.64	(267)
16	cis-[Fe(CO)cp(PPh$_2$)]$_2$	0.52	1.65	(332)
17	trans-[Fe(CO)cp(PPh$_2$)]$_2$	0.52	1.66	(332)
18	[Fe(CO)cp(SPh)]$_2$	0.61	1.67	(267)
19	[Fe(CO)cp(SMe)]$_2$	0.56	1.65	(267, 333)
20	Fe(CO)$_2$cpI	0.48	1.83	(333)
21	[Fe(CO)$_2$cp]$_2$	0.48	1.89	(333)
22	Fe(CO)$_2$cpCl	0.49	1.88	(333)
23	Fe(CO)$_2$cpSnPh$_3$	0.35	1.82	(149)
24	Fe(CO)cp(PPh$_3$)SnPh$_3$	0.46	1.84	(149)
25	Fe(CO)cp(AsPh$_3$)SnPh$_3$	0.53	1.90	(149)
26	Fe(CO)cp(SbPh$_3$)SnPh$_3$	0.55	1.90	(149)
27	Fe(CO)cp(PPh$_2$CF$_3$)SnPh$_3$	0.46	1.83	(149)
28	Fe(CO)cp(PMe$_2$Ph)SnPh$_3$	0.43	1.71	(149)
29	PPh$_3$Fe(CO)$_4$	0.17	2.54	(133)
30	P(OEt)$_3$Fe(CO)$_4$	0.13	2.31	(133)
31	AcNpFe(CO)$_4$	0.24	1.78	(133)
32	MaAFe(CO)$_4$	0.27	1.41	(133)
33	[π-AllylFe(CO)$_4$]$^+$BF$_4^-$	0.32	1.01	(133)
34	f_6fosFe(CO)$_3$	0.20	2.34	(147)
35	diphosFe(CO)$_3$	0.19	2.12	(147)
36	$ffos$Fe(CO)$_4$	0.21	2.80	(147)
37	diphosFe$_2$(CO)$_8$	0.16	2.47	(147)
38	diarsFe$_2$(CO)$_8$	0.19	2.68	(147)
39	(cp)$_2$Fe	0.77	2.39	(208, 382, 561)
40	cpFe(C$_5$H$_4$)COCH$_3$	0.71	2.27	(533)
41	cpFe(C$_5$H$_4$)CH$_3$	0.77	2.39	(382)

TABLE LXII—continued

Code No.[b]	Compound[c]	C.S.[d]	Q.S.	Ref.
42	cpFe(C$_5$H$_4$)Cl	0.67	2.42	(533)
43	[cpFe(C$_5$H$_4$)]$_2$CH$^+$BF$_4^-$	0.79	2.11	(325)
44	[(cp)$_2$Fe]$^+$Br$^-$	0.70	~0.2	(561)
45	[(cp)$_2$Fe]$^+$BF$_4^-$	0.62	0.65	(533)
46	[Me$_4$N]$_2$[Fe(C$_2$B$_9$H$_{11}$)$_2$]	0.56	2.80	(60)
47	(Me$_4$N)[Fe(C$_2$B$_9$H$_{11}$)$_2$]	0.50	0.64	(60)
48	(cp)Fe(C$_2$B$_9$H$_{11}$)[f]	0.61	0.53	(325)
49	[BrFe(NO)$_2$]$_2$	0.67	1.76	(146)
50	[IFe(NO)$_2$]$_2$	0.60	1.68	(146)
51	(C$_5$H$_{11}$N)Fe(NO)$_2$Br	0.60	1.12	(146)
52	Ph$_3$PFe(NO)$_2$Br	0.50	1.02	(146)
53	Ph$_3$AsFe(NO)$_2$Br	0.52	1.28	(146)
54	Ph$_3$SbFe(NO)$_2$Br	0.55	1.49	(146)

[a] Data given in mm/sec at 80°K unless otherwise noted.
[b] Code number will be preceded by table number in the text.
[c] AcNp = acenaphthalene; MaA = maleic anhydride;
$f_6 f$os = Ph$_2$PC͞ = C͞P(Ph)$_2$(CF$_2$)$_2$CF$_2$; diphos = Ph$_2$PCH$_2$CH$_2$PPh$_2$;
diars = Ph$_2$AsCH$_2$CH$_2$AsPh$_2$; and ffos = Ph$_2$P͞C = CP(Ph)$_2$(CF$_2$)C͞F$_2$.
[d] Relative to sodium nitroprusside.
[e] At room temperature.
[f] Temperature = 140°K.

valencies. Although the agreement between predicted and observed values was quite good for a number of compounds, the p.c.s. value for cp varied markedly, and the Fe^{II} low-spin work discussed previously strongly indicates that the C.S. values for compounds containing CO are not additive. Also, it appears that the coordination number and valency of the iron should be a constant for a meaningful application of the p.c.s. treatment.

From a general structural point of view, Greenwood and co-workers have made several important empirical generalizations (224, 267). First, five-coordinate Fe^0 compounds (Table LXII, compounds 1–3, 12) have much larger quadrupole splittings than four-coordinate compounds (Table LXII, compound 4) and six-coordinate Fe^0 compounds (Table LXII compounds 6, 8, and 9), while seven-coordinate values (Table LXII, compounds 6 and 7) are also generally small. A lone pair is considered in compounds 8 and 9 to occupy an effective ligand site and complete the octahedral coordination of the metal. Greenwood et al. also noted that the C.S. value is reduced as the anionic charge on the metal cluster increases, e.g., for the series $Fe_2(CO)_9$, $Fe_2(CO)_8H^-$, and $Fe_2(CO)_8^{2-}$, the C.S. decreases from 0.42 to 0.33 to 0.18 mm/sec. It is also generally true that the C.S. increases as the coordination number increases. Thus, the C.S. for $Na_2Fe(CO)_4$, $Fe(CO)_5$, and $Fe_2(CO)_9$ are 0.08, 0.17, and 0.39 mm/sec, respectively.

From compounds 8, 9, and 14–19 in Table LXII, it is apparent that replacement of S by P in a bridging position lowers the C.S. (267). This is consistent with P being a stronger σ donor than S. However, the Q.S. for compounds 14–19 are all somewhat surprisingly within experimental error.

In the $Fe(CO)LcpX$ (L = CO, PPh_3, $AsPh_3$ etc.; X = Cl^-, I^-, $SnPh_3^-$) compounds, the C.S. has a small and measurable trend, but the Q.S. values are again remarkably constant. The trend in C.S. values is consistent with an increase in σ + π in the order (149) $SbPh_3 < AsPh_3 < PPh_3 \sim PPh_2CF_3 < PMe_2Ph < CO$.

Perhaps the largest variations in Q.S. within a series of compounds is given by compounds 29–33 in Table LXII (133). An increase in C.S. is paralleled by a large decrease in Q.S. These C.S. observations are consistent with σ + π decreasing in the order $P(OEt)_3 > PPh_3 > AcNp > MaA > \pi$ allyl, although an adequate explanation for the Q.S. trend has not yet been proposed.

In compounds 34–38 and other similar complexes (147), the $LFe(CO)_4$ derivatives give larger Q.S. than the corresponding $L_2Fe(CO)_3$ derivatives. In contrast, other similar compounds show the reverse trend (133, 333). Possible explanations of these trends have been discussed (147), but the signs of the Q.S. should be determined for a meaningful rationalization.

Very interesting studies on π-cyclopentadienyl and analogous carborane species have appeared (Table LXII, compounds 39–48) (60, 132, 154, 174, 514, 561). The very large splitting in $(cp)_2Fe$ and its almost complete collapse in $(cp_2Fe)^+$ species has been a subject of considerable discussion. The sign of the Q.S. in $(cp)_2Fe$ has been found to be positive using both the magnetic field technique (132) and single crystals (174). Using molecular orbital calculations (154, 514), this large positive splitting has been attributed mainly to an electron localized in d_{xy}. The removal of this electron on going to $[(cp)_2Fe]^+$ fortuitously collapses the splitting. The analogous carbollyl complexes (Table LXII, compounds 46–48) give slightly larger splittings than their cyclopentadienyl counterparts, indicating that the three B atoms in the C_2B_3 face of the carborane icosahedron have very similar bonding characteristics to those of the replaced carbon atoms. The smaller C.S. in the boranes indicates greater s electron density at the Fe nucleus in the carboranes than π-cp compounds. This has been attributed to the stronger π-acceptor properties of the boranes (60).

Finally, in this section, we mention a few of the Fe^{-I} compounds whose spectra have been recorded (146). The order of C.S. gives a measure of $(\sigma + \pi)$ for these ligands; i.e., $C_5H_5N < Ph_3Sb < Ph_3As < Ph_3P$. Fe^{-I} compounds have larger Q.S. values than their Fe^{-II} analogs because of the $q_{C.F.}$ term in Fe^{-I} compounds (146). Rather surprisingly, Fe^{-I} compounds have a larger C.S. than Fe^{-II} compounds. This is probably due to the fact that on going from Fe^{-II} to Fe^{-I} we have replaced strong $\sigma + \pi$ ligands such as CO by weak $\sigma + \pi$ ligands such as Cl and Br (146).

5. Fe Intermediate Spins

A number of interesting papers have appeared which discuss the Mössbauer parameters of intermediate spin ($S = 1$) Fe^{II} compounds and $S = \frac{3}{2}$ Fe^{III} compounds. A selection of the available data is given in Table LXIII. Much of the interest in the bis(N,N-disubstituted dithiocarbamate)iron(III) compounds (Table LXIII, compounds 8–11) has centered around the interpretation of magnetic and relaxation phenomena of these compounds [(486, 563, 565) and references in (564)]. These papers have been of considerable interest and importance, but as discussed at the beginning of this chapter, they are beyond the scope of this article.

The first six compounds in Table LXIII have magnetic moments at 295°K of about 3.9 B.M., and this has been attributed (372, 375) to two unpaired electrons with a contribution of ~1.0 B.M. from the second-order Zeeman effect. The Mössbauer spectra, however, are of very little use in characterizing the spin state of these compounds. Thus, both the C.S. and Q.S. are very similar to those of Fe^{II} low-spin compounds, although the quadrupole splittings are rather small. The magnetic data, and other spectral data strongly suggests that these compounds do have the $S = 1$ configuration.

FePc has long been of great interest. The magnetic moment of this complex is much above that expected for two unpaired electrons [see references in (158)]. Both an $S = 1$ state and a thermal admixture of $S = 2$ and $S = 0$ states are possible explanations for this unusual magnetic moment. The complex obeys a Curie–Weiss law and gives an almost temperature-independent susceptibility (384). The temperature independence of the Q.S. (155, 179, 410) down to 4°K indicates that there are no low-lying excited states within about 4 cm^{-1} of the ground state (410). More recently, Dale et al. (155, 158) have obtained magnetic data down to 1.25°K and shown that the susceptibility is virtually independent of temperature between 1.25° and 20°K. Their results are compatible with the iron atom having an orbital singlet with a real spin triplet state. This $S = 1$ state is split by second-order spin-orbit coupling into a singlet ground state and a doublet state at 70 cm^{-1}. Mössbauer spectra in large magnetic fields *(155)* show that the field gradient is positive. Most of the very large Q.S. can be attributed to the strong in-plane covalent bonding *(155)*, and reasonable agreement between predicted and observed values has been obtained *(155, 179)*, using the M.O. calculations used by Zerner *et al.* *(575)* for Fe^{II} porphin complexes, which should be closely analogous to FePc.

The pentacoordinate dithiocarbamate complexes have an orbitally degenerate spin quartet ground term as indicated by the magnetic

TABLE LXIII

Mössbauer Parameters for Intermediate Spin Fe^{II} and Fe^{III} Compounds[a]

Code No.[b]	Compound	C.S.[i]	Q.S.	Ref.
1	[Fe(phen)$_2$OX]·5H$_2$O[c,d]	+0.58	0.21	*(372, 375)*
2	[Fe(phen)$_2$mal]·7H$_2$O	+0.52	0.18	*(372, 375)*
3	[Fe(phen)$_2$F$_2$]·4H$_2$O	+0.55	0.16	*(372, 375)*
4	[Fe(dip)$_2$OX]·3H$_2$O[d]	+0.52	0.26	*(372, 375)*
5	[Fe(4,7-dmph)$_2$OX]·4H$_2$O[d]	+0.54	0.21	*(372, 375)*
6	[Fe(4,7-dmph)$_2$mal]·7H$_2$O	+0.52	0.21	*(372, 375)*
7	FePc[d]	+0.78	2.67	*(155, 347, 410)*
		+0.83[e]	2.67[e]	*(155, 179, 410)*
8	(R$_2$NCS$_2$)$_2$FeCl[h]	~+0.64[g]	2.64	*(212)*
			2.68[f]	*(564)*
9	(R$_2$NCS$_2$)$_2$FeBr[h]	~+0.64[g]	2.82	*(212)*
			2.88[f]	*(564)*
10	(R$_2$NCS$_2$)$_2$FeI[h]	~+0.64[g]	2.92	*(212)*
11	(R$_2$NCS$_2$)$_2$FeSCN[h]	~+0.64[g]	2.56	*(212)*

[a] Data given in mm/sec at 80°K unless otherwise noted.
[b] Code number will be preceded by table number in the text.
[c] At room temperature.
[d] phen = Phenanthroline; dip = 2,2'-dipyridyl; 4,7-dmph = 4,7-dimethyl-1,10-phenanthroline; Pc = phthalocyanine; ox = oxalate; mal = malonate.
[e] At ~4°K.
[f] At 1.2°K.
[g] These C.S. values are averages for all R and X values for the compounds (R$_2$NCS$_2$)$_2$FeX *(212)*.
[h] R = CH$_3$, C$_2$H$_5$, C$_6$H$_{12}$, C$_6$H$_{11}$, C$_8$H$_{14}$, etc. *[212]*.
[i] Values quoted relative to nitroprusside.

susceptibility, ESR, and Mössbauer results [(563, 564) and references]. Most of the compounds are simple paramagnets, although $[(C_2H_5)_2NCS_2]_2FeCl$ has shown magnetic ordering at low temperatures (563). The large positive Q.S. values (564) cannot be readily explained because the q_{lat} contribution to the EFG cannot be calculated. It is likely (564) that an appreciable q_{val} contribution arises from the ground quartet term. The variations in Q.S. (Table LXIII, compounds 8–11) could be due to the variation in q_{lat} (212, 564).

G. OTHER MÖSSBAUER ISOTOPES

In the preceding parts of this chapter we have discussed Mössbauer results for ^{57}Fe, ^{119}Sn, ^{99}Ru, ^{193}Ir, ^{129}I, and ^{127}I compounds. Despite the fact that there are over thirty isotopes which exhibit this effect, we restrict the discussion in this final section to four isotopes which have yielded useful and interesting chemical information: ^{197}Au, ^{121}Sb, ^{125}Te, and ^{129}Xe. Interesting chemical information has also been obtained on many rare earth isotopes [for a review see (427)] and ^{237}Np [for a review see (458)], while other isotopes such as ^{61}Ni (214, 518), and many heavy isotopes such as ^{177}Hf, ^{181}Ta, ^{182}W, ^{186}Os, and ^{195}Pt have not yet yielded useful chemical information [for a review see (270, 512)]. A recent paper on ^{182}W Mössbauer (33) indicates that useful chemical information should be readily obtained from tungsten compounds.

1. ^{197}Au

Mössbauer spectra of ^{197}Au compounds were first reported by Roberts (491) and Shirley (511). Although the natural linewidth of the 77.3 keV ($\frac{1}{2} \to \frac{3}{2}$) resonance is about 50 times that of ^{57}Fe, the changes in C.S. and Q.S. are two to three times those observed for ^{57}Fe compounds. The sign of $\delta R/R$ has recently been shown to be positive (490), and the quadrupole moment of the ground nuclear state has the value +0.586 barns (67). Because of the liquid He temperatures required to obtain reasonable spectra, and the very short half-life (~18 hr) of the ^{195}Pt precursor, very little quantitative data had been reported until very recently. Three recent papers (49, 108, 222) demonstrate the considerable utility of Mössbauer parameters for discussing structure and bonding in Au compounds. A selection of data (49, 108, 222) for Au^I and Au^{III} compounds is given in Tables LXIV and LXV, respectively.

a. *Center Shifts.* Au^I compounds contain linear L–Au–L' units in the first coordination sphere of the gold atom [see references in (49)]. The electron configuration of Au^{+1} ion is $5d^{10}$, and the ligands donate electrons to the $6s6p$ Au hybrid orbitals. Since the C.S. is usually more

TABLE LXIV

MÖSSBAUER PARAMETERS FOR SOME RECENTLY MEASURED Au[I] COMPOUNDS[a]

Code No.[b]	Compound	C.S.[c]	Q.S.	Ref.
1	AuCl	~−1.4	~4.6	(49, 108, 222)
2	AuBr	−1.47	4.23	(222)
3	AuI	−1.28	~4.1	(49, 108, 222)
4	AuCN	+2.11	8.16	(49, 108, 222)
5	As$(C_6H_5)_4$Au$(N_3)_2$	+1.43	6.84	(49)
6	KAu(CN)$_2$	+3.12	10.12	(49)
7	$(C_{10}H_{12})$AuCl	+0.83	6.04	(49)
8	$(C_{18}H_{36})$AuCl	+0.91	6.41	(49)
9	$(C_{16}H_{32})$AuCl	+1.03	6.29	(49)
10	Me$_2$SAuCl	+1.26	6.42	(108)
11	C$_5$H$_5$NAuCl	+1.7	6.4	(108)
12	Ph$_3$AsAuCl	+1.92	7.00	(108)
13	Ph$_3$PAuCl	+2.96	7.47	(108)
14	$(C_6F_5Ph_2P)$AuCl	+2.93	7.87	(108)
15	Ph$_3$PAuN$_3$	+3.3	8.4	(108)
16	Ph$_3$PAu(OCOMe)	+3.3	7.6	(108)
17	Ph$_3$PAuCN	+3.9	10.5	(108)
18	Ph$_3$PAuMe	+4.93	10.35	(108)

[a] At 4.2°K(mm/sec). In many cases, averages of the available recent data (see references) are given.
[b] Code number will be preceded by table number in text.
[c] Relative to the source Au/Pt at 4.2°K.

sensitive to the s electron augmentation than p electron augmentation and $\delta R/R$ is positive, we would expect that an increase in the σ-donor characteristics of the two ligands would increase $[\Psi(O)_s]^2$ and the C.S. π-accepting ligands would decrease the $5d$ electron density and also result in an increase in the C.S. However, π-bonding effects are usually not considered to be very important in Au compounds.

Au[III] compounds are known to be square-planar [see references in (49)] and the bonding involves dsp^2 hybrid orbitals on the gold. Assuming again that the C.S. is most sensitive to $6s$ orbital augmentation, the C.S. should again increase with increased σ-donor characteristics of the ligands. In addition, it might be expected that Au[III] would have larger C.S. values than Au[I] because of the smaller d electron density in Au[III] compounds and resultant increase in $[\Psi(O)_s]^2$. However, the increased shielding of s electrons from a larger $6p$ electron density in Au[III] compounds would tend to offset the d-electron density effect.

TABLE LXV
Mössbauer Parameters for Some Recently Measured Au^{III} Compounds[a]

Code No.[b]	Compound	C.S.[c]	Q.S.	Ref.
1	AuF$_3$	−1.07	2.74	(222)
2	AuCl$_3$	+0.57	0.75	(222)
		+0.83	—	(108)
3	AuBr$_3$	+0.79	1.27	(222)
		+0.48	—	(49)
		+0.18	1.8	(108)
4	[BrF$_2$]$^+$[AuF$_4$]$^-$	−0.69	1.82	(222)
5	MAuF$_4$[d]	+0.03	<1.0	(49, 108)
6	KAuI$_4$	+0.43	1.28	(49)
7	KAuBr$_4$	+0.64	1.13	(49, 222)
8	KAuCl$_4$·2H$_2$O	+0.71	~1.3	(49, 108, 222)
9	Ph$_4$AsAuCl$_4$	+1.09	1.88	(108)
10	KAu(SCN)$_4$	+1.63	2.04	(49)
11	As(C$_6$H$_5$)$_4$Au(N$_3$)$_4$	+1.66	2.89	(49)
12	Na$_3$AuO$_3$	+2.45	3.02	(49)
13	KAu(CN)$_2$X$_2$[d]	~+2.7	~5.5	(49, 222)
14	KAu(CN)$_4$	+4.12	6.93	(49, 222)
15	Ph$_3$PAuCl$_3$	2.06	3.25	(108)
16	C$_5$H$_5$NAuCl$_3$	1.45	—	(108)
17	Me$_2$SAuCl$_3$	1.26	2.20	(108)
18	p-MeC$_6$H$_4$NCAuCl$_3$	0.75	2.00	(108)

[a] At 4.2°K(mm/sec). In many cases, averages of the available recent data (see references) are given.
[b] Code number will be preceded by table number in text.
[c] Relative to the source Au/Pt at 4.2°K.
[d] M = alkali metal; X = Cl, Br, I.

In agreement with the above considerations, it is apparent from the data in Tables LXIV and LXV that the C.S. values for both Au^I and Au^{III} compounds increase as the σ-donor characteristics of the ligands increase. Thus, for Au^I halides, azide, and cyanide (Table LXIV, compounds 1–5) the C.S. becomes more positive in the order Cl$^-$ < Br$^-$ < I$^-$ < N$_3^-$ < CN$^-$. Similarly, in the LAuCl compounds (Table LXIV, compounds 10–14), the C.S. increases in the order L = Me$_2$S < C$_5$H$_5$N < Ph$_3$As < Ph$_3$P ~ (C$_6$F$_5$)Ph$_2$P. Again, in the Ph$_3$P AuX series (Table LXIV, compounds 15–18), the C.S. increases in the order X = N$_3^-$ ~ OCOMe$^-$ < CN$^-$ < Me. The order of these C.S. values agrees with generally recognized bonding characteristics, and it is apparent that the C.S. should be useful for placing ligands in a σ-bonding order.

For the Au^{III} compounds, a similar order of C.S. values is found (Table LXV) (except for the halides), and the order of center shifts correlates very well with the spectrochemical ranking of ligands (49). A similar correlation has been noted previously in this chapter for Fe^{II} low-spin compounds (44). Thus, in the AuX_4^- series, (Table LXV, compounds 5–8, 10–12, 14), the C.S. values increase in the order $F^- < I^- < Br^- < Cl^- < SCN^- \sim N_3^- < O^{2-} < CN^-$. For the $LAuCl_3$ compounds (Table LXV, compounds 15–17), the order of C.S. is identical to that of the Au^I compounds, i.e., $L = Me_2S < C_5H_5N < Ph_3P$, indicating again that Ph_3P is the strongest σ donor in this series.

It is interesting to note that the C.S. value for $KAu(CN)_2X_2$ (X = Cl, Br, I) compounds ($\sim$ +2.7 mm/sec) is close to the arithmetic mean of the C.S. values for $KAuX_4$ compounds ($\sim$0.9 mm/sec) and that for $KAu(CN)_4$ ($\sim$4.1 mm/sec). It is possible then that a partial center shift approach would be useful for predicting and rationalizing Au^{III} C.S. data. It is also noticeable from Tables LXIV and LXV that Au^{III} compounds generally have more positive C.S. values than their Au^I analogs, consistent with considerations outlined above. Thus, the C.S. of $KAu(CN)_4 > KAu(CN)_2$ and the C.S. of $AuCl_3 > AuCl$. However, this generalization is not always true; for example, $C_5H_5NAuCl_3$ has a smaller C.S. than C_5H_5NAuCl. It is often difficult then to distinguish Au^I from Au^{III} on the basis of C.S. alone (49, 108, 222).

b. *Quadrupole Splittings*. The quadrupole splittings also show a large range of values, and they can also be interpreted qualitatively using the simple bonding considerations outlined in the previous section. The signs of the quadrupole splittings have not been determined, but they can be predicted with considerable confidence. For Au^I compounds, the Z EFG axis lies through the bond axes, and σ bonding involves donation of electrons to the $6s$ and $6p_z$ orbitals. V_{ZZ} then should be negative, and since Q is $+ve$, e^2qQ for Au^I compounds is negative. Since the $6s$ population should be proportional to the $6p$ population, then an increase in σ-donor power of a ligand should increase the magnitude of the Q.S., while also increasing the C.S. Inspection of Table LXIV shows that an increase in C.S. is paralleled usually by an increase in Q.S. Indeed, Faltens and Shirley (222) have fit a Q.S.–C.S. plot for compounds 1–4 and 6 (Table LXIV) to a straight line with the equation:

$$C.S. = 0.872 Q.S. - 0.474 (cm/sec) \qquad (88)$$

For Au^{III}, the situation is slightly more complicated. The Z EFG axis lies along the 4-fold symmetry axis. The $d_{x^2-y^2}$ hole in a Au^{3+} ion would then produce a negative e^2qQ since there is a concentration of electron density (two d_{z^2} electrons) along the Z EFG axis. Covalent bonding, however, to the $5d_{x^2-y^2}$, $6p_x$ and $6p_y$ orbitals produces a positive contribution to the field gradient, and calculations show (222) that, except for the most ionic Au^{III} compounds, the latter contribution should dominate and give a positive e^2qQ. Thus, for the more ionic auric fluorides (Table LXV, compounds 1, 4, and 5), the sign of e^2qQ is taken to be negative, whereas for all the other compounds, the sign of e^2qQ is taken as positive (222). Supporting these assignments, a plot of C.S. versus Q.S. for a number of auric compounds (222) is reasonably linear with an equation:

$$\text{C.S.} = 0.532 \text{ Q.S.} + 0.016 \text{ (cm/sec)} \tag{89}$$

Faltens and Shirley noted that the C.S. and Q.S. taken together could be used to determine the oxidation state of gold unambiguously.

Like the C.S. values, the Q.S. for $KAu(CN)_2X_2$ compounds (~5.5 mm/sec) is intermediate between that of the AuX_4^- species (~1.5 mm/sec) and the $Au(CN)_4^-$ species (~6.9 mm/sec). The Q.S. value for trans-$Au(CN)_2X_2^-$ compounds would be expected to be considerably larger than the average value of the two end members because of the large η in $Au(CN)_2X_2^-$ (222). It would appear that partial quadrupole splittings could be derived and used successfully for such compounds.

Faltens and Shirley (222) have attempted to interpret the C.S. and Q.S. variations more quantitatively with very little success. Fortunately, it seems that even the qualitative interpretation is of considerable bonding and structural use in these gold compounds.

2. ^{121}Sb

A large number of Sb Mössbauer spectra have been recorded using the 37.2 keV $\frac{7}{2}+ \to \frac{5}{2}+$ transition. The majority of the data for Sb^{III} and Sb^V stoichiometric compounds containing one oxidation state of Sb are given in Tables LXVI and LXVII. Spectra of simple compounds can be obtained at 80°K, but liquid He temperatures are desirable to obtain good spectra and to extract more meaningful center shifts and quadrupole splittings (524). The quadrupole splittings are never well resolved, and most of the discrepancies in C.S. values are probably due to fitting one Lorentzian to an asymmetric peak.

The C.S. values are very sensitive both to the oxidation state of Sb, and the type of ligand. This is due to the very large value of $\delta \langle R^2 \rangle / \langle R^2 \rangle (\rho)$ (Table LXVIII) for [121] Sb, in comparison with other Mössbauer nuclei discussed in this chapter. The values in Table LXVIII were obtained in two ways. In the first method, (498, 510) experimentally determined C.S. values for tin, antimony, tellurium, iodine, and xenon have been compared with atomic Hartree–Fock density calculations to give values

TABLE LXVI

MÖSSBAUER PARAMETERS FOR Sb^{III} COMPOUNDS[a]

Compound	C.S.[b]	e^2qQ	Ref.
SbF_3	−14.6	+19.6	(72, 496, 522)
$SbCl_3$[c]	−13.8	+12.2	(72, 496, 522)
$SbBr_3$[c]	−13.9	+9.4	(72, 522)
SbI_3[c]	−15.9	—	(72, 522)
Sb_2O_3	−11.4	~18.0	(62, 72, 496, 522)
Sb_2S_3	−14.4	—	(62, 72, 522)
Sb_2Te_3	−15.3	—	(62)
$Co(NH_3)_6SbCl_6$	−19.7	—	(63)
K_3SbCl_6	−18.2	—	(63)
Cs_3SbCl_6	−18.1	—	(63)
$(NH_4)_3SbCl_6$	−17.2	—	(63)
$(NH_4)_2SbCl_5$	−15.2	—	(63)
$Sb^{III}{}_{in}Rb_4[Sb^{III}Sb^VCl_{12}]$	−19.6	—	(63)
Ph_3Sb	−9.69	+17.5	(389)
$(p\text{-}ClC_6H_4)_3Sb$	−9.3	—	(307)
$(p\text{-}CH_3OC_6H_4)_3Sb$	−9.0	—	(307)

[a] Data given in mm/sec with both source and absorber at 80°K except for Ph_3Sb. This spectrum was recorded with both source and absorber at 4.2°K.

[b] Center shift values are quoted relative to $^{121}SnO_2$ or $Ca^{121}SnO_3$, taking these compounds to have identical C.S. values. The C.S. of InSb w.r.t. from these sources is -8.5 ± 0.1 mm/sec.

[c] Russian work (377, 513) has yielded substantially more negative values for the C.S. For example, the C.S. values (377) for $SbCl_3$, $SbBr_3$, and SbI_3 are −15.5, −15.85, and −16.5 mm/sec, respectively.

of $\delta\langle R^2\rangle/\langle R^2\rangle$ for each nucleus. In the second, the center shifts for isoelectronic pairs of compounds were compared to obtain ratios of $\delta\langle R^2\rangle/\langle R^2\rangle$ for adjacent nuclei. Taking the value of ρ for ^{119}Sn to be 2.4×10^{-4}, as deduced by a comparison of center shifts with calculated charge densities (496), the values of ρ for the other Mössbauer nuclei could be determined. It is important to note that ρ is negative for ^{121}Sb. The very large differences in C.S. between Sb^{III} compounds (C.S. ranges from −9 to −19 mm/sec) and Sb^V compounds (C.S. ranges from −7 to +4 mm/sec) have been very useful in detecting the existence of Sb^{III} and Sb^V in mixed oxides and sulfides of Sb (62, 71, 388), as well as in complex chlorides such as $Rb_4[Sb_2Cl_{12}]$ (63). The large range of C.S. values within one oxidation state (Tables LXVI and LXVII) has been useful in obtaining structural and bonding information (vide infra). Because ρ is negative, an increase in $[\psi(O)_s]^2$ decreases the C.S. Sb^V, then,

TABLE LXVII
MÖSSBAUER PARAMETERS FOR Sb^V COMPOUNDS[a]

Compound	C.S.[c]	e^2qQ	Ref.
SbF_5	+2.23	—	(72, 522)
$SbCl_5$	−3.12	−4.4	(72, 522)
	−3.5	—	(513)
Sb_2O_5	+1.2	−4.3	(72, 496, 522)
$RbSbCl_6$	−2.7	—	(63)
$HSbCl_6 \cdot XH_2O$	−3.0	—	(513)
$NaSbF_6$	+1.7	—	(496, 513)
$KSbF_6$	+3.8	—	(496)
$NaSb(OH)_6$	+0.5	—	(513)
Ph_5Sb	−4.6	—	(389)
Ph_4SbF	−4.5	—	(389)
Ph_4SbF[b]	−4.56	−7.2	(389)
Ph_4SbCl[b]	−5.26	−6.0	(389)
Ph_4SbBr[b]	−5.52	−6.8	(389)
Ph_4SbNO_3[b]	−5.49	−6.4	(389)
Ph_3SbF_2[b]	−4.69	−22.0	(389, 523)
Ph_3SbCl_2[b]	−6.02	−20.6	(389, 523)
Ph_3SbBr_2[b]	−6.32	−19.8	(389, 523)
Ph_3SbI_2[b]	−6.72	−18.1	(389, 523)
$(PhCH_2)_3SbCl_2$[b]	−5.86	−23.0	(389)
$(CH_3)_3SbCl_2$[b]	−6.11	−24.0	(389, 523)
$(CH_3)_3SbBr_2$[b]	−6.40	−22.1	(389, 523)
Ph_4SbClO_4	−5.9	—	(389)

[a] Data given in mm/sec with both source and absorber at 80°K unless otherwise noted.
[b] Spectra recorded with both source and absorber at 4.2°K.
[c] Center shift values are quoted relative to $^{121}SnO_2$ or $Ca\ ^{121}SnO_3$ (^{121}Sb), taking these two compounds to have identical C.S. values. The center shift of InSb w.r.t. from these sources is -8.5 ± 0.1 mm/sec.

has a more positive C.S. than Sb^{III}. In contrast, because ^{119}Sn has a positive ρ, Sn^{IV} has more negative C.S. values than Sn^{II} (Section III, A). The quadrupole splittings are never well resolved; only for the largest Q.S. values for spectra at 4.2°K (such as the spectra of $R_{5-n}SbX_n$ (R = Ph, Me; X = F^-, Cl^-, Br^-, I^-; $n = 0-2$) can more than two of the eight lines be visually resolved. As a result, the quadrupole splittings (if detectable) normally have large errors. From spectra of the alkyl and aryl halides, the ratio of quadrupole moments Q_{ex}/Q_{gr} has been determined to be 1.34 ± 0.01 (523), but the value of Q_{gr} is not nearly as well defined. Ruby et al. (496) concluded from a number of widely varying Q_{gr} values and their calculated value (496), that $Q_{gr} = -0.28 \pm 0.1$ barns. It should be noted here that both Q_{ex} and Q_{gr} have negative values. Thus, as for ^{119}Sn, an excess charge density along the Z EFG direction gives a negative q, but a positive e^2qQ.

TABLE LXVIII

VALUES OF $\rho = \delta\langle R^2\rangle/\langle R^2\rangle \times 10^4$ FOR VARIOUS NUCLEI[a]

	From comparison of isoelectronic structures	From density calibration
^{119}Sn	—	2.4
^{121}Sb	−14.6	−17.0
^{125}Te	1.9	—
^{127}I	−4.8	−5.6
^{129}I	6.2	6.6
^{129}Xe	0.55	0.66

[a] From Refs. (498, 510).

a. Experimental Data. For Sb^{III} halides and oxide, a linear relationship has been noted between C.S. values and the difference in Pauling electronegativity (ΔX) between Sb and the ligand (72, 522). The least-squares fit to the data gives:

$$\text{C.S.} = -18.3 + 4.35\Delta X \quad (90)$$

SbF$_3$ lies well off this line. A plot of C.S. versus Q.S. for SbI$_3$, SbBr$_3$, SbCl$_3$, and Sb$_2$O$_3$ gives another straight line (72).

$$\text{C.S. (mm/sec)} = -18.0 + 0.011 e^2qQ \text{ (MH}_z\text{)} \quad (91)$$

These correlations indicate that a "bare" Sb^{3+} ion should have a C.S. of about −18 mm/sec. As the electronegativity of the ligand increases, $[\Psi(O)_s]^2$ decreases, causing the C.S. to become more positive. The C.S. values for Ph$_3$Sb and related compounds, using the above correlation Eq. (90), are somewhat anomalous, since the electropositive Ph group might be expected to increase $[\Psi(O)_s]^2$ more than in SbI$_3$, and give a very negative C.S. value. In fact, the most positive values for Sb^{III} compounds are observed. The structures of SbI$_3$ and Ph$_3$Sb are, however, not analogous. SbI$_3$ is essentially octahedral (542), whereas Ph$_3$Sb is trigonal-pyramidal with C–Sb–C bond angles of ∼113° (101), implying that Ph$_3$Sb has σ bonds which are essentially sp^3 hybrids with sufficient p_z electron density in the lone pair to give the large positive Q.S. (389).

It has also been suggested that there may be some resonance between Sb and Ph groups (*148, 389*) which permits some of the s electron density of the lone pair to be dissipated into the aromatic rings and give the unexpected C.S. values.

Some of the C.S. values for the $SbCl_6^{3-}$ species in Table LXVI are more negative than the -18 mm/sec suggested (*72, 522*) as the C.S. value for a "bare" Sb^{3+} ion from Eqs. (90) and (91). These very negative values imply (*63*) that the $5s$ pair is not being used greatly in the bonding to the six chlorides and/or that the value of -18 mm/sec is inaccurate and that the C.S. value for a bare Sb^{3+} is considerably more negative than this value.

As for Sb^{III} compounds, it is apparent from Table LXVII that the C.S. values for Sb^{V} compounds become more positive as the electronegativity of the ligands increases. The center shifts for analogous compounds vary as follows: $SbCl_5 < Sb_2O_5 < SbF_5$; $SbCl_6^- < SbF_6^-$; $Ph_4SbBr < Ph_4SbCl < Ph_4SbF$; and $Ph_3SnI_2 < Ph_3SbBr_2 < Ph_3SbCl_2 < Ph_3SbF_2$. Also, the C.S. values for the Ph_3SbX_2 compounds are more negative than for the Ph_4SbX compounds. It should be noted here that Gukasyan and Shpinel (*307*) have quoted substantially more negative values for the C.S. of Ph_3SbF_2 and Ph_3SbCl_2 than those given in Table LXVII. This discrepancy is probably due to fitting one peak to the asymmetric spectrum. The above trends in C.S. are again consistent with the increased withdrawal of s electron density as the electronegativity of the ligand increases.

Considerable structural and bonding information can be obtained for the $Ph_{5-n}SbX_n$ species, and this series of spectra are the most interesting Sb spectra yet reported (*389*). The sign and magnitude of the Q.S. values for the R_3SbX_2 species is expected from the known trigonal-bipyramidal structure (*72, 469, 522*) and the signs and magnitudes of the Q.S. for the isoelectronic Sn compounds (Table XXII). In fact, it should be possible to use the isoelectronic Sn and Sb compounds along with Eq. (51) to derive an accurate value of $Q_{119_{Sn}}/Q_{121_{Sb}}$. The positive q values observed (negative e^2qQ) are expected from the excess negative charge lying in the XY plane due to the strong σ-donor properties of the R groups. As with Sn compounds (Table XX), as the electronegativity of the halide decreases (from F to I), the e^2qQ values become more negative. It is also apparent that the methyl compounds have larger Q.S. values than the phenyl compounds $[Ph_3SbCl_2 = -20.6$ mm/sec; $(CH_3)_3SbCl_2 = -24.0$ mm/sec] indicating that methyl is a better σ donor than phenyl. Once again, the same trend is noticed in the isoelectronic Sn^{IV} species. Thus, $[Me_3SnCl_2]^-$ and $[Ph_3SnCl_2]^-$ (Table XXII) have Q.S. values of -3.31 and -3.02 mm/sec, respectively, and the p.q.s. value for methyl is substantially more negative than phenyl. The C.S. values for the methyl compounds are slightly more negative than those for the phenyl compounds. Again, this is consistent with Me being a better donor than Ph, and once again the Sn^{IV} compounds show the analogous trend (Table XX).

The Ph_4SbX compounds are of some considerable structural interest. Although recent X-ray structures of Ph_4SbOH (*52*) and Ph_4SbOCH_3 (*509*) have shown five-coordinate trigonal-bipyramidal coordination about Sb, the Ph_4Sb group has often been taken as ionic, i.e., Ph_4Sb^+. Recent infrared evidence has been shown to be consistent with this cation being present (*396*). A tetrahedral species would be expected to give little or no quadrupole splitting, and little or no variation in C.S. as the counterion is varied. In contrast, the C.S. varies from −4.56 (Ph_4SbCl) to −5.9 mm/sec (Ph_4SbClO_4), and the Q.S. values of ∼7 mm/sec (except for Ph_4SbClO_4) approach those expected from a point charge model taking the Sb–X bond to be identical in character in both R_3SbX_2 and R_4SbX compounds. Taking structures 8 and 10 in Table IV, and assuming $[L]^{tbe} \equiv [L]^{tba}$, the ratio of Q.S. for $R_3SbCl_2 : R_4SbCl$ is expected to be greater than 2 : 1 in comparison to the ∼3 : 1 observed. The quadrupole splittings strongly indicate that except for Ph_4SbClO_4, a strong covalent bond is formed between the Sb and X ligand, and that these compounds are five-coordinate in the solid state. The narrow single-line spectrum of Ph_4SbClO_4, the large absorption, and the negative C.S. all are consistent with this compound being formulated as $Ph_4Sb^+(ClO_4)^-$ (*389*). Solution conductance data indicate that whereas Ph_4SbF and Ph_4SbCl are essentially undissociated, Ph_4SbNO_3 could be a 1 : 1 electrolyte (*389*) in solution.

The Mössbauer spectra of SbF_5 and $SbCl_5$ have not been very useful for structural elucidation (*72, 522*). For example, two proposed structures for the low temperature phase of $SbCl_5$ are: a dimeric structure with Sb in six coordination, and an ionic $[SbCl_4]^+[SbCl_6]^-$ structure (*430, 508*). The Mössbauer spectrum of $SbCl_5$ is consistent with essentially octahedral coordination about Sb. A two-peak fit gave an equally good fit to the data (*72*) and the ionic structure cannot be ruled out from this spectrum.

3. ^{125}Te

Several papers have appeared which discuss the Mössbauer spectra of Te compounds using the 35.48 KeV $\frac{3}{2}+ \rightarrow \frac{1}{2}+$ transition. The majority of data for tellurium halide and oxygen complexes are given in Tables LXIX and LXX. All spectra have been obtained at 80°K, but many absorbers have been enriched to obtain reasonable spectra. Because of the very large line widths (~5.3 mm/sec), the quadrupole splittings are usually not well resolved. More seriously, however, agreement of C.S. values among various workers has been extremely poor (Tables LXIX and LXX), and this has made it extremely difficult to obtain meaningful chemical bonding information, or for that matter, the sign and magnitude of $\delta R/R$. For example, as summarized by Gibb et al. (266), Russian work

TABLE LXIX

Mössbauer Parameters for Te Halides and Halide Complexes[a]

Compound	C.S.[c]	Q.S.	Ref.
TeF_4	+0.4	7.0	(546)
	+0.4	6.8	(547)
$TeCl_4$	+1.2	4.0	(361)
	+2.7	5.4	(545)
$TeBr_4$	+1.1	3.8	(361)
	+1.6	5.0	(545)
TeI_4	+1.0	3.0	(547)
	+1.8	~4	(361)
	+1.0	6.0	(546)
M_2TeCl_6[b]	+1.95	0.0	(266)
	+1.4	0.0	(85, 513)
	+1.8	0.0	(545)
	+1.70[d]	0.0	(306)
M_2TeBr_6[b]	+1.74	0.0	(266)
	+1.7	0.0	(85, 513)
	+2.2	0.0	(545)
	+1.56[d]		(306)
M_2TeI_6[b]	+1.59	0.0	(266)
	+2.0	0.0	(85, 513)
	+1.27[d]		(306)
TeF_6^{2-}?	+0.0	0.0	(513)
	+0.97	5.6	(306)
$MTeF_5$[b]	+0.89	5.98	(266)
$TeCl_2$	+0.5	6.5	(361)
	+1.3	6.5	(545)
$TeBr_2$	+0.6	6.3	(545)

[a] Data given in mm/sec at 80°K.
[b] M = NH_4, Rb, Cs, K, NMe_4.
[c] Relative to $^{125}I/Cu$ as source.
[d] These values were quoted relative to ZnTe. ZnTe has a C.S. very close to zero relative to $^{125}I/Cu$ source (266, 361).

(*85, 513*) for the hexahalogen complexes of tellurium, M_2TeX_6 (X = Cl, Br, I), showed that the C.S. of these complexes increased in the order Cl < Br < I, i.e., in the same order for the hexahalogen complexes of Sn indicating that like Sn, $\delta R/R$ is positive. However, more recent and precise work (*266, 306*) shows that the trend is just the opposite. The C.S. varies in the order I < Br < Cl. The explanation of this trend will be examined shortly. However, it has been concluded (*266, 306, 361*) that $\delta R/R$ is positive for ^{125}Te.

Calculations of the quadrupole moment of the excited state of ^{125}Te (*99, 548*) indicate that Q is negative (*99*) and that its magnitude is about 0.2 barns.

Some information of chemical interest can be obtained from the results in Tables LXIX and LXX. For example, Te^{VI} has substantially more negative C.S. values than Te^{IV}, and this difference has been used by Erickson and Maddock (*219*) to calculate the amount of TeO$_2$ in α-TeO$_3$. The lack of quadrupole splittings in the M_2TeX_6 species is consistent with the known regular octahedral structure of some of the anions (*78, 162, 266*). The far-infrared spectra of these anions also indicate octahedral symmetry (*301*). In contrast, the compounds MTeF$_5$ (M = NH$_4^+$, Cs$^+$) have a large quadrupole splitting, consistent with the square-pyramidal structure deduced from IR and Raman spectroscopy (*302*). The agreement between the recently reported parameters for supposed TeF$_6^{2-}$ species (*306*) and the parameters for the TeF$_5^-$ species (*266*) strongly suggests that Gukasyan et al. (*306*) measured spectra of TeF$_5^-$ and not TeF$_6^{2-}$ (*266*).

The trend in C.S. values for the M_2TeX_6 series is of considerable bonding interest. The structural data described above indicate that the lone pair in Te is stereochemically inactive and, thus, has a very high s character. The very positive C.S. values for this species are consistent with this observation. Also, the trend in C.S. I < Br < Cl (opposite to that for Sn species) suggests that little s character is involved in the Te–X bonds. If the bonding mainly involves Te p electrons, then an electronegative ligand would cause an increase in $[\Psi(O)_s]^2$ (the lone pair contracts from deshielding) (*266*), and the C.S. decreases as is observed for Xe (*454*) and iodine (*445*) compounds having regular 90° and 180° bond angles.

Any other bonding information is very difficult to extract from present data because of the large discrepancies in C.S. values between workers. Erickson and Maddock (*219*) have suggested that agreement is improved between workers if TeO$_2$ is used as the C.S. standard. However, even if all the data are converted to TeO$_2$, large discrepancies are still observed.

4. ^{129}Xe

The Perlows have published several interesting papers on the Mössbauer spectra of Xe compounds using the 39.58 keV $\frac{3}{2}+ \rightarrow \frac{1}{2}+$ transition in ^{129}Xe. Much of the interest in this work has stemmed from the "production" of novel xenon compounds from their iodine precursors. This aspect of the work has been outlined in Section III, E of this chapter. In addition, however, useful bonding information has been

TABLE LXX

Mössbauer Parameters for Te Oxygen Compounds[a]

Compound	C.S.[c]	Q.S.	Ref.
Te(OH)$_6$ (monoclinic)	−0.98	0	(266)
	−1.01	0	(219)
	−1.15	0	(361)
	−0.8	0	(545)
BaH$_4$TeO$_6$	−0.87	0	(266)
Na$_2$H$_4$TeO$_6$	−0.94	0	(219)
Na$_2$TeO$_4$·2H$_2$O	−0.95	0	(266)
Na$_2$TeO$_4$	−1.26	0	(219)
MTeO$_4$[b]	~−0.9	0	(266)
	−0.9	0	(361, 545)
H$_2$TeO$_3$	+1.3	7.7	(545)
Na$_2$TeO$_3$	+0.42	6.65	(266)
	+0.35	5.94	(219)
	+0.22	5.78	(361)
SrTeO$_3$	+0.66	5.97	(266)
TeO$_2$	+0.91	6.25	(266)
	+0.91[d]	6.63	(219)
	+0.72	6.54	(361)
	+1.3	6.8	(545)
	+0.78	7.3	(546)
Te$_2$O$_4$	+0.89	6.65	(266)
Te$_2$O$_4$·HNO$_3$	+1.7	6.0	(545)
β-TeO$_3$	+1.01	—	(219)
α-TeO$_3$	−0.90	0	(219)
	−1.07	2.6	(361)
	−0.1	0	(545)

[a] Data given in mm/sec at 80°K
[b] M = K$_2$, Ca, Sr, Co, Ni.
[c] Relative to ^{125}I/Cu as source.
[d] This value and all other C.S. values from Ref. (219) are taken after fixing this TeO$_2$ value at +0.91 mm/sec.

obtained using the Townes–Dailey Q.S. approach outlined for iodine in Section IV,D. A summary of most of the xenon work is given by Perlow (449).

Although some xenon–oxygen compounds have been studied (449), examination of the data for xenon halides (Table LXXI) will illustrate the useful bonding information obtainable from xenon Mössbauer.

For ^{129}Xe, ρ is positive and comparatively small (Table LXVIII), giving rise to rather small changes in C.S. with rather large errors (Table

TABLE LXXI

Mössbauer Parameters for Xenon Halides[a]

Substance[b]	C.S.[c]	$\tfrac{1}{2}e^2qQ$ (mm/sec)[d]	e^2qQ (Mc/sec)[d]	U_p	h_p	Electron transfer/ bond
XeF$_4$	+0.40 ± 0.04	41.04	(+)2620	1.50	3.00	0.75
[XeCl$_4$]	+0.25 ± 0.08	25.62	(+)1640	0.94	1.88	0.47
XeF$_2$	+0.10 ± 0.12	39.00	(−)2490	1.43	1.43	0.72
[XeCl$_2$]	+0.17 ± 0.08	28.20	(−)1800	1.03	1.03	0.52
[XeBr$_2$]	−0.03 ± 0.07	22.2	(−)1415	0.81	0.81	0.41

[a] From Ref. (454). Data given in mm/sec at 4.2°K.
[b] Those compounds in brackets were prepared from the corresponding I compound and used as sources. For example, XeCl$_4$ was prepared from KICl$_4$.
[c] Relative to xenon clathrate at 4.2°K.
[d] e^2qQ for one p_z electron in ^{129}Xe is +1742 Mc/sec or +54.6 mm/sec, taking Eγ = 39.58 keV.

LXXI). An increase in $[\Psi(O)_s]^2$ should thus increase the C.S. The quadrupole moment of ^{129}Xe has been determined to be $Q = -0.41 \pm 0.06$ barns (448, 449). The e^2qQ value for one Xe $5p_z$ electron is equal to +1740 Mc/sec or +54.6 mm/sec (448). It should be recognized that q is negative for one p_z electron, but the negative sign of Q gives a positive e^2qQ for an excess of negative charge along the Z EFG direction.

XeF$_2$ is known to be linear, and the linear and square-planar structures of ICl$_2^-$ and ICl$_4^-$ are probably preserved in XeCl$_2$ and XeCl$_4$ [see references in (453)]. The center shifts and quadrupole splittings are consistent with pure p bonding, perhaps expected from the 90° and 180° bond angles in linear and square-planar structures (454). On the assumption of pure p bonding, we would expect that the XeX$_4$ compounds would have a more positive C.S. than the XeX$_2$ compounds owing to the lower p-electron density and the resulting higher $[\Psi(O)_s]^2$ owing to deshielding in XeX$_4$. This order is found experimentally (Table LXXI). Also, on the basis of pure p bonding we would expect the order of C.S. values to be $F^- > Cl^- > Br^-$, and this is generally true, although the differences are very small and are often within the errors. This same trend holds for the analogous iodine compounds (Section IV,D). Appreciable s character in the bonds would lead to the opposite trend in C.S. values, i.e., $F^- < Cl^- < Br^-$. More negative C.S. values for the xenon oxides (453) implies that the Xe–O bonding involves appreciable s electron density on the xenon atom—as has been concluded for iodine–oxygen compounds [(445) and Section IV,D].

Since e^2qQ for one p_z electron has been determined, Eq. (53) can be used along with known Q.S. values to derive U_p, and these are given in Table LXXI. The signs of e^2qQ cannot be determined directly [although the Goldanskii–Karyagin asymmetry is helpful in assigning the sign of e^2qQ (449)], but it is apparent that the signs of the square-planar and linear compounds will be the same as those for the isoelectronic and isostructural iodine species (Table LII). The square-planar tetrahalides have a positive e^2qQ ($q = -ve$), and the linear dihalides a negative e^2qQ ($q = +ve$). In the square-planar compounds, the bonding involves the p_x and p_y orbitals and $h_p = 2U_p$, while in the dihalides, $h_p = U_p$. The assumption of pure p bonding can be tested using the C.S. data, which can be thought of as measuring h_p directly [for example, in the ^{129}I case, see Eq. (73)]. Unfortunately, the xenon C.S. values are not precisely known, but a plot of h_p (from Q.S.) versus C.S. gives a reasonable straight line through the origin. The change in center shift per p electron hole (0.13 mm/sec) was obtained from the slope of this line.

The Mössbauer data suggests, then, that as for the isoelectronic iodine compounds, the bonding from xenon to the halides involves mainly xenon p electrons. The p-electron transfer/bond given in the last column of Table LXXI is consistent with this interpretation. The two chlorides and the two fluorides display very similar electron transfers per bond (449).

ACKNOWLEDGMENTS

We are very grateful to Dr. M. G. Clark for helpful comments.

REFERENCES

1. Ablov, A. V., Belozerskii, G. N., Goldanskii, V. I., Makarov, E. F., Trukhtanov, V. A., and Khrapov, V. V., *Dokl. Phys. Chem.* **151**, 712 (1963) (Russ.: *Dokl. Akad. Nauk SSSR, Phys. Chem. Sect.* p. 1352).*
2. Ablov, A. V., Bersuker, I. B., and Goldanskii, V. I., *Dokl. Phys. Chem.* **152**, 934 (1963) (Russ., p. 1391).

* Hereafter the Russian journals will be designated by "Russ." and the page number will follow.

3. Ablov, A. V., Goldanskii, V. I., Stukan, R. A., and Makarov, E. F., *Dokl. Phys. Chem.* **170**, 565 (1966) (Russ. p. 128).
4. Alcock, N. W., and Timms, R. E., *J. Chem. Soc.*, *A* 1876 (1968).
5. Aleksandrov, A. Yu., Delyagin, N. N., Mitrofanov, K. P., Polak, L. S., and Shpinel, V. S., *Dokl. Phys. Chem.* **148**, 1 (1963) (Russ. p. 126).
6. Aleksandrov, A. Yu., Delyagin, N. N., Mitrofanov, K. P., Polak, L. S., and Shpinel, V. S., *Sov. Phys. JETP* **16**, 879 (1963).
7. Aleksandrov, A. Yu., Dorfman, Ya. G., Lependina, O. L., Mitrofanov, K. P., Plotnikova, M. V., Polak, L. S., Temkin, A. Ya., and Shpinel, V. S., *Russ. J. Phys. Chem.* **38**, 1185 (1964).
8. Aleksandrov, A. Yu., Mitrofanov, K. P., Okhlobystin, O. Yu., Polak, L. S., and Shpinel, V. S., *Dokl. Phys. Chem.* **153**, 974 (1963) (Russ. p. 370).
9. Aleksandrov, A. Yu., Okhlobystin, O. Yu., Polak, L. S., and Shpinel, V. S., *Dokl. Phys. Chem.* **157**, 768 (1964) (Russ. p. 934).
10. Ali, K. M., Cunningham, D., Frazer, M. J., Donaldson, J. D., and Senior, B. J., *J. Chem. Soc.*, *A* 2836 (1969).
11. Alti, G. De., Galasso, V., and Bigotto, A., *Inorg. Chim. Acta* **3**, 527 (1969).
12. Alti, G. De., Galasso, V., Bigotto, A., and Costa, G., *Inorg. Chim. Acta* **3**, 533 (1969).
13. Alyea, E. C., Bradley, D. C., Copperthwaite, R. G., Sales, K. D., Fitzsimmons, B. W., and Johnson, C. E., *Chem. Commun.* 1715 (1970)
14. "An Introduction to Mössbauer Spectroscopy" (L. May, ed.). Plenum, New York, 1971.
15. Araujo, F. T. de, Dufresne, A., Lima, G. C. de, and Knudsen, J. M., *Chem. Phys. Lett.* **7**, 333 (1970).
16. Archer, E. M., and van Schalkwyk, T. G. D., *Acta Crystallogr.* **6**, 88 (1953).
17. Artman, J. O., *in* "Mössbauer Effect Methodology" (I. J. Gruverman, ed.). Vol. 7. Plenum, New York, 1971.
18. Attridge, C. J., *Organometal. Chem. Rev. Sect. A* **5**, 323 (1970).
19. Axtmann, R. C., Hazony, Y., and Hurley, J. W. Jr., *Chem. Phys. Lett.* **2**, 673 (1968).
20. Bancroft, G. M., *Phys. Lett. A* **26**, 17 (1967).
21. Bancroft, G. M., *Chem. Geol.* **5**, 255 (1970).
22. Bancroft, G. M., *Chem. Phys. Lett.* **10**, 449 (1971).
23. Bancroft, G. M., and Burns, R. G., *1966 Int. Mineral Assoc. Pap. Proc., 5th Gen. Meeting Mineral Soc., London* 36 (1968).
24. Bancroft, G. M., and Burns, R. G., *Mineral Soc. Amer. Spec. Pap.* **2**, 137 (1969).
25. Bancroft, G. M., Burns, R. G., and Howie, R. A., *Nature (London)* **213**, 1221 (1967).
26. Bancroft, G. M., Burns, R. G., and Maddock, A. G., *Amer. Mineral.* **52**, 1009 (1967).
27. Bancroft, G. M., Burns, R. G., and Stone, A. J., *Geochim. Cosmochim. Acta* **32**, 547 (1968).
28. Bancroft, G. M., Butler, K. D., Dale, B., and Rake, A. T., *J. Chem. Soc., Dalton Trans.* (1972). In press.
29. Bancroft, G. M., Butler, K. D., and Rake, A. T., *J. Organometal. Chem.* (1972). **34**, 137 (1972).
30. Bancroft, G. M., Dharmawardena, K. G., and Maddock, A. G., *J. Chem. Soc., A* 2914 (1969).

31. Bancroft, G. M., Dharmawardena, K. G., and Maddock, A. G., *Inorg. Chem.* **9**, 223 (1970).
32. Bancroft, G. M., Dharmawardena, K. G., and Stone, A. J., *Chem Commun.* 6 (1971).
33. Bancroft, G. M., Garrod, R. E. B., and Maddock, A. G., *Inorg. Nucl. Chem. Lett.* **7**, 1157 (1971).
34. Bancroft, G. M., Garrod, R. E. B., and Maddock, A. G., *J. Chem. Soc.*, A 3165 (1971).
35. Bancroft, G. M., Garrod, R. E. B., and Maddock, A. G., unpublished observations.
36. Bancroft, G. M., Garrod, R. E. B., Maddock, A. G., Mays, M. J., and Prater, B. E., *Chem. Commun.* 201 (1970).
37. Bancroft, G. M., Garrod, R. E. B., Maddock, A. G., Mays, M. J., and Prater, B. E., *J. Amer. Chem. Soc.* **94**, 647 (1972).
38. Bancroft, G. M., Garrod, R. E. B., Poliakoff, M., and Turner, J. J., unpublished observations.
39. Bancroft, G. M., and Libbey, E. T., unpublished results.
40. Bancroft, G. M., Maddock, A. G., Burns, R. G., and Strens, R. G. J., *Nature (London)* **212**, 913 (1966).
41. Bancroft, G. M., Maddock, A. G., Ong, W. K., and Prince, R. H., *J. Chem. Soc.*, A 723 (1966).
42. Bancroft, G. M., Maddock, A. G., Ong, W. K., Prince, R. H., and Stone, A. J., *J. Chem. Soc.*, A 1966 (1967).
43. Bancroft, G. M., Maddock, A. G., and Randl, R. P., *J. Chem. Soc.*, A 2939 (1968).
44. Bancroft, G. M., Mays, M. J., and Prater, B. E., *J. Chem. Soc.*, A 956 (1970).
45. Bancroft, G. M., Mays, M. J., Prater, B. E., and Stefanini, F. P., *J. Chem. Soc.*, A 2146 (1970).
46. Baranovskii, V. I., Sergeev, V. P., and Dzevitskii, B. E., *Dokl. Phys. Chem.* **184**, 55 (1969) (Russ. p. 632).
47. Barber, M., and Swift, P., *Chem. Commun.* 1338 (1970).
48. Bargeron, C. B., and Drickamer, H. G., *J. Chem. Phys.* **55**, 3471 (1971).
49. Bartunik, H. D., Potzel, W., Mössbauer, R. L., and Kaindl, G., *Z. Phys.* **240**, 1 (1970).
50. Bearden, A. J., Marsh, H. S., and Zuckerman, J. J., *Inorg. Chem.* **5**, 1260 (1966).
51. Beattie, I. R., and McQuillan, G. P., *J. Chem. Soc.*, A 1519 (1963).
52. Beauchamp, A. L., Bennett, M. J., and Cotton, F. A., *J. Amer. Chem. Soc.* **91**, 297 (1969).
53. Beckett, R., Heath, G. A., Hoskins, B. F., Kelly, B. P., Martin, R. L., Roos, I. A. G., and Weickhardt, P. L., *Inorg. Nucl. Chem. Lett.* **6**, 257 (1970).
54. Bergen, A. Van den, Murray, K. S., West, B. O., and Buckley, A. N., *J. Chem. Soc.*, A 2051 (1969).
55. Berrett, R. R., and Fitzsimmons, B. W., *J. Chem. Soc.*, A 525 (1967).
56. Berrett, R. R., Fitzsimmons, B. W., and Owusu, A. A., *J. Chem. Soc.*, A 1575 (1968).
57. Bilevich, K. A., Goldanskii, V. I., Rochev, V. Ya., and Khrapov, V. V., *Izv. Akad. Nauk SSSR Ser. Khim.* 1583 (1969) (Russ. p. 1705).
58. Birchall, T., *Can. J. Chem.* **47**, 1351 (1969).
59. Birchall, T., *Can. J. Chem.* **47**, 4563 (1969).

60. Birchall, T., and Drummond, I., *Inorg. Chem.* **10**, 399 (1971).
61. Birchall, T., and Greenwood, N. N., *J. Chem. Soc.*, A 286 (1969).
62. Birchall, T. and Valle B. D., *Chem. Commun.* 675 (1970).
63. Birchall, T., Valle, B. D., Martineau, E., and Milne, J. B., *J. Chem. Soc.*, A 1855 (1971).
64. Bird, S. R. A., Donaldson, J. D., Holding, A. F. LeC., Senior, B. J., and Tricker, M. J., *J. Chem Soc.*, A 1616 (1971).
65. Bird, S. R. A., Donaldson, J. D., Keppie, S. A., and Lappert, M. F., *J. Chem. Soc.*, A 1311 (1971).
66. Biryukov, B. P., Unisimov, K. N., Struchkov, Yu. T., Kolobova, N. E., Osipoua, O. P., and Zakharova, M. Ya., *J. Struct. Chem.* (*USSR*) **8**, 554 (1967).
67. Blackman, A. G., Landman, D. A., and Lurio, A., *Phys. Rev.* **161**, 60 (1967).
68. Bloembergen, N., and Sorokin, P. P., *Phys. Rev.* **110**, 865 (1958).
69. Blom, E. A., Penfold, B. R., and Robinson, W. T., *J. Chem. Soc.*, A 913 (1969).
70. Bokii, N. G., Zakharova, G. N., and Struchkov, Yu. T., *J. Struct. Chem.* (*USSR*) **11**, 828 (1970) (Russ. p. 895).
71. Bowen, L. H., Garrou, P. E., and Long, G. G., *J. Inorg. Nucl. Chem.* **33**, 953 (1971).
72. Bowen, L. H., Stevens, J. G., and Long, G. G., *J. Chem. Phys.* **51**, 2010 (1969).
73. Boylan, M. J., Nelson, S. M., and Deeney, F. A., *J. Chem. Soc.*, A 976 (1971).
74. Boyle, A. J. F., Bunbury, D. St. P., and Edwards, C., *Proc. Phys. Soc.* **79**, 416 (1962).
75. Boyle, A. J. F., and Hall, H. E., *Rep. Progr. Phys.* **25**, 441 (1962).
76. Brady, P. R., Duncan, J. F., and Mok, K. F., *Proc. Roy. Soc. Ser.* A **287**, 343 (1965).
77. Brändén, C. I., *Acta Chem. Scand.* **17**, 759 (1963).
78. Brown, I. D., *Can. J. Chem.* **42**, 2758 (1964).
79. Bryan, R. F., *J. Amer. Chem. Soc.* **86**, 733 (1964).
80. Bryan, R. F., *J. Chem. Soc.*, A 696 (1968).
81. Bryuchova, E. V., Semin, G. K., Goldanskii, V.I., and Khrapov, V. V., *Chem. Commun.* 491 (1968).
82. Bryukhanov, V. A., Delyagin, N. N., Kuzmin, R. N., and Shpinel, V. S., *Sov. Phys. JETP* **19**, 1344 (1964).
83. Bryukhanov, V. A., Goldanskii, V. I., Delyagin, N. N., Makarov, E. F., and Shpinel, V. S., *Sov. Phys. JETP* **15**, 443 (1962).
84. Bryukhanov, V. A., Goldanskii, V. I., Delyagin, N. N., Korytko, L. A., Makarov, E. F., Suzdalev, I. P., and Shpinel, V. S., *Sov. Phys. JETP* **16**, 321 (1963).
85. Bryukhanov, V. A., Iofa, B. Z., Opalenko, A. A., and Shpinel, V. S. *Zh. Neorg. Khim.* **12**, 1985 (1967).
86. Buckley, A. N., Rumbold, B. D., Wilson, G. V. H., and Murray, K. S., *J. Chem. Soc.*, A 2298 (1970).
87. Buckley, A. N., Wilson, G. V. H., and Murray, K. S., *Solid State Commun.* **7**, 471 (1969).
88. Bukshpan, S., *J. Chem. Phys.* **48**, 4242 (1968).
89. Bukshpan, S., Goldstein, C., and Sonnino, T., *J. Chem. Phys.* **49**, 5477 (1968).
90. Bukshpan, S., Goldstein, C., Soriano, J., and Shamir, J., *J. Chem. Phys.* **51**, 3976 (1969).
91. Bukshpan, S., and Herber, R. H., *J. Chem. Phys.* **46**, 3375 (1967).
92. Bukshpan, S., Soriano, J., and Shamir, J., *Chem. Phys. Lett.* **4**, 241 (1969).

93. Buskshpan, S., and Sonnino, T., *J. Chem. Phys.* **48**, 4442 (1968).
94. Burbridge, C. D., and Goodgame, D. M. L., *J. Chem. Soc., A* 694 (1967).
95. Burbridge, C. D., and Goodgame, D. M. L., *J. Chem. Soc., A* 1074 (1968).
96. Burbridge, C. D., and Goodgame, D. M. L., *Inorg. Chim. Acta* **4**, 231 (1970).
97. Burbridge, C. D., Goodgame, D. M. L., and Goodgame, M., *J. Chem. Soc., A* 349 (1967).
98. Burns, G. R., *Inorg. Chem.* **7**, 277 (1968).
99. Buyrn, A. B., and Grodzins, L., *Bull. Amer. Phys. Soc.* **8**, 43 (1963).
100. Cameron, J. A., Keszthelyi, L., Nagy, G., and Kacsoh, L., *Chem. Phys. Lett.* **8**, 628 (1971).
101. Campbell, I. G. M., *J. Chem. Soc.* 3116 (1955).
102. Carty, A. J., Efratry, A., Ng, T. W., and Birchall, T., *Inorg. Chem.* **9**, 1263 (1970).
103. Carty, A. J., Hinsperger, T., Mihichuk, L., and Sharma, H. D., *Inorg. Chem.* **9**, 2573 (1970).
104. Carty, A. J., Ng, T. W., Carter, W., Palenik, G. J., and Birchall, T., *Chem. Commun.*, 1101 (1969).
105. Casabella, P. A., and Bray, P. J., *J. Chem. Phys.*, **28**, 1182 (1958).
106. Cassidy, J. E., Moser, W., Donaldson, J. D., Jelen, A., and Nicholson, D. G., *J. Chem. Soc., A* 173 (1970).
107. Champion, A. R., Vaughan, R. W., and Drickamer, H. G., *J. Chem. Phys.* **47**, 2583 (1967).
108. Charlton, J. S., and Nichols, D. I., *J. Chem. Soc. A* 1484 (1970).
109. Chatt, J., and Hayter, R. G., *J. Chem. Soc.* 5507 (1961).
110. "Chemical Applications of Mössbauer Spectroscopy" (V. I. Goldanskii and R. H. Herber, eds.). Academic Press, New York, 1968.
111. *Chem. Soc. Spec. Publ. No.* **11** (1958); *No.* **18** (1965).
112. Cheng, H. S., and Herber, R. H., *Inorg. Chem.* **9**, 1686 (1970).
113. Cheng, H. S., and Herber, R. H., *Inorg. Chem.* **10**, 1315 (1971).
114. Chivers, T., and Sams, J. R., *Chem. Commun.* 249 (1969).
115. Chivers, T., and Sams, J. R., *J. Chem. Soc., A* 928 (1970).
116. Chow, Y. M., *Inorg. Chem.* **9**, 794 (1970).
117. Christe, K. O., and Sawodny, W., *Inorg. Chem.* **6**, 1783 (1967).
118. Clark, H. C., O'Brien, R. J., and Trotter, J., *J. Chem. Soc.* 2332 (1964).
119. Clark, M. G., *Mol. Phys.* **20**, 257 (1971).
120. Clark, M. G., *J. Chem. Phys.* **54**, 697 (1971).
121. Clark, M. G., Bancroft, G. M., and Stone, A. J., *J. Chem. Phys.* **47**, 4250 (1967).
122. Clark, M. G., Maddock, A. G., and Platt, R. H., *J. Chem. Soc., Dalton Trans.* 281 (1972).
123. Clark, R. J. H., Davies, A. G., and Puddephatt, R. J., *J. Chem. Soc., A* 1828 (1968).
124. Clark, R. J. H., Maresca, L., and Smith, P. J., *J. Chem. Soc., A* 2687 (1970).
125. Clausen, C. A., and Good, M. L., *in* "Mössbauer Effect Methodology" (I. J. Gruverman, ed.), Vol. 4. Plenum, New York, 1968.
126. Clausen, C. A., and Good, M. L., *Inorg. Chem.* **9**, 220 (1970).
127. Clausen, C. A., and Good, M. L., *Inorg. Chem.* **9**, 817 (1970).
128. Clausen, C. A., Prados, R. A., and Good, M. L., *Chem. Commun.* 1188 (1969).
129. Clausen, C. A., Prados, R. A., and Good, M. L., *J. Amer. Chem. Soc.* **92**, 7483 (1970).

130. Clausen, C. A., Prados, R. A., and Good, M. L., *Chem. Phys. Lett.* **8**, 565 (1971).
131. Cody, V., and Corey, E. R., *J. Organomet. Chem.* **19**, 359 (1969).
132. Collins, R. L., *J. Chem. Phys.* **42**, 1072 (1965).
133. Collins, R. L., and Pettit, R., *J. Chem. Phys.* **39**, 3433 (1963).
134. Collins, R. L., and Pettit, R., *J. Amer. Chem. Soc.* **85**, 2332 (1965).
135. Collins, R. L., Pettit, R., and Baker, W. A., *J. Inorg. Nucl. Chem.* **28**, 1001 (1966).
136. Collins, R. L., and Travis, J. C., in "Mössbauer Effect Methodology" (I. J. Gruverman, ed.), Vol. 3. Plenum, New York, 1967.
137. Cordey Hayes, M., *J. Inorg. Nucl. Chem.* **26**, 915 (1964).
138. Cordey Hayes, M., *J. Inorg. Nucl. Chem.* **26**, 2306 (1964).
139. Cordey Hayes, M., in "Chemical Applications of Mössbauer Spectroscopy" (R. H. Herber and V. I. Goldanskii, eds.), p. 314. Academic Press, New York, 1968.
140. Cordey Hayes, M., Kemmitt, R. D. W., Peacock, R. D., and Rimmer, G. D., *J. Inorg. Nucl. Chem.* **31**, 1515 (1969).
141. Cordey Hayes, M., Peacock, R. D., and Vucelic, M., *J. Inorg. Nucl. Chem.* **29**, 1177 (1967).
142. Costa, N. L., Danon, J., and Xavier, R. M., *J. Phys. Chem. Solids* **23**, 1783 (1962).
143. Cotton, F. A., and Edwards, W. T., *J. Amer. Chem. Soc.* **91**, 843 (1969).
144. Cowan, M., and Gordy, W., *Phys. Rev.* **104**, 551 (1956).
145. Cox, M., Darken, J., Fitzsimmons, B. W., Smith, A. W., Larkworthy, L. G., and Rogers, K. A., *Chem. Commun.* 105 (1970).
146. Crow, J. P., Cullen, W. R., Herring, F. G., Sams, J. R., and Tapping, R. L., *Inorg. Chem.* **10**, 1616 (1971).
147. Cullen, W. R., Harbourne, D. A., Liengme, B. V., and Sams, J. R., *Inorg. Chem.* **8**, 1464 (1969).
148. Cullen, W. R., and Hochstrasser, R. M., *J. Mol. Spectrosc.* **5**, 118 (1960).
149. Cullen, W. R., Sams, J. R., and Thompson, J. A. J., *Inorg. Chem.* **10**, 843 (1971).
150. Cullen, W. R., Sams, J. R., and Waldman, M. C., *Inorg. Chem.* **9**, 1682 (1970).
151. Cullen, W. R., and Waldman, M. C., *Can. J. Chem.* **47**, 3093 (1969).
152. Cunningham, D., Frazer, M. J., and Donaldson, J. D., *J. Chem. Soc.*, A 2049 (1971).
153. Da Costa, M. I., Fraga, E. F. R., and Sonnino, T., *J. Chem. Phys.* **52**, 1611 (1970).
154. Dahl, J. P. and Ballhausen, C. J., *Kgl. Dan. Vidensk. Selsk. Mat. Fys. Medd.* **33**, No. 5 (1961).
155. Dale, B. W., Williams, R. J. P., Edwards, P. R., and Johnson, C. E., *J. Chem. Phys.* **49**, 3445 (1968).
156. Dale, B. W., Williams, R. J. P., Edwards, P. R., and Johnson, C. E., *Trans. Faraday Soc.* **64**, 620 (1968).
157. Dale, B. W., Williams, R. J. P., Edwards, P. R., and Johnson, C. E., *Trans. Faraday Soc.* **64**, 3011 (1968).
158. Dale, B. W., Williams, R. J. P., Johnson, C. E., and Thorp, T. L., *J. Chem. Phys.* **49**, 3441 (1968).
159. Dalton, R. F., and Jones, K., *Inorg. Nucl. Chem. Lett.* **5**, 785 (1969).

160. Danon, J., in "Chemical Applications of Mössbauer Spectroscopy" (V. I. Goldanskii and R. H. Herber, eds.), p. 159. Academic Press, New York, 1968.
161. Danon, J., and Iannarella, L., *J. Chem. Phys.* **47**, 382 (1967).
162. Das, A. K., and Brown, I. D., *Can. J. Chem.* **44**, 939 (1966).
163. Das, T. P., and Hahn, E. L., "Nuclear Quadrupole Resonance Spectroscopy." Academic Press, New York, 1958.
164. Davies, A. G., personal communication.
165. Davies, A. G., and Donaldson, J. D., *J. Chem. Soc.*, A 946 (1968).
166. Davies, A. G., Donaldson, J. D., and Simpson, W. B., *J. Chem. Soc.*, A 417 (1969).
167. Davies, A. G., Milledge, H. J., Puxley, D. C., and Smith, P. J., *J. Chem. Soc.*, A 2862 (1970).
168. Davies, A. G., Smith, L., and Smith, P. J., *J. Organomet. Chem.* **23**, 135 (1970).
169. Davies, A. G., Smith, L., Smith, P. J., and McFarlane, W., *J. Organomet. Chem.* **29**, 245 (1971).
170. Debye, N. W. G., Fenton, D. E., Ulrich, S. E., and Zuckerman, J. J., *J. Organomet. Chem.* **28**, 339 (1971).
171. Debye, N. W. G., Rosenberg, E., and Zuckerman, J. J., *J. Amer. Chem. Soc.* **90**, 3234 (1968).
172. Deer, W. A., Howie, R. A., and Zussman, J., "Rock Forming Minerals." Longmans, Green, New York, 1963.
173. Dehmelt, H. G., *Naturwissenschaften* **37**, 398 (1950).
174. Dehn, J. T., and Mulay, L. N., *J. Inorg. Nucl. Chem.* **31**, 3103 (1969).
175. DeVoe, J. R., and Spijkerman, J. J., *Mössbauer Spectrometry, Anal. Chem. Ann. Rev.* **42**, 366R (1970) and previous years.
176. Devooght, J., Gielen, M., and Lejeune, S., *J. Organomet. Chem.* **21**, 333 (1970).
177. deVries, J. L. K. F., Trooster, J. M., and de Boer, E., *Chem. Commun.* 604 (1970).
178. deVries, J. L. K. F., Trooster, J. M., and de Boer, E., *Inorg. Chem* **10**, 81 (1971).
179. Dezsi, I., Balizs, A., Molnar, B., Gorobchenko, V. D., and Lukashevich, I. I., *J. Inorg. Nucl. Chem.* **31**, 1661 (1969).
180. Dezsi, I., Keszthelyi, L., Molnar, B., and Pocs, L., *Phys. Lett.* **18**, 28 (1965).
181. Dollase, W. A., *Amer. Mineral.* **56**, 447 (1971).
182. Donaldson, J. D., *Progr. Inorg. Chem.* **8**, 287 (1967).
183. Donaldson, J. D., Filmore, E. J., and Tricker, M. J., *J. Chem. Soc.*, A 1109 (1971).
184. Donaldson, J. D. and Jelen, A., *J. Chem. Soc.*, A 1448 (1968).
185. Donaldson, J. D., and Nicholson, D. G., *J. Chem. Soc.*, A 145 (1970).
186. Donaldson, J. D., and Nicholson, D. G., *Inorg. Nucl. Chem. Lett.* **6**, 151 (1970).
187. Donaldson, J. D., Nicholson, D. G., and Senior, B. J., *J. Chem. Soc.*, A 2928 (1968).
188. Donaldson, J. D., and Oteng, R., *Inorg. Nucl. Chem. Lett.* **3**, 163 (1967).
189. Donaldson, J. D., Oteng, R., and Senior, B. J., *Chem. Commun.* 618 (1965).
190. Donaldson, J. D., and Senior, B. J., *J. Chem. Soc.*, A 1796 (1966).
191. Donaldson, J. D., and Senior, B. J., *J. Chem. Soc.*, A 1798 (1966).
192. Donaldson, J. D., and Senior, B. J., *J. Inorg. Nucl. Chem.* **31**, 881 (1969).
193. Donaldson, J. D., and Senior, B. J., *J. Chem. Soc.*, A 2358 (1969).

194. Doskey, M. A., and Curran, C., *Inorg. Chim. Acta* **3**, 169 (1969).
195. Dosser, R. J., Eilbeck, W. J., Underhill, A. E., Edwards, P. R., and Johnson, C. E., *J. Chem. Soc.*, A 810 (1969).
196. Drickamer, H. G., Lewis, G. K., and Fung, S. C., *Science* **163**, 885 (1969).
197. Dulaney, G. W., and Clifford, A. F., *in* "Mössbauer Effect Methodology" (I. J. Gruverman, ed.), p. 65. Vol. 5. Plenum, New York, 1970.
198. Duncan, J. F., and Golding, R. M., *Quart. Rev., Chem. Soc.* **19**, 36 (1965).
199. Duncan, J. F., Golding, R. M., and Mok, K. F., *J. Inorg. Nucl. Chem.* **28**, 1114 (1966).
200. Dundon, R. W., and Walter, L. S., *Earth Planet. Sci. Lett.* **2**, 372 (1967).
201. Edwards, P. R., and Johnson, C. E., *J. Chem. Phys.* **49**, 211 (1968).
202. Edwards, P. R., Johnson, C. E., and Williams, R. J. P., *J. Chem. Phys.* **47**, 2074 (1967).
203. Ehrlich, B. S., and Kaplan, M., *J. Chem. Phys.* **50**, 2041 (1969).
204. Ehrlich, B. S., and Kaplan, M., *J. Chem. Phys.* **51**, 603 (1969).
205. Ehrlich, B. S., and Kaplan, M., *J. Chem. Phys.* **54**, 612 (1971).
206. Emerson, G. F., Mahler, J. E., Pettit, R., and Collins, R., *J. Amer. Chem. Soc.* **86**, 3590 (1964).
207. Ensling, J., Gütlich, P., Hassellbach, K. M., and Fitzsimmons, B. W., *J. Chem. Soc.*, A 1940 (1971).
208. Epstein, L. M., *J. Chem. Phys.* **36**, 2731 (1962).
209. Epstein, L. M., *J. Chem. Phys.* **40**, 435 (1964).
210. Epstein, L. M., and Straub, D. K., *Inorg. Chem.* **4**, 1551 (1965).
211. Epstein, L. M., and Straub, D. K., *Inorg. Chem.* **8**, 453 (1969).
212. Epstein, L. M., and Straub, D. K., *Inorg. Chem.* **8**, 560 (1969).
213. Epstein, L. M., and Straub, D. K., *Inorg. Chem.* **8**, 784 (1969).
214. Erich, U., Frölich, F., Gütlich, P., and Webb, G. A., *Inorg. Nucl. Chem. Lett.* **5**, 855 (1969).
215. Erickson, N. E., Ph.D. Thesis, Columbia Univ., New York, New York, 1964.
216. Erickson, N. E., *in* "Mössbauer Effect and Its Applications in Chemistry" (R. F. Gould, ed.). Amer. Chem. Soc. Publ., Washington, D.C., 1967. No. 68.
217. Erickson, N. E., *Chem. Commun.* 1349 (1970).
218. Erickson, N. E., and Fairhall, A. W., *Inorg. Chem.* **4**, 1320 (1965).
219. Erickson, N. E., and Maddock, A. G., *J. Chem. Soc.*, A 1665 (1970).
220. Erickson, N. E., and Sutin, N., *Inorg. Chem.* **5**, 1834 (1966).
221. Evans, B. J., Ghose, S., and Hafner, S. S., *J. Geol.* **75**, 306 (1967).
222. Faltens, M. O., and Shirley, D. A., *J. Chem. Phys.* **53**, 4249 (1970).
223. Farmery, K., Kilner, M., Greatrex, R., and Greenwood, N. N., *Chem. Commun.* 593 (1968).
224. Farmery, K., Kilner, M., Greatrex, R., and Greenwood, N. N., *J. Chem. Soc.*, A 2339 (1969).
225. Fenger, J., Maddock, A. G., and Siekierska, K. E., *J. Chem. Soc.*, A 3255 (1970).
226. Fenton, D. E., and Zuckerman, J. J., *J. Amer. Chem. Soc.* **90**, 6226 (1968).
227. Fenton, D. E., and Zuckerman, J. J., *Inorg. Chem.* **8**, 1771 (1969).
228. Finger, L. W., *Mineral. Soc. Amer. Spec. Pap.* **2**, 95 (1969).
229. Finger, L. W., *Amer. Mineral.* **55**, 300 (1970).
230. Fischer, D. C., and Drickamer, H. G., *J. Chem. Phys.* **54**, 4825 (1971).
231. Fischer, K. F., *Amer. Mineral.* **51**, 814 (1966).
232. Fitzsimmons, B. W., *J. Chem. Soc.*, A 3235 (1970).

233. Fitzsimmons, B. W., Owusu, A. A., Seeley, N. J., and Smith, A. W., *J. Chem. Soc., A* 935 (1970).
234. Fitzsimmons, B. W., Seeley, N. J., and Smith, A. W., *J. Chem. Soc., A* 143 (1969).
235. Fluck, E., *Advan. Inorg. Chem. Radiochem.* **6**, 433 (1964).
236. Fluck, E., *in* "Chemical Applications of Mössbauer Spectroscopy" (R. H. Herber and V. I. Goldanskii, eds.), Chap. 4. Academic Press, New York, 1968.
237. Fluck, E., and Brauch, K. F., *Z. Anorg. Chem.* **364**, 107 (1969).
238. Fluck, E., Kerler, W., and Neuwirth, W., *Angew. Chem. Int. Ed., Engl.* **2**, 277 (1963).
239. Fluck, E., and Kuhn, P., *Z. Anorg. Chem.* **350**, 263 (1967).
240. Ford, B. F. E., Liengme, B. V., and Sams, J. R., *J. Organomet. Chem.* **19**, 53 (1969).
241. Ford, B. F. E., and Sams, J. R., *J. Organomet. Chem.* **21**, 345 (1970).
242. Ford, B. F. E., Sams, J. R., Goel, R. G., and Ridley, D. R., *J. Inorg. Nucl. Chem.* **33**, 23 (1971).
243. Forder, R. A., and Sheldrick, G. M., *J. Organomet. Chem.* **21**, 115 (1970).
244. Forder, R. A., and Sheldrick, G. M., *J. Organomet. Chem.* **22**, 611 (1970).
245. Forster, D., and Goodgame, D. M. L, *J. Chem. Soc.* 262 (1965).
246. Frank, E., and Abeledo, C., *Inorg. Chem.* **5**, 1453 (1966).
247. Frauenfelder, H., "The Mössbauer Effect." Benjamin, New York, 1962.
248. Friedt, J. M., *J. Inorg. Nucl. Chem.* **32**, 431 (1970).
249. Friedt, J. M., and Adloff, J. P., *C. R. Acad. Sci. Ser. C* **264**, 1356 (1967).
250. Fung, S. C., and Drickamer, H. G., *J. Chem. Phys.* **51**, 4353 (1969).
251. Fung, S. C., and Drickamer, H. G., *J. Chem. Phys.* **51**, 4360 (1969).
252. Gabriel, J. R., and Ruby, S. L., *Nucl. Instrum. Methods* **36**, 23 (1965).
253. Gallagher, P. K., and Kurkjian, C. R., *Inorg. Chem.* **5**, 214 (1966).
254. Gancedo, R. R., Maddock, A. G., and Platt, R. H., unpublished observations.
255. Garg, A. N., and Goel, P. S., *Inorg. Chem.* **10**, 1344 (1971).
256. Garg, A. N., Shukla, P. N., and Goel, P. S., *Chem. Phys. Lett.* **7**, 494 (1970).
257. Garrod, R. E. B., Platt, R. H., and Sams, J. R., *Inorg. Chem.* **10**, 424 (1971).
258. Gassenheimer, B., and Herber, R. H., *Inorg. Chem.* **8**, 1120 (1969).
259. Gerloch, M., Lewis, J., Mabbs, F. E., and Richards, A., *J. Chem. Soc., A* 112 (1968) and references.
260. Gerloch, M., and Mabbs, F. E., *J. Chem. Soc., A* 1598 (1967).
261. Gerloch, M., and Mabbs, F. E., *J. Chem. Soc., A* 1900 (1967).
262. Ghose, S., *Acta Crystallogr.* **14**, 622 (1961).
263. Ghose, S., *Z. Kristallogr. Kristallgeometrie, Kristallphys. Kristallchem.* **122**, 81 (1965).
264. Ghose, S., and Hafner, S. S., *Z. Kristallogr., Kristallgeometrie, Kristallphys., Kristallchem.* **125**, 157 (1967).
265. Gibb, T. C., *J. Chem. Soc., A* 2503 (1970).
266. Gibb, T. C., Greatrex, R., Greenwood, N. N., and Sarma, A. C., *J. Chem. Soc., A* 212 (1970).
267. Gibb, T. C., Greatrex, R., Greenwood, N. N., and Thompson, D. T., *J. Chem. Soc., A* 1663 (1967).
268. Gibb, T. C., and Greenwood, N. N., *J. Chem. Soc.*, 6989 (1965).
269. Gibb, T. C., and Greenwood, N. N., *J. Chem. Soc., A* 43 (1966).
270. Gibb, T. C., and Greenwood, N. N., "Mössbauer Spectroscopy." Chapman & Hall, London, 1971.

271. Gibb, T. C., Goodman, B. A., and Greenwood, N. N., *Chem. Commun.* 774 (1970).
272. Glentworth, P., Nichols, A. J., Large, N. R., and Bullock, R. J., *Chem. Commun.* 206 (1971).
273. Goldanskii, V. I., *At. Energy Rev.* **1**, 3 (1963).
274. Goldanskii, V. I., "The Mössbauer Effect and Its Applications in Chemistry." Consultants Bureau, New York, 1964.
275. Goldanskii, V. I., Borshagovskii, B. V., Makarov, E. F., Stukan, R. A., Anisimov, K. N., Kolobova, N. E., and Shripkin, V. V., *Theor. Exp. Chem.* (*USSR*) **3**, 275 (1967) (Russ. p. 478).
276. Goldanskii, V. I., Gorodinskii, G. M., Karyagin, S. V., Korytko, L. A., Krizhanskii, L. M., Makarov, E. F., Suzdalev, I. P., and Khrapov, V. V., *Dokl. Phys. Chem.* **147**, 766 (1962) (Russ. p. 127).
277. Goldanskii, V. I., Khrapov, V. V., Stukan, R. A., *Organometal. Chem. Rev. Sect. A* **4**, 225 (1969).
278. Goldanskii, V. I., Makarov, E. F., Stukan, R. A., Sumarokova, T. N., Trukhtanov, V. A., and Khrapov, V. V., *Dokl. Phys. Chem.* **156**, 474 (1964) (Russ. p. 400).
279. Goldanskii, V. I., Makarov, E. F., Stukan, R. A., Trukhtanov, V. A., and Khrapov, V. V., *Dokl. Phys. Chem.* **151**, 598 (1963) (Russ. p. 357).
280. Goldanskii, V. I., Rochev, V. Ya., and Khrapov, V. V., *Dokl. Phys. Chem.* **156**, 571 (1964) (Russ. p. 909).
281. Golding, R. M., *Mol. Phys.* **12**, 13 (1967).
282. Golding, R. M., "Applied Wave Mechanics." Van Nostrand, Princeton, New Jersey, 1969.
283. Golding, R. M., Jackson, F., and Sinn, E., *Theor. Chim. Acta* **15**, 123 (1969).
284. Golding, R. M., Mok, K. F., and Duncan, J. F., *Inorg. Chem.* **5**, 774 (1966).
285. Goldstein, C., and Pasternak, M., *Phys. Rev.* **177**, 481 (1969).
286. Goldstein, M., and Unsworth, W. D., *Spectrochim. Acta Part A*, **27**, 1055 (1971).
287. Goodgame, D. M. L., and Machado, A. A. S. C., *Chem. Commun.* 1420 (1969).
288. Goodman, B. A., Greatrex, R., and Greenwood, N. N., *J. Chem. Soc., A* 1868 (1971).
289. Goodman, B. A., and Greenwood, N. N., *Chem. Commun.* 1105 (1969).
290. Goodman, B. A., and Greenwood, N. N., *J. Chem. Soc., A* 1862 (1971).
291. Goodman, B. A., Greenwood, N. N., Jaura, K. L., and Sharma, K. K., *J. Chem. Soc., A* 1865 (1971).
292. Gordy, W., *Discuss. Faraday Soc.*, **19**, 14 (1955).
293. Greatrex, R., and Greenwood, N. N., *Discuss. Farad. Soc.* **47**, 126 (1969).
294. Greatrex, R., Greenwood, N. N., and Kaspi, P., *J. Chem. Soc. A* 1873 (1971).
295. Greatrex, R., Greenwood, N. N., and Pauson, P. L., *J. Organomet. Chem.* **13**, 533 (1968).
296. Greene, P. T., and Bryan, R. F., *J. Chem. Soc., A* 1696 (1970); 2549 (1971).
297. Greenwood, N. N., *Chem. Brit.* **3**, 56 (1967).
298. Greenwood, N. N., in "Spectroscopic Properties of Inorganic and Organometallic Compounds," Chem. Soc., Specialist Periodical Rep. 1967 and annually afterwards.
299. Greenwood, N. N., Perkins, P. G., and Wall, D. H., *Symp. Farad. Soc.* **1**, 51 (1967).
300. Greenwood, N. N., and Ruddick, J. N. R., *J. Chem. Soc., A* 1679 (1967).
301. Greenwood, N. N., and Straughan, B. P., *J. Chem. Soc., A* 962 (1966).

302. Greenwood, N. N., Sarma, A. C., and Straughan, B. P., *J. Chem. Soc.*, A 1446 (1966).
303. Greenwood, N. N., and Timnick, A., *J. Chem. Soc.*, A 676 (1971).
304. Gregory, N. W., *J. Amer. Chem. Soc.* **73**, 472 (1951).
305. Grover, J. E., and Orville, P. M., *Geochim. Cosmochim. Acta* **33**, 205 (1969).
306. Gukasyan, S. E., Iofa, B. Z., Karasev, A. N., Semenov, S. I., and Shpinel, V. S., *Phys. Status Solidi* **37**, 91 (1970).
307. Gukasyan, S. E., and Shpinel, V. S., *Phys. Status Solidi* **29**, 49 (1968).
308. Gütlich, P., and Hassellbach, K. M., *Angew Chem.* **81**, 627 (1969).
309. Gütlich, P., Odar, S., Fitzsimmons, B. W., and Erickson, N. E., *Radiochim. Acta* **10**, 147 (1968).
310. Hafemeister, D. W., DePasquali, G., and DeWaard, H., *Phys. Rev.* **135**, B1089 (1964).
311. Hafner, S. S., Virgo, D., and Warburton, D., *Proc. Apollo 12 Lunar Sci., Conf.* 1972).
312. Hall, D., Slater, J. H., Fitzsimmons, B. W., and Wade, K., *J. Chem. Soc.*, A 800 (1971).
313. Hall, L. H., Spijkerman, J. J., and Lambert, J. L., *J. Amer. Chem. Soc.* 2044 (1968).
314. Halsey, M. J., and Pritchard, A. M., *J. Chem. Soc.*, A 2878 (1968).
315. Hansson, A., and Brunge, O., private communication quoted by I. Lindquist, "Inorganic Adduct Molecules of Oxo Compounds." Academic Press, New York, New York, 1963.
316. Harmon, K. M., Hesse, L. L., Klemann, L.P., Kocher, C.W., McKinley, S. V., and Young, A. E., *Inorg. Chem.* **8**. 1054 (1969).
317. Harrison, P. G., and Zuckerman, J. J., *J. Amer. Chem. Soc.* **91**, 6885 (1969).
318. Harrison, P. G., and Zuckerman, J. J., *Inorg. Chem.* **9**, 175 (1970).
319. Hazony, Y., *J. Chem. Phys.* **45**, 2664 (1966).
320. Hazony, Y., and Axtmann, R. C., *Chem. Phys. Lett.* **8**, 571 (1971).
321. Hazony, Y., Axtmann, R. C., and Hurley, J. W., Jr., *Chem. Phys. Lett.* **2**, 440 (1968).
322. Hazony, Y., and Herber, R. H., *J. Inorg. Nucl. Chem.* **33**, 961 (1971).
323. Herber, R. H., "Applications of the Mössbauer Effect in Chemistry and Solid State Physics," Int. At. Energy Agency, Tech. Dept., No. 50, p. 130. Vienna, 1965.
324. Herber, R. H., *Progr. Inorg. Chem.* **8**, 1 (1966).
325. Herber, R. H., *Inorg. Chem.* **8**, 174 (1969).
326. Herber, R. H., *J. Chem. Phys.* **54**, 3755 (1971).
327. Herber, R. H., and Chandra, S., *J. Chem. Phys.* **52**, 6045 (1970).
328. Herber, R. H., and Chandra, S., *J. Chem. Phys.* **54**, 1847 (1971).
329. Herber, R. H., Chandra, S., and Hazony, Y., *J. Chem. Phys.* **53**, 3330 (1970).
330. Herber, R. H., and Cheng, H. S., *Inorg. Chem.* **8**, 2145 (1969).
331. Herber, R. H., and Goscinny, Y., *Inorg. Chem.* **7**, 1293 (1968).
332. Herber, R. H., and Hayter, R. G., *J. Amer. Chem. Soc.* **86**, 301 (1964).
333. Herber, R. H., King, R. B., and Wertheim, G. K., *Inorg. Chem.* **3**, 101 (1964).
334. Herber, R. H., and Parisi, G. I., *Inorg. Chem.* **5**, 769 (1966).
335. Herber, R. H., and Stapfer, C. H., *Inorg. Nucl. Chem. Lett.* **7**, 617 (1971).
336. Herber, R. H., Stöckler, H. A., and Reichle, W. T., *J. Chem. Phys.* **42**, 2447 (1965).
337. Hermodsson, Y., *Acta Crystallogr.* **13**, 656 (1960).

338. Hill, J. C., Drago, R. S., and Herber, R. H., *J. Amer. Chem. Soc.* **91**, 1644 (1969).
339. Hinze, J., and Jaffé, H. H., *J. Amer. Chem. Soc.* **84**, 540 (1962).
340. Hinze, J., and Jaffé, H. H., *J. Phys. Chem.* **67**, 1501 (1963).
341. Hofmann, W., *Z. Kristallogr.* **92**, 161 (1935).
342. Hogben, M. G., Gay, R. S., and Graham, W. A. G., *J. Amer. Chem. Soc.* **88**, 3457 (1966): Hogben, M. G., and Graham, W. A. G., *ibid.* **91**, 283 (1969); Hogben, M. G., Gay, R. S., Oliver, A. J., Thompson, J. A. J., and Graham, W. A. G., *ibid.* **91**, 291 (1969).
343. Hogg, C. S., and Meads, R. E., *Mineral. Mag.* **37**, 606 (1970).
344. Hoppe, R., and Dähne, W., *Naturwissenschaften* **49**, 254 (1962).
345. Hoy, G. R., and Chandra, S., *J. Chem. Phys.* **47**, 961 (1967).
346. Huang, H. H., and Hui, K. M., *J. Organomet. Chem.* **2**, 288 (1964).
347. Hudson, A., and Whitfield, H. J., *Inorg. Chem.* **6**, 1120 (1967).
348. Hulme, R., *J. Chem. Soc.* 1524 (1963).
349. Ichiba, S., Mishima, M., Sakai, H., and Negita, H., *Bull. Chem. Soc. Jap.* **41**, 49 (1968).
350. Ichiba, S., Sakai, H., Negita, H., and Maeda, Y., *J. Chem. Phys.* **54**, 1627 (1971).
351. Ingalls, R., *Phys. Rev.* **133 A**, 787 (1964).
352. Ingalls, R., Costan, C. J., DePasquali, G., Drickamer, H. G., and Pinajian, J. J., *J. Chem. Phys.* **45**, 1057 (1966).
353. Isaacs, N. W., and Kennard, C. H. L., *J. Chem. Soc.*, A 1257 (1970).
354. Jaccarino, V., King, J. G., Satten, R. A., and Stroke, H. H., *Phys. Rev.* **94**, 1798 (1954).
355. Janssen, M. J., Luijken, J. G. A., and Van Der Kerk, G. J. M., *Rec. Trav. Chim. Pays Bas.* **82**, 90 (1963).
356. Jesson, J. P., Weiher, J. F., and Trofimenko, S., *J. Chem. Phys.* **48**, 2058 (1968).
357. Johnson, C. E., *Symp. Faraday Soc.* **1**, 1 (1967).
358. Jones, C. H. W., and Warren, J. L., *J. Chem. Phys.* **53**, 1740 (1970).
359. Jones, M. T., *Inorg. Chem.* **6**, 1249 (1967).
360. Josephson, B. D., *Phys. Rev. Lett.* **4**, 341 (1960).
361. Jung, P., and Triftshauser, W., *Phys. Rev.* **175**, 512 (1968).
362. Kaindl, G., Potzel, W., Wagner, F. E., Zahn, U., and Mössbauer, R. L., *Z. Phys.* **226**, 103 (1969).
363. Kamenar, B., and Grdenic, D., *J. Inorg. Nucl. Chem.* **24**, 1039 (1962).
364. Karasyov, A. N., Kolobova, N. E., Polak, L. S., Shpinel, V.S., and Anisimov, K. N., *Teor. Eksp. Khim. Akad. Nauk Ukr.* (*Engl.*) **2**, 126 (1966).
365. Karyagin, S. V., *Sov. Phys. Dokl.* **148**, 110 (1963) (Russ. p. 1102); *Sov. Phys. Solid State* **5**, 1552 (1964) (Russ. p. 2128); *ibid.* **8**, 1387 (1966) (Russ. p. 1739).
366. Kasai, N., Yasuda, K., and Okawara, R., *J. Organomet. Chem.* **3**, 172 (1965).
367. Khrapov, V. V., Candidate Dissertation, Inst. Chem. Phys. Acad. Sci. USSR, Moscow, 1965.
368. Khrapov, V. V., Goldanskii, V. I., Prokof'ev, A. K., and Kostyanovskii, R. G. *J. Gen. Chem. USSR* **37**, 1 (1967) (Russ. p. 3).
369. Kistner, O. C., and Sunyar, A. W., *Phys. Rev. Lett.* **4**, 412 (1960).
370. Knight, J., and Mays, M. J., *J. Chem. Soc.*, A 654 (1970).
371. Kojima, S., Tsukada, K., Ogawa, S., and Shimauchi, A., *J. Chem. Phys.* **23**, 1963 (1955).

372. Konig, E., Hüfner, S., Steichele, E., and Madeja, K., Z. Naturforsch. A **22**, 1543 (1967).
373. Konig, E., and Kremer, S., Chem. Phys. Lett. **8**, 312 (1971).
374. Konig, E., and Madeja, K., Inorg. Chem. **6**, 48 (1967).
375. Konig, E., and Madeja, K., Inorg. Chem. **7**, 1848 (1968).
376. Korecz, L., and Burger, K., J. Inorg. Nucl. Chem. **30**, 781 (1968).
377. Kothekar, V., Iofa, B. Z., Semenov, S. I., and Shpinel, V. S., Sov. Phys. JETP **28**, 86 (1969).
378. Kriegsmann, V. H., and Geissler, H., Z. Anorg. Chem. **323**, 170 (1963).
379. Kriegsmann, V. H., and Pischtschan, S., Z. Anorg. Chem. **308**, 212 (1961).
380. Krizhanskii, L. M., Okhlobystin, O. Yu., Popov, A. V., and Rogozev, B. I., Dokl. Phys. Chem. **160**, 142 (1965) (Russ. p. 1121).
381. Lees, J. K., and Flinn, P. A., J. Chem. Phys. **48**, 882 (1968).
382. Lesikar, A. V., J. Chem. Phys. **40**, 2746 (1964).
383. Leung, K. L., and Herber, R. H., Inorg. Chem. **10**, 1020 (1971).
384. Lever, A. B. P., J. Chem. Soc. 1821 (1965).
385. Lewis, G. K., and Drickamer, H. G., J. Chem. Phys. **49**, 3782 (1968).
386. Lewis, G. K., and Drickamer, H. G., Proc. Nat. Acad. Sci. U.S. **61**, 414 (1968).
387. Lewis, J., Mabbs, F. E., and Richards, A., J. Chem. Soc., A 1014 (1967).
388. Long, G. G., Stevens, J. G., and Bowen, L. H., Inorg. Nucl. Chem. Lett. **5**, 799 (1969).
389. Long, G. G., Stevens, J. G., Tullbane, R. J., and Bowen, L. H., J. Amer. Chem. Soc. **92**, 4230 (1970).
390. Long, G. L., and Whitney, D. L., J. Inorg. Nucl. Chem. **33**, 1196 (1971).
391. Lucken, E. A. C., "Nuclear Quadrupole Coupling Constants." Academic Press, New York, 1969.
392. Ludwig, G. W., J. Chem. Phys. **25**, 159 (1956).
393. McClure, D. S., Advan. Chem. Co-ord. Compounds 498 (1961).
394. McDonald, R. R., Larson, A. C., and Cromer, D. T., Acta Crystallogr. **17**, 1104 (1964).
395. McGrady, M. M., and Tobias, R. S., J. Amer. Chem. Soc. **87**, 1909 (1965).
396. MacKay, K. M., Sowerby, D. B., and Young, W. C., Spectrochim. Acta Part A **24**, 611 (1968).
397. McWhinnie, W. R., Poller, R. C., and Thevarasa, M., J. Chem. Soc., A 1671 (1967).
398. Maddock, A. G., "Mössbauer Spectroscopy in the Study of Nuclear Reactions in Solids," MTP Publications, 1972. Butterworths, London, and University Park Press, State College, Pennsylvania.
399. Maddock, A. G., Medeiros, L. O., and Bancroft, G. M., Chem. Commun. 1067 (1967).
400. Maddock, A. G., and Platt, R. H., J. Chem. Soc., A 1191 (1971).
401. Maddock, A. G., and Platt, R. H., J. Chem. Phys. **55**, 1490 (1971).
402. Maddock, A. G., and Platt, R. H., unpublished observations.
403. Malathi, N., and Puri, S. P., J. Phys. Soc. Jap. **29**, 108 (1970).
404. Matas, J., and Zemcik, T., Phys. Lett. **19**, 111 (1965).
405. Matthews, C. K., J. Inorg. Nucl. Chem. **31**, 2853 (1969).
406. Mazak, R. A., and Collins, R. L., J. Chem. Phys. **51**, 3220 (1969).
407. Melson, G. A., Stokley, P. F., and Bryan, R. F., J. Chem. Soc., A 2247 (1970).
408. Menes, M., and Bolef, D. I., J. Phys. Chem. Solids **19**, 79 (1961).
409. Moore, W. J., and Pauling, L., J. Amer. Chem. Soc. **63**, 1392 (1941).

410. Moss, T. H., and Robinson, A. B., *Inorg. Chem.* **7**, 1692 (1968).
411. "Mössbauer Effect Data Index" (A. H. Muir, K. J. Ando, and H. M. Coogan, eds.) Wiley, New York, 1966.
412. "Mössbauer Effect Methodology" (I. J. Gruverman, ed.), Vols. 1–7. Plenum, New York, 1965–1971.
413. Mössbauer, R. L., *Naturwissenschaften* **45**, 538 (1958).
414. Mueller, R. F., *Mineral Soc. Amer. Spec. Paper* **2**, 83 (1969).
415. Mullins, F. P., *Can. J. Chem.* **48**, 1677 (1970).
416. Mullins, F. P., *Can. J. Chem.* **49**, 2719 (1971).
417. Mullins, M. A., and Curran, C., *Inorg. Chem.* **6**, 2017 (1967).
418. Mullins, M. A., and Curran, C., *Inorg. Chem.* **7**, 2584 (1968).
419. Murin, A. N., Lure, B. G., and Grushko, Yu. S., *Fiz. Tverd. Tela* **9**, 1820 (1967); *Chem. Abstr.* **67**, 69,093 (1967).
420. Naik, D. V., and Curran, C., personal communication, presented at the Int. Conf. Organometal. Chem. Moscow (1971).
421. Naik, D. V., and Curran, C., *Inorg. Chem.* **10**, 1017 (1971).
422. Nath, A., Harpold, M., Klein, M. P., and Kundig, W., *Chem. Phys. Lett.* **2**, 471 (1968).
423. Nath, A., Klein, M. P., Kundig, W., and Lichtenstein, D., in "Mössbauer Effect Methodology" (I. J. Gruverman, ed.), Vol. 5, p. 163. Plenum, New York, 1970.
424. Nesmeyanov, A. N., Goldanskii, V. I., Khrapov, V. V., Rochev, V. Ya., Kravtsov, D. N., and Rokhlina, E. M., *Bull. Acad. Sci. USSR* **4**, 763 (1968).
425. Nozik, A. J., and Kaplan, M., *J. Chem. Phys.* **47**, 2960 (1967).
426. O'Connor, J. E., and Corey, E. R., *Inorg. Chem.* **6**, 968 (1967).
427. Ofer, S., Nowik, I., and Cohen, S. G., in "Chemical Applications of Mössbauer Spectroscopy" (V. I. Goldanskii and R. H. Herber, eds.), p. 426. Academic Press, New York, 1968.
428. Okawara, R., Webster, D. E., and Rochow, E. G., *J. Amer. Chem. Soc.* **82**, 3287 (1960).
429. Okazaki, A., and Ueda, I., *J. Phys. Soc. Jap.* **11**, 470 (1956).
430. Olie, K., Smitskamp, C. C., and Gerding, H., *Inorg. Nucl. Chem. Lett.* **4**, 129 (1968).
431. Onaka, S., Sasaki, Y., and Sano, H., *Bull. Chem. Soc. Jap.* **44**, 726 (1971).
432. Oosterhuis, W. T., and Lang, G., *J. Chem. Phys.* **50**, 4381 (1969).
433. O'Rourke, M., and Curran, C., *J. Amer. Chem. Soc.* **92**, 1501 (1970).
434. Osborn, J. A., Gillard, R. D., and Wilkinson, G., *J. Chem. Soc.* 3168 (1964).
435. Panyushkin, V. N., DePasquali, G., and Drickamer, H. G., *J. Chem. Phys.* **51**, 3305 (1969).
436. Papike, J. J., and Clark, J. R., *Amer. Mineral.* **53**, 1156 (1968).
437. Parish, R. V., *Progr. Inorg. Chem.* **15**, 101 (1972).
438. Parish, R. V., and Johnson, C. E., *Chem. Phys. Lett.* **6**, 239 (1970).
439. Parish, R. V., and Johnson, C. E., *J. Chem. Soc., A* 1906 (1971).
440. Parish, R. V., and Platt, R. H., *J. Chem. Soc., A* 2145 (1969).
441. Parish, R. V., and Platt, R. H., *Inorg. Chim. Acta.* **4**, 589 (1970).
442. Parish, R. V., and Platt, R. H., *Inorg. Chim. Acta* **4**, 65 (1970).
443. Pasternak, M., Simopoulos, A., and Hazony, Y., *Phys. Rev., A* **140**, 1892 (1965).
444. Pasternak, M., and Sonnino, T., *Phys. Rev.* **164**, 384 (1967).
445. Pasternak, M., and Sonnino, T., *J. Chem. Phys.* **48**, 1997 (1968).

446. Pasternak, M., and Sonnino, T., *J. Chem. Phys.* **48**, 2009 (1968).
447. Pelah, I., and Ruby, S. L., *J. Chem. Phys.* **51**, 383 (1969).
448. Perlow, G. J., *Phys. Rev.* **135**, B1102 (1964).
449. Perlow, G. J., in "Chemical Applications of Mössbauer Spectroscopy" (V. I. Goldanskii and R. H. Herber, eds.), p. 377. Academic Press, New York, 1968.
450. Perlow, G. J., and Perlow, M. R., *Rev. Mod. Phys.* **36**, 353 (1964).
451. Perlow, G. J., and Perlow, M. R., *J. Chem. Phys.* **41**, 1157 (1964).
452. Perlow, G. J., and Perlow, M. R., *J. Chem. Phys.* **45**, 2193 (1966).
453. Perlow, G. J., and Perlow, M. R., *J. Chem. Phys.* **48**, 955 (1968).
454. Perlow, G. J., and Yoshida, H., *J. Chem. Phys.* **49**, 1474 (1968).
455. Petrides, D., Mullins, F. P., and Curran, C., *Inorg. Chem.* **9**, 1270 (1970).
456. Petrov, A. A., Rogozev, B. I., Krizhanskii, L. M., and Zaugorodnii, V. S., *J. Gen. Chem. USSR* **38**, 1151 (1968) (Russ. p. 1196).
457. Philip, J., Mullins, M. A., and Curran, C., *Inorg. Chem.* **7**, 1895 (1968).
458. Pillinger, W. L., and Stone, J. A., in "Mössbauer Effect Methodology" (I. J. Gruverman, ed.), Vol. 4, p. 217. Plenum, New York.
459. Poder, C., and Sams, J. R., *J. Organometal. Chem.* **19**, 67 (1969).
460. Poeth, T. P., Harrison, P. G., Long, T. V., Willeford, B. R., and Zuckerman, J. J., *Inorg. Chem.* **10**, 522 (1971).
461. Poliakoff, M., and Turner, J. J., *J. Chem. Soc.*, A 654 (1971).
462. Poller, R. C., and Ruddick, J. N. R., *J. Chem. Soc.*, A 2273 (1969).
463. Poller, R. C., Ruddick, J. N. R., and Spillman, J. A., *Chem. Commun.* 680 (1970).
464. Poller, R. C., Ruddick, J. N. R., Taylor, B., and Toley, D. L. B., *J. Organometal. Chem.* **24**, 341 (1970).
465. Poller, R. C., Ruddick, J. N. R., Thevarasa, M., and McWhinnie, W. R., *J. Chem. Soc.*, A 2327 (1969).
466. Poller, R. C., and Toley, D. L. B., *J. Chem. Soc.*, A 1578 (1967).
467. Poller, R. C., and Toley, D. L. B., *J. Chem. Soc.*, A 2035 (1967).
468. Poller, R. C., and Toley, D. L. B., *J. Inorg. Nucl. Chem.* **31**, 2973 (1969).
469. Polynova, T. W., and Porai-Koshits, M. A., *J. Struct. Chem.* **7**, 691 (1966).
470. Potzel, W., Wagner, F. E., Zahn, U., Mössbauer, R. L., and Danon, J., *Z. Phys.* **240**, 306 (1970).
471. Potzel, W., Wagner, F. E., Mössbauer, R. L., Kaindl, G., and Selzter, H. E., *Z. Phys.* **241**, 179 (1971).
472. Prados, R. A., Ph.D. Thesis, Louisiana State Univ., Baton Rouge, Louisiana. Kindly communicated by M. L. Good.
473. Prados, R. A., Clausen, C. A., and Good, M. L., Abstr. Amer. Chem. Soc. 1971 Spring meeting, Los Angeles.
474. "Pyroxenes and Amphiboles; Crystal Chemistry and Phase Petrology." *Min. Soc. Amer. Spec. Paper No.* **2**, (1969).
475. Reddy, J. M., Knox, K., and Robin, M. B., *J. Chem. Phys.* **40**, 1082 (1964).
476. Reiff, W. M., *J. Chem. Phys.* **54**, 4718 (1971).
477. Reiff, W. M., *Chem. Phys. Lett.* **8**, 297 (1971).
478. Reiff, W. M., *Inorg. Chim. Acta*. In press.
479. Reiff, W. M., *Coord. Chem. Rev.* In press.
480. Reiff, W. M., Baker, W. A., and Erickson, N. E., *J. Amer. Chem. Soc.* **90**, 4794 (1968).
481. Reiff, W. M., Long, G. J., and Baker, W. A., *J. Amer. Chem. Soc.* **90**, 6347 (1968).

482. Renovitch, G. A., and Baker, W. A., *J. Amer. Chem. Soc.* **89**, 6377 (1967).
483. Rentzeperis, P. J., *Z. Kristallogr.* **117**, 431 (1962).
484. Richards, R. R., and Gregory, N. W., *J. Phys. Chem.* **69**, 239 (1965).
485. Rickards, R., Johnson, C. E., and Hill, H. A. O., *J. Chem. Phys.* **48**, 5231 (1968).
486. Rickards, R., Johnson, C. E., and Hill, H. A. O., *Trans. Faraday Soc.*, **65**, 2847 (1969).
487. Rickards, R., Johnson, C. E., and Hill, H. A. O., *J.Chem. Phys.* **51**, 846 (1969).
488. Rickards, R., Johnson, C. E., and Hill, H. A. O., *J. Chem. Phys.* **53**, 3118 (1970).
489. Rickards, R., Johnson, C. E., and Hill, H. A. O., *J. Chem. Soc. A* 797 (1971).
490. Roberts, L. D., Patterson, D. O., Thomson, J. O., and Levey, R. P., *Phys. Rev.* **179**, 656 (1969).
491. Roberts, L. D., Pomerance, H., Thomson, J. O., and Dam, C. F., *Bull. Amer. Phys. Soc.* **7**, 565 (1962).
492. Robin, M. B., *J. Chem. Phys.* **40**, 3369 (1964).
493. Robinson, H., Dehmelt, H. G., and Gordy, W., *J. Chem. Phys.* **22**, 511 (1954).
494. Rogozev, B. I., Zavgorodnii, V. S., Krizhanskii, L. M., and Petrov, A. A., *J. Gen. Chem. USSR* **38**, 1999 (1968) (Russ. p. 2064).
495. Rother, P., Wagner, F., and Zahn, U., *Radiochim. Acta* **11**, 203 (1969).
496. Ruby, S. L., Kalvius, G. M., Beard, G. B., and Snyder, R. E., *Phys. Rev.* **159**, 239 (1967).
497. Ruby, S. L., and Selig, H., *Phys. Rev.* **147**, 348 (1966).
498. Ruby, S. L., and Shenoy, G. K., *Phys. Rev.* **186**, 326 (1969).
499. Rundle, R. E., and Olson, D. H., *Inorg. Chem.* **3**, 596 (1964).
500. Sano, H., and Hashimoto, F., *Bull. Chem. Soc. Jap.* **38**, 684 (1965).
501. Sano, H., and Kono, H., *Bull. Chem. Soc. Jap.* **38**, 1228 (1965).
502. Sasane, A., Nakamura, D., and Kubo, M., *J. Phys. Chem.* **71**, 3249 (1967).
503. Saxena, S. K., and Ghose, S., *Amer. Mineral.* **56**, 532 (1971).
504. Schilt, A. A., *Inorg. Chem.* **3**, 1323 (1964).
505. Schlemper, E. O., *Inorg. Chem.* **6**, 2012 (1967).
506. Schlemper, E. O., and Britton, D., *Inorg. Chem.* **5**, 507 (1966).
507. Schlemper, E. O., and Hamilton, W. C., *Inorg. Chem.* **5**, 995 (1966).
508. Schneider, R. F., and Di Lorenzo, J. V., *J. Chem. Phys.* **47**, 2343 (1967).
509. Shen, K., McEwen, W. E., LaPlaca, S. J., Hamilton, W. C., and Wolf, A. P., *J. Amer. Chem. Soc.* **90**, 1718 (1968).
510. Shenoy, G. K., and Ruby, S. L., in "Mössbauer Effect Methodology" (I. J. Gruverman, ed.), Vol. 5, p. 77. Plenum, New York, 1970.
511. Shirley, D. A., *Rev. Mod. Phys.* **36**, 339 (1964).
512. Shirley, D. A., in "Chemical Applications of Mössbauer Spectroscopy" (V. I. Goldanskii and R. H. Herber, eds.), p. 504. Academic Press, New York, 1968.
513. Shpinel, V. S., Bryukhanov, V. A., Kothekar, V., Iofa, B. Z., and Senov, S. I., *Symp. Faraday Soc.* **1**, 69 (1967).
514. Shustorovich, E. M., and Djatkina, M. E., *Dokl. Akad. Nauk. SSSR (Engl.)* **128**, 1234 (1959).
515. Simopoulos, A., Wickman, H. H., Kostikas, A., and Petrides, D., *Chem. Phys. Lett.* **7**, 615 (1970).
516. Smith, P. J., *Organometal. Chem. Rev. Sect. A* **5**, 373 (1970).
517. Spiess, H. W., Haas, H., and Hartmann, H., *J. Chem. Phys.* **50**, 3057 (1969).

518. Spijkerman, J. J., *Symp. Faraday Soc.* **1**, 134 (1967).
519. Spijkerman, J. J., Hall, L. H., and Lambert, J. L., *J. Amer. Chem. Soc.* **90**, 2039 (1968).
520. Stapfer, C. H., Leung, K. L., and Herber, R. H., *Inorg. Chem.* **9**, 970 (1970).
521. Stein, L., in "Halogen Chemistry" (V. Gutman, ed.), Vol. 1, p. 133 and refs. quoted therein. Academic Press, New York, 1967.
522. Stevens, J. G., and Bowen, L. H., in Mössbauer Effect Methodology" (I. J. Gruverman, ed.), Vol. 5, p. 27. Plenum, New York, 1970.
523. Stevens, J. G., and Ruby, S. L., *Phys. Lett.* **32 A**, 91 (1970).
524. Stevens, P. J., D. Phil. Thesis, Oxford Univ., Oxford, 1964.
525. Stöckler, H. A., and Sano, H., *Phys. Lett. A* **25**, 550 (1967).
526. Stöckler, H. A., and Sano, H., *Chem. Phys. Lett.* **2**, 448 (1968).
527. Stöckler, H. A., and Sano, H., *Phys. Rev.* **165**, 406 (1968).
528. Stöckler, H. A., and Sano, H., *Trans. Faraday Soc.* **64**, 577 (1968).
529. Stöckler, H. A., and Sano, H., *Chem. Commun.* 954 (1969).
530. Stöckler, H. A., Sano, H., and Herber, R. H., *J. Chem. Phys.* **45**, 1182 (1966).
531. Stöckler, H. A., Sano, H., and Herber, R. H., *J. Chem. Phys.* **47**, 1567 (1967).
532. Stukan, R. A., Goldanskii, V. I., Makarov, E. F., and Rukhadze, E. G., *J. Struct. Chem. (USSR)* **8**, 239 (1965).
533. Stukan, R. A., Gulein, S. P., Nesmeyanov, A. N., Goldanskii, V. I., and Makarov, E. F., *Teor. Eksp. Khim.* **2**, 805 (1966).
534. Sukhoverkhov, V. F., and Dzevitskii, B. E., *Dokl. Akad. Nauk SSSR Ser. Fiz.* **177**, 1089 (1967) (Russ. p. 6111).
535. Taft, R. W., in "Steric Effects in Organic Chemistry" (M. S. Newman, ed.), p. 556. Wiley, New York, 1956.
536. "The Mössbauer Effect and Its Applications in Chemistry." (R. F. Gould, ed.). Amer. Chem. Soc., Washington, D.C., 1967.
537. Thomas, K., Osborn, J. A., Powell, A. R., and Wilkinson, G., *J. Chem. Soc., A* 1801 (1968).
538. Thompson, J. B., *Amer. Mineral.* **54**, 341 (1969); **55**, 528 (1970).
539. Tinkham, M., *Proc. Roy. Soc., Ser. A* **236**, 549 (1956).
540. Townes, C. H., and Dailey, B. P., *J. Chem. Phys.* **17**, 782 (1949).
541. Triftshauser, W., and Craig, P. P., *Phys. Rev.* **162**, 274 (1967).
542. Trotter, J., and Zobel, T., *Z. Kristallogr.* **123**, 67 (1966).
543. Tsai, J. H., Flynn, J. J., and Boer, F. P., *Chem. Commun.* 702 (1967).
544. Uchida, T., *Bull. Chem. Soc. Jap.* **40**, 2244 (1967).
545. Unland, M. L., *J. Chem. Phys.* **49**, 4514 (1968).
546. Violet, C. E., in "The Mössbauer Effect and Its Applications in Chemistry" (R. F. Gould, ed.), p. 147. Amer. Chem. Soc., Washington, D.C., 1967.
547. Violet, C. E., and Booth, R., *Phys. Rev.* **144**, 225 (1966).
548. Violet, C. E., Booth, R., and Wooten, F., *Phys. Lett.* **5**, 230 (1963).
549. Virgo, D., and Hafner, S. S., *Mineral Soc. Amer. Spec. Paper* **2**, 67 (1969).
550. Virgo, D., and Hafner, S. S., *Amer. Mineral.* **55**, 201 (1970).
551. Vucelic, M., *Croat. Chem. Acta* **40**, 255 (1968).
552. Wagner, F., and Zahn, U., *Z. Phys.* **233**, 1 (1970).
553. Walker, L. R., Wertheim, G. K., and Jaccarino, V., *Phys. Rev. Lett.* **6**, 98 (1961).
554. Watanabe, N., and Niki, E., *Bull. Chem. Soc. Jap.* **43**, 3034 (1970).
555. Watson, R. E., and Freeman, A. J., *Phys. Rev.* **120**, 1125 (1960).
556. Wedd, R. W. J., and Sams, J. R., *Can. J. Chem.* **48**, 71 (1970).

557. Wei, C. H., and Dahl, L. F., *J. Amer. Chem. Soc.* **91**, 1351 (1969).
558. Wells, W. L., and Brown, T. L., *J. Organomet. Chem.* **11**, 271 (1968).
559. Wertheim, G. K., "Mössbauer Effect: Principles and Applications." Academic Press, New York, 1964.
560. Wertheim, G. K., and Buchanan, D. N. E., *Chem. Phys. Lett.* **3**, 87 (1969).
561. Wertheim, G. K., and Herber, R. H., *J. Chem. Phys.* **38**, 2106 (1963).
562. Wickman, H. H., and Silverthorn, W. E., *Inorg. Chem.* **10**, 2333 (1971).
563. Wickman, H. H., Trozzolo, A. M., Williams, H. J., Hull, G. W., and Merrett, F. R., *Phys. Rev.* **155**, 563 (1967).
564. Wickman, H. H., and Trozzolo, A. M., *Inorg. Chem.* **7**, 63 (1968).
565. Wickman, H. H., and Wagner, C. F., *J. Chem. Phys.* **51**, 435 (1969).
566. Williams, D. E., and Kocher, C. W., *J. Chem. Phys.* **52**, 1480 (1970).
567. Williams, D. E., and Kocher, C. W., *J. Chem. Phys.* **55**, 1491 (1971).
568. Williams, P. G. L., and Bancroft, G. M., "Mössbauer Effect Methodology" (I. J. Gruverman, ed.), Vol. 7, p. 39. Plenum, New York, 1971.
569. Wynter, C. I., and Chandler, L., *Bull. Chem. Soc. Jap.* **43**, 2115 (1970).
570. Wynter, C. I., Hill, J., Bledsoe, W., Shenoy, G. K., and Ruby, S. L., *J. Chem. Phys.* **50**, 3872 (1969).
571. Yamasaki, R. S., and Cornwell, C. D., *J. Chem. Phys.* **30**, 1265 (1959).
572. Yeats, P. A., Ford, B. F. E., Sams, J. R., and Aubke, F., *Chem. Commun.* 791 (1969).
573. Yeats, P. A., Poh, B. L., Ford, B. F. E., Sams, J. R., and Aubke, F., *J. Chem. Soc., A* 2188 (1970).
574. Yeats, P. A., Sams, J. R., and Aubke, F., *Inorg. Chem.* **9**, 740 (1970).
575. Zerner, M., Gouterman, M., and Kobayashi, H., *Theor. Chim. Acta* **6**, 363 (1966).
576. Zhdonov, Y. A., and Minkin, V. I., "Correlation Analysis in Organic Chemistry." Izd. Rostovsk Univ., 1961.
577. Zuckerman, J. J., *Advan. Organomet. Chem.* **9**, 21 (1970).

METAL ALKOXIDES AND DIALKYLAMIDES

D. C. Bradley

Department of Chemistry, Queen Mary College, London, England

I. Introduction 259
II. Metal Alkoxides 260
   A. Alkali Metal Alkoxides 260
   B. Alkoxides of Beryllium, Magnesium, Zinc, and the Alkaline Earths 264
   C. Aluminum and Gallium Alkoxides 266
   D. Transition Metal Alkoxides 272
   E. Alkoxides of Lanthanides and Actinides 290
   F. Double Alkoxides 293
   G. Metal Trialkylsilyloxides 295
III. Metal Dialkylamides 298
   A. The Preparation of Metal Dialkylamides 298
   B. Chemical Properties of Metal Dialkylamides 301
   C. Physical Properties of Metal Dialkylamides 302
   References 316

I. Introduction

Since the metal alkoxides $M(OR)_x$ were first comprehensively reviewed in 1960 (1), considerable progress has been maintained and the next major review in 1967 (2) quoted over three hundred references. The most significant advances have involved the chemistry of the transition metal alkoxides (3) with the emphasis on the ligand field aspects of the alkoxo group (e.g., electronic spectra, magnetism, etc.) and on X-ray crystallographic and NMR structural determinations. Industrial applications (4) have been concerned with metal alkoxides as components of soluble Ziegler–Natta catalysts for olefin polymerization and also as sources for the production of pure metal oxides.

The metal dialkylamides $M(NR_2)_x$ are of special interest in that they contain covalent metal–nitrogen bonds and occupy a position between the metal alkoxides and the metal alkyls. The field has not been fully reviewed hitherto. However, excellent reviews have recently appeared on titanium dialkylamides (5) and the reactivity of metal–dialkylamido bonds (6). A concise review of recent work on transition metal dialkylamides has also been given (6). It is clear that this is a rapidly developing field and that metal dialkylamides are synthetic reagents showing considerable versatility.

II. Metal Alkoxides

In this chapter we shall restrict discussion to the fully substituted alkoxides $M(OR)_x$, with an occasional reference to metal oxide alkoxides $MO_y(OR)_{(x-2y)}$, bearing in mind that the latter class of compounds was recently reviewed (7). This excludes from consideration a considerable range of metal compounds containing alkoxo groups, in addition to one or more other types of ligand, viz., $M(OR)_x X_y L_z$ (where X = anionic ligand and L = neutral ligand). In most of these compounds the alkoxo groups play only a subsidiary role in determining the nature of the molecule, whereas in the alkoxides $M(OR)_x$ the alkoxo group is unambiguously the key ligand.

The methods of preparation of metal alkoxides are well established (1, 2, 4) and have recently been reviewed (8). It is considered unnecessary to treat this aspect under a separate heading. Similarly, the basic physical and chemical properties of the metal alkoxides are also firmly established (1, 2, 4) and these do not require separate consideration other than in a brief restatement in this introduction. Thus, we may deal with the metal alkoxides in a different fashion from previous reviews and concentrate on areas of recent development of the subject.

To summarize rather briefly, the chief chemical properties of metal alkoxides are their characteristic ease of hydrolysis (except for a few notable exceptions) and reactivity with hydroxylic molecules. Physical properties involve a balance between the tendency of the metal to expand its coordination number by utilizing the bridging property of alkoxo groups and the opposition to this by the steric effect of the alkyl group. This results in a wide spectrum of properties ranging from insoluble, nonvolatile, polymeric solids to volatile, monomeric liquids. Some notable progress has been made in the elucidation of the structures of the intermediate oligomeric compounds $[M(OR)_x]_n$.

A. ALKALI METAL ALKOXIDES

With the current resurgence of interest in the chemistry of the alkali metals it is timely to review the present status of the alkali metal alkoxides. The common assumption that all alkali metal compounds are ionic has been critically reexamined and it is noteworthy that several years ago the covalent nature of alkali metal zirconium alkoxides $MZr_2(OR)_9$ (M = Li, Na, or K) was reported (9). Wheatley (10) determined the crystal structure of LiOMe and showed it to be an infinite two-dimensional polymer with a layer structure. The environment of the lithium atoms

(Li–O = 1.95 Å) was a squashed tetrahedron (OLiO = 101.7° and 131.1°), while the oxygens were each bonded to four lithiums and a methyl group in a tetragonal-pyramidal configuration with the methyl group taking the apical position. In classical terms the CH_3O^- oxygen has only three pairs of electrons for coordination to four lithium ions, therefore the covalency must involve delocalized bonding typical of electron-deficient structures. Bains (*11*) showed that lithium alkoxides become more soluble (in nonpolar solvents), less polymeric, and more volatile in the following order $LiOEt < LiOPr^i < LiOBu^t$. Thus, the *tert*-butoxide sublimed at 110°/0.1 mm and it was tetrameric in benzene and hexameric in cyclohexane indicative of covalent nature. The proton NMR spectrum of $(LiOBu^t)_n$ showed a single line ($\tau = 8.473$; CCl_4 solution) shifted downfield from *tert*-butanol ($\tau = 8.778$) indicating either that all ligands are equivalent or that nonequivalent ligands are engaged in rapid exchange. Independent work by Golovanov *et al.* (*12*) also showed the *tert*-butoxide to be hexameric ($n = 6.4 \pm 0.7$) in cyclohexane, but a higher degree of polymerization was observed ($n = 9.4 \pm 0.9$) in benzene. In order to clarify the behavior in benzene and to check the proposal by Simonov *et al.* (*13*) that polymeric species exist in the vapor (from the similarity of the infrared spectra of the vapor at 180° and the solid) further work was carried out by Hartwell and Brown (*14*). These authors found that solutions in benzene were hexameric and that the mass spectrum at 130° gave fragment ions of the hexamer [e.g., $Li_6(OBu^t)_5(OCMe_2)^+$ and $Li_6(OBu^t)_5{}^+$]. They proposed a structure for $(LiOBu^t)_6$ based on a distorted octahedral cluster of lithiums with triple bridging *tert*-butoxo groups occupying six faces and resulting in D_{3d} symmetry with three-coordinated lithium. Presumably the steric effect of the *tert*-butyl groups prevents the formation of a layer structure similar to $(LiOMe)_\infty$ in which the lithiums are four-coordinated.

X-Ray crystallographic analysis of the tetrameric *tert*-butoxides $[MOBu^t]_4$ (M = K, Rb, or Cs) has shown (*15*) a "cubane" type of structure similar to that exhibited by thallous alkoxides. In the structure (Fig. 1) a tetrahedral cluster of metal atoms is bonded by triple-bridging alkoxo groups whose oxygen atoms occupy positions outside each face of the tetrahedron. The metal is thus three-coordinated and presumably is prevented from attaining the four-coordination found in LiOMe by the steric repulsion of the *tert*-butyl groups. Broad line ^1H NMR studies suggested free rotation of the *tert*-butyl groups at temperatures $> -20°$, but at lower temperatures the rotation is frozen out. Sodium *tert*-butoxide gave crystals of lower symmetry. Weiss (*16*) has also shown that KOMe crystals are tetragonal, and the infinite polymer adopts an interesting double layer structure (Fig. 2) containing two parallel sheets

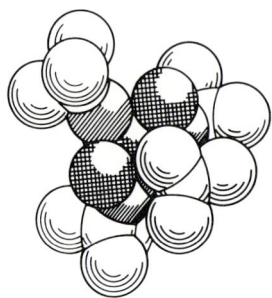

Fig. 1. The [MOBu^t]$_4$ cubane structure. Cross-hatched, M; hatched, O.

Fig. 2. The structure of [KOMe]. Hatched, K; stippled, O; plain, CH$_3$.

of potassium atoms and two sheets of oxygen atoms with the methyl groups above and below these sheets. Each potassium is five-coordinated to oxygen and each oxygen is six-coordinated (5-potassiums and 1-methyl). On the other hand, sodium methoxide (17) is isostructural with lithium methoxide, although partial hydrolysis [e.g., to Na$_3$(OH)(OMe)$_2$] leads to the more compact potassium methoxide structure.

Conductance measurements on sodium alkoxides dissolved in their parent alcohols (18) showed a marked decrease in the order NaOMe > NaOEt > NaOPr^i ≫ NaOBu^t, with the latter being practically nonconducting. This confirms the covalent nature of sodium tert-butoxide, but it would be preferable to study these compounds in the same noninteracting solvent before attempting to compare their relative tendencies toward ionization.

Some interesting results have recently been obtained with alkali metal derivatives of fluorinated alcohols. A novel synthetic method was employed to obtain the trifluoromethoxides of the higher alkali metals (19).

$$\text{MF} + \text{COF}_2 \rightleftharpoons \text{MOCF}_3 \quad \quad (1)$$
$$(\text{M} = \text{K, Rb, Cs})$$

The reaction is reversible, but in anhydrous acetonitrile the metal trifluoromethoxides are preferentially formed as stable solids. The reaction was reversed by heating *in vacuo* and thermolysis studies indicated the following order of stability: $CsOCF_3 > RbOCF_3 > KOCF_3$. Further work (*20*) has shown that other perfluoroalkoxides [Eqs. (2) and (3)] may be obtained by suitable choice of perfluorocarbonyl compounds.

$$MF + R_fCOF \rightleftharpoons MOCF_2R_f \qquad (2)$$
$$(M = Rb, Cs; R_f = CF_3, C_2F_5, C_3F_7)$$

$$MF + (CF_3)_2CO \rightleftharpoons MOCF(CF_3)_2 \qquad (3)$$
$$(M = Rb, Cs)$$

However, these perfluoroalkoxides were less stable than the trifluoromethoxides, although they were considerably more soluble in polar solvents and enabled ^{19}F NMR measurements to be obtained. The thermal instability of these derivatives of primary and secondary perfluoroalkoxides is probably related more to the ease of α-fluorine elimination rather than ionic character, and a dramatic increase in stability was found by Dear *et al.* (*21*) for the derivatives of perfluoro-*tert*-butanol. For example, both the lithium and sodium compounds could be melted and distilled under atmospheric pressure [b.p. $LiOC_4F_9$, 218°; $NaOC_4F_9$, 232°] while the potassium compound (decomposes >220°) could be sublimed *in vacuo* (140°/0.2 mm). The compounds $MOC(CF_3)_3$ (M = Li, Na, K) were all soluble in ether, acetone, and acetonitrile and could be recrystallized from benzene. Sodium perfluoro-*tert*-butoxide gave a complex mass spectrum (65°), which included fragment ions of the tetramer species, and it crystallizes in the cubic system ($a = 18.32$ Å) with 32 molecules in the unit cell. It is tempting to speculate on the possibility that the structure may involve tetrameric units having the cubane structure exhibited by the *tert*-butoxides. In concluding this section on alkali metal alkoxides it is relevant to note that an X-ray crystal analysis (*22*) has confirmed the tetrameric structure of thallium(I) alkoxides since the methoxide has a distorted "cubane" configuration. Raman and infrared spectra have been obtained for $Tl_4(OR)_4$ (R = Et, Pr^n) and a normal coordinate analysis has been carried out for the $Tl_4(OC)_4$ core of the structure (*23*). The high intensity of the lowest frequency Raman bands was indicative of metal–metal bonding and a metal–metal stretching force constant of 0.26 mdyne/Å was deduced.

B. Alkoxides of Beryllium, Magnesium, Zinc, and the Alkaline Earths

Although the alkyl metal(II) alkoxides, RMOR′, studied by Coates and co-workers do not come strictly within the terms of reference of this chapter, their interesting chemistry and relevance to the alkoxides justifies some comment. Of particular interest are the tetrameric species such as [MeBeOPr^i]$_4$ (*24*), [EtMgOBu^t]$_4$ (*25*), [MeZnOMe]$_4$ (*26*), and [MeCdOEt]$_4$ (*27*), which are all believed to have the "cubane" structure (Fig. 3) found by X-ray crystal analysis of [MeZnOMe]$_4$ (*28*). The struc-

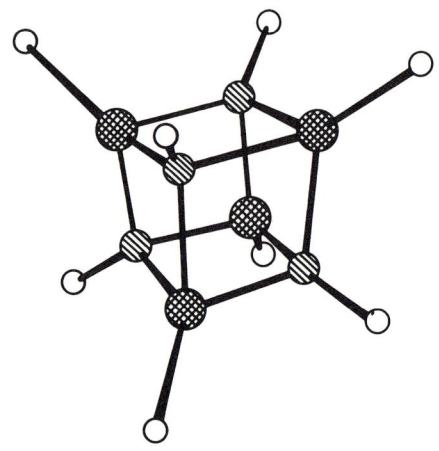

FIG. 3. The [MeZnOMe]$_4$ cubane structure. Cross-hatched, Zn; hatched, O; plain, CH$_3$.

ture is clearly determined by the triple-bridging alkoxo groups which occupy the four faces of the metal atom tetrahedron. The relationship of these structures to those of the tetrameric alkali metal and thallium(I) alkoxides is striking. It is also noteworthy that in these compounds the Group II metals are attaining four-coordination.

Beryllium alkoxides Be(OR)$_2$ of the lower aliphatic alcohols are insoluble, nonvolatile compounds, which are probably infinite polymers with tetrahedrally coordinated beryllium. Mehrotra (*2, 29*) found that alcoholysis of beryllium alkoxides was very slow. The methoxide and ethoxide have been included in an infrared study of Group II alkoxides (*30*). Splitting of the ν_{CO} band was taken to indicate appreciable covalent character in the BeO bonds. Steric hindrance of bulky alkyl groups has a profound effect on intermolecular bonding (1), and Coates and Fishwick (24) found that di-*tert*-butoxyberyllium prepared by the interesting route [Eq. (4)] was a trimer $Be_3(OBu^t)_6$. The compound (m.p. 112°) was soluble

$$Me_2Be \xrightarrow[\text{or}]{Bu^tOH} (MeBeOBu^t)_4 \xrightarrow{Me_2CO} Be_3(OBu^t)_6 \quad (4)$$
$$\quad\quad\quad Me_2CO$$

in hydrocarbon solvents and could be sublimed *in vacuo* ($100°/10^{-3}$ mm). Its ^1H NMR spectrum in benzene gave a single line ($\tau = 8.56$), but in perdeuteriomethylcyclohexane there were two lines ($\tau = 8.60, 8.75$) in a 2:1 ratio and the spectrum was unchanged from 33°–100°. A similar spectrum ($\tau = 8.62, 8.78; 2:1$) was given in carbon tetrachloride and the data are consistent with structure (I) involving a tetrahedrally coordinated beryllium in the middle with two three-coordinated berylliums

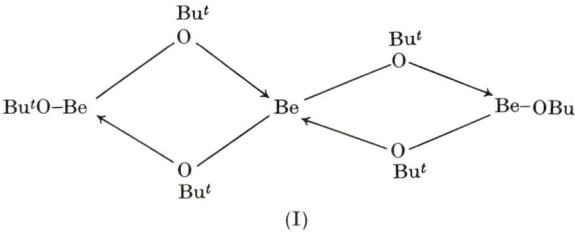

(I)

at the ends. In view of the preference for the tetrameric cubane structure by RBeOR' species (24), it would not have been surprising if $Be(OBu^t)_2$ had adopted a similar structure with four bridging and four terminal alkoxo groups. However, it appears that such a structure is prevented by steric hindrance, and it is noteworthy that $(Bu^tBeOBu^t)_2$ is dimeric. Using the more bulky triethylcarbinol produced the dimeric derivative $Be_2(OCEt_3)_4$, which was more volatile (sublimes at $50°-60°/10^{-3}$ mm) and no doubt contains three-coordinated beryllium. Its ^1H NMR spectrum showed only one type of alkoxide species and it appears that rapid exchange takes place between terminal and bridging ligands in contrast to behavior of species (I).

The lower aliphatic alkoxides of magnesium $Mg(OR)_2$ are also insoluble and nonvolatile. Infrared studies have been carried out on the methoxide and ethoxide (30–32) and interpreted as indicating a considerable degree of covalency in the MgO bonds compared with alkaline earth alkoxides which appear to be more ionic in character. Magnesium methoxide is sparingly soluble in methanol and an unstable solvate can readily be crystallized, but an attempt to determine the crystal structure of $Mg(OMe)_2$ was unsuccessful because the unsolvated compound was amorphous (*31*). Bryce-Smith and Wakefield (*33*) reported that Mg(OPr^i)$_2$ obtained by the action of Pr^{i}OH on Bu^{n}MgOPr^i was initially a porous rubbery material which on standing changed to a friable solid. Coates et al. (*34*) found that Mg(OBu^t)$_2$ was insoluble in ether, but it reacted with magnesium bromide to produce the dimer Mg$_2$(OBu^t)$_2$-Br$_2$(Et$_2$O)$_2$ which contains four-coordinated magnesium in a structure involving *tert*-butoxo bridges.

Since zinc alkoxides Zn(OR)$_2$ are insoluble, nonvolatile compounds (including the *tert*-butoxide) various preparative methods have been devised. Using the dialkylzinc compound as a starting material, the dialkoxides have been obtained by controlled oxygenation (*35*) or by prolonged reaction with excess alcohol (*36*). Talalaeva et al. (*37*) also prepared Zn(OBu^t)$_2$ by means of the reaction of zinc chloride with lithium *tert*-butoxide in ether solution, whereas Mehrotra and Arora (*38*) obtained Zn(OMe)$_2$ and Zn(OPr^i)$_2$ by using the lithium alkoxide in its parent alcohol. They also prepared alkoxides by means of the alcoholysis reaction provided some soluble lithium–zinc double alkoxide was present as a catalyst. It appears, therefore, that zinc alkoxides are all highly polymeric compounds, but the coordination number of the zinc is not known.

X-Ray diffraction analysis of the alkaline earth dimethoxides M(OMe)$_2$ (M = Ca, Sr, Ba) revealed hexagonal layer lattices (CdI$_2$) indicative of six-coordination with all methoxo groups triple-bridging through edge-shared octahedra (*31, 39*). The corresponding ethoxides proved to exhibit a similar structure (*40*). Infrared studies were also carried out on the methoxides and ethoxides (*30–32*), and it was concluded that the ionic character in the M–O bonds increased from Be to Ba with the alkaline earths being predominantly ionic.

C. Aluminum and Gallium Alkoxides

Aluminum alkoxides have been known since 1876 and have found important uses, yet detailed structural knowledge was lacking until recently and there are still many unsolved problems. Physical data from the earlier literature are unreliable owing to insufficient precautions being taken to prevent hydrolysis and also to a lack of knowledge of the "aging" phenomenon (slow change of the state of aggregation) of certain alkoxides.

For example, aluminum trimethoxide is a solid which is commonly believed to be nonvolatile, but Bradley and Faktor (*41*) found that the pure compound sublimed readily at 240° in a high vacuum. Wilhoit et al. (*42*) reinvestigated the alleged polymorphism of aluminum triethoxide and found only one form (m.p. ~ 140°) which was sparingly soluble contrary to earlier reports. Aluminum triisopropoxide, which is the most frequently used derivative, is obtained as a liquid by distillation and may remain supercooled for a long time, although the solid eventually formed has a high melting point. A recent report (43) identifies three forms of solid Al(OPr^i)$_3$, a low melting material (m.p. 29.5°) produced by solidification of the liquid (previously heated to 130°–140°) at 0°, the usually obtained compound (m.p. 118°) produced by cooling the liquid at room temperature or by crystallization from isopropanol, and a high melting modification (m.p. ~137°) obtained by keeping the liquid at 60°–65°.

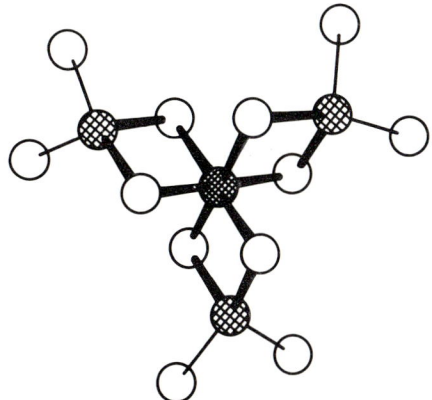

FIG. 4. The structure of [Al(OR)$_3$]$_4$. Fine cross-hatched, Al (octahedral); hatched, Al (tetrahedral); plain, OR.

The sec-butoxide appears to be the only commonly used alkoxide of aluminum which is a liquid at room temperatures.

Mehrotra (44, 45) made a systematic study of a wide range of aluminum alkoxides and established that straight-chain alkoxides were tetramers Al$_4$(OR)$_{12}$, the freshly distilled isopropoxide was a trimer which slowly transformed to a tetramer, and the tert-butoxide was a dimer Al$_2$(OBu^t)$_6$. Fieggen (43) found that the low melting aluminum isopropoxide was trimeric and the higher melting forms were tetrameric. Cyclic structures involving tetrahedral aluminum were first proposed for the trimers and tetramers, but it was suggested by Bradley (46), in order to explain the "aging" phenomenon, that the tetramer might involve a unique structure (Fig. 4) with a central octahedral aluminum and three peripheral tetrahedral aluminum atoms.

The structure (Fig. 4) was confirmed for solutions of the tetramer by Shiner et al. (47) using ^{1}H NMR, which gave three doublets for the CH_3 protons in a 1:1:2 ratio, because the methyl groups in the bridging isopropoxo groups are nonequivalent. Kleinschmidt (48) obtained a ^{27}Al NMR spectrum of the tetramer. A broad peak and a sharper peak were found in the ratio 3:1, which also agrees with the proposed structure, since the central octahedral aluminum which experiences a higher symmetry electric field than the tetrahedral aluminums should give the sharper signal. Recent ^{1}H NMR work by Oliver and Worrall (49) has convincingly shown that the asymmetric central aluminum in the tetrameric $Al_4(OCH_2R)_{12}$ (R = C_6H_5, 4-Cl–C_6H_4) causes nonequivalence of the methylene protons with consequent appearance of an AB quartet ($J_{AB} = 11$ Hz). In both compounds the methylenes in the bridging groups gave well-defined quartets ($\delta_{AB} = 22.5$ and 21.3 Hz, respectively) at 60 MHz, but the terminal groups gave unresolved singlets. However, at 220 MHz the terminal group methylenes also gave AB quartets ($\delta_{AB} = 0.5$ and 6 Hz converted to 60 MHz equivalents). The larger δ_{AB} values for the bridging group methylenes were ascribed to their closer proximity to the asymmetric center.

Conclusive evidence that tetrameric $Al_4(OPr^i)_{12}$ maintains its integrity in the vapor phase was reported by Fieggen et al. (50) from mass spectral studies of $Al_4(OPr^i)_{12}$ and $Al_4(OCDMe_2)_{12}$. The highest mass fragment was due to $Al_4(OPr^i)_{11}(OCHMe)^+$, but more intense peaks were found corresponding to $Al_4(OPr^i)_{11}{}^+$, $Al_4(OPr^i)_{10}(OCMe_2)^+$, $Al_4(OPr^i)_{10}(OCHMeCH_2)^+$, and $Al_4O(OPr^i)_9{}^+$. Fragment ions containing three and two aluminums were also present which could have arisen from trimer and dimer species or by fragmentation of the tetramer species. Metastable peaks corresponding to the loss of CH_3CHO, $CH_3CH:CH_2$, and $Pr_2{}^iO$ were found, and a plausible fragmentation pattern was deduced. These mass spectral data show that the initial vapor evaporated from $Al_4(OPr^i)_{12}$ at 118° contains predominantly tetrameric species, but Mehrotra (45) reported that vapor density measurements showed the vapor to be dimeric. It is therefore desirable to carry out further mass spectral studies to determine whether the composition of vapor depends on its thermal history (e.g. equilibrium studies). This is particularly important in interpreting thermodynamic data such as entropies of vaporization (44, 45, 51). By means of reaction calorimetry Wilson (52) has derived a value for the enthalpy of formation of tetrameric aluminum isopropoxide, $\Delta H^0{}_f[Al_4(OPr^i)_{12}](C) = -5149.5$ kJ·mole^{-1}.

In their NMR studies Shiner et al. (47) examined the supercooled liquid aluminum isopropoxide, which they confirmed was essentially

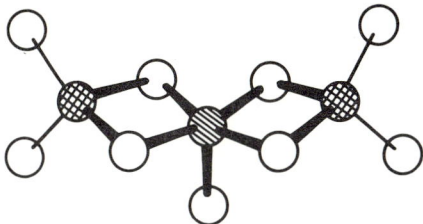

FIG. 5. A structure for [Al(OR)₃]₃. Hatched, Al (five-coordinated); cross-hatched, Al (tetrahedral); plain, OR.

trimeric in solution. Solutions of the trimer in various solvents gave a simple spectrum showing a single isopropoxide species with a chemical shift close to that observed for the terminal groups of the tetramer. At low temperatures the trimer signals broadened and, in some cases, split into two (approximately species 1:2), in accordance with the requirements of a cyclic trimer (II). Since the addition of isopropanol

$$\begin{array}{c} \text{Pr}^i\text{O} \quad \text{OPr}^i \\ \text{Pr}^i \diagdown \text{Al} \diagdown \text{Pr}^i \\ \text{O} \quad \text{O} \\ | \quad \downarrow \\ \text{Pr}^i\text{O}—\text{Al} \quad \text{Al}—\text{OPr}^i \\ \text{Pr}^i\text{O} \diagup \diagdown \text{O} \diagup \diagdown \text{OPr}^i \\ | \\ \text{Pr}^i \end{array}$$

(II)

demonstrated that alcoholysis of the trimer was relatively slow (separate signals for trimer and alcohol), it was concluded that the fast exchange of terminal and bridging groups in the trimer was an intramolecular process assisted by the coordinatively unsaturated nature of the four-coordinated aluminums. In the tetramer (Fig. 4) the very slow terminal-bridging exchange is understandably blocked by the presence of the central octahedral aluminum, whereas a cyclic tetramer would be expected to give rapid exchange. Shiner et al. (47) also noted that solutions of the trimer exhibited a rate of transformation to the tetramer which was much slower than in the molten trimer, and they determined the half-life of the trimer in the melt at 22° to be ca. 50 hr. They also noted that an "aged" sample of the trimer showed weak signals due to yet another species besides the tetramer.

Kleinschmidt (48) proposed an alternative structure for the trimer (Fig. 5) involving a central five-coordinated aluminum bridged to two four-coordinated aluminums, and Fieggen (43) has found ^{1}H NMR evidence for chloroform solutions of the trimer, which supports this structure.

The best established dimeric aluminum alkoxide is $Al_2(OBu^t)_6$ (44), whose structure (III) was confirmed by ^{1}H NMR (47, 53). The mixed

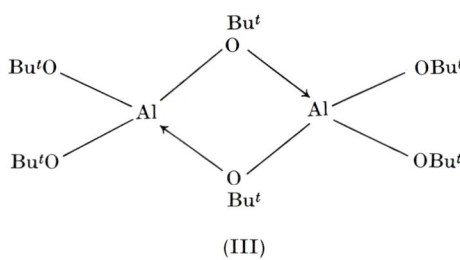

(III)

alkoxide $Al(OPr^i)(OBu^t)_2$ is also dimeric (53), and recent ^{1}H NMR studies (54) suggest that it has the interesting unsymmetrical structure (IV), which contains one asymmetric aluminum atom. However, the

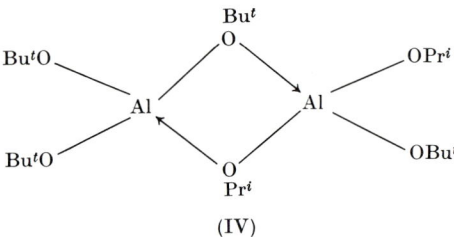

(IV)

spectrum at 60 MHz is poorly resolved and could be interpreted in terms of a mixture of dimer species differing in the constitution of the bridge system. It is clearly desirable to obtain spectra at 220 MHz in more than one solvent to obtain more definitive information. The mass spectrum of this compound (54) gave numerous fragment ions containing two aluminum atoms, and the proposed fragmentation pattern was consistent with structure (IV).

Other dimeric species recently reported are $Al_2(OCH_2CCl_3)_6$ (55) and $Al_2(OCH_2CH_2Cl)_6$ (56) which are of interest because of the ease with which the bridge is cleaved by donor molecules (L) to form complexes $Al(OR)_3L$.

The infrared spectra have been obtained for several aluminum alkoxides [see Mehrotra (2) for earlier references], but they have not been particularly helpful from a structural viewpoint. Fieggen (43) has compared the infrared and Raman spectra of the tetrameric and trimeric forms of aluminum isopropoxide and has found that they are consistent with the proposed structures.

Aluminum alkoxides are commonly used as catalysts in the Meerwein–Ponndorf–Verley and Tishchenko reactions (*1, 46*), and the nature of the alkoxide has an important bearing on the mechanism of the reaction. Bearing in mind the complex nature of aluminum alkoxides, Bains and Bradley (*57*) studied the kinetics of the system depicted by Eq. (5).

$$Me_2CO + Al_2(OPr^{i*})_2(OBu^t)_4 \rightleftharpoons Me_2{}^*CO + Al_2(OPr^i)_2(OBu^t)_4 \quad (5)$$

The course of the reaction was followed by using ^{14}C-labeled isopropoxo groups, but the kinetics at different temperatures proved to be unexpectedly complex and cast doubts on the accepted mechanism of the Meerwein–Ponndorf–Verley reaction. However, it had been assumed that the dimer structure involved isopropoxide bridges, but recent NMR work (*54*) suggests that structure (IV) is present with both bridging and terminal isopropoxide groups. It is also possible that more than one dimer species is present and the proportions may vary with temperature; hence, the system may not be as simple as was originally envisaged.

Even more compelling evidence refuting the accepted mechanism of the Meerwein–Ponndorf–Verley reaction was forthcoming from the ^{1}H NMR studies of Shiner and Whittaker (*58*). They showed that acetone did not break down the polymeric structure of either $Al_3(OPr^i)_9$ or $Al_4(OPr^i)_{12}$, and there was no evidence for a monomeric activated complex $(Pr^iO)_3Al(Me_2CO)$. In the reaction of acetophenone with the isopropoxides [Eqs. (6) and (7)] the rate-determining step was the alcoholysis [Eq. (7)], since the rate of formation of acetone was much

$$C_6H_5COCH_3 + \frac{1}{n}[Al(OPr^i)_3]_n \rightarrow \frac{1}{n}\{Al[OCH(CH_3)C_6H_5](OPr^i)_2\}_n + Me_2CO \quad (6)$$

$$Pr^iOH + \frac{1}{n}\{Al[OCH(CH_3)C_6H_5](OPr^i)_2\}_n \rightarrow \frac{1}{n}[Al(OPr^i)_3]_n + C_6H_5(CH_3)CHOH \quad (7)$$

greater than that of α-phenylethanol. It was also demonstrated that the trimer $Al_3(OPr^i)_9$ reacted much faster than the tetramer $Al_4(OPr^i)_{12}$.

Recent work on the kinetics of the Tishchenko reaction of acetaldehyde with aluminum isopropoxide catalysts (*59*) showed the formation of ethyl acetate to be first-order in catalyst and first-order in acetaldehyde. Surprisingly, the rate constant for the trimer as catalyst was lower than for the tetramer, but isopropyl acetate was also formed and there may be complications owing to simultaneous Meerwein–Ponndorf–Verley reactions.

Alkoxides of gallium(III) have recently been prepared (*60, 61*) and found to undergo the typical reactions of metal alkoxides.

Gallium trimethoxide is a white, insoluble solid which can be sublimed in vacuo (275°–280°/0.4 mm). Other normal alkoxides Ga(OR)$_3$ (R = Et, Pr^n, Bu^n) are more volatile and are tetrameric in solution like the corresponding aluminum derivatives. However, the isopropoxide, a viscous liquid (b.p. 120°/1.0 mm), surprisingly is dimeric. The crystalline dimeric *tert*-butoxide has been shown by ^{1}H NMR *(54)* to have the same structure as Al$_2$(OBu^t)$_6$ (III).

D. TRANSITION METAL ALKOXIDES

Alkoxides of the transition metals have received considerable attention during the past 20 years, and the basic features of the subject have been dealt with in earlier reviews *(1, 2, 46)*. A recent review *(3)* has concentrated on developments over the past 3–5 years, so this account will be restricted to a few highlights and it is not intended to deal with all reported work.

1. d^0 Systems

It is convenient to deal with d^0 systems separately since crystal field effects are absent. Also, these alkoxides are all diamagnetic and the metals are present in their highest oxidation states.

All these compounds are very rapidly hydrolyzed. Dealing first with the metal methoxides, the following compounds are known M(OMe)$_4$ (M = Ti, Zn, Hf), VO(OMe)$_3$, M(OMe)$_5$ (M = Nb, Ta), and W(OMe)$_6$. The tetramethoxides are generally insoluble powdery solids of low volatility owing to their polymeric nature, whereas the dimeric pentamethoxides and monomeric hexamethoxide are soluble and much more volatile. A soluble tetrameric form of Ti(OMe)$_4$ was discovered by Dunn *(62)*, and it was shown by X-ray crystal analysis *(63)* to have the interesting structure shown in Fig. 6. This has the same M$_4$O$_{16}$ framework as crystalline Ti$_4$(OEt)$_{16}$ *(64)* and Ti$_4$(OMe)$_4$(OEt)$_{12}$ *(65)*, and this is clearly of fundamental importance to our understanding of the structural chemistry of metal alkoxides. These units are centrosymmetric and contain octahedrally coordinated titanium in a group of four edge-sharing octahedra. Within each tetrameric unit there are two alkoxo groups which are triple-bridging (i.e., Ti$_3$OR), four which are double-bridging (Ti$_2$OR), and ten nonbridging groups. The latter are disposed in groups of either two or three per titanium atom in mutually cis positions. There are two different environments for the titanium atoms. Ti(1) and Ti(1') are bonded to three nonbridging and three bridging oxygens, whereas Ti(2) and Ti(2') are bonded to two nonbridging and four bridging.

A detailed analysis of the bond lengths is limited by a peculiar uncertainty in the positions of methoxo groups (4) and (5), which the authors ascribed to partial hydrolysis and replacement by hydroxo groups. However, it does seem clear that a trans influence is operating in this system. Thus, ignoring groups (4) and (5), the shortest Ti–O bond lengths are found to involve the nonbridging methoxo groups [(1), (2), and (3)], which are trans to the triple-bridging methoxo groups

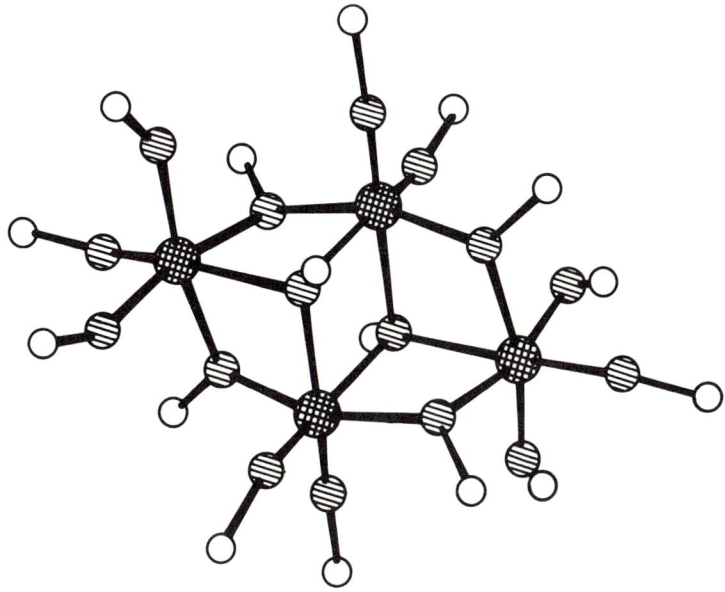

FIG. 6. The structure of [Ti(OMe)$_4$]$_4$. Cross-hatched, Ti; hatched, O; plain, CH$_3$.

[(8) and (8')]: Ti(2)–O(1) = 1.79; Ti(2)–O(2) = 1.78; and Ti(1)–O(3) = 1.83 Å. Although this may be largely a σ-bonding effect, it is significant that all the valence electrons on the triple-bridging oxygens are involved in σ bonds and cannot contribute to π bonding, whereas the trans nonbridging oxygens have lone pairs available for π donation to vacant titanium d orbitals. Moreover this π-donation hypothesis is consistent with the wide angles: Ti(2)–Ô(1)–C(1) = 161; Ti(2)–O(2)–C(2) = 152; and Ti(1)–Ô(3)–C(3) = 140°; and the short C–O bond lengths C(1)–O(1) = 1.38; C(2)–O(2) = 1.39; and C(3)–O(3) = 1.38 Å. Since each titanium atom shares 12 electrons by virtue of σ bonding, it needs a share of six π electrons to achieve the 18-electron configuration. In Ti(1) this could arise if the nonbridging oxygens (3), (4), and (5) each contributed a π-electron pair by interacting with the three vacant d orbitals. Unfortunately, only Ti(1)–O(3) is known accurately, but it is considerably shorter than the trans bond Ti(1)–O(8) (2.20 Å). In Ti(2) four π electrons could be donated by the nonbridging oxygens (1) and (2) producing short bonds and resulting in the long bonds in the trans position, viz, Ti(2)–O(8′) = 2.15 and Ti(2)–O(8) = 2.13 Å. The other two π electrons could be contributed by the bridging oxygens (6) and (7), which are equidistant with Ti–O = 1.96 Å [significantly shorter than the bonds between these oxygens and titanium (1)]. Consistent with this view the double-bridging oxygens have intermediate C–O bond lengths [C(6)–O(6) = 1.45 and C(7)–O(7) = 1.44 Å] compared with the short nonbridging (1.38 Å) and long triple-bridging [C(8)–O(8) = 1.50 Å] bond lengths and also intermediate Ti–Ô–C angles (117°–124°).

Thus, the π-bonding hypothesis explains not only the mutually cis configuration of nonbridging methoxo groups, but also the relative order of bond lengths and bond angles for all the ligands in the tetrameric molecule. A NMR spectrum of titanium methoxide gave four broad peaks in the ratio 1:2:3:2, which were assigned on the basis of the tetrameric structure *(66)*.

Before completing the discussion of titanium tetramethoxide it should be mentioned that Winter *et al.* have shown that the insoluble form (A) has a different X-ray powder pattern from the soluble species (B) *(67)*, and there are also significant differences in their infrared spectra in the C–O and Ti–O regions *(68)*. It was implied that the insoluble form (A) does not contain the triple-bridging methoxo groups characteristic of the tetrameric units in (B) and, hence, would inevitably be more highly polymeric.

The crystal structure of VO(OMe)$_3$ was recently determined by Caughlan *et al.* *(69)*, and it is instructive to compare it with Ti(OMe)$_4$ and consider why it has such a different structure. In fact, it is composed of dimeric units V$_2$O$_2$(OMe)$_6$ (Fig. 7), which are linked by weak bridges at O(10) and O(9′) to generate a linear polymer down the c-axis of the crystal. Hence, the overall structure is that of an infinite linear polymer involving edge-sharing distorted octahedra. The hypotheses used in interpreting the Ti$_4$(OMe)$_{16}$ structure may be applied equally well to the [VO(OMe)$_3$]$_\infty$ structure. Inspection of the dimer unit reveals two types of nonbridging groups, the V–O at O(7) and (O8) and the terminal methoxo groups at O(5) and O(6), and it is significant that each V=O is cis to the terminal methoxo group and trans to the very weak interdimer bridging methoxo group. With six σ bonds and one π bond each vanadium has a share in 14 electrons and requires four more π electrons to achieve the 18-electron configuration. The V=O oxygen could donate a lone pair to give some triple bond character $V^-\equiv O^+$ and the other lone-pair π donation could be contributed by the terminal methoxo group. The terminal methoxo groups thus have shorter V–O distances (1.74 Å) than the weakly bridging methoxo groups [O(9) and O(10); V–O = 1.84 and 1.86 Å] which, in turn, are shorter than the intradimer bridges [O(3) and O(4); V–O = 1.96 and 2.03 Å trans to O(10) and O(9) and V–O = 2.04 and 2.05 Å trans to O(5) and O(6)], whereas the longest bonds are given by the weak interdimer bridges [O(9') and O(10'); V–O = 2.30 and 2.25 Å] which are trans to the very short V=O bonds.

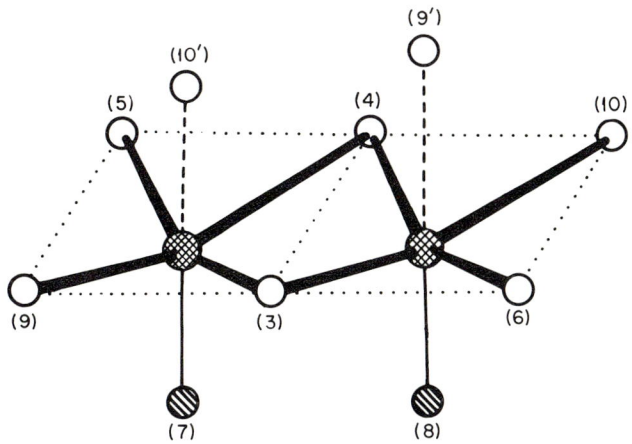

FIG. 7. The dimer unit in $[VO(OMe)_3]_x$; Cross-hatched, V; hatched, 0 in V=O group; plain, O in OMe group.

Unfortunately, there are no X-ray structures available on the dimeric methoxides $M_2(OMe)_{10}$ (M = Nb, Ta), but their structures were solved using variable temperature ^1H NMR measurements in solution (70). At low temperatures (ca. −60°) three signals were obtained in the ratio 2:2:1 corresponding to the two distinguishable types of terminal groups and the bridging group required by the structure in Fig. 8. Once again there is octahedral coordination with edge-sharing by bridging methoxo groups, and the structural relationship between this dimer and the $Ti_4(OMe)_{16}$ tetramer is evident. Applying the same arguments as before and in anticipation of an X-ray structural analysis, it is tempting to predict that the terminal methoxo groups [O(3), O(3'), O(4), and O(4')] trans to the bridging oxygens [O(5) and O(5')] will have slightly shorter M–O bond lengths than the other type of terminal groups which are paired off in trans positions. Thus, each metal atom shares 12 electrons

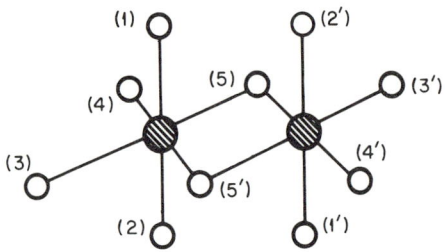

FIG. 8. The dimer [M(OR)₅]₂. Hatched, Nb or Ta; plain, OR.

from σ bonding and requires six π electrons to attain the 18-electron configuration. Assuming that the bridging oxygens form longer M–O bonds and are less effective π donors than the terminal oxygens, then the terminal oxygens trans to them [O(3) and O(4)] will each donate a pair of electrons to a vacant d orbital, leaving one pair to be donated from the two trans terminal oxygens [O(1) and O(2)] into the remaining vacant d orbital. Since O(1) and O(2) are directly competing with each other both as σ and π donors, whereas O(3) and O(4) are only competing with the weaker bridging oxygens, it follows that M–O(3) and M–O(4) should be slightly shorter than M–O(1) and M–O(2), which, in turn, should be shorter than M–O(5) and M–O(5'). At higher temperatures the NMR signals of terminal methoxo group protons coalesced as intramolecular exchange occurred, and the activation energy for exchange for $Ta_2(OMe)_{10}$ was derived as $E_a = 8.6 \pm 0.5$ kcal·mole⁻¹ from the temperature dependence of line broadening. At still higher temperatures further broadening occurred as terminal groups exchanged with bridging groups ($E_a = 10.4 \pm 0.4$ kcal·mole⁻¹) and eventually the spectrum collapsed into one sharp signal. This work was independently confirmed by Riess and Pfalzgraf (71), who also showed by NMR measurements that the complex $Nb(OMe)_5(C_5H_5N)$ was reversibly formed in solution. The formation of weak complexes $Ta(OR)_5(C_5H_5N)$ had earlier been proposed by Bradley et al. (72) to account for the monomeric behavior of tantalum alkoxides in pyridine.

Finally we mention $W(OMe)_6$, which was obtained by methanolysis of $W(NMe_2)_6$ (73). It is a monomeric, white crystalline solid which sublimes in vacuo (50°–60°/10⁻⁴ mm) and is undoubtedly an octahedral molecule.

Metal ethoxides have received considerable attention, and recent work has clarified the structural status of titanium ethoxide, which had been a controversial subject for some time. A major advance came with the advent of the X-ray crystal analysis by Ibers (64), which demonstrated the presence of tetramers (same Ti_4O_{16} configuration as Fig. 6) in the solid state. The positions of the carbon atoms were not reported, but it was stated that the ethyl groups took up positions which gave a cylindrical structure to the tetrameric molecule. The titanium–oxygen bond lengths reflected the nature of the coordination of the oxygen atom, i.e., Ti_3OEt, 2.23; Ti_2OEt, 2.03; and TiOEt, 1.77 Å. The same tetrameric Ti_4O_{16} framework was also found in the mixed alkoxide $[Ti(OMe)(OEt)_3]_4$ (65), which had Ti–O bond lengths in the range 1.6–2.4 Å and the closest Ti...Ti distances 3.3–3.5 Å. Unfortunately, the positions of the carbon atoms were not given, but it would be very interesting to know which sites were occupied by the methoxo groups. Titanium tetraethoxide has interesting physical properties. It is very soluble in organic

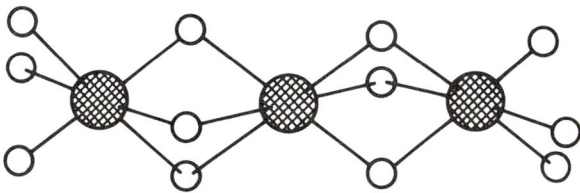

FIG. 9. A structure for $[Ti(OR)_4]_3$. Cross-hatched, Ti (octahedral); plain, OR.

solvents and is usually obtained as a supercooled liquid by vacuum distillation. Over a period of several months at room temperature it gradually solidifies to the tetrameric crystalline form, but in solution it is predominantly, if not entirely, trimeric. Bradley and Holloway (74, 75) claimed that earlier work suggesting a decreasing degree of polymerization with decreasing concentration was erroneous because of the sensitivity of molecular weight measurements to traces of water and the difficulty in keeping benzene rigorously dry. Using exceptional precautions to avoid hydrolysis they found by cryoscopic measurements that the compound was trimeric over a wide concentration range. These results have since been confirmed by the completely independent technique of light scattering (76), and the problem which remains is to determine the structure of the trimer. The chief candidates are shown in Figs. 9–12. The linear structure (Fig. 9) based on face-sharing octahedral units was originally proposed by Caughlan et al. (77) and much favored by Bradley (1, 46). The alternative structures (Figs. 10 and 11) were proposed by Martin and Winter (78). In Fig. 10 the structure involves a triangular arrangement of titanium atoms with edge-sharing trigonal prisms, while in Fig. 11 another triangular arrangement involves five-

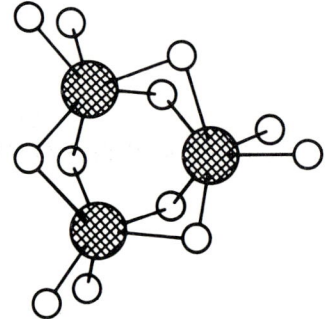

FIG. 10. A structure for [Ti(OR)$_4$]$_3$. Cross-hatched, Ti (trigonal-prismatic); plain, OR.

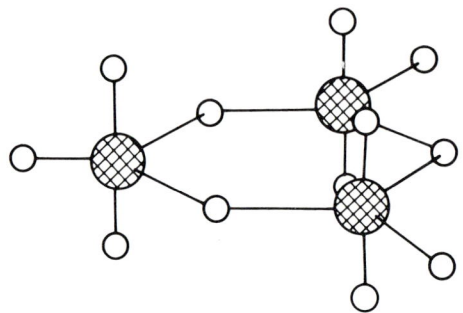

FIG. 11. A structure for [Ti(OR)$_4$]$_3$. Cross-hatched, Ti (trigonal-bipyramidal); plain, OR.

coordinated titanium (trigonal bipyramid) with corners sharing. Since titanium is predominantly octahedral in six-coordinated systems, Fig. 10 seems highly unlikely and, although five-coordination is a distinct possibility, it is difficult to imagine why the trimer structure (Fig. 11) would have preference over an edge-shared dimer. The structure in Fig. 12 (66) has a linear arrangement of titaniums with the central one octahedrally coordinated and the outer ones five-coordinated in an edge-sharing arrangement of two trigonal bipyramids and an octahedron. This is the structure which is clearly favored by the results of Russo and Nelson (76). They calculated values for the molecular polarizability anisotropy for the three structures [Fig. 9 (9.0×10^{-48} cm^6), Fig. 11 (14.0×10^{-48} cm^6), and Fig. 12 (111.6×10^{-48} cm^6)]. The average value obtained from six concentrations in cyclohexane (0.03–0.60 M) and three in carbon tetrachloride (0.10–0.60 M) was $111.5 \pm 2 \times 10^{-48}$ cm^6, so even allowing generously for errors in the calculated values there is no doubt that the structures in Figs. 9 and 11 can be eliminated. These authors also observed that below 800 cm^{-1} there are no coincident bands in the infrared and Raman spectra which suggests a centrosymmetrical structure which again rules out Fig. 11. Bradley and Westlake (79) had earlier attempted to assign structures of metal alkoxides by determining the ratio of terminal to bridging groups from band intensities of the C–O stretching frequencies. For $Ti_3(OEt)_{12}$ they found a ratio of 2:1, which agreed with none of the structures in Figs. 9–11 which were then being considered. However, it is now seen to be in agreement with the structure in Fig. 12. Attempts have also been made to assign the

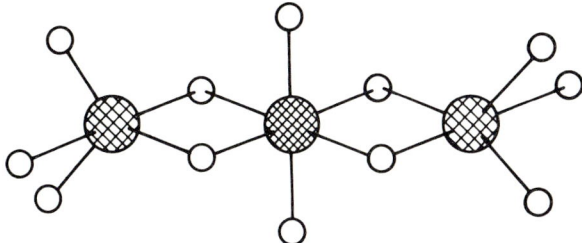

FIG. 12. A structure for $[Ti(OR)_4]_3$. Fine cross-hatched, Ti (octahedral); cross-hatched, Ti (five-co-ordinated); plain, OR.

$Ti_3(OEt)_{12}$ structure by ^{1}H NMR studies (72, 75, 80). Owing to the fluxional nature of the molecule the spectrum at room temperature or above shows only one quartet (CH_2 protons) and one triplet (CH_3 protons) due to rapid intramolecular exchange of terminal and bridging alkoxo groups. At lower temperatures broadening and splitting are observed, but have been interpreted in different ways. Weingarten and Van Wazer (66) considered that the unequal splitting of the quartet at −48° in toluene ruled out Figs. 9 and 10 and they favored the structure in Fig. 10 with the 3:1 ratio of terminal:bridging, although they did not completely rule out Fig. 12. In a more detailed variable temperature study in toluene and carbon disulfide, Bradley and Holloway (80) observed some additional features. In both solvents it was noted that the new peaks increased in intensity at the expense of the original peaks as the temperature was lowered, and it suggested that a new species was developing rather than that terminal-bridging peaks of one species were being resolved. Below −60° the spectra were broadened owing to viscosity effects, and it was not possible to make any definite structural assignments. Nevertheless, it was found that the ethoxide readily crystallized from CS_2 at low temperatures, whereas the liquid trimer transforms to the solid tetramer very slowly at room temperatures. It was therefore proposed that the low-temperature NMR spectra were showing the transformation in solution of the trimer to the tetramer. Similar measurements on the trimeric n-propoxide and n-butoxide in CS_2 showed no splitting even down to $-90°$, although the lines were broadened, and this indicates very rapid terminal-bridging exchange in the trimer (*80*). This behavior is readily understood for the structure in Fig. 12 containing two five-coordinated and one six-coordinated titanium. Since the enthalpy of formation of an ethoxo bridge was estimated by various methods to be ca. -11 kcal·mole^{-1} (*66*), it seems unlikely that the terminal-bridging exchange for the trimer would involve a simple bridge-dissociation mechanism.

Assuming that the trimer has the structure shown in Fig. 12, a reappraisal of the structural implications of the theory of Bradley *et al.* (*7*) for the soluble hydrolysis products of $Ti_3(OEt)_{12}$ is called for. On the basis of ebullioscopic molecular weight studies a series of polymers $[Ti_{3(x+1)}O_{4x}(OEt)_{4(x+3)}]$ ($x = 0, 1, 2, 3, \ldots \infty$) was postulated based on a condensation process involving the octahedral trimers (Fig. 9) as structural units. In addition, a crystalline product corresponding to $x = 1$ $[Ti_6O_4(OEt)_{16}]$ was isolated. An X-ray crystal analysis of this compound (*81*) has shown that a heptameric unit Ti_7O_{24} (Fig. 13) is present, and it is clearly patterned on the tetramer unit Ti_4O_{16} of the crystalline tetraethoxide. Unfortunately the positions of the carbon atoms were not located, so the details of the structure are not known, but the proposed formula $Ti_7O_5(OEt)_{19}$ is impossible for quadrivalent titanium and the formula is presumably either $Ti_7O_4(OEt)_{20}$ or $Ti_7O_4(OH)(OEt)_{19}$. It is possible to reconcile the difference between the species in solution and in the crystal by considering that the species in solution contains some five-coordinated titanium [e.g., a species $Ti_6O_4(OEt)_{16}$ involving two five-coordinated titaniums and four octahedral titaniums can be visualized], and rearrangement occurs to give an entirely octahedral system in the crystal.

The pentaethoxides of niobium and tantalum were shown by variable temperature ^1H NMR studies to have the dimeric structure (Fig. 8) involving edge-sharing octahedra (*70*), and hydrolysis studies (*7*) indicated the presence of polymers in the series $M_{2(x+1)}O_{3x}(OR)_{2(2x+5)}$ based on a condensation process of dimer units. The only crystalline hydrolysis product which has so far been obtained was shown by X-ray analysis to be $Nb_8O_{10}(OEt)_{20}$ (*82*). This has the interesting cage structure shown in Fig. 14, which is based on octahedral niobium. It may be visualized as comprising two Nb_3O_{13} (three edge-sharing octahedra)

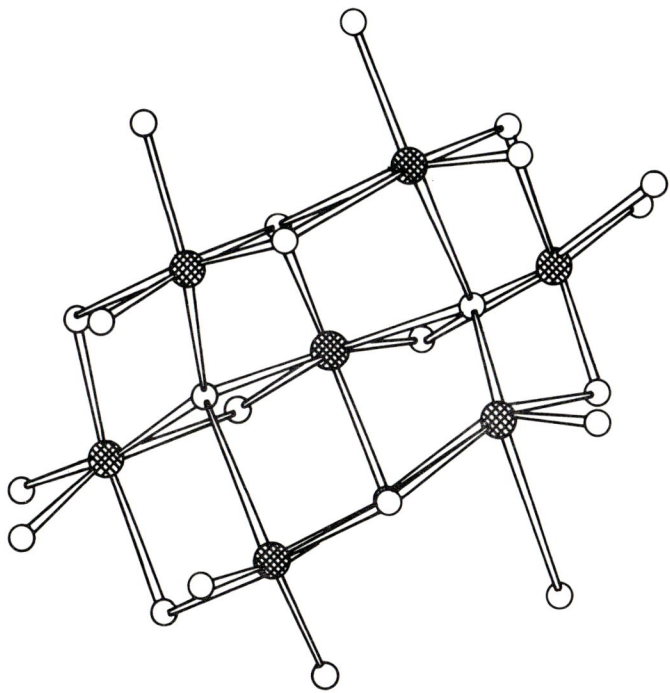

FIG. 13. The [Ti$_7$O$_{24}$] unit. Cross-hatched, Ti; plain, O.

units linked by two corner-sharing octahedra. It is noteworthy that there are no Nb=O bonds [compare VO(OMe)$_3$], and consequently all the oxo oxygens are in bridging positions (two triple-bridging and eight double-bridging), whereas the nonbridging ethoxo groups are arranged in cis pairs (except for the two niobiums which have only a single nonbridging group). It is significant that all the polymeric six-coordinated methoxide and ethoxide structures which are known with certainty involve edge-sharing octahedra in contrast to the adjacent face-sharing depicted for Fig. 9. It was pointed out by Russo and Nelson (76) that the face-sharing structure would bring the titanium atoms too close together (ca. 2.3 Å compared with Ti...Ti of 3 Å in the edge-shared structures). To conclude this section on the ethoxides we note the ^{1}H NMR and calorimetric studies on the "scrambling" reactions of Ti(OEt)$_4$ with Ti(OR)$_4$ (R = Pr^i, Bu^t), Ti(NMe$_2$)$_4$, and TiCl$_4$ (66) and also the thermochemical work from which the standard heat of formation ΔH_f^0[Ti(OEt)$_4$, (c)] = -349 ± 1.4 kcal·mole^{-1} and the average Ti–O bond dissociation energy $\bar{D} = 101 \pm 2.1$ kcal·mole^{-1} were derived (83). Other titanium

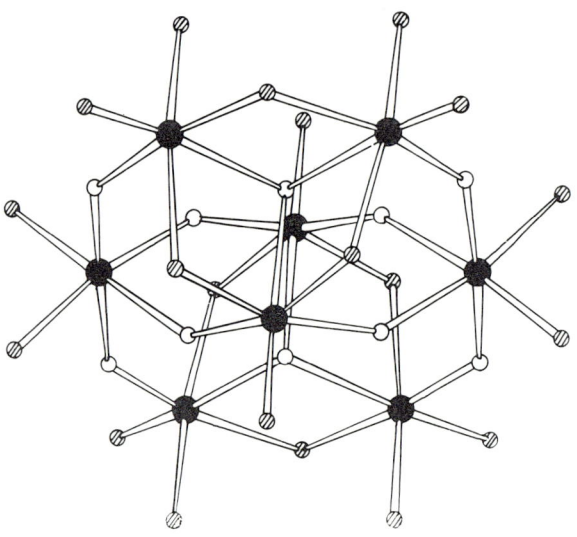

FIG. 14. The structure of [Nb$_8$O$_{10}$(OEt)$_{20}$]. Solid, Nb; plain, O; hatched, OEt.

alkoxides were similarly studied and all gave $\bar{D}$ (Ti–O) values in the range 101–109 kcal·mole^{-1} (84).

Cryoscopic molecular weight determinations and ^{1}H NMR studies have been carried out on the metal isopropoxides (75, 80). Steric hindrance of the branched alkyl groups opposes intermolecular bonding, and titanium isopropoxide is monomeric in solution and more volatile than the normal alkoxides. Nevertheless, it was shown by NMR measurements that polymerization takes place at lower temperatures. Zirconium isopropoxide ($\bar{n} = 3.57 \pm 0.08$) shows a small but significant difference in degree of polymerization from hafnium isopropoxide ($\bar{n} = 3.33 \pm 0.09$), and it appears that trimers and tetramers are present, although no change in molecular weight with concentration was detected. A single type of isopropoxo group was found in the NMR spectrum of the freshly distilled compounds indicating rapid terminal-bridging exchange in the trimer species, but "aged" samples had three additional small sets of doublets believed to be due to tetramer species. The extra peaks coalesced into the main doublet at higher temperatures. The pentaisopropoxides of niobium and tantalum gave interesting ^{1}H NMR spectra in solution (70). Distinct resonances were found for the monomer M(OPr^i)$_5$ (a single type of isopropoxide) and the dimer M$_2$(OPr^i)$_{10}$ (two types of isopropoxide in 4:1 ratio), and the variation of intensities of monomer and dimer peaks with temperature and concentration fitted the mass law relationship for an equilibrium.

$$M_2(OPr^i)_{10} \rightleftharpoons 2M(OPr^i)_5 \qquad (8)$$

At a given temperature and concentration the niobium compound had a lower dimer concentration than tantalum. From the temperature dependence of the equilibrium constant for Eq. (8), the enthalpy of the dimerization process was derived as $Nb_2(OPr^i)_{10}$, $\Delta H = 16.3$ kcal·mole^{-1}, and $Ta_2(OPr^i)_{10}$, $\Delta H = 17.0 \pm 1.5$ kcal·mole^{-1}. Addition of isopropanol caused the removal of the monomer peak which coalesced with that of the alcohol and a reduction in intensity of the dimer peaks owing to some solvation.

$$M_2(OPr^i)_{10} + 2Pr^iOH \rightleftharpoons 2M(OPr^i)_5(Pr^iOH) \qquad (9)$$

Evidently the dimer exchanges isopropoxo groups slowly with isopropanol, whereas the solvated monomer undergoes a rapid intramolecular ligand exchange and a rapid intermolecular exchange with the alcohol.

Steric hindrance is very marked in the *tert*-butoxo group, and the *tert*-butoxides of titanium, zirconium, hafnium, niobium, and tantalum are all monomeric and volatile. For the tetra-*tert*-butoxides the ^{1}H NMR studies (75) showed that $Ti(OBu^t)_4$ remained monomeric in solution down to $-50°$, whereas $M(OBu^t)_4$ (M = Zr, Hf) showed some broadening and a slight shift to low field, which indicated some polymerization at $-50°$. Addition of *tert*-butanol showed that exchange with $Ti(OBu^t)_4$ was slow, whereas coalescence of alkoxide and alcohol signals showed rapid exchange for $Zr(OBu^t)_4$ and $Hf(OBu^t)_4$. The penta-*tert*-butoxides $M(OBu^t)_5$ (M = Nb, Ta) remained monomeric in solution down to $-100°$, and intramolecular ligand exchange was too fast to allow structural identification (i.e., distinction between trigonal bipyramid and square pyramid). Addition of *tert*-butanol gave a single coalesced signal indicating rapid exchange.

To summarize this brief account of the d^0 alkoxides of Groups IV and V, it appears that in the polymeric species involving edge-shared octahedral structures the intramolecular exchange of alkoxo groups is relatively slow as is the intermolecular exchange with the parent alcohol. For species involving some or all of the metal atoms in lower coordination numbers the exchange processes appear to be much more rapid.

2. *d^n Systems*

The d^n systems ($n = 1$–9) are of interest because of the intervention of crystal field effects. From an important survey of the polymeric dimethoxides $M(OMe)_2$ (M = Cr, Mn, Fe, Co, Ni, Cu), in which polymeric structures involving octahedral coordination were found, Martin et al. (85) deduced from the reflectance spectra and magnetism that the methoxo group exerted a crystal field splitting similar to that of water in these high-spin systems. Infrared spectra in the M–O region (600–200 cm^{-1}) showed trends which could be related to ligand field effects on the M–O vibrations (86). It is convenient to deal with the d^n metal alkoxides according to the number of d electrons (n).

d^1 *Metal Alkoxides.* Titanium trimethoxide was isolated as a highly reactive (to air), insoluble, yellow-green solid (85). It would be expected to have octahedral Ti(III), but the reflectance spectrum gave a "d–d" transition at 10,000 cm^{-1} and the compound was diamagnetic indicating strong metal–metal interactions. Lappert and Sanger (87) have recently reported that triethoxotitanium(III) is a diamagnetic, dark blue-green crystalline tetramer $Ti_4(OEt)_{12}$. A number of tetraalkoxovanadium(IV) compounds have been prepared (3). The trimethoxo compound is a soluble trimeric derivative $V_3(OMe)_{12}$, and it is paramagnetic with μ_{eff} varying slightly with temperature (1.79 at 289°K to 1.70 at 123°K) (88). The data could be fitted to a model involving a distorted octahedral configuration with the $^2T_{2g}$ state split by >1000 cm^{-1} leading to an orbital singlet ground state. This was confirmed by ESR studies (89), which showed a broad signal ($g = 1.955 \pm 0.005$) in the solid at room temperature with no hyperfine structure. However, it should be borne in mind that the actual structure of the trimer may be analogous to $Ti_3(OEt)_{12}$ (Fig. 12) with both six-coordinated and five-coordinated metal atoms. Vanadium tetraethoxide has received further study (88). It is a dimeric solid which obeys the Curie law with $\mu_{eff} = 1.69$ (123°–289°K). Electronic spectra showed two "d–d" transitions at 6000 and 14,200 cm^{-1}, which could be assigned to five-coordinated vanadium (trigonal-bipyramidal). The broadness of the bands suggested splitting of the $^2E''$ and $^2E'$ states by a lowering in symmetry from D_{3h} to C_{2v}, in accordance with an edge-sharing bridged structure for the dimer $V_2(OEt)_8$. Also an ESR signal was given by the solid at room temperature ($g = 1.954$) and in solution there was a partial resolution of the ^{51}V hyperfine coupling (89). The tetra-*tert*-butoxovanadium(IV) is a bright blue, volatile, monomeric liquid, and its electronic absorption spectrum gave a broad asymmetric band which was resolved by gaussian analysis into two "d–d" transitions at 10,930 and 13,900 cm^{-1} (89). Earlier work had shown that $V(OBu^t)_4$ gave at room temperature an ESR signal ($\langle g \rangle = 1.964$; $\langle a \rangle = 0.0064$ cm^{-1}) with ^{51}V hyperfine splitting, and frozen solutions ($-196°$) gave resolved anisotropic spectra ($g_\| = 1.940$; $g_\perp = 1.984$; $A_\| = 0.0125$; $A_\perp = 0.0036$ cm^{-1}) (90). A molecular orbital treatment gave a reduced spin–orbit coupling constant $\lambda = 156$ cm^{-1}), and the coefficients of the wave functions suggested a moderate degree of covalency owing to participation of the d orbitals. These ESR data were confirmed ($\langle g \rangle = 1.962$) (*89*) and found to be in agreement with the magnetic susceptibility data ($\mu_{\text{eff}} = 1.69 \pm 0.03$ independent of temperature) (*88*) and all the data point to a distorted tetrahedral configuration (D_{2d} point group) with a $^2B_1(d_{x^2-y^2}$ orbital) ground state (Fig. 15). The d–d transitions then correspond to $^2B_1 \to {}^2B_2$ (10,930 cm^{-1}) and $^2B_1 \to {}^2E$ (13,900 cm^{-1}). The distortion of the d^1 tetrahedral system (^2E ground state) could be attributed to the Jahn–Teller effect, but similar distortions of d^2 systems have been found which cannot be

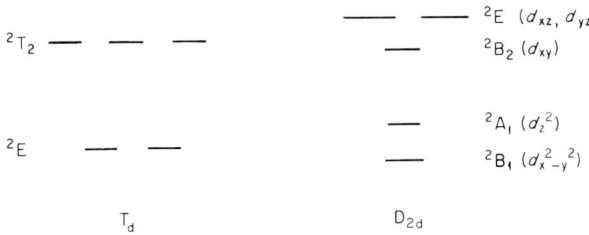

FIG. 15. Splitting of d-orbital energy levels.

so explained, and Bradley and Chisholm (*91*) prefer an explanation based on the consequences of covalent bonding. Thermochemical measurements (*92*) gave a value of $D(\text{V–O}) = 87.5$ kcal·mole^{-1} for V(OBu^t)$_4$.

Some alkoxo derivatives of the d^1 species Nb(IV) and W(V) have been isolated by Brubaker and co-workers. Ethanolic solutions of NbCl$_5$ saturated with HCl were electrolytically reduced to Nb(IV) and the red, diamagnetic, dimeric complex Nb$_2$Cl$_2$(OEt)$_6$(C$_5$H$_5$N)$_2$ was obtained by addition of pyridine. Treatment of this complex with NaOEt gave Nb(OEt)$_4$ as a dark red oil (b.p. ca. 160°/0.002 mm) which had a shoulder at 26,300 cm^{-1}, which may be due to a "d–d" transition (*93*). The infrared spectrum in the region 1000–1150 cm^{-1} (mainly C–O stretching vibrations) was very similar to that of the polymeric ethoxo compounds Ti$_3$(OEt)$_{12}$, Nb$_2$(OEt)$_{10}$, and Ta$_2$(OEt)$_{10}$ (*79*), and together with its low volatility and diamagnetic nature suggests that Nb(OEt)$_4$ is polymeric, but a molecular weight determination was not carried out. The pentaethoxotungsten(V) was obtained by addition of NaOEt to an ethanolic solution of WCl$_5$ (*94*). It was a brown, diamagnetic, liquid which showed only one type of ethoxo group in the ^{1}H NMR spectrum down to $-80°$. This behavior is consistent with a fluxional metal–metal bonded dimer or a very labile ethoxo-bridged polymer, but no molecular weight measurements were made.

d^2 *Metal Alkoxides.* Bradley et al. (*95*) have recently reported fully on the physico-chemical properties of the interesting volatile, monomeric $Cr(OR^t)_4$ ($R^t = tert$-alkyl) compounds. Previous calorimetric work (*92*) had shown that $Cr(OBu^t)_4$ has considerable thermodynamic stability: $-\Delta H_f^0[Cr(OBu^t)_4, (g)] = -305$ kcal·mole^{-1} and average bond dissociation energy $\bar{D}(Cr-O) = 73$ kcal·mole^{-1}; and it is clear that covalent Cr(IV) compounds are quite stable in nonaqueous systems. A parent ion $Cr(OBu^t)_4^+$ was obtained in the mass spectrum, together with several interesting chromium-containing fragment ions and metastable peaks (*95*). The characteristic bright blue color of these compounds is due to absorption in the 15,000 cm^{-1} region, and the complex spectrum was interpreted in terms of the three d–d transitions expected for a tetrahedral d^2 complex: $^3T_2(F) \leftarrow {}^3A_2(F)$ (9100); $^3T_1(F) \leftarrow {}^3A_2$ (15,200); $^3T_1(P) \leftarrow {}^3A_2$ (25,000 cm^{-1}) with $10Dq = 9430$ cm^{-1} and $B = 795$ cm^{-1}. However, the lowest energy bands were doublets (8700, 9500 and 13,700, 15,750 cm^{-1}) suggesting a lowering in symmetry to D_{2d}, which would split the T terms. The magnetic susceptibility data showed practically Curie law behavior with $\mu_{eff} = 2.80 \pm 0.03$ independent of temperature, and ESR signals at 10°K in frozen toluene solution (broad absorption $g \sim 4$, sharp absorption $g = 1.962$) agreed with an orbital singlet ground state ($^3B_1$ in D_{2d} symmetry) and a distorted tetrahedral configuration with zero-field splitting. Attempts to prepare other Cr(IV) alkoxo compounds showed that only the tertiary alkoxides were stable, since primary and secondary alcohols were oxidized and led to the formation of stable Cr(III)(OR)$_3$ derivatives. During the spectral work on $Cr(OBu^t)_4$, it was noted that dilute solutions (cyclohexane) of the Cr(IV) compound were perceptibly less readily hydrolyzed than the corresponding d^0 compound $Ti(OBu^t)_4$, and it was suggested that the presence of d electrons in the chromium(IV) compound impeded nucleophilic attack by water molecules (*95*). Wiberg and Foster (*96*) using ^{18}O-labeled water in a medium of acetic (91%) and perchloric (0.23 M) acids showed that in the hydrolysis of $Cr(OBu^t)_4$ the Cr–O bonds underwent cleavage. They also observed that $Cr(OBu^t)_4$ dissolved in styrene was stable for a week showing that butoxo radicals were not produced, thus confirming earlier work of Hagihara and Yamazaki (*97*). The latter authors also showed that polymerization of styrene was initiated by the addition of methanol to $Cr(OBu^t)_4$, which thus produced radicals during the reaction leading to Cr(OMe)$_3$, Bu^{t}OH, and formaldehyde.

The d^2 vanadium(III) alkoxides V(OR)$_3$ (R = Me, Et) were obtained as insoluble, nonvolatile, green solids, which were very sensitive to oxidation and presumably are polymeric (98). A number of green vanadium(III) chloride methoxide complexes $VCl(OMe)_2$, $VCl(OMe)_2(MeOH)$, $VCl(OMe)_2(Me_2CO)$, and $VCl_2(OMe)(MeOH)_2$ were prepared by Kakos and Winter (99), who studied spectral and magnetic properties. The electronic reflectance spectra were interpreted basically in terms of octahedral V(III): $^3T_{2g} \leftarrow {}^3T_{1g}$ (15,500); $^3T_{1g}(P) \leftarrow {}^3T_{1g}$ (25,000); $^3A_{2g} \leftarrow {}^3T_{1g}$ (35,700 cm^{-1}) with $10Dq = 17,400$ cm^{-1} and $B = 745$ cm^{-1}; but splitting of the bands suggested a lowering of symmetry. The anomalously low magnetic moments were accounted for by spin interactions of neighboring metal atoms in trimeric or tetrameric clusters.

d^3 *Metal Alkoxides.* The d^3 Cr(III) alkoxides have been known for many years, but only recently have spectral and magnetic studies been carried out (85, 100). The trimethoxide and triethoxide are insoluble, nonvolatile, pale green solids whose reflectance spectra agreed with ligand field predictions for an octahedral d^3 ion with $10Dq \sim 17,000$ cm^{-1}. X-Ray powder photographs and infrared and Raman spectra for $Cr(OMe)_3$ were interpreted on the basis of a layer type lattice (hexagonal symmetry) involving edge-sharing methoxo-bridged octahedra (100). Magnetic susceptibility measurements gave anomalously low moments owing to antiferromagnetic interactions and extrapolation of the χ^{-1} vs. T data (Curie–Weiss law) gave very large θ values (270°–320°), which then gave μ_{eff} close to the d^3 spin-only value of 3.88 (85, 100). Superexchange involving the alkoxo bridges was invoked as a likely mechanism for the antiferromagnetic behavior.

By using bulky alkoxo groups it was shown that chromium can be prevented from achieving octahedral coordination. Thus, chromium tri-*tert*-butoxide is soluble in organic solvents and is extremely easily oxidized to Cr(IV) (92, 95), and it is most probably dimeric with bridging alkoxo groups and distorted tetrahedral Cr(III) (101).

The compound $LiCr(OBu^t)_4$ is pink and insoluble with probably an infinite linear structure with alkoxo bridges and alternating lithium and chromium atoms (102). The electronic spectrum was assigned on the basis of tetrahedral d^3 Cr(III): $^4T_2(F) \leftarrow {}^4T_1(F)$ (10,000); $^4T_1(P) \leftarrow {}^4T_1(F)$ (17,400); $^4A_2 \leftarrow {}^4T_1$ (19,400 cm^{-1}) with $10Dq = 11,200$ cm^{-1} and $B = 560$ cm^{-1}. The magnetic susceptibility measurements showed that the magnetic moment varied slightly with temperature ($\mu_{eff}^{298} = 3.58$; $\mu_{eff}^{98} = 3.43$) as expected for a system with a T ground state. The ease of oxidation of tetrahedral d^3 Cr(III) can be ascribed to the instability resulting from placing one electron in a higher energy d orbital, whereas both electrons in tetrahedral d^2 Cr(IV) are in lower energy orbitals.

By contrast molybdenum tri-*tert*-butoxide was found to be a reactive orange solid, which could be sublimed *in vacuo* (100°/10^{-3} mm) (73).

The compound was dimeric in benzene and gave a strong parent ion $Mo_2(OBu^t)_6{}^+$ in the mass spectrum together with many strong fragment ion peaks containing two metal atoms. It was also diamagnetic and the infrared and ^1H NMR spectra showed no evidence for bridging alkoxo groups, and it was therefore concluded that the structure entailed a multiple metal–metal bond.

d^4–d^9 *Metal Alkoxides.* An example of a d^4 metal alkoxide is the purple $Cr(OMe)_2$, which is a polymeric, insoluble, nonvolatile solid (*85*). The reflectance spectrum (transitions at 18,200 and 22,200 cm^{-1}) showed evidence of a tetragonally distorted octahedral configuration for the metal, and magnetic measurements showed strong antiferromagnetic interaction ($\mu_{eff} = 5.16$; $\theta = 160°$). There are two examples of d^5 metal alkoxides $Mn(OMe)_2$ and $Fe(OR)_3$. The manganese dimethoxide is a pale pink, insoluble solid whose magnetic susceptibility conforms to the Curie–Weiss law ($\mu_{eff} = 5.96$; $\theta = 35°$), and the reflectance spectrum (spin-forbidden transitions at: 18,500, 24,400, 27,070, 29,000, and 31,300 cm^{-1}) was assigned to octahedral Mn(II). Thus, a polymeric methoxo-bridged octahedral structure is envisaged.

It had been known for several years that ferric alkoxides were trimeric (*103*) and, thus, differed from the tetrameric aluminum alkoxides, but in neither case should there be crystal field effects. Martin and co-workers (*104*) found interesting magnetic susceptibility results for the $Fe_3(OR)_9$ compounds. At room temperature the magnetic moments per iron atom of the normal alkoxides (R = Me, Et, Bu^n) were $\mu_{eff} = 4.51$–4.35, compared to 5.9 required for a d^5 high-spin system. Moreover, the same value was found in solution independent of concentration, showing that the magnetic anomaly was a property of the trimeric cluster and not due to general lattice interactions. The magnetic moments decreased with temperature ($\mu_{eff}^{90} = 3.4$) and the Curie–Weiss law was obeyed ($\theta \sim 200°K$). The magnetic moment of $Fe_3(OBu_n)_9$ decreased with increasing degree of hydrolysis, but increased with thermal decomposition (>80°). A broad band at 11,000 cm^{-1} was observed in the electronic diffuse reflectance spectrum. Further work (*105*) showed that the magnetic susceptibility data were well represented by a model involving cooperative spin coupling of the d^5 ($^6A_1$) metal atoms in an equilateral triangular configuration with an isotropic coupling constant $J = -10$ cm^{-1}. The 216-fold spin degeneracy of the Fe_3 unit is partially split by the spin–spin interactions to give eight spin levels characterized by the spin quantum numbers $S^1 = \frac{1}{2}, \frac{3}{2}, \frac{5}{2}, \ldots, \frac{15}{2}$; and the magnetic field splits each level into $2S^1 + 1$ sublevels. At temperatures approaching the absolute zero only the lowest level ($S^1 = \frac{1}{2}$) is occupied, and $\mu_{eff}^2 = 1$ (B.M.)^2 per Fe_3 unit. At higher temperatures the higher levels become populated leading ultimately to $\mu_{\text{eff}}^2 = 105$ (B.M.)^2 per Fe$_3$ unit, and it was shown that the Weiss temperature $\theta = 35J/3k$. It was noted that quite a small J value ($-15°$K) could cause a dramatic temperature variation of the magnetic susceptibility of the trinuclear cluster. It was suggested that a cyclic structure of tetrahedral iron atoms with alkoxo bridges (D$_{3h}$ symmetry) was most probable and that a superexchange mechanism might operate via the bridging ligands.

The only representative of high-spin d^6 metal alkoxides is Fe(OMe)$_2$, a dark green compound, which obeys the Curie law ($\mu_{\text{eff}} = 5.14$). Its reflectance spectrum ($^5E_g \leftarrow {}^5T_{2g}$; 10,000 cm^{-1}) meets the requirements of an octahedrally coordinating ferrous ion, and it probably has a high polymeric methoxo-bridged edge-sharing octahedral structure (*85*). Cobalt(II) methoxide is an insoluble purple solid, which shows Curie–Weiss law behavior ($\mu_{\text{eff}} = 5.46$; $\theta = 15°$), and its reflectance spectrum [$^4T_{2g}$(F) $\leftarrow {}^4T_{1g}$(F) (9500); ^2E_g(G) $\leftarrow {}^4T_{1g}$ (12,000); $^4A_{2g}$(F) $\leftarrow {}^4T_{1g}$ (12,000); $^4A_{2g}$(F) $\leftarrow {}^4T_{1g}$ (17,900); $^4T_{1g}$(P) $\leftarrow {}^4T_{1g}$ (21,000 cm^{-1})] was assigned in terms of octahedral cobalt(II) (d^7 high-spin) (*85*). Nickel(II) methoxide is an insoluble, pale green solid, which showed Curie law behavior ($\mu_{\text{eff}} = 3.38$), and its reflectance spectrum [$^3T_{2g}$(F) $\leftarrow {}^3A_{2g}$(F) (8700); $^3T_{1g}$(F) $\leftarrow {}^3A_{2g}$ (14,500); $^3T_{1g}$(P) $\leftarrow {}^3A_{2g}$ (25,000 cm^{-1})] was assigned to octahedral nickel(II) (*85*). Infrared bands (420 and 275 cm^{-1}) were assigned to nickel–oxygen stretching vibrations (*86*). Further work by Krüger and Winter (*106*) led to the isolation of a number of methoxohalonickel(II) compounds, e.g., Ni(OMe)Cl; Ni(OMe)Cl(MeOH), Ni(OMe)Cl(MeOH)$_2$, Ni$_3$(OMe)$_4$Cl$_2$, Ni$_3$(OMe)$_5$Cl, etc., and their reflectance spectra were analyzed in terms of octahedral Ni(II). The 10Dq values increased steadily as chloride was replaced by methoxide in the above series [i.e., 7900, 8100, 8200, 8200, and 8300, respectively, and for Ni(OMe)$_2$ 8500 and NiCl$_2$ 7200 cm^{-1}] as would be expected. These compounds exhibited anomalous magnetic properties with μ_{eff} increasing with a decrease in temperature, and the data were ingeniously fitted to a structural model involving "cubane" Ni$_4$(OMe)$_4$ clusters with the octahedral coordination of each Ni(II) completed by additional ligands in terminal and bridging positions. The magnetic interactions corresponded to ferromagnetic interactions between the four nickel atoms in the cluster (J values +4 to +16 cm^{-1}).

Finally we mention work carried out on the d^9 Cu(II) alkoxides. Brubaker and Wicholas (*107*) found that at room temperature the magnetic moments for Cu(II) in Cu(OMe)$_2$ and Cu(OEt)$_2$ were abnormally low and they suggested a highly polymeric structure with tetragonally distorted octahedral copper. Martin *et al.* (*85, 108*) measured the magnetic susceptibility of Cu(OMe)$_2$ over the range 80°–350°K; they found a maximum around 260°K and pronounced antiferromagnetic behavior. They favored a linear chain model for polymeric Cu(OMe)$_2$ analogous in structure to CuCl$_2$. The J value for antiferromagnetic spin–spin interactions was calculated to be -191 cm^{-1} and is comparable with values found for copper(II) carboxylates. The dimeric methoxo-β-diketonato compound Cu$_2$(OMe)$_2$(acac)$_2$ gave a much larger $J = -725$ cm^{-1}, whereas Cu(OMe)Cl showed completely different magnetic behavior (Curie law behavior, $\mu_{\text{eff}} = 2.0$), which was interpreted in terms of pairs of copper atoms interacting to give a triplet ground state.

It is noteworthy that all the dimethoxides M(OMe)$_2$ (M = Cr, Mn, Fe, Co, Ni, Cu) were found to have highly polymeric structures involving octahedral or distorted octahedral metal atoms, and it will be very interesting to obtain complete crystal structures in order to get a better understanding of the mechanism of ferromagnetic and antiferromagnetic interactions of neighboring paramagnetic atoms. Applying Bradley's structural theory (109) for the minimum polymer size modified in the light of recent X-ray work to restrict octahedral structures to edge-sharing, an infinite two-dimensional layer lattice (CdX$_2$) would be expected involving each oxygen in bridging to three metals and with the alkyl groups taking up peripheral positions. For the octahedral Cr(OMe)$_3$ a layer lattice is possible with each oxygen bridging only two metals, but a more compact unit would arise if triple bridging occurred. The smaller polymer based on edge-sharing MO$_6$ octahedra would be M$_{14}$(OR)$_{42}$ with 16 triple-bridging oxygens, 10 double-bridging, and 16 nonbridging.

E. ALKOXIDES OF LANTHANIDES AND ACTINIDES

Lanthanum trialkoxides were first reported over ten years ago (110), but it is only in recent years that lanthanide alkoxide chemistry has received serious attention. Not surprisingly, it has been found that they have similar properties to the metal alkoxides in general.

There are three main preparative routes to the lanthanide alkoxides. For the insoluble methoxides the method of Bradley and Faktor (110) can be applied. This involves the addition of lithium methoxide to the methanolic metal chloride solution. The insoluble metal methoxide is filtered off and washed with methanol to remove lithium chloride.

$$3\text{LiOMe} + \text{LnCl}_3 \rightarrow \text{Ln(OMe)}_3\downarrow + 3\text{LiCl} \qquad (10)$$

Other alkoxides can be prepared from the methoxide by alcoholysis, but the reaction is very slow and tedious.

$$\text{Ln(OMe)}_3 + 3\text{ROH} \rightarrow \text{Ln(OR)}_3 + 3\text{MeOH} \qquad (11)$$

The method of Mehrotra et al. (111) is of more general application. This involves preparation of the triisopropoxide by the action of sodium isopropoxide on the lanthanide trichloride isopropanolate and then conversion to other alkoxides by alcoholysis.

$$\text{LnCl}_3 + 3\text{NaOPr}^i \rightarrow \text{Ln(OPr}^i)_3 + 3\text{NaCl} \qquad (12)$$

The method has been applied to the preparation of alkoxides of lanthanum, praseodymium and neodymium (*111*), samarium (*112*), gadolinium, erbium, and ytterbium (*113*). Another method of general applicability is due to Mazdiyasni et al. (*114, 115*), who showed that the lanthanide isopropoxides could be obtained from the reaction of the metal and the alcohol catalyzed by mercuric salts.

$$\text{Ln} + 3\text{Pr}^i\text{OH} \rightarrow \text{Ln(OPr}^i)_3 + 3/2\text{H}_2 \qquad (13)$$

The lanthanide methoxides and ethoxides are insoluble solids, but the higher alkoxides are soluble in typical organic solvents. The lanthanide triisopropoxides sublime at 200°–300°/0.1 mm and it is clear that the lanthanide trialkoxides are polymeric (*110*), but conflicting results have been reported for molecular weight determinations. For example, it was reported that $\text{Ln(OPr}^i)_3$ (Ln = La, Pr, Nd) were monomeric in benzene (*111, 116*), whereas Mazdiyasni et al. (*114*) had proposed dimeric structures and samarium isopropoxide was found to be tetrameric (*112*). However, the ^{1}H NMR spectra of ytterium, lanthanum, and lutetium isopropoxides show that polymeric species are present, and preliminary mass spectral studies have indicated fragment ions of dimeric and tetrameric species (*115*). In view of the existence of $\text{LaAl}_3(\text{OPr}^i)_{12}$ (*117*) and the NMR evidence for polymeric structures, it seems more probable that Sankla and Kapoor (*112*) are correct in finding a tetrameric samarium isopropoxide and it is reasonable to suppose that a tetramer structure similar to that of $\text{Al}_4(\text{OPr}^i)_{12}$ (Fig. 4) may apply to a number of lanthanide isopropoxides. Brown and Mazdiyasni (*115*) have presented infrared spectral data (350–3000 cm^{-1}), visible and ultraviolet spectra, and thermogravimetric data for the series of lanthanide isopropoxides. Typical lanthanide chemistry is also apparent in the oxidation states exhibited by the alkoxides. Thus, although the trialkoxides may be isolated in all cases, it is noteworthy that cerium(IV) gives stable tetraalkoxides (*118–120*) and europium(II) is listed as forming a diisopropoxide (*115*).

Rather less is known about actinide alkoxides because, in addition to problems of hydrolysis and oxidation, there is often the special problem of high radioactivity to contend with. Thorium(IV) alkoxides were prepared by Bradley and co-workers (*121*), who studied the effect of steric hindrance of the alkyl group on the degree of polymerization and volatility. The tetraisopropoxide ($\bar{n} = 3.8$) was practically tetrameric and could be sublimed (200°/0.1 mm), whereas the tetra-*tert*-butoxide ($\bar{n} = 3.4$) was nearer to trimeric and more volatile (160°/0.1 mm), while the derivative of triethylcarbinol was monomeric. Uranium alkoxides have received much attention. Earlier synthetic work was carried out by Gilman and co-workers (*122–124*) during a search for volatile uranium compounds suitable for the separation of uranium isotopes. It was found that uranium(IV) tetraalkoxides were extremely susceptible to oxidation and were readily converted to uranium(V) pentaalkoxides.

$$U(OEt)_4 + \tfrac{1}{2}Br_2 + NaOEt \rightarrow U(OEt)_5 + NaBr \qquad (14)$$

With more powerful oxidizing agents (e.g., dibenzoyl peroxide), the uranium(VI) hexaalkoxide was obtained.

$$U(OEt)_5 + NaOEt + \tfrac{1}{2}(C_6H_5CO)_2O_2 \rightarrow U(OR)_6 + C_6H_5CO_2Na \qquad (15)$$

The stability of uranium(V) alkoxides toward disproportionation was demonstrated in two ways. They could be distilled *in vacuo* and they could be prepared by a reproportionation from the uranium(IV) and uranium(VI) alkoxides.

$$U(OR)_4 + U(OR)_6 \rightarrow 2U(OR)_5 \qquad (16)$$

This is yet another example [cf. Cr(IV)] of the stability of an intermediate valence in the covalent state where the aquocation in the same oxidation state is very unstable.

Later work by Bradley and co-workers showed that for the uranium-(V) alkoxides the methoxide was trimeric $U_3(OMe)_{15}$, but the other normal alkoxides were dimeric $U_2(OR)_{10}$ (*125, 126*). With very bulky highly branched alkyl groups it was possible to synthesize monomeric uranium(V) alkoxides whereas the penta-*tert*-butoxide gave monomeric 1:1 complexes $U(OBu^t)_5 \cdot L$ (L = Bu^tOH, C_5H_5N) (*127*). In addition to uranium(VI) hexaalkoxides $U(OR)_6$, it has proved possible to prepare uranyl alkoxides such as the insoluble yellow $UO_2(OMe)_2(MeOH)$ (*128*). With *tert*-butanol the latter compound underwent an unprecedented reaction which produced some of the volatile, deep red crystalline $U(OBu^t)_6$, which was perceptibly resistant toward hydrolysis (*128*). The effect of steric hindrance of the alkyl groups on the degree of polymerization and volatility of uranium(IV) tetraalkoxides was also studied (*129*), and it was shown that uranium tetraalkoxides were significantly more volatile than the corresponding thorium compounds. The extreme ease of oxidation of uranium(IV) alkoxides precluded the isolation of the tetra-*tert*-butoxide.

Recently some spectroscopic studies have been made on $U_2(OEt)_{10}$. Bradley and Westlake (*79*) assigned various bands to C–O and U–O stretching vibrations and deduced from measurements of the intensities of terminal and bridging species that the structure of the dimer was probably the edge-shared octahedral one exhibited by niobium and tantalum alkoxides (Fig. 8). Karraker (*130*) has assigned the electronic absorption spectrum (which shows weak but narrow bands at 5405, 5680, 6622, 6934, 10,200, 11,690, and 14,490 cm^{-1}) to f–f transitions for a distorted octahedral f^1 system with a spin–orbit coupling constant of 1905 cm^{-1}. By comparing this spectrum with that of $UCl_5(SOCl_2)$ it was shown that the f–f transitions of the U(V) (f^1) system are sensitive to ligand field strength and indicated that the f orbital extended into the bonding region. Karraker et al. (*131*) have confirmed the edge-shared octahedral structure for the dimer by ^{1}H NMR and a low magnetic moment ($\mu_{eff} = 1.12$) was reported from a susceptibility determination at room temperature, but variable temperature studies are clearly required for any meaningful discussion.

Samulski and Karraker (*132*) have synthesized $Np(OMe)_4$ and $Np(OEt)_4$ from reactions involving $NpCl_4$ and the appropriate lithium alkoxide. The neptunium(IV) tetraalkoxides were nonvolatile, red-brown solids, which doubtless are polymeric. Electronic absorption spectra and infrared spectra were obtained for the tetraethoxide, which was soluble in carbon tetrachloride. Although the green quinquevalent neptunium compound $NpBr(OEt)_4$ was obtained by bromination of $Np(OEt)_4$, attempts to obtain Np(V) or Np(VI) alkoxides were unsuccessful.

Since $PuCl_4$ was unstable, Bradley and co-workers (*133*) used the complex Pu(IV) chloride, $(C_5H_5NH)_2PuCl_6$, as a starting material for preparing plutonium(IV) tetraalkoxides. The tetraisopropoxide was isolated as a soluble green solid, which could be sublimed (220°/0.05 mm) and formed a crystalline solvate $Pu(OPr^i)_4 \cdot Pr^iOH$ analogous to those of Zr, Hf, and Ce(IV). The tetra-*tert*-butoxide $Pu(OBu^t)_4$ was more volatile (sublimes 112°/0.05 mm), and it was pointed out that in contrast to the behavior of uranium(IV) alkoxides those of plutonium(IV) were not readily oxidized by air.

F. Double Alkoxides

The double alkoxides (derivatives containing two different metals) were first studied systematically by Meerwein and Bersin (*134*) who obtained a wide range of compounds: e.g., $KLi(OPr^i)_2$; $K_2Be(OEt)_4$; $Na_2Mg(OPr^n)_4$; $KZn(OMe)_3$; $LiAl(OR)_4$; $NaFe(OEt)_4$; $MAl_2(OEt)_8$ (M = Ca, Mg, Co, Ni, Cu); $Zn_3Al(OR)_9$; $ZnAl_2(OR)_8$; $NaHSn(OEt)_6$; $KH_3[Ti(OBu^n)_6]_2$; $NaH[Zr(OEt)_6]$; and $NaSb(OEt)_6$. They considered these compounds to be alkoxo salts derived from a basic alkoxide and an acidic alkoxide, and they showed that in several cases the "acidic" alkoxide could be titrated to a sharp end point with sodium alkoxides. However, it is now evident that double alkoxides will be formed generally from polymeric metal alkoxides, although the stability of the double alkoxides may vary considerably depending on the nature of the two metals and the alkoxo group. In fact, it would not be surprising to find multiple alkoxides, e.g., $M_1M_2M_3M_4(OR)_x$, and a triple alkoxide $K_2Sn^{II}Sn_2^{IV}(OEt)_{12}$ was listed by Meerwein and Bersin (134). They also drew attention to the fact that the double alkoxides $MgAl_2(OR)_8$ and $CaAl_2(OR)_8$ (R = Et, Pr^i) could be distilled in vacuo, whereas the calcium and magnesium alkoxides individually were infusible and nonvolatile. Later work by Wardlaw and co-workers (135) on the reactions of zirconium tetrachloride with sodium alkoxides and on titrations of $Zr(OR)_4$ with NaOR failed to confirm the existence of $NaH[Zr(OR)_6]$ (134), but showed that stable compounds of the type $NaZr_2(OR)_9$ could be obtained. The alkali metal–zirconium alkoxides $MZr_2(OR)_9$ (M = Li, Na, K) were soluble in organic solvent and could be distilled in vacuo, and they are clearly substantially covalent in nature. Similarly, tin(IV) gave the double alkoxide $NaSn_2(POr^i)_9$ (136). An interesting volatile uranium(IV) aluminum alkoxide $UAl_4(OPr^i)_{16}$ was reported by Albers et al. (137), and Gilman et al. (138) have obtained double alkoxides of uranium(V): $NaU(OEt)_6$; $CaU_2(OEt)_{12}$; and $U_3Al(OEt)_{18}$. Recently Mehrotra and Agarwal (117) have reported the preparation of volatile double alkoxides of aluminum and lanthanides $MAl_3(OPr^i)_{12}$ (M = La, Pr), and they suggested a structure analogous to that of the tetrameric $Al_4(OPr^i)_{12}$ (Fig. 4) with the central octahedral aluminum replaced by the lanthanide. The ^{1}H NMR spectrum of $LaAl_3(OPr^i)_{12}$ showed only one type of isopropoxo group down to $-60°$, indicating a rapid intramolecular exchange of bridging and terminal ligands (54), and the mass spectrum gave the parent ion $LaAl_3(OPr^i)_{12}^+$ and several fragment ions and metastables including a very intense peak for $LaAl_3(OPr^i)_{11}^+$. Using a pH titration method involving lithium methoxide and metal chlorides in anhydrous methanol, Gut (139) obtained evidence for the presence of several anionic species: $B(OMe)_4^-$, $Al(OMe)_4^-$, $Ti_2(OMe)_9^-$, $Nb(OMe)_6^-$, and $Ta(OMe)_6^-$. Mehrotra and Agrawal (140) have confirmed the existence of $MZr_2(OR)_9$ (M = Li, Na, K) and showed that other double alkoxides such as $Li_2Zr_3(OEt)_{14}$, $Na_2Zr_3(OPr^i)_{14}$, and $MZr(OBu^t)_5$ (M = Li, Na, K) could also be isolated. They were all soluble in organic solvents and could be sublimed in vacuo. It is apparent from this short review that the field of multiple metal alkoxides offers scope for further interesting research.

G. Metal Trialkylsilyloxides

The metal trialkylsilyloxides $M(OSiR_3)_x$ are a special case of the general class of compounds containing the heterosiloxane group Si–O–M. They are of interest in revealing the effect on physicochemical properties of replacing the carbinol carbon of the tertiary alkoxo group OCR_3 by silicon. The larger silicon atom should reduce the steric effect of the alkyl groups, since they are placed further away from the central atom, although the silicon atom will exert more shielding than the carbon atom. Electronic effects are twofold and mutually opposed. Thus, silicon is less electronegative than carbon, and this should lead to the $^-OSiR_3$ ion being a better electron donor than $^-OCR_3$, but this is opposed by the fact that silicon has vacant d orbitals which may withdraw electron density from the oxygen by $p\pi$–$d\pi$ bonding. The field has received much attention during the past decade mainly by Andrianov and co-workers, Bradley and co-workers, and Schmidt and Schmidbaur and their co-workers, and derivatives of a wide range of metals are known. Since recent comprehensive reviews are available (7, 141, 142), we shall deal here only with the highlights and very recent publications.

The metal trialkylsilyloxides generally have good thermal stability and are perceptibly less readily hydrolyzed than the alkoxides. The least stable are $Cr^{VI}O_2(OSiMe_3)_2$ (dangerously explosive), $Hg(OSiMe_3)_2$, and $[Au^{III}(OSiMe_3)_3]_2$ (decomposes at room temperature) (142). A novel method of thermal decomposition was exhibited by bistrimethylsilyoxymercury(II) involving intramolecular transmethylation (142).

$$[Hg(OSiMe_3)_2]_n \rightarrow n MeHg(OSiMe_3) + (Me_2SiO)_n \qquad (17)$$

A similar behavior was shown by the zinc compounds, but not by cadmium. The methyltrimethylsilyloxymetal(II) compounds $[MeM(OSiMe_3)_4]$ (M = Zn, Cd, Hg) are all tetrameric in the crystalline state and have the interesting cubane structure (Fig. 3) analogous to that of $(MeZnOMe)_4$ (142). The same structure is exhibited by $[MeBe(OSiMe_3)]_4$ (143). Zeitler and Brown (144) showed that $Ti[OSi(C_6H_5)_3]_4$ had exceptionally high thermal and hydrolytic stability, while Chamberland et al. found that $VO[OSi(C_6H_5)_3]_3$ could be purified by washing with water (145).

Such p_π–d_π bonding between silicon and oxygen as is present in R_3SiO groups does not inhibit the oxygen from acting as a bridging group. Thus, the compounds $[MeM(OSiMe_3)]_4$ contain trimethylsilyloxy groups in triple-bridging configurations, and in the alkali metal compounds $[MOSiMe_3]_4$ (M = K, Rb, Cs) (146) the cubane structure is present analogous to that (Fig. 1) exhibited by the *tert*-butoxides (15). Mass spectra showed the presence of tetrameric fragment ions $M_4(OSiMe_3)_3(OSiMe_2)^+$ (M = K, Rb, Cs), whereas the lithium and sodium derivatives gave hexameric fragment ions $M_6(OSiMe_3)_5(OSiMe_2)^+$ (M = Li, Na) (146). Earlier work by Bradley and Thomas (147) had shown that crystalline tetrakistrimethylsilyloxyzirconium (m.p. 152°; sublimes at 135°/0.1 mm) was dimeric $[Zr(OSiMe_3)_4]_2$, presumably with five-coordinated zirconium, in marked contrast to the very volatile, monomeric liquid $Zr(OCMe_3)_4$. It was also noted that the pentakistrialkylsilyloxyuranium(V) compounds $[U(OSiR_3)_5]_n$ were more polymeric than the corresponding tertiary alkoxides, but the hexakis derivatives $U(OSiR_3)_6$ were all monomeric (148). Schmidbaur and Schmidt (149–151) have shown that tristrimethylsilyloxymetal(III) compounds $[M(OSiMe_3)_3]_2$ (M = Al, Ga, Fe) are dimeric and yield saltlike double compounds, e.g., $M[Fe(OSiMe_3)_4]$ (M = Li, Na, K, SbMe_4), containing tetrahedrally coordinated metals (142). A novel complex $Fe(OSiMe_3)_3(Me_3NO)$ (yellow crystals; m.p. 86°–90°; (b.p. 145°–148°/ 1 mm.) was obtained by treating the dimeric $[Fe(OSiMe_3)_3]_2$ with trimethylamine oxide (142). Recently, Shiotani and Schmidbaur (152) have succeeded in isolating gold (I) trimethylsilyl oxide stabilized by phosphine- or arsine-donor ligands.

$$AuCl \cdot L + NaOSiMe_3 \rightarrow Au(OSiMe_3)L + NaCl \qquad (18)$$
$$(L = Me_3P, \phi_3P, \phi_3As)$$

The trimethylphosphine complex $Au(OSiMe_3)(PMe_3)$ was a soluble, colorless crystalline compound which was monomeric and could be sublimed *in vacuo*. It gave a parent molecular ion in the mass spectrum, and at temperatures above 190° it dissociated to a gold mirror, hexamethyldisiloxane, trimethylphosphine oxide, and trimethylphosphine. The trimethylarsine adduct was less stable. These interesting new compounds are examples of univalent gold exhibiting the coordination number of two.

The first examples of lanthanide trimethylsilyloxides were recently reported by Batwara and Mehrotra (153).

$$Ln(OPr^i)_3 + nMe_3SiO_2CMe \rightarrow Ln(OPr^i)_{(3-n)}(OSiMe_3)_n + nPr^iO_2CMe \qquad (19)$$
$$(Ln = Gd, Er; n = 1, 2, 3)$$

These tristrimethylsilyloxy derivatives of gadolinium and erbium had an average degree of polymerization *ca.* 3.5 and were thus more polymerized than the corresponding *tert*-butoxides which were closer to trimeric.

Several investigators have pointed out that the metal trialkylsilyl oxides are less susceptible to hydrolysis than the corresponding alkoxides.

Bradley and Prevedorou-Demas have studied the controlled hydrolysis of trialkylsilyloxy derivatives of titanium, zirconium, and tantalum (7) in dioxane solution and the thermal stability and degree of polymerization of the hydrolysis products. The titanium compound $Ti(OSiMe_3)_4$ was notably resistant toward hydrolysis, but the initial products of hydrolysis were unstable owing to disproportionation to the parent compound and more highly condensed products (154).

$$2Ti(OSiMe_3)_4 + H_2O \rightarrow Ti_2O(OSiMe_3)_6 + 2Me_3SiOH \quad (20)$$

$$3Ti_2O(OSiMe_3)_6 \rightarrow 4Ti(OSiMe_3)_4 + \tfrac{1}{4}[Ti_8O_{12}(OSiMe_3)_8] \quad (21)$$

The solid polymer $Ti_8O_{12}(OSiMe_3)_8$ was soluble in cyclohexane and a cubic cage was proposed for the Ti_8O_{12} unit with the tetrahedral coordination of each titanium being completed by a pendant trimethylsilyloxy group. The more highly polymeric products obtained at higher degrees of hydrolysis were formulated in terms of Ti–O–Ti cross-linking of octamer units. The initial products of hydrolysis of $[Zr(OSiMe_3)_4]_2$ conformed to a polymer system reminiscent of the zirconium oxide alkoxides (7) based on octahedrally coordinated zirconium (155). This is understandable since the steric constraint on coordination expansion exerted by the bulky Me_3SiO groups is relieved by the removal of these groups by hydrolysis. However, the degrees of polymerization of the residual more highly condensed polymers remaining after thermal disproportionation appeared to require the presence of octameric units analogous to those of titanium (155). The hydrolysis products from $Ta(OSiMe_3)_5$ could all be related to structures involving five-coordinated tantalum (156), whereas the behavior of $Ti(OSiEt_3)_4$ (157, 158) was significantly different from $Ti(OSiMe_3)_4$. In the hydrolysis of $[Al(OSiMe_3)_3]_2$ it was noted that a small proportion of OH groups was present, but the polymers formed were interpreted on the basis of Al–O–Al cross-linking of dimeric units (159). A preliminary account has recently appeared (160) on the preparation and properties of dialkylsilanediol derivatives of transition metals. Thermolysis of the gummy polymer $[TiO_2(OSiEt_2)_2]_n$ gave an insoluble nonvolatile residue $[(TiO_2)_x(OSiEt_2)_y]$, and a titanium-containing distillate corresponding to $TiO_2(OSiEt_2)_{10}$ was obtained. The distillate may be a mixture of $(Et_2SiO)_3$, $(Et_2SiO)_4$, and $TiO_2(OSiEt_2)_x$ ($x = 2$–4), since it appeared to disproportionate readily to the volatile cyclic diethyldisiloxanes leaving a polymeric residue $[TiO_2(OSiEt_2)_4]_n$. However, fragment ions corresponding to species of the general formula $TiO_2(OSiEt_2)_x$ ($x = 4$–10) were clearly observed in mass spectrometric studies (160). A recent account of the comprehensive studies by Andrianov and co-workers on the elementoorganosiloxanes has appeared in review form (161).

III. Metal Dialkylamides

Although metal dialkylamides $M(NR_2)_x$ (M = a metal of valency x; R = an alkyl group) have been known for many years, the subject experienced a rather slow development until recently. In a review covering amino derivatives of metals and metalloids in 1965 by Jones and Lappert (*162*), twenty-eight elements were listed as known to form dialkylamido derivatives, and it was noteworthy that only a few transition metals were featured. Since that time the list has been extended, the reactivity of metal–nitrogen bonds has been explored, and some crystal structures have been elucidated, but much still remains to be investigated in this fascinating field.

Considering covalently bonded dialkylamido groups, there are three distinct possibilities:

$$\begin{array}{ccc} R\diagdown\overset{\cdot\cdot}{N}\!-\!M & R\diagdown\overset{+}{N}\diagup M^{-\frac{1}{2}} & R\diagdown\overset{+}{N}\!=\!\overset{-}{M} \\ R\diagup\;\;| & R\diagup\;\diagdown M^{-\frac{1}{2}} & R\diagup \\ R & & \\ (\text{V}) & (\text{VI}) & (\text{VII}) \end{array}$$

In structure (V) is depicted a metal–nitrogen σ bond with a pyramidal nitrogen containing a basic lone pair of electrons. If steric factors allow, this lone pair may be involved in donation to another metal giving rise to the dialkylamido bridge (VI). If bridging is precluded by steric factors, the nitrogen lone pair may engage in π bonding giving rise to a trigonal-planar nitrogen as shown in structure (VII).

A. The Preparation of Metal Dialkylamides

There are three main preparative procedures for synthesizing metal dialkylamides. Some derivatives can be obtained by treating the metal hydride with the secondary amine.

$$MH_x + xR_2NH \rightarrow M(NR_2)_x + xH_2 \tag{22}$$

Derivatives of aluminum hydride were used by Ruff (*163*) to prepare aluminum dialkylamides.

$$LiAlH_4 + 4Me_2NH \rightarrow LiAl(NMe_2)_4 + 4H_2 \tag{23}$$
$$3LiAl(MNe_2)_4 + AlCl_3 \rightarrow 2Al_2(NMe_2)_6 + 3LiCl \tag{24}$$
$$AlH_3(NMe_3) + 3Pr^i_2NH \rightarrow Al(NPr^i_2)_3 + Me_3N + 3H_2 \tag{25}$$

The second method involves the reaction between the secondary amine and a metal alkyl.

$$MR_x + xR_2NH \rightarrow M(NR_2)_x + xRH \tag{26}$$

This method was used by Coates and Glockling (*164*) to prepare the trimeric beryllium bisdimethylamide $Be_3(NMe_2)_6$. It is also the most convenient method for obtaining lithium dialkylamides since *n*-butyllithium is commercially available.

The third method involves the reaction of a metal halide (usually the chloride) with an alkali metal (usually Li or Na) dialkylamide, e.g.,

$$MCl_x + xLiNR_2 \rightarrow M(NR_2)_x + xLiCl \tag{27}$$

The first transition metal derivative synthesized was $Ti[N(C_6H_5)_2]_4$, which was prepared from $TiCl_4$ and potassium diphenylamide (*165*). The only uranium derivative prepared to date is $U(NEt_2)_4$, which was obtained from UCl_4 and lithium diethylamide (*122*). In recent years tetradialkylamides of titanium (*166*), vanadium (*167*), chromium (*168, 169*), zirconium (*166*), niobium (*170*), hafnium (*171*), thorium (*171*), and tin (*167*), pentadialkylamides of niobium (*170*) and tantalum (*172*), and the hexadimethylamide of tungsten (*173*) have all been obtained by means of metal chloride/lithium dialkylamide reactions. However, the reaction of metal chloride and lithium dialkylamide does not always proceed according to the requirements of Eq. (27) (*3*). For example, in the case of $NbCl_5$, the bulkier dialkylamides give rise to tetradialkylamidoniobium(IV) compounds (*170*), whereas the same ligands with $TaCl_5$ gave the monoalkylimidotrisdialkylamidotantalum(V) compounds $RN=Ta(NR_2)_3$ (*172*). This behavior was ascribed to steric factors which led to the instability of $M(NR_2)_5$ derivatives, but the different modes of breakdown obviously reflect differences in electronic factors (redox properties) between niobium and tantalum. Further complications occurred when chlorides of Mn(II), Fe(III), Co(II), and Ni(II) were treated with $LiNEt_2$ and no diethylamido derivatives were isolated. Instead the interesting unsymmetrical nitrogen chelate (VIII) was obtained from a reaction which is summarized by Eq. (28) (*174, 175*).

(VIII)

$$7CoCl_2 + 14LiNEt_2 \rightarrow CoN_4C_{16}H_{30} + 6Co + 14LiCl + 10Et_2NH \tag{28}$$

In some reactions reduction of the metal to a lower oxidation state occurs even with the less bulky groups such as NMe_2. Thus, the major product from the $MoCl_5/LiNMe_2$ reaction was a polymeric Mo(III) dimethylamide $[Mo(NMe_2)_3]_x$, which afforded the volatile $Mo(NMe_2)_4$ by disproportionation (176). Similarly the $WCl_6/LiNMe_2$ reaction gave as the major product polymeric W(III) dimethylamide $[W(NMe_2)_3]_x$ with only a small yield of $W(NMe_2)_6$ (173). Mass spectral studies suggested that some $W_2(NMe_2)_{10}$ was formed in the disproportionation of $[W(NMe_2)_3]_x$ (177). Disproportionation was also a characteristic feature of the trisdialkylamido derivatives of titanium (178), vanadium (178), and chromium (168, 169).

$$MCl_3 + 3LiNEt_2 \rightarrow M(NEt_2)_3 + 3LiCl \quad (29)$$

$$2M(NEt_2)_3 \rightarrow M(NEt_2)_4 + M(NEt_2)_2 \quad (30)$$
$$(M = Ti, V, Cr)$$

However, by using the very bulky ligand NPr^i_2 it was possible to stabilize trisdiisopropylamidochromium(III) as a volatile, monomeric derivative containing three-coordinated chromium (168, 179). Bürger and Wannagat have prepared a number of interesting bistrimethylsilylamido derivatives $M[N(SiMe_3)_2]_x$ using the reaction of the sodium derivative $NaN(SiMe_3)_2$ with a metal halide. Thus, the tris derivatives of Cr(III) (180) and Fe(III) (181), the bis derivatives of Mn(II) (180), Co(II) (181), nickel(II) (180), and the zinc subgroup (182), and the mono derivative of Cu(I) (180), were isolated by the general method:

$$MX_x + xNaN(SiMe_3)_2 \rightarrow M[N(SiMe_3)_2]_x + xNaCl \quad (31)$$

Recently the tris derivatives of Ti(III) and V(III) $M[N(SiMe_3)_2]_3$ (M = Ti, V) were obtained by Bradley and Copperthwaite (183), using the reaction of $LiN(SiMe_3)_2$ with the five-coordinated metal trichloride complexes $MCl_3(Me_3N)_2$. Also, the first lanthanide(III) tris derivatives have been obtained using the lithium derivatives of diisopropylamine and bistrimethylsilylamine (184).

In some cases {e.g., $Cr[N(SiMe_3)_2]_2$ (T.II.F.)$_2$ (185), $Mn[N(SiMe_3)_2]_2$ (T.H.F.) (186)} solvent molecules (T.H.F. = tetrahydrofuran) remained coordinated to the metal silylamide and in other reactions involving phosphine complexes of metal chlorides the phosphine remained coordinated {e.g., $Me_3P \cdot AuN(SiMe_3)_2$ (152), $(C_6H_5)_3P \cdot Co[N(SiMe_3)_2]_2$ and $[(C_6H_5)_3P]_2NiN(SiMe_3)_2$ (187)}.

In addition to the three main preparative methods dealt with above, there is a novel method recently reported by Ashby and Kovar (188) for the synthesis of aluminum trisdiethylamide and the diethylaminoalanes.

$$Al + 3/2H_2 + nEt_2NH \rightarrow H_{(3-n)}Al(NEt_2)_n + nH_2 \quad (32)$$
$$(n = 1, 2, 3)$$

The reaction was carried out at moderate temperatures and pressures in benzene solution and careful control of the conditions led to the formation of H_2AlNEt_2, $HAl(NEt_2)_2$, or $Al(NEt_2)_3$. A 91% yield of $Al(NEt_2)_3$ was obtained using diethylamine as solvent at 150° under 3000 psig of hydrogen for 4 hr. This direct synthesis of aluminum diethylamide from the metal may well have commercial applications.

Another procedure of limited application is the aminolysis or transamination reaction (*166*, *170*, *172*).

$$M(NR_2)_x + yHNR_2' \rightarrow M(NR_2')_y(NR_2)_{(x-y)} + yHNR_2 \qquad (33)$$

The extent of this reaction is often limited by steric hindrance.

An unusual reaction which was reported to produce the dimethylamide of Na, K, Cu(II), Cd, Hg(II), and Al involved prolonged heating of the metal cyanide (Na, K) or acetate (Na, K, Cu, Cd, Hg, Al) with anhydrous dimethylformamide (*189*).

In this review we are primarily concerned with the fully substituted metal dialkylamides $M(NR_2)_x$, but reference to the synthesis of "mixed ligand compounds" $TiX_n(NR_2)_{(4-n)}$ ($n = 1, 2, 3$; X = alkyl, aryl, cyclopentadienyl, halide, alkoxide, etc.) may be found in the comprehensive account by Bürger and Neese (*5*). Most of the reactions described for Ti(IV) will be applicable to other polyvalent metals.

B. CHEMICAL PROPERTIES OF METAL DIALKYLAMIDES

Most of the metal dialkylamides are very readily hydrolyzed to the amine and metal oxide or hydroxide. This is a special case of the more general reactivity of the metal dialkylamides with molecules HL containing active hydrogen.

$$M(NR_2)_x + yHL \rightarrow MLy(NR_2)_{(x-y)} + yHNR_2 \qquad (34)$$
$$(L = \text{halogen, OH, OR, etc.})$$

Some interesting tetrakis-Schiff base complexes TiL_4 (*190*) and ZrL_4 (*191*) were thus obtained. The scope of this reaction has been very systematically explored by Lappert and co-workers with special reference to the reactivity of the Sn–NR$_2$ bonds, but some reactions of $M(NMe_2)_4$ (M = Ti, Zr, Hf) and $Ti(NR_2)_3$ have been included (*162*, *192–196*). Streitwieser and co-workers (*197*) have utilized the reactivity of lithium and cesium cyclohexylamides to determine the acidity of hydrocarbons.

Another type of reaction is ligand exchange or metathetical reaction.

$$M(NR_2)_4 + M'X_4 \rightarrow MX(NR_2)_3 + M'X_3(NR_2) \qquad (35)$$
$$\rightarrow MX_2(NR_2)_2 + M'X_2(NR_2)_2 \qquad (36)$$
$$\rightarrow MX_3(NR_2) + M'X(NR_2)_3 \qquad (37)$$
$$\rightarrow MX_4 + M'(NR_2)_4 \qquad (38)$$

The equilibria involved for M = M' = Ti and X = halogen or OR, have been studied by Weingarten and Van Wazer (66), and this type of reaction has been well documented for titanium derivatives by Bürger and Neese (5) and for organotin dialkylamides by George and Lappert (198).

The third type of reaction involves insertion of an unsaturated group between the metal–nitrogen bond. The N,N-dialkyl dithiocarbamates of the early transition metals were first prepared by insertion of carbon disulfide into metal dialkylamides (199).

$$M(NR_2)_x + xCS_2 \rightarrow M(S_2CNR_2)_x \qquad (39)$$

The generalized insertion reaction [Eq. (40)] has been thoroughly and systematically explored by Lappert and co-workers (162, 200), who

$$L_xM(NR_2)_y + yA{=}BC \rightarrow L_xM[AB(NR_2)C]_y \qquad (40)$$

demonstrated the great synthetic value of this reaction. Nitriles undergo a variety of reactions including polymerization (200–202).

Dialkylamides of transition metals in lower valencies [e.g., Ti(III), V(III), V(IV), Cr(III), Nb(IV), and Mo(IV)] and some of the bistrimethylsilylamides are extraordinarily sensitive to oxygen and demand exceptional precautions in handling these compounds for spectroscopic and other physical measurements. It is possible that peroxo or superoxo compounds are initially formed by addition of molecular oxygen, but the instability of such derivatives poses problems and hazards in characterizing them. Thus, under certain conditions the highly reactive chromium(III) trisdiisopropylamide $Cr(NPr^i{}_2)_3$ formed a 1:1 compound $Cr(O_2)(NPr^i{}_2)_3$, which may be a peroxochromium(V) derivative (203), but at a low temperature the uptake of oxygen corresponds to the formation of $CrO_3(NPr^i{}_2)_3$, a dangerously explosive compound (204). However, with nitric oxide the reaction with $Cr(NPr^i{}_2)_3$ gave rise to the diamagnetic $Cr(NO)(NPr^i{}_2)_3$, which is a rare example of four-coordinated chromium(II) (205). The stability of the Cr–NO system was evident from reactions with *tert*-butanol, which gave derivatives such as $Cr(NO)(OBu^t)(NPr^i{}_2)_2$ and $Cr(NO)(OBu^t)_3$ without loss of nitric oxide.

It has been found that metal dialkylamides $M(NR_2)_4$ (M = Ti, Zr, Hf) undergo "addition reactions" with metal carbonyls $M'(CO)_x$ (M' = Cr, Fe, Ni) to form intractable compounds which were formulated as $\{M(NR_2)_4[M'(CO)_x]_2\}$ (206).

C. Physical Properties of Metal Dialkylamides

The metal dialkylamides are reasonably volatile compounds which may be sublimed or distilled *in vacuo*, and they are usually soluble in nonreactive organic solvents. This has enabled molecular weights, absorption spectra, and NMR and ESR spectra to be determined in solution. Some of these compounds are crystalline and a few X-ray crystal analyses have been carried out.

The dialkylamido group may act as a bridging ligand as shown in structure (VI), and oligomeric species are found for the alkali metals, beryllium and aluminum. In the case of quadrivalent, quinquevalent, and sexivalent metals the dialkylamides are usually monomeric owing to steric hindrance which prevents coordination polymerization. The zirconium (hafnium) tetradimethylamide is a borderline case since it showed some polymerization in solution (number average degree of polymerization = 1.22) (166), which was confirmed by variable temperature NMR studies (171) and low-temperature infrared spectra (207).

1. The Nature of the Metal–Nitrogen Bond

It is sometimes assumed that the high chemical reactivity of metal dialkylamides must be due to weak metal–nitrogen bonds with a tendency to ionic character $\overset{+}{M}\overset{-}{N}R_2$. This does not necessarily follow, and it is well established that some metal complexes containing strongly covalently bonded ligands are also reactive or labile owing to the availability of vacant low-energy orbitals in the metal which can facilitate interactions with a nucleophilic reagent. The observation of strong infrared- and Raman-active metal–nitrogen bands is indicative of substantial covalent character in the polyvalent metal dialkylamides. The ESR spectra of vanadium(IV) and niobium(IV) dialkylamides may also be interpreted as suggesting the presence of covalent metal–nitrogen bonds. In the few X-ray structures which have been completed, it is noteworthy that the nonbridging dialkylamido groups invariably have trigonal-planar nitrogens implying π-donor characteristics in the dialkylamido group as shown in structure (VII). In addition, the electronic absorption spectra of transition metal dialkylamides show a substantial ligand-field splitting energy for dialkylamido groups. Altogether these physical data give a picture of the metal–nitrogen bond as being substantially covalent and directional in character.

Thermochemical studies by Bradley and Hillyer (84) on $Ti(NEt_2)_4$ led to a value of the standard heat of formation $\Delta H_f^0[Ti(NEt_2)_4, (liq.)] = -116 \pm 2.0$ kcal/mole with an estimated $\Delta H_f^0[Ti(NEt_2)_4, (g)] = -100 \pm 2.3$ kcal/mole. Uncertainty in the value of $\Delta H_f^0[Et_2N, (g)]$ precluded an accurate estimate of the average bond dissociation energy, but the value $\bar{D}(Ti-N) \sim 73$ kcal/mole was derived. This shows that the metal–nitrogen bonds are moderately strong in $Ti(NEt_2)_4$. However, a preliminary report on $Me_3Sn-NMe_2$ gave the Sn–N bond energy as ~40 kcal/mole *(162)*.

Mass spectral studies gave independent evidence of the considerable thermodynamic stability of transition metal dialkylamides. Thus, parent ion peaks were observed for $Ti(NMe_2)_4^+$, $V(NMe_2)_4^+$, $Zr(NMe_2)_4^+$, $Hf(NMe_2)_4^+$, $Nb(NMe_2)_5^+$, and $Ta(NMe_2)_5^+$ *(207)*, $Cr(NEt_2)_4^+$ *(169)*, $Nb(NEt_2)_4^+$ *(208)*, $Mo(NMe_2)_4^+$ *(176)*, $W(NMe_2)_6^+$ *(173)*, $Al_2(NMe_2)_6^+$ and $Al(NPr_2^i)_3^+$ *(209)*, in addition to many metal-containing fragment ions, metastable peaks, and doubly charged metal-containing species. Parent ion peaks were also obtained in the mass spectra of $M[N(SiMe_3)_2]_3$ (M = Ti, V, Cr, Fe) *(183)*.

2. *Infrared and Raman Spectra of Metal Dialkylamides*

The infrared and Raman spectra for $Ti(NMe_2)_4$ were first reported by Bürger *et al.* *(210)*. The spectra were interpreted on the basis of a tetrahedral (T_d) TiN_4 skeleton with a strong infrared band at 590 cm^{-1} assigned to the stretching mode $\nu_{as}TiN_4(F_2)$ and a polarized Raman band at 532 cm^{-1} assigned to $\nu_s TiN_4(A_1)$. Other characteristic strong infrared absorptions were assigned to ligand vibrations: δCH_3 (1249 cm^{-1}), $\nu_s NC_2$ (945 cm^{-1}). Bürger and Sawodny *(211)* also reported infrared and Raman spectra for the series $M(NMe_2)_4$ (M = Si, Ge, Sn) and calculated the MN_4 and NC_2 force constants. It is noteworthy that the force constants for $Sn(NMe_2)_4$ (Sn–N = 3.110; C–N = 4.351 mdyne/Å) and $Ti(NMe_2)_4$ (Ti–N = 3.111; C–N = 4.355 mdyne/Å) were virtually identical, although the frequencies differed significantly. An interesting compilation of data on Ti–N frequencies for compounds of the type $Ti(NR_2)_xX_{(4-x)}$ [where R = Me, Et; X = Cl, Br, I, Me, Et; and x = 4, 3, 2, 1], is available in the review by Bürger and Neese *(5)*. Bradley and Gitlitz *(212)* obtained infrared spectra on a range of dialkylamides of Ti(IV), V(IV), Zr(IV), Nb(V), Nb(IV), Hf(IV), Ta(V), Th(IV), and Ta=NR(NR$_2$)$_3$. The data were interpreted in terms of the dialkylamido groups acting as π donors with significant contributions of the type $\overset{+}{M}=\overset{-}{N}R_2$ to the metal–nitrogen bonds. Although the tetrahedral molecules $M(NR_2)_4$ should give only one infrared-active M–N stretching frequency the five-coordinated species $M(NMe_2)_5$ (M = Nb, Ta) should exhibit either two (A_2'', E' in D_{3h}) for a trigonal bipyramid or three (2A, E in C_{4v}) for a tetragonal pyramid. Both niobium and tantalum pentadimethyl-amides gave only one band in the M–N stretching region in cyclohexane solutions, and it was concluded that the other bands were too close in frequency to be resolved. Similarly, only one Zr–N band was observed for $Zr(NMe_2)_4$, although it was known to contain some polymeric species.

Further work using mulled samples of the metal dimethylamides has revealed more spectral details (176, 207). Thus, the infrared spectrum Ti(NMe$_2$)$_4$ gave a single symmetrical Ti–N band, but V(NMe$_2$)$_4$ and Mo(NMe$_2$)$_4$ gave M–N bands exhibiting shoulders on the high-frequency side indicative of a lowering of symmetry to D_{2d}. The pentadimethylamides of Nb and Ta each showed two partially resolved M–N bands and, taken in conjunction with their Raman spectra, suggested a trigonal-bipyramidal configuration. At low temperatures (~100°K) most of the M–N bands shifted (5–10 cm^{-1}) to lower frequencies, but for Ti(NMe$_2$)$_4$, V(NMe$_2$)$_4$, Mo(NMe$_2$)$_4$, Nb(NMe$_2$)$_5$, Ta(NMe$_2$)$_5$, and W(NMe$_2$)$_6$ there were no major changes. With Zr(NMe$_2$)$_4$ and Hf(NMe$_2$)$_4$ some important changes occurred on lowering the temperature, which could be explained on the assumption that the degree of polymerization increased markedly at lower temperatures. The infrared and Raman spectra of W(NMe$_2$)$_6$ (W–N stretching bands at 555 cm^{-1} Raman and 545 cm^{-1} infrared) were consistent with the known octahedral structure of this compound (173).

Infrared and Raman spectra have also been obtained for some of the bistrimethylsilylamido derivatives M[N(SiMe$_3$)$_2$]$_x$. Bürger et al. (213) assigned the spectra of Be[N(SiMe$_3$)$_2$]$_2$ in terms of a linear two-coordinated beryllium compound with the D_{2d} configuration for the Si$_2$NBeNSi$_2$ framework. Force constant calculations gave Be–N, 2.957 and Si–N, 3.395 mdyne/Å. Similarly the infrared and Raman spectra of M[N(SiMe$_3$)$_2$]$_2$ (M = Zn, Cd, Hg) were also assigned on the basis of linear monomeric molecules (182). The monomeric Co[N(SiMe$_3$)$_2$]$_2$ gave a very similar infrared spectrum to those of the zinc, cadmium, and mercury derivatives (214).

3. NMR Spectra of Metal Dialkylamides

On the basis of its proton NMR spectrum (215), one form of Be(NMe$_2$)$_2$ was given the linear trimeric structure (IX). The NMR spectra have also

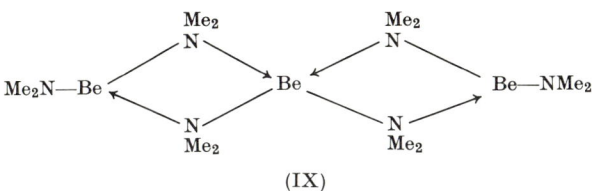

(IX)

been reported for some dialkylaminoberyllium alkyls by Coates et al. (216, 217).

The proton NMR spectrum of Al$_2$(NMe$_2$)$_6$ also showed distinct peaks corresponding to terminal and bridging NMe$_2$ groups in accordance with structure (X). These peaks showed no tendency to coalesce even up to 150°C, indicating a relatively strong nitrogen bridge (*209*). The monomeric derivative Al(NPr$^i_2$)$_3$ showed only a single type of dialkylamide group.

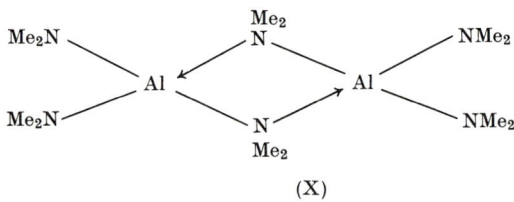

(X)

Moedritzer (*218*) obtained the chemical shifts for NMe$_2$ protons in Si(NMe$_2$)$_4$, Ge(NMe$_2$)$_4$, P(NMe$_2$)$_3$, As(NMe$_2$)$_3$, Sb(NMe$_2$)$_3$, and Ti(NMe$_2$)$_4$ on the neat liquids. Bradley and Gitlitz (*219*) reported chemical shift data for a number of transition metal dialkylamides in cyclohexane solution, but could find no correlation with other properties of the metals. However, in the case of Zr(NMe$_2$)$_4$, they found evidence for polymerization at low temperatures. Proton NMR data are also given by Bürger and Neese (*5*) for a range of derivatives Ti(NR$_2$)$_x$X$_{(4-x)}$ ($x = 3, 2, 1$; X = R, Cl, Br, SEt). Single NMR peaks were also found for the diamagnetic, monomeric species Mo(NMe$_2$)$_4$ (*176*) and W(NMe$_2$)$_6$ (*173*).

The proton NMR spectra of Be[N(SiMe$_3$)$_2$]$_2$ (*213*) and M[N(SiMe$_3$)$_2$]$_2$ (M = Zn, Cd, Hg) (*182*) all gave single peaks only a few hertz downfield (2–6 Hz) from T.M.S., but satellites due to proton coupling to ^{29}Si and ^{13}C enabled J_{H-Si} (6–7 Hz) and J_{H-C} (116–119 Hz) to be determined.

4. *Paramagnetic Metal Dialkylamides* (d–d *Bands, Magnetism, and ESR Spectra*)

Some of the transition metal dialkylamides are paramagnetic owing to the presence of unpaired d electrons, and these give rise to measurable ligand-field and magnetic effects.

a. Four-Coordinated Derivatives. The compounds V(NR$_2$)$_4$(d^1), Nb(NR$_2$)$_4$(d^1), Cr(NEt$_2$)$_4$(d^2), and Mo(NMe$_2$)$_4$(d^2) are all monomeric and all except Mo(NMe$_2$)$_4$ are paramagnetic. Electron spin resonance studies (*89, 220*) on V(NMe$_2$)$_4$ and V(NEt$_2$)$_4$ showed conclusively that these molecules have a distorted (D_{2d}) tetrahedral VN$_4$ structure. Thus, a regular tetrahedral d^1 species would have a degenerate ground state (^2E) and an ESR signal would not be expected at room temperature. However, both V(NMe$_2$)$_4$ (*89*) and V(NEt$_2$)$_4$ (*220*) gave strong signals at room temperature suggestive of an orbitally nondegenerate ground state and in frozen solution at $-150°C$ the anisotropic g values ($g_\parallel < g_\perp$) corresponded to axial symmetry (D_{2d}) with the electron occupying the $d_{x^2-y^2}$ orbital (see Fig. 15). The two partially resolved d–d bands at ~17,500 and ~13,300 cm^{-1} may be assigned to the transitions $^2E \leftarrow {}^2B_1$ and $^2B_2 \leftarrow {}^2B_1$ (see Fig. 15), respectively (*89*). The $3d^1$ electron thus benefits from a significant amount of ligand-field stabilization energy. At low temperature (77°K) the higher frequency band in V(NEt$_2$)$_4$ was split into a doublet ($\lambda\lambda_{max}$ 18,700 and 16,600 cm^{-1}). This may be due to the Jahn–Teller effect on the ^2E excited state. The magnetic susceptibility of V(NEt$_2$)$_4$ obeyed the Curie–Weiss law (small θ value) and gave the magnetic moment $\mu_{eff} = 1.70 \pm 0.02$ independent of temperature as expected for the $^2B_1$ ground state (*89*). This agreed well with the value (1.71) calculated from the g-values. Holloway et al. (*220*), following and refining the methods used by Kokoszka et al. (*90*) for V(OBu^t)$_4$, calculated the molecular orbital parameters for V(NEt$_2$)$_4$ from the values of $g_\parallel$, $A_\parallel$, $g_\perp$, and $A_\perp$ determined by electron spin resonance (the V^{51} nucleus with spin $I = 7/2$ gives an eight-line spectrum owing to electron-nuclear hyperfine coupling). According to these calculations the $d_{xy}(B_2)$ orbital appeared to be most affected by covalency, but from the energy level diagram (Fig. 15) the d_{xz} and d_{yz} orbitals are most involved. Apart from this discrepancy the electronic spectra and the ESR spectra both suggest that a considerable amount of covalency is involved in these molecules.

Bradley and Chisholm (*208*) have studied the Nb(NR$_2$)$_4$ compounds since these involve $4d^1$ systems. Each compound had a "d–d" band around 18,000–21,000 cm^{-1}, which was assigned to the $^2E \leftarrow {}^2B_1$ transition and at low temperature (77°K) this band split into two partially resolved bands (20,400 and 18,200 cm^{-1}). This behavior was similar to that observed with V(NR$_2$)$_4$ compounds. However, the Nb(NR$_2$)$_4$ spectra showed no bands due to the $^2B_2 \leftarrow {}^2B_1$ transition, presumably because the separation in energy of ^2E and $^2B_2$ states was much greater for niobium(IV) than vanadium(IV) and, thus, the $^2B_2 \leftarrow {}^2B_1$ symmetry-forbidden transition was too weak to be observed. The ESR spectra of Nb(NR$_2$)$_4$ compounds showed well-resolved ten-line spectra (Nb^{93}; $I = \frac{9}{2}$) at room temperature, whereas frozen solutions (toluene; 123°K) gave clear evidence for a D_{2d} distortion with $d_{x^2-y^2}(b_1)$ ground state ($g_\parallel < g_\perp$). Nevertheless, the magnetic susceptibilities all gave anomalously low values ($\mu_{eff} \sim 0.8$), although the Curie–Weiss law was obeyed.

Interesting results were obtained with the $3d^2$ and $4d^2$ systems Cr(NR$_2$)$_4$ and Mo(NR$_2$)$_4$. The chromium(IV) compounds gave an intense ($\epsilon_m = 1200$) d–d transition at 13,700 cm^{-1}, which was provisionally assigned to $^3A_2 \leftarrow {}^3B_1$ in D_{2d} symmetry. Magnetic susceptibilities conformed to the Curie–Weiss law with small θ values and gave $\mu_{eff} \sim 2.80$

independent of temperature (*169*). The molybdenum(IV) compounds $Mo(NR_2)_4$ (R = Me, Et) were diamagnetic and gave strong "*d–d*" bands [$Mo(NMe_2)_4$, broad doublet 21,740 and 19,600 cm^{-1}; $Mo(NEt_2)_4$, broad band at 18,500 cm^{-1}] assigned to $^1E \leftarrow {}^1A_1$ (in D_{2d}) transitions which are symmetry-allowed. $Mo(NMe_2)_4$ gave weaker shoulders at 14,300 and 10,500 cm^{-1} assigned to $^1A_2 \leftarrow {}^1A_1$ and $^1B_1 \leftarrow {}^1A_1$ transitions, but all bands gave further splitting into unresolved doublets at low temperature (*176*). Referring to the single-electron energy level diagram in D_{2d} (Fig. 15), it appears that in $Mo(NR_2)_4$ the separation between the $d_{x^2-y^2}$ and higher energy orbitals is sufficient to cause spin pairing. It was pointed out (*176*) that in D_{2d} symmetry the $M(NR_2)_4$ molecule with the conformation giving minimum interligand steric hindrance would allow all *d* orbitals except $d_{x^2-y^2}$ to be involved in both σ and π bonding. Thus, the spin pairing in $Mo(NR_2)_4$ may be considered as a consequence of strong covalent bonding.

   b. Three-Coordinated Derivatives. Some preliminary reports have appeared on the spectra and magnetic properties of the trigonally coordinated transition metals in compounds such as $M[N(SiMe_3)_2]_3$ (M = Ti, V, Cr, Fe) and $Cr(NPr^i_2)_3$. The d^1 compound $Ti[N(SiMe_3)_2]_3$ gave an ESR signal at room temperature indicative of a $^2A'_1$ ground state, and frozen solutions (135°K) exhibited *g*-anisotropy which confirmed axial symmetry (D_{3h}) for the TiN_3 framework (*183*). Magnetic susceptibility data and electronic spectra have also been obtained (*186*). The vanadium compound (d^2) did not give an ESR signal, although it is paramagnetic and this behavior is also consistent with the trigonal coordination of the metal (*183*). The d^3 compounds $Cr(NPr^i_2)_3$ and $Cr[N(SiMe_3)_2]_3$ both gave magnetic susceptibilities corresponding to spin-only magnetic moments ($\mu_{eff} = 3.80$) independent of temperature (*179*). Neither compound in solution at room temperature gave an ESR signal, but in frozen solutions at 130°K a spectrum was obtained with $g_\parallel = 2.0$ and $g_\perp = 4.0$ corresponding to an axially symmetric (D_{3h}) system with a large zero-field splitting (*221, 222*). The iron(III) compound $Fe[N(SiMe_3)_2]_3$ obeyed the Curie–Weiss law (small θ) and gave a temperature-independent magnetic moment ($\mu_{eff} = 5.91$) corresponding to a high-spin d^5 species. The ESR spectra on a powder and on oriented single crystals ($g_\parallel = 2.007$; $g_\perp = 6.021$) showed that a large zero-field splitting was present in this axially symmetric compound (*222, 223*). An interesting Mössbauer spectrum was also obtained (*223*). At 77°K a typical two-line spectrum was obtained ($\delta = 0.43$ mm sec^{-1}) with a large quadrupole splitting ($\Delta E = 5.12$ mm·sec^{-1}). At 4.2°K a five-line spectrum appeared which on application of a weak magnetic field gave a fully resolved six-line spectrum. It was deduced that the principal component of the electric field gradient tensor was positive. The electronic absorption spectrum of $Fe[N(SiMe_3)_2]_3$ gave bands which were rather strong for spin-forbidden transitions associated with the $^6A_1'$ ground state, but the two d–d transitions at 16,100 and 20,000 cm^{-1} were assigned to $(^4A_1'', {}^4A_2'') \leftarrow {}^6A_1'$ and $^4E' \leftarrow {}^6A_1'$, respectively. Assignments of the electronic spectra of the d^1, d^2, and d^3 trigonal compounds have also been made on a crystal field model for D_{3h} symmetry (224).

c. *Two-Coordinated Derivatives.* The cobalt(II) derivative $Co[N(SiMe_3)_2]_2$ (181) was shown by Bradley and Fisher (214) to be monomeric in solution, and it gave the same electronic absorption spectrum in solution, liquid, or crystalline states. The spectrum was therefore assigned by analogy with gaseous $CoCl_2$ for a linear two-coordinated ($D_{\infty h}$) species. Its magnetic susceptibility obeyed the Curie law with $\mu_{eff} = 4.83$ independent of temperature.

5. *Structures of Metal Dialkylamides Determined by X-Ray or Electron Diffraction*

Structures of various metal dialkylamides have recently been determined and important data on M–N bond lengths and the bond angles in these molecules are being collected.

The single crystal X-ray analysis of $W(NMe_2)_6$ showed (Fig. 16) the expected octahedral WN_6 framework (W–N = 2.032 ± 0.025 Å) (173). A feature of special interest was the shape and conformation of the

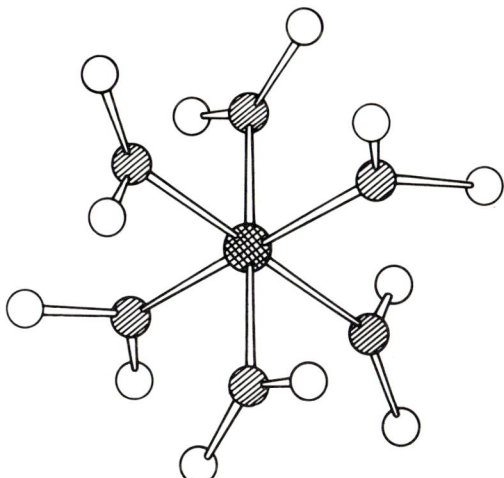

Fig. 16. The structure of $[W(NMe_2)_6]$. Cross-hatched, W; hatched, N; plain, Me.

dimethylamide ligands. The symmetry point group of the heavy atom framework $W(NC_2)_6$ was T_h due to extensive planarity (e.g., C_2NWNC_2 units are all coplanar). Thus, the nitrogen atoms are trigonal planar rather than pyramidal, and this is consistent with delocalized nitrogen-to-tungsten π bonding involving molecular orbitals of T_{2g} symmetry. An 18-electron valency group can thus be accorded to the d^0 tungsten(VI) atom, but lack of comparable data on W(VI)–N bond lengths renders uncertain an assessment of π character from the observed W–N bond length. Another point of interest was the small CŃC (104.92°) angle

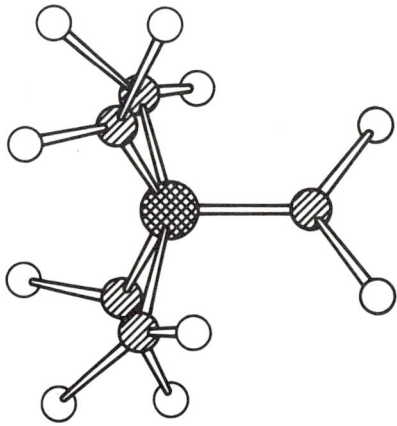

FIG. 17. The structure of [Nb(NMe2)5]. Cross-hatched, Nb; hatched, N; plain, Me.

of the ligand and the unexpectedly long C–N (1.516 Å) bonds. It thus appears that the W–N bond has more s-character (approx. sp hybridization) than the N–C bonds (approx. sp^3 hybridization). If the W–N and N–C bonds all involved sp^2-hybrid nitrogen σ orbitals the N–C bond length should be close to 1.485 ± 0.002 Å (225) and the CŃC angle 120°, whereas the observed CŃC is close to the tetrahedral angle and the N–C bond length is close to the $N(sp^3)$–$C(sp^3)$ value of 1.51 Å. The structure also shows that the central atom is well shielded by the symmetrical array of twelve CH_3 groups and this, in conjunction with the 18-electron valency group for tungsten, explains the relatively inert character of this compound to attack by nucleophilic reagents.

Heath and Hursthouse (226) have published a preliminary account of the structures of the five-coordinated metal dialkylamides $Nb(NMe_2)_5$ and $Nb(piperidide)_5$. In both compounds the structure of the partial framework $Nb(NC_2)_5$ is the same and approximates to a distorted tetragonal pyramid (Fig. 17). The Nb–N(1) bond occupies the axial position (it is a 2-fold axis), and it is significantly shorter (1.977 ± 0.017 Å) than the basal Nb–N bonds (average 2.042 ± 0.015 Å). The nitrogens are again trigonal planar implying that the nitrogen is acting as a π donor to the niobium(V) d^0 atom. However, it was pointed out that the shape of the ligand could be due to intramolecular congestion. It is interesting that both $Nb(NMe_2)_5$ and $Nb(piperidide)_5$ adopt this structure with one Nb–N bond shorter than the other four. It may be that the short bond Nb–N(1) is due to the presence of a localized π bond involving one vacant metal d orbital and the nitrogen p_π orbital, while the longer

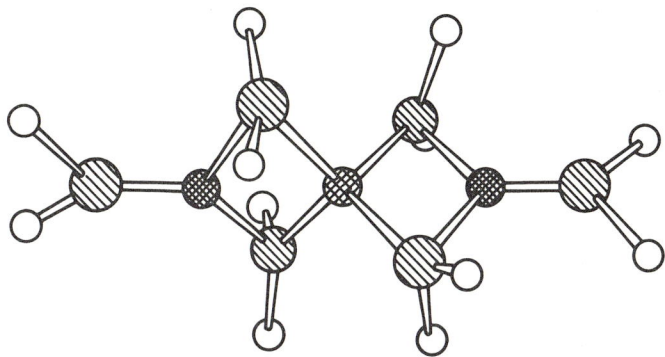

FIG. 18. The structure of $[Be(NMe_2)_2]_3$. Fine cross-hatched, Be (tetrahedral); cross-hatched, Be (trigonal); hatched, N; plain, Me.

bonds are due to delocalized partial π bonds involving only two d orbitals with the remaining four nitrogens.

Vilkov et al. (227) have determined the structure of $Sn(NMe_2)_4$ in the vapor state by electron diffraction. The SnN_4 framework was tetrahedral (Sn–N = 2.045 ± 0.060 Å; NŜnN = 109.5°) and the nitrogen atoms were close to trigonal planar (CÑC = 119 ± 3; SnNC = 117.5 ± 1.5°). The calculations were based on a C_{2v} point group for the molecules and gave C–N = 1.450 ± 0.045 Å and C–H = 1.10 ± 0.10 Å.

The structure of the trimer $Be_3(NMe_2)_6$ was determined by X-ray diffraction analysis by Atwood and Stucky (228). The molecule (Fig. 18) has a linear configuration of three berylliums with the central one being four-coordinated by bridging dimethylamide groups and the outer ones three-coordinated giving a symmetry point group of D_{2d}. A number of interesting features are apparent in this structure. For example, the terminal dimethylamide groups contain trigonal-planar nitrogens with CÑC = 103.6° and a short Be–N bond length (1.560 Å). This is consistent with a localized beryllium–nitrogen π bond involving the three-coordinated beryllium atom and the terminal nitrogen atom. The bridging dimethylamides are unequally bonded to the two types of beryllium. Thus, $N_{bridge}-Be_{terminal} = 1.61$ Å and $N_{bridge}\hat{Be}N_{bridge} = 102.9°$, whereas $N_{bridge}-Be_{central} = 1.76$ Å and $N_{bridge}\hat{Be}_{central}N_{bridge} = 92.3°$. The distorted tetrahedral central beryllium atom has slightly longer bonds to nitrogen than the trigonal terminal berylliums as would be expected. The bridging dimethylamides have $C\hat{N}C = 108.2°$, which is much closer to the tetrahedral angle than in the terminal ligands, but the C–N distances (terminal = 1.52; bridging = 1.53 Å) are practically the same. Presumably interligand intramolecular steric interactions prevent the development of an infinite linear polymer with all-tetrahedral beryllium.

Although methylamino metal derivatives $MeNH-ML_x$ are not strictly dialkylamido metal compounds, it seems relevant to note the very interesting example of stereoisomerism reported for the trimeric compounds $(Me_2AlNHMe)_3$ (229). Both forms were obtained from the reaction of Al_2Me_6 with methylamine and were separated by fractional sublimation. The stable less volatile form (I) (m.p. 110°) was rhombohedral and had the Al_3N_3 ring in the chair conformation with all N-methyl groups in equatorial positions. The metastable form (II) gave monoclinic crystals with the Al_3N_3 ring in a skew-boat conformation with two N-methyl groups equatorial and the third one axial. Interestingly, the trimer $[Me_2AlN(CH_2)_2]_3$ also adopted the skew-boat conformation (230). A preliminary report on the crystal structure of the tetramer $(C_6H_5AlNC_6H_5)_4$ showed the presence of a cubane configuration of the Al_4N_4 framework reminiscent of the metal–oxygen cubanes $(RMOR)_4$ (Figs. 1 and 3).

Some structures have recently been determined for bistrimethylsilylamido metal compounds. The lithium derivative behaves as a dimer in solution, but it was found to be trimeric in the crystalline state (143). X-Ray crystal analysis gave the ring structure for Li_3N_3 (Fig. 19) with Li–N = 2.00, Si–N = 1.72, and Si–C = 1.89 Å; $N\hat{L}iN = 148°$, $Li\hat{N}Li = 92°$, and SiNSi = 118°. For the parent amine $(Me_3Si)_2NH$ it was shown by electron diffraction (231) that Si–N = 1.735 Å and $Si\hat{N}Si = 125.5°$.

The beryllium compound was shown by Bürger et al. (213) to be monomeric in solution and thus contained two-coordinated beryllium. The structure of this compound in the vapor state was determined by Clark and Haaland (232) by electron diffraction. As shown in Fig. 20, the $Si_2NBeNSi_2$ framework has D_{2d} symmetry consistent with the presence of $p_\pi-p_\pi$ bonding involving vacant acceptor p orbitals on the beryllium. Although this π bonding (ligand-to-metal) is in competition with donation of the nitrogen lone pair into vacant d orbitals on the silicons, the Si–N bond length (1.726 Å) is very similar to that in

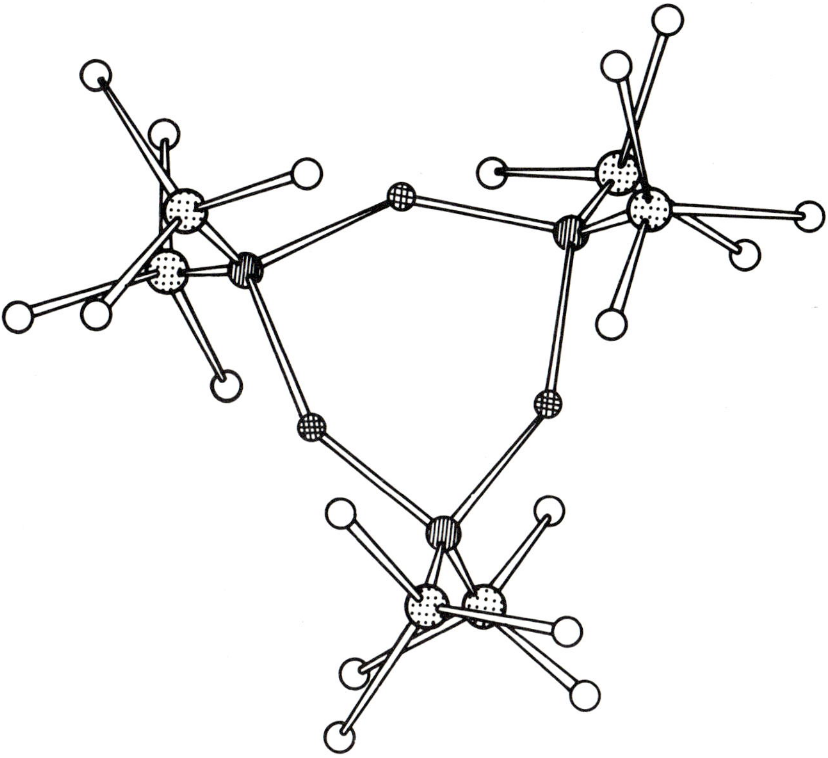

Fig. 19. The structure of [LiN(SiMe$_3$)$_2$]$_3$. Cross-hatched, Li; hatched, N; stippled, Si; plain, Me.

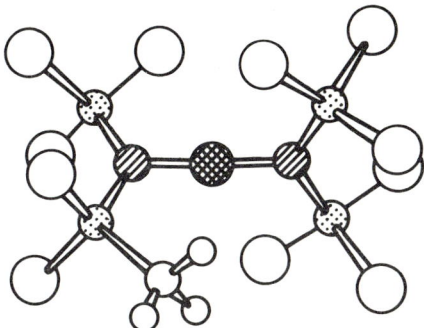

Fig. 20. The structure of {Be[N(SiMe$_3$)$_2$]$_2$}. Cross-hatched, Be; hatched, N; stippled, Si; plain, Me.

(Me₃Si)₂NH. However, the Be–N bond length (1.566 Å) is very close to that found (1.56 Å) in the terminally bonded Be–NMe₂ group in Be₃(NMe₂)₆, and this suggests that significant beryllium–nitrogen π bonding is present. The SiN̂Si (129.3°) is wider than in (Me₃Si)₂NH or [LiN(SiMe₃)₂]₃. Owing to intramolecular congestion involving methyl groups, there was rotation of SiMe₃ groups about the Si–N bonds resulting in a conformation corresponding to D_2 symmetry, although the D_{2d} symmetry of the Si₂NBeNSi₂ unit was preserved.

The first three-coordinated metal silylamide structure was reported by Bradley *et al.* (*233*), who determined the structure of Fe[N(SiMe₃)₂]₃ by single crystal X-ray analysis. As expected the FeN₃ unit was trigonal planar (Fig. 21) with Fe–N = 1.918 ± 0.004 Å and the nitrogens were also trigonally planar, but each FeNSi₂ plane made a dihedral angle of 49° with the FeN₃ plane. Thus, the Fe(NSi₂)₃ framework had the symmetry point group D_3. The ligand geometry, Si–N = 1.731 ± 0.003 Å and SiN̂Si = 121.24°, was rather similar to that found in the free amine and in the lithium and beryllium derivatives. The chromium derivative Cr[N(SiMe₃)₂]₃ had the same structure as, indeed, have all the transition metal trissilylamides (*234*).

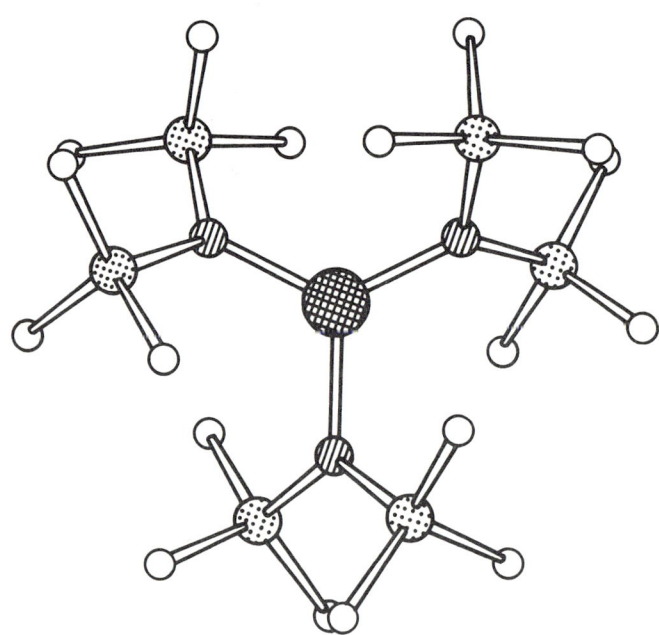

Fig. 21. The structure of {Fe[N(SiMe₃)₂]₃}. Cross-hatched, Fe; hatched, N; stippled, Si; plain, Me.

The only nontransition metal trissilylamide structure so far reported is that of the aluminum compound Al[N(SiMe$_3$)$_2$]$_3$ (235). This compound is isostructural with the transition metal derivatives and, although it has shorter metal–nitrogen bonds (Al–N = 1.78 ± 0.02 Å), it has practically the same dihedral angle (50°) between MNSi$_2$ and MN$_3$ planes. Also the ligand geometry (Si–N = 1.75 ± 0.01 Å; SiÑSi = 118 ± 1°) appears to be the same within a wider uncertainty than that found in the other silylamides. It is very difficult to assess the degree of π bonding in the metal–nitrogen bonds in the silylamides, because no authentic single σ-bond distances are known for these metals with such low coordination numbers. The planarity of the MNSi$_2$ groups does not necessarily prove that nitrogen is donating π electrons to the metal because nitrogen-to-silicon π donation would have the same effect (viz., in the free amine). Further light has been shed on this fascinating problem by the determination of the structure of chromium trisdiisopropylamide, which was also known to be monomeric in solution (179). Single-crystal X-ray analysis (221) showed that the structure (Fig. 22) was trigonal with the Cr(NC$_2$)$_3$ approximating to D_3 symmetry as in the metal trissilylamides. The CrNC$_2$ units were planar suggesting nitrogen-to-metal π-electron donation and the ligand planes made dihedral angles of ~70° with the CrN$_3$ plane. The short Cr–N bond distance (1.87 Å) also suggested that some metal–nitrogen π bonding was involved. Further X-ray structural analyses of crystalline metal dialkylamides may be expected in the near future.

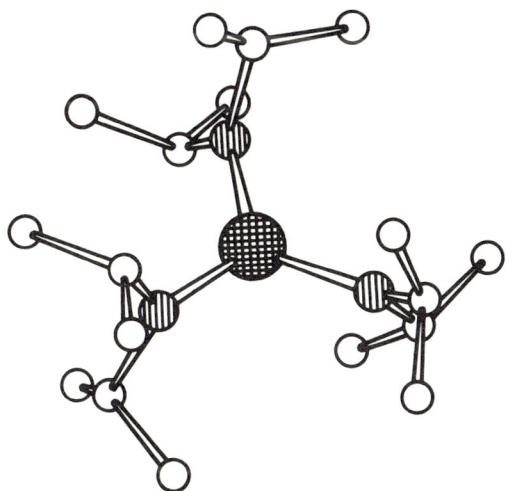

FIG. 22. The structure of [Cr(NPr^i$_2$)$_3$]. Cross-hatched, Cr; hatched, N; plain, C.

References

1. Bradley, D. C., *Progr. Inorg. Chem.* **2**, 303–361 (1960).
2. Mehrotra, R. C., *Inorg. Chim. Acta Rev.* **1**, 99 (1967).
3. Bradley, D. C., and Fisher, K. J., *in* "M.T.P. International Review of Science, General Chemistry of the Transition Metals" (D. W. A. Sharp, ed.), Vol. 5, Part I, pp. 65–91. Butterworths, London, 1972.
4. Bradley, D. C., Alkoxides, *in* "Enciclopedia della Chimica," Utet/Sansoni Edizioni Scientifiche. In press.
5. Bürger, H., and Neese, H. J., *Chimica* **24**, 209 (1970).
6. Lappert, M. F., and Prokai, B., *Advan. Organomet. Chem.* **5**, 225 (1967).
7. Bradley, D. C., *Coord. Chem. Rev.* **2**, 299 (1967).
8. Bradley, D. C., *in* "Preparative Inorganic Reactions" (W. Jolly, ed.), Vol. 2, pp. 169–186. Wiley, New York, 1965.
9. Bartley, W. G., and Wardlaw, W., *J. Chem. Soc.* 422 (1958).
10. Wheatley, P. J., *J. Chem. Soc.* 4270 (1961).
11. Bains, M. S., *Can. J. Chem.* **42**, 945 (1964).
12. Golovanov, I. B., Simonov, A. P., Priskunov, A. K., Talalaeva, T. V., Tsareva, G. W., and Kocheshkov, K. A., *Dokl. Akad. Nauk. SSSR* **149**, 835 (1963).
13. Simonov, A. P., Shigorin, D. N., Talalaeva, T. V., and Kocheshkov, K. A., *Izv. Akad. Nauk. SSSR. Otd. Khim. Nauk* 1126 (1962).
14. Hartwell, G. E., and Brown, T. L., *Inorg. Chem.* **5**, 1257 (1966).
15. Weiss, E., Alsdorf, H., and Kühr, H., *Angew. Chem. Int. Ed. Engl.* **6**, 801 (1967).
16. Weiss, E., *Helv. Chim. Acta* **46**, 2051 (1963).
17. Weiss, E., and Alsdorf, H., *Z. Anorg. Allg. Chem.* **372**, 206 (1970).
18. Mehrotra, R. C., and Agrawal, M. M., *J. Chem. Soc.* 1026 (1967).
19. Bradley, D. C., Redwood, M. E., and Willis, C. J., *Proc. Chem. Soc.* 416 (1964); Redwood, M. E., and Willis, C. J., *Can. J. Chem.* **43**, 1893 (1965).
20. Redwood, M. E., and Willis, C. J., *Can. J. Chem.* **45**, 389 (1967).
21. Dear, R. E. A., Fox, W. B., Fredericks, R. J., Gilbert, E. E., and Huggins, D. K., *Inorg. Chem.* **9**, 2590 (1970).
22. Dahl, L. F., Davis, G. L., Wampler, D. L., and West, K., *J. Inorg. Nucl. Chem.* **24**, 357 (1962).
23. Matoni, V. A., and Spiro, T. G., *Inorg. Chem.* **7**, 193 (1968).
24. Coates, G. E., and Fishwick, A. H., *J. Chem. Soc. A* 477 (1968).
25. Coates, G. E., and Ridley, D., *Chem. Commun.* 560 (1966).
26. Coates, G. E., and Ridley, D., *J. Chem. Soc. A* 1064 (1966).
27. Coates, G. E., and Lauder, A., *J. Chem. Soc. A* 264 (1966).
28. Shearer, H. M. M., and Spencer, C. B., *Chem. Commun.* 194 (1966).
29. Mehrotra, R. C., and Arora, M., *Indian J. Chem.* **7**, 399 (1969).
30. Grigor'ev, A. I., and Turova, N. Ya., *Dokl. Akad. Nauk. SSSR* **162**, 98 (1965).
31. Lutz, H. D., *Z. Anorg. Allg. Chem.* **353**, 207 (1967).
32. Lutz, H. D., *Z. Anorg. Allg. Chem.* **356**, 132 (1968).
33. Bryce-Smith, D., and Wakefield, B. J., *J. Chem. Soc.* 2483 (1964).
34. Coates, G. E., Heslop, J. A., Redwood, M. E., and Ridley, D., *J. Chem. Soc. A* 1118 (1968).
35. Abraham, M. H., *J. Chem. Soc.* 4130 (1960).

36. Coates, G. E., and Roberts, P. D., *J. Chem. Soc. A* 1233 (1967).
37. Talalaeva, T. V., Zenina, G. V., and Kocheshkov, K. A., *Dokl. Akad. Nauk SSSR* **171**, 122 (1966).
38. Mehrotra, R. C., and Arora, M., *Z. Anorg. Allg. Chem.* **370**, 300 (1969).
39. Turova, N. Ya., Popovkin, B. A., and Novoselova, A. V., *Dokl. Akad. Nauk SSSR* **167**, 604 (1966).
40. Turova, N. Ya., Popovkin, B. A., and Novoselova, A. V., *Izv. Akad. Nauk SSSR, Neorg. Mater.* **3**, 1435 (1967).
41. Bradley, D. C., and Faktor, M. M., *Nature (London)* **184**, 55 (1959).
42. Wilhoit, R. C., Burton, J. R., Kuo, F.-T., Huang, S.-R., and Viguesnel, K., *J. Inorg. Nucl. Chem.* **24**, 851 (1962).
43. Fieggen, W., Doctoral Thesis, Univ. of Amsterdam, 1970.
44. Mehrotra, R. C., *J. Indian. Chem. Soc.* **30**, 585 (1953).
45. Mehrotra, R. C., *J. Indian. Chem. Soc.* **31**, 85 (1954).
46. Bradley, D. C., *Advan. Chem. Ser.* **23**, 10 (1959).
47. Shiner, V. J., Whittaker, D., and Fernandez, V. P., *J. Amer. Chem. Soc.* **85**, 2318 (1963).
48. Kleinschmidt, D. C., Ph.D. Thesis, Indiana Univ., Bloomington, Indiana, 1967 [reported in Fieggen (*43*)].
49. Oliver, J. G., and Worrall, I. J., *J. Chem. Soc. A* 1389 (1970) and earlier references therein.
50. Fieggen, W., Gerding, H., and Nibbering, N. M. M., *Rec. Trav. Chim.* **87**, 377 (1968).
51. Wilhoit, R. C., *J. Phys. Chem.* **61**, 114 (1957).
52. Wilson, J. W., *J. Chem. Soc. A* 981 (1971).
53. Bains, M. S., *Can. J. Chem.* **40**, 381 (1962).
54. Oliver, J. G., and Worrall, I. J., *J. Chem. Soc. A* 845 (1970).
55. Saegusa, T., and Veshima, T., *Inorg. Chem.* **6**, 1679 (1967).
56. Paul, R. C., Makhini, H. S., and Chadha, S. L., *Chem. Ind. (London)* 829 (1970).
57. Bains, M. S., and Bradley, D. C., *Chem. Ind. (London)* 1032 (1961).
58. Shiner, V. J., and Whittaker, D., *J. Amer. Chem. Soc.* **85**, 2337 (1963).
59. Ogata, Y., Kawasaki, A., and Kishi, I., *Tetrahedron* **23**, 825 (1967).
60. Mehrotra, R. C., and Mehrotra, R. K., *Current Sci.* **8**, 241 (1964).
61. Bindal, S. R., Mathur, V. K., and Mehrotra, R. C., *J. Chem. Soc. A* 863 (1969).
62. Dunn, P., *Aust. J. Appl. Sci.* **10**, 458 (1959).
63. Wright, D. A., and Williams, D. A., *Acta Crystallogr. Sect. B* **24**, 1107 (1968).
64. Ibers, J. A., *Nature (London)* **197**, 686 (1963).
65. Witters, R. D., and Caughlan, C. N., *Nature (London)* **205**, 1312 (1965).
66. Weingarten, H., and Van Wazer, J. R., *J. Amer. Chem. Soc.* **87**, 724 (1965).
67. Adams, R. W., and Winter, G., *Aust. J. Chem.* **20**, 171 (1967).
68. Kakos, G. A., and Winter, G., *Aust. J. Chem.* **21**, 793 (1968).
69. Caughlan, C. N., Smith, H. M., and Watenpaugh, K., *Inorg. Chem.* **5**, 2131 (1966).
70. Bradley, D. C., and Holloway, C. E., *J. Chem. Soc. A* 219 (1968).
71. Riess, J. G., and Pfalzgraf, L. G., *Bull. Soc. Chim.* 2401 (1968); Pfalzgraf, L. G., and Reiss, J. G., *ibid.* 4348 (1968).
72. Bradley, D. C., Wardlaw, W., and Whitley, A., *J. Chem. Soc.* 5 (1956).
73. Bradley, D. C., and Chisholm, M. H., to be published.
74. Bradley, D. C., and Holloway, C. E., *Inorg. Chem.* **3**, 1163 (1964).
75. Bradley, D. C., and Holloway, C. E., *J. Chem. Soc. A* 1316 (1968).

76. Russo, W. R., and Nelson, W. H., *J. Amer. Chem. Soc.* **92**, 1521 (1970).
77. Caughlan, C. N., Smith, H. S., Katz, W., Hodgson, W., and Crowe, R. W., *J. Amer. Chem. Soc.* **73**, 5652 (1951).
78. Martin, R. L., and Winter, G., *J. Chem. Soc.* 2947 (1961).
79. Bradley, D. C., and Westlake, A. H., "Proceedings of the Symposium on Coordination Chemistry, Tihany, Hungary" (M. Beck, ed.), pp. 309–315. Publ. House Hung. Acad. Sci., Budapest, 1965.
80. Bradley, D. C., and Holloway, C. E., *Proc. Paint Res. Inst. (Off. Dig.)* **37**, 487 (1965).
81. Watenpaugh, K., and Caughlan, C. N., *Chem. Commun.* 76 (1967).
82. Bradley, D. C., Hursthouse, M. B., and Rodesiler, P. F., *Chem. Commun.* 1112 (1968).
83. Bradley, D. C., and Hillyer, M. J., *Trans. Faraday Soc.* **62**, 2367 (1966).
84. Bradley, D. C., and Hillyer, M. J., *Trans. Faraday Soc.* **62**, 2374 (1966).
85. Adams, R. W., Bishop, E., Martin, R. L., and Winter, G., *Aust. J. Chem.* **19**, 207 (1966).
86. Adams, R. W., Martin, R. L., and Winter, G., *Aust. J. Chem.* **20**, 773 (1967).
87. Lappert, M. F., and Sanger, A. R., *J. Chem. Soc. A* 1314 (1971).
88. Alyea, E. C., and Bradley, D. C., *J. Chem. Soc. A* 2330 (1969).
89. Bradley, D. C., Moss, R. H., and Sales, K. D., *Chem. Commun.* 1255 (1969).
90. Kokoszka, G. F., Allen, H. C., and Gordon, G., *Inorg. Chem.* **5**, 91 (1966).
91. Bradley, D. C., and Chisholm, M. H., *J. Chem. Soc. A* 2741 (1971).
92. Bradley, D. C., and Hillyer, M. J., *Trans. Faraday Soc.* **62**, 2382 (1966).
93. Wentworth, R. A. D., and Brubaker, C. H., *Inorg. Chem.* **3**, 47 (1964).
94. Reagan, W. J., and Brubaker, C. H., *Inorg. Chem.* **9**, 827 (1970).
95. Alyea, E. C., Basi, J. S., Bradley, D. C., and Chisholm, M. H., *J. Chem. Soc. A* 772 (1971).
96. Wiberg, K. B., and Foster, G., *Chem. Ind. (London)* 108 (1961).
97. Hagihara, M., and Yamasaki, H., *Nippon Kagaku Zasshi* **81**, 822 (1960).
98. Bradley, D. C., and Mehta, M. L., *Can. J. Chem.* **40**, 1710 (1962).
99. Kakos, G. A., and Winter, G., *Aust. J. Chem.* **23**, 15 (1970).
100. Brown, D. A., Cunningham, D., and Glass, W. K., *J. Chem. Soc. A* 1563 (1968).
101. Alyea, E. C., Basi, J. S., and Bradley, D. C., unpublished results.
102. Alyea, E. C., Basi, J. S., Bradley, D. C., and Chisholm, M. H., *Chem. Commun.* 495 (1968).
103. Bradley, D. C., Multani, R. K., and Wardlaw, W., *J. Chem. Soc.* 126 (1958).
104. Adams, R. W., Martin, R. L., and Winter, G., *Aust. J. Chem.* **19**, 363 (1966).
105. Adams, R. W., Barraclough, C. G., Martin, R. L., and Winter, G., *Inorg. Chem.* **5**, 346 (1966).
106. Krüger, A. G., and Winter, G., *Aust. J. Chem.* **23**, 1 (1970).
107. Brubaker, C. H., and Wicholas, M., *J. Inorg. Nucl. Chem.* **27**, 59 (1965).
108. Adams, R. W., Barraclough, C. G., Martin, R. L., and Winter, G., *Aust. J. Chem.* **30**, 2351 (1971).
109. Bradley, D. C., *Nature (London)* 1211 (1958).
110. Bradley, D. C., and Faktor, M. M., *Chem. Ind. (London)* 1332 (1958).
111. Misra, S. N., Misra, T. N., Kapoor, R. N., and Mehrotra, R. C., *Chem. Ind. (London)* 120 (1963).
112. Sankhla, B. S., Misra, S. N., and Kapoor, R. N., *Chem. Ind. (London)* 382 (1965); Sankhla, B. S., and Kapoor, R. N., *Austr. J. Chem.* **20**, 2013 (1967).

113. Batwara, J. M., Tripathi, U. D., Mehrotra, R. K., and Mehrotra, R. C., *Chem. Ind. (London)* 1379 (1966); Tripathi, U. D., Batwara, J. M., and Mehrotra, R. C., *J. Chem. Soc. A* 991 (1967).
114. Mazdiyasni, K. S., Lynch, C. T., and Smith, J. S., *Inorg. Chem.* **5**, 342 (1966).
115. Brown, L. M., and Mazdiyasni, K. S., *Inorg. Chem.* **9**, 2783 (1970).
116. Misra, S. N., Misra, T. N., and Mehrotra, R. C., *Aust. J. Chem.* **21**, 797 (1968).
117. Mehrotra, R. C., and Agrawal, M. M., *Chem. Commun.* 469 (1968).
118. Bradley, D. C., Chatterjee, A. K., and Wardlaw, W., *J. Chem. Soc.* 2260 (1956).
119. Bradley, D. C., Chatterjee, A. K., and Wardlaw, W., *J. Chem. Soc.* 3469 (1956).
120. Bradley, D. C., Chatterjee, A. K., and Wardlaw, W., *J. Chem. Soc.* 2600 (1957).
121. Bradley, D. C., Saad, M. A., and Wardlaw, W., *J. Chem. Soc.* 1091, 3488 (1954).
122. Gilman, H., Jones, R. G., Karmas, G., and Martin, G. A., *J. Amer. Chem. Soc.* **78**, 4285 (1956).
123. Gilman, H., Jones, R. G., Bindschadler, E., Karmas, G., and Yoeman, F. A., *J. Amer. Chem. Soc.* **78**, 4287 (1956).
124. Jones, R. G., Bindschadler, E., Blume, D., Karmas, G., Martin, G. A., Thirtle, J. R., Yoeman, F. A., and Gilman, H., *J. Amer. Chem. Soc.* **78**, 7030 (1956).
125. Bradley, D. C., Chakravarti, B. N., and Chatterjee, A. K., *J. Inorg. Nucl. Chem.* **3**, 367 (1957).
126. Bradley, D. C., and Chatterjee, A. K., *J. Inorg. Nucl. Chem.* **4**, 279 (1957).
127. Bradley, D. C., Kapoor, R. N., and Smith, B. C., *J. Chem. Soc.* 1023 (1963).
128. Bradley, D. C., Chatterjee, Amar K., and Chatterjee, Amiya K., *Proc. Chem. Soc.* 260 (1957); *J. Inorg. Nucl. Chem.* **12**, 71 (1959).
129. Bradley, D. C., Kapoor, R. N., and Smith, B. C., *J. Inorg. Nucl. Chem.* **24**, 863 (1962).
130. Karraker, D. G., *Inorg. Chem.* **3**, 1618 (1964).
131. Karraker, D. G., Siddal, T. H., and Stewart, W. E., *J. Inorg. Nucl. Chem.* **31**, 711 (1969).
132. Samulski, E. T., and Karraker, D. G., *J. Inorg. Nucl. Chem.* **29**, 993 (1967).
133. Bradley, D. C., Harder, B., and Hudswell, F., *J. Chem. Soc.* 3318 (1957).
134. Meerwein, H., and Bersin, T., *Ann.* **475**, 113 (1929).
135. Bradley, D. C., and Wardlaw, W., *J. Chem. Soc.* 280 (1951); Bartley, W. G., and Wardlaw, W., *ibid.* 421 (1958).
136. Bradley, D. C., Caldwell, E. W., and Wardlaw, W., *J. Chem. Soc.* 4775 (1957).
137. Albers, H., Deutsch, M., Krastinak, W., and Von Osten, H., *Chem. Ber.* **85**, 267 (1952).
138. Jones, R. G., Bindschadler, E., Blume, D., Martin, G. A., Thirtle, J. R., and Gilman, H., *J. Amer. Chem. Soc.* **78**, 6027 (1956).
139. Gut, R., *Helv. Chim. Acta* **47**, 2262 (1964).
140. Mehrotra, R. C., and Agrawal, M. M., *J. Chem. Soc. A* 1026 (1967).
141. Schmidbaur, H., *Angew. Chem. Int. Ed. Engl.* **4**, 201 (1965).
142. Schindler, F., and Schmidbaur, H., *Angew. Chem. Int. Ed. Engl.* **6**, 683 (1967).
143. Mootz, D., Zinnius, A., and Böttcher, B., *Angew. Chem. Int. Ed. Engl.* **8**, 378 (1969).
144. Zeitler, V. A., and Brown, C. A., *J. Phys. Chem.* **61**, 1174 (1957).
145. Chamberlain, M. M., Jabs, G. A., and Wayland, B. B., *J. Org. Chem.* **27**, 3321 (1962).

146. Weiss, E., Hoffmann, K., and Grützmacher, H. F., *Chem. Ber.* **103**, 1190 (1970).
147. Bradley, D. C., and Thomas, I. M., *J. Chem. Soc.* 3404 (1959).
148. Bradley, D. C., Kapoor, R. N., and Smith, B. C., *J. Chem. Soc.* 204 (1963).
149. Schmidbaur, H., and Schmidt, M., *Angew. Chem. Int. Ed.* **1**, 328 (1962).
150. Schmidbaur, H., *Chem. Ber.* **96**, 2692 (1963).
151. Schmidbaur, H., *Chem. Ber.* **97**, 836 (1964).
152. Shiotani, A., and Schmidbaur, H., *J. Amer. Chem. Soc.* **92**, 7003 (1970).
153. Batwara, J. M., and Mehrotra, R. C., *J. Inorg. Nucl. Chem.* **32**, 411 (1970).
154. Bradley, D. C., and Prevedorou-Demas, C., *Can. J. Chem.* **41**, 629 (1963).
155. Bradley, D. C., and Prevedorou-Demas, C., *J. Chem. Soc.* 1580 (1964).
156. Bradley, D. C., and Prevedorou-Demas, C., *J. Chem. Soc. A* 1139 (1966).
157. Bradley, D. C., and Prevedorou-Demas, C., *J. Chem. Soc. A* 43 (1967).
158. Bradley, D. C., Lorimer, J. W., and Prevedorou-Demas, C., *Can. J. Chem.* **47**, 4113 (1969).
159. Bradley, D. C., Lorimer, J. W., and Prevedorou-Demas, C., *Can. J. Chem.* **49**, 2310 (1971).
160. Bradley, D. C., and Prevedorou-Demas, C., *Chem. Ind. (London)* 1659 (1970).
161. Andrianov, K. A., *Inorg. Macromol. Rev.* **1**, 33 (1970).
162. Jones, K., and Lappert, M. F., *J. Organomet. Chem.* **3**, 295 (1965).
163. Ruff, J. K., *J. Amer. Chem. Soc.* **83**, 2835 (1961).
164. Coates, G. E., and Glockling, F., *J. Chem. Soc.* 22 (1954).
165. Dermer, D. C., and Fernelius, W. C., *Z. Anorg. Chem.* **221**, 83 (1935).
166. Bradley, D. C., and Thomas, I. M., *J. Chem. Soc.* 3857 (1960).
167. Thomas, I. M., *Can. J. Chem.* **39**, 1386 (1961).
168. Basi, J. S., and Bradley, D. C., *Proc. Chem. Soc.* 305 (1963).
169. Basi, J. S., Bradley, D. C., and Chisholm, M. H., *J. Chem. Soc. A* 1433 (1971).
170. Bradley, D. C., and Thomas, I. M., *Can. J. Chem.* **40**, 449 (1962).
171. Bradley, D. C., and Gitlitz, M. H., *J. Chem. Soc. A* 980 (1969).
172. Bradley, D. C., and Thomas, I. M., *Can. J. Chem.* **40**, 1355 (1962).
173. Bradley, D. C., Chisholm, M. H., Heath, C. E., and Hursthouse, M. B., *Chem. Commun.* 1261 (1969).
174. Bradley, D. C., Bonnett, R., and Fisher, K. J., *Chem. Commun.* 886 (1968).
175. Bradley, D. C., Bonnett, R., Fisher, K. J., and Rendall, I. F., *J. Chem. Soc. A* 1622 (1971).
176. Bradley, D. C., and Chisholm, M. H., *J. Chem. Soc. A* 2741 (1971).
177. Bradley, D. C., and Chisholm, M. H., unpublished work.
178. Alyea, E. C., Bradley, D. C., Lappert, M. F., and Sanger, A. R., *Chem. Commun.* 1064 (1969).
179. Alyea, E. C., Basi, J. S., Bradley, D. C., and Chisholm, M. H., *Chem. Commun.* 495 (1968).
180. Bürger, H., and Wannagat, U., *Monatsch. Chem.* **95**, 1099 (1964).
181. Bürger, H., and Wannagat, U., *Monatsch. Chem.* **94**, 1007 (1963).
182. Bürger, H., Sawodny, W., and Wannagat, U., *J. Organomet. Chem.* **3**, 113 (1965).
183. Bradley, D. C., and Copperthwaite, R. G., *Chem. Commun.* 764 (1971); Alyea, E. C., Bradley, D. C., and Copperthwaite, R. G., *J. Chem. Soc. Dalton Trans.* 1580 (1972).
184. Bradley, D. C., Ghotra, J. S., and Hart, F. A., *Chem. Commun.* 349 (1972).
185. Bradley, D. C., and Newing, C. W., unpublished work; Bradley, D. C., Hursthouse, M. B., Newing, C. W., and Welch, A. J., *Chem. Commun.* 567 (1972).

186. Bradley, D. C., and Copperthwaite, R. G., unpublished work.
187. Bradley, D. C., Hursthouse, M. B., Smallwood, R. J., and Welch, A. J., *Chem. Commun.* 872 (1972).
188. Ashby, E. C., and Kovar, R., *J. Organomet. Chem.* **22** C34 (1970).
189. Paul, R. C., and Sreenathan, B. R., *Indian J. Chem.* **4**, 382 (1966).
190. Bradley, D. C., Hursthouse, M. B., and Rendall, I. F., *Chem. Commun.* 672 (1969).
191. Bradley, D. C., Hursthouse, M. B., and Rendall, I. F., *Chem. Commun.* 368 (1970).
192. Chandra, G., and Lappert, M. F., *J. Chem. Soc. A* 1940 (1968).
193. Jenkins, A. D., Lappert, M. F., and Srivastava, R. C., *J. Organomet. Chem.* **23**, 165 (1970).
194. Cardin, D. J., Keppie, S. A., and Lappert, M. F., *J. Chem. Soc. A* 2594 (1970).
195. Lappert, M. F., and Sanger, A. R., *J. Chem. Soc. A* 874 (1971).
196. Lappert, M. F., and Sanger, A. R., *J. Chem. Soc. A* 1314 (1971).
197. Streitwieser, A., Brauman, J. I., Hammons, J. H., and Pudjaatmaka, A. H., *J. Amer. Chem. Soc.* **87**, 384 (1965); Streitwieser, A., Hammons, J. H., Ciuffarin, E., and Brauman, J. I., *J. Amer. Chem. Soc.* **89**, 59 (1967); Streitwieser, A., Ciuffarin, E., and Hammons, J. H., *ibid.* **89**, 63 (1967).
198. George, T. A., and Lappert, M. F., *Chem. Commun.* 463 (1966); see also Chandra, G., George, T. A., and Lappert, M. F., *J. Chem. Soc. C* 2569 (1969).
199. Bradley, D. C., and Gitlitz, M. H., *Chem. Commun.* 289 (1965); *J. Chem. Soc. A* 1152 (1969).
200. Chandra, G., Jenkins, A. D., Lappert, M. F., and Srivastava, R. C., *J. Chem. Soc. A* 2550 (1970).
201. Bradley, D. C., and Ganorkar, M. C., *Chem. Ind. (London)* 1521 (1968).
202. Jenkins, A. D., Lappert, M. F., and Srivastava, R. C., *Polymer. Lett.* **6**, 865 (1968).
203. Bradley, D. C., Newing, C. W., Chien, J. C. W., and Kruse, W., *Chem. Commun.* 1177 (1970).
204. Bradley, D. C., and Newing, C. W., unpublished work.
205. Bradley, D. C., and Newing, C. W., *Chem. Commun.* 219 (1970).
206. Bradley, D. C., Charalambous, J., and Jain, S., *Chem. Ind. (London)* 1730 (1965).
207. Bradley, D. C., and Chisholm, M. H., unpublished work.
208. Bradley, D. C., and Chisholm, M. H., *J. Chem. Soc. A* 1511 (1971).
209. Bradley, D. C., and Kinsella, E., unpublished work.
210. Bürger, H., Stammreich, H., and Sans, Th. Teixeira, *Monatsch. Chem.* **97**, 1276 (1966).
211. Bürger, H., and Sawodny, W., *Spectrochim. Acta Part A* **23**, 2841 (1967).
212. Bradley, D. C., and Gitlitz, M. H., *Nature (London)* **218**, 353 (1968); *J. Chem. Soc. A* 980 (1969).
213. Bürger, H., Forker, C., and Goubeau, J., *Monatsch. Chem.* **96**, 597 (1965).
214. Bradley, D. C., and Fisher, K. J., *J. Amer. Chem. Soc.* **93**, 2058 (1971).
215. Fetter, N. R., and Peters, F. M., *Can. J. Chem.* **43**, 1884 (1965).
216. Bell, N. A., Coates, G. E., and Emsley, J. W., *J. Chem. Soc. A* 49 (1966).
217. Coates, G. E., and Fishwick, A. H., *J. Chem. Soc. A* 1199 (1967).
218. Moedritzer, K., *Inorg. Chem.* **3**, 609 (1964).
219. Bradley, D. C., and Gitlitz, M. H., *J. Chem. Soc. A* 980 (1969).
220. Holloway, C. E., Mabbs, F. E., and Smail, W. R., *J. Chem. Soc. A* 2980 (1968).

221. Bradley, D. C., Hursthouse, M. B., and Newing, C. W., *Chem. Commun.* 411 (1971).
222. Bradley, D. C., Copperthwaite, R. G., Cotton, S. A., Gibson, J. F., and Sales, K. D., Paper presented at Autumn Meeting of the Chemical Society, York, September, 1971.
223. Alyea, E. C., Bradley, D. C., Copperthwaite, R. G., Sales, K. D., Fitzsimmons, B. W., and Johnson, C. E., *Chem. Commun.* 1715 (1970).
224. Alyea, E. C., Bradley, D. C., Copperthwaite, R. G., and Sales, K. D., *J. Chem. Soc., Dalton Trans.* In press.
225. Camerman, A., *Can. J. Chem.* **48**, 179 (1970).
226. Heath, C., and Hursthouse, M. B., *Chem. Commun.* 143 (1971).
227. Vilkov, L. V., Tarasenko, N. A., and Prokof'ev, A. K., *Zh. Strukt. Khim.* **11**, 129 (1970).
228. Atwood, J. L., and Stucky, G. D., *Chem. Commun.* 1169 (1967); *J. Amer. Chem. Soc.* **91**, 4426 (1969).
229. Gosling, K., McLaughlin, G. M., Sim, G. A., and Smith, J. D., *Chem. Commun.* 1617 (1970).
230. Atwood, J. L., and Stucky, G. D., *J. Amer. Chem. Soc.* **92**, 285 (1970).
231. Robiette, A. G., Sheldrick, G. M., Sheldrick, W. S., Beagley, B., Cruickshank, D. W. J., Monaghan, J. J., Aylett, B. J., and Ellis, I. A., *Chem. Commun.* 909 (1968).
232. Clark, A. H., and Haaland, A., *Chem. Commun.* 912 (1969).
233. Bradley, D. C., Hursthouse, M. B., and Rodesiler, P. F., *Chem. Commun.* 14 (1969).
234. Heath, C. E., and Hursthouse, M. B., unpublished work.
235. Sheldrick, G. M., and Sheldrick, W. S., *J. Chem. Soc. A* 2279 (1969).

FLUOROALICYCLIC DERIVATIVES OF METALS AND METALLOIDS

W. R. Cullen

Chemistry Department, University of British Columbia, Vancouver, British Columbia, Canada

I. Introduction 323
II. Preparative Methods 324
   A. Carbene and Carbenoid Additions 324
   B. Cycloaddition Reactions 327
   C. Hydride Additions 328
   D. Metal–Fluoride Additions 331
   E. Metal–Alkyl and Metal–Aryl Additions 332
   F. Reactions with Compounds Containing Metal–Metal Bonds . . 333
   G. Oxidative Addition Reactions 335
   H. Direct Reaction with a Metal 337
   I. Exchange Reactions 338
   J. Preparation by Modification of Existing Fluoroalicyclic Derivatives 340
III. Other Chemical Properties 342
IV. Physical Properties 345
V. Coordination Complexes 346
   A. Monoligate Complexes 347
   B. Biligate Monometallic (Chelate) Complexes 351
   C. Biligate Bimetallic Complexes 356
   D. Triligate Bimetallic Complexes 363
   E. Decomposition Products of Coordination Complexes . . . 366
   References 368

I. Introduction

This article will be concerned with the preparation and properties of alicyclic fluorocarbon (fluoroalicyclic) derivatives of metals and metalloids. The discussion will be restricted to compounds containing the fluoroalicyclic ring sigma-bonded to elements other than carbon, nitrogen, oxygen, and the halogens, although inevitably the chemistry of some compounds from this latter group will need to be described. Complexes in which the metal or metalloid is part of a ring such as (**1**) (*160*) are excluded as are π complexes of fluoroalicyclic olefins which some authors regard as metallocyclopropanes as in (**2**) (*131*).*

* The mode of bonding of olefins lies between the extremes represented by

$$\begin{matrix} C \\ \| \\ C \end{matrix} \to M \quad \text{and} \quad \begin{matrix} C \\ | \\ C \end{matrix} \!\!\!\!\bigtriangledown\!\!\!\! M$$

(*130*). The latter is preferred by some workers (e.g., *24*, *131*) to describe the fluoro-olefin case.

```
    (C₆H₅)₃  CF₃                        F  F
       Sb   /  \CF₃                      |  |
   OC\  |  C=C                (C₆H₅)₃P\    /C—C—F
      Rh      /                        Pt      |  |
   Cl/  |  C=C\                (C₆H₅)₃P/   \C—C—F
       Sb       \CF₃                      |  |
    (C₆H₅)₃  CF₃                        F  F
        (1)                                (2)
```

The general class of sigma-bonded fluorocarbon derivatives of metals and metalloids has been known for some time (5, 114). Most of the early work was concerned with the chemistry of trifluoroiodomethane and the preparation of trifluoromethyl derivatives such as $(CF_3)_2Hg$ (115). Since then fluoroaliphatic derivatives of many transition metals have been prepared (28, 208) and some interest has been shown in studying vinylic and acetylenic derivatives (27, 28, 43, 63, 89, 203).

Some of the material presented in this review has been described previously (62). The present article is an updating of this work and, in particular, gives an account of the coordination complexes derived from fluoroalicyclic-bridged ditertiary phosphines and arsines.

II. Preparative Methods

A. Carbene and Carbenoid Additions

The addition of a carbene (R_2C:) to a carbon–carbon double bond is a well-known source of cyclopropanes (143, 153), and addition to a triple bond is a less extensively investigated route to cyclopropenes (20, 98, 143, 153, 195). Trimethyl(trifluoromethyl)tin acts as a source of difluorocarbene when heated in the gas phase at ~150°C (45). The carbene so produced adds stereospecifically to cis- and trans-butene-2, so it is probably in the singlet state (87, 153). It also adds to vinylic and acetylenic derivatives of the Group IV and V elements to give the corresponding cyclopropene or cyclopropane derivative (81, 87). Again these additions are probably stereospecific.

$$R_nMC{\equiv}CR_f + (CH_3)_3SnCF_3 \xrightarrow{150°} \underset{F_2}{\overset{R_nM\quad R_f}{\triangledown}} + (CH_3)_3SnF \qquad (1)$$

R = mainly CH_3, M = As, Si, Ge, Sn
R_f = mainly CF_3, also C_2F_5, and $CF(CF_3)_2$

$trans$-$(CH_3)_2AsC(CF_3)$=$C(CF_3)H$ + $(CH_3)_3SnCF_3$ $\xrightarrow{150°}$

$$\underset{F_2}{\underset{|}{CF_3}}\overset{(CH_3)_2As}{\underset{}{\bigtriangledown}}\overset{CF_3}{\underset{}{H}} + (CH_3)_3SnF \quad (2)$$

The yields in the thermal reactions [reactions (1) and (2)] are quite high, e.g., 84% in the case of $(CH_3)_3GeC=C(CF_3)CF_2$ but, not surprisingly, they are poor when addition of the carbene to a bisacetylide

$(CH_3)_2Ge(C\equiv CCF_3)_2 + 2(CH_3)_3SnCF_3$ $\xrightarrow{150°}$

$$CF_3 \bigtriangledown_{F_2} \overset{(CH_3)_2}{\underset{}{Ge}} \bigtriangledown_{F_2} CF_3 + 2(CH_3)_3SnF \quad (3)$$

is attempted. Further reaction of the cyclopropene products of reaction (1) with the carbene source does not seem to occur, even though Mahler (*161*) found that a bicyclobutane results when hexafluorobutyne-2 is heated with tris(trifluoromethyl)phosphorus difluoride. The phosphorus fluoride is another carbene source.

$$CF_3C\equiv CCF_3 \xrightarrow{CF_2} \overset{CF_3 \quad CF_3}{\underset{F_2}{\bigtriangledown}} \xrightarrow{CF_2} CF_3-\overset{F_2}{\underset{F_2}{\diamondsuit}}-CF_3 \quad (4)$$

Other cyclopropyl derivatives, $(C_2H_5)_3MCHCH_2CF_2$, have been obtained by reacting vinyl derivatives with a CF_2 source. In this investigation the carbene was generated by treating the trifluoromethyltin compound with sodium iodide (*198*).

$$(CH_3)_3SnCF_3 + I^- \xrightarrow[80°]{DME} (CH_3)_3SnI + CF_3^- \quad (5)$$

$$CF_3^- \longrightarrow CF_2 + I^-$$

Seyferth and his co-workers have also extensively investigated mercurials as carbene sources (*194*), and have prepared dichloro- and chlorofluorocyclopropanes by either heating the mercurial with the Group IV vinyl derivative or by using an iodide displacement method as described in reaction (5) (*196, 197*).

$$R_3MCH{=}CH_2 + C_6H_5HgCCl_2Br \longrightarrow \underset{Cl_2}{\overset{R_3M\quad H}{\triangle}}_{H\quad H} + C_6H_5HgBr \qquad (6)$$

$$R = C_2H_5$$
$$M = Si, Ge, Sn$$

$$(CH_3)_3SiCH{=}CH_2 + C_6H_5HgCCl_2F \longrightarrow \underset{FCl}{\overset{(CH_3)_3Si\quad H}{\triangle}}_{H\quad H} + C_6H_5HgCl \qquad (7)$$

The reaction of nonfluorinated diazomethanes with unsaturated metallic species has received scant attention (*94*). In the fluorocarbon field it has been found that bis(trifluoromethyl)diazomethane, $(CF_3)_2CN_2$ (*129*), reacts with alkynyl derivatives to afford cyclopropenes (*79, 88*).

$$(CF_3)_2CN_2 + (CH_3)_3GeC{\equiv}CCF_3 \longrightarrow \underset{(CF_3)_2}{\overset{(CH_3)_3Ge\quad CF_3}{\triangle}} + N_2 \qquad (8)$$

$$(CH_3)_3SiC{\equiv}CSi(CH_3)_3 + (CF_3)_2CN_2 \longrightarrow \underset{(CF_3)_2}{\overset{(CH_3)_3Si\quad Si(CH_3)_3}{\triangle}} + N_2 \qquad (9)$$

The product of reaction (8) can also be prepared by reacting the alkyne with bis(trifluoromethyl)diazirine, $(CF_3)_2\overset{\frown}{C-N{=}N}$ (*88*). The yield of the bistrimethylsilyl derivative, the only known cyclopropene with two metalloid substituents, is low. It is possible that the free carbene $[(CF_3)_2C:]$ is involved in these reactions. An alternative path would allow 1,3-addition to form an isopyrazole which would then eliminate nitrogen, either thermally or photo-chemically, to give the cyclopropene. This is found in the thermal reaction of the diazomethane with hexafluorobutyne-2 (*88*).

$$CF_3C{\equiv}CCF_3 + (CF_3)_2\bar{C}{-}N{=}\overset{+}{N} \longrightarrow \underset{(CF_3)_2}{\overset{CF_3\quad CF_3}{\underset{N{\diagup}N}{\bigtriangleup}}} + \underset{(CF_3)_2}{\overset{CF_3\quad CF_3}{\triangle}} \qquad (10)$$

$$\underset{-N_2}{\xrightarrow{300°}}$$

B. Cycloaddition Reactions

1. The [2 + 2] Reaction

Although the concerted [2 + 2] cycloaddition reaction of olefins to give a cyclobutane is disallowed (218), the reaction can still take place via a diradical intermediate (7, 181, 187, 206). So far only silicon derivatives have been prepared by this procedure (21, 144, 179, 207).

$$\begin{array}{c}>C{=}C< \\ + \\ >C{=}C<\end{array} \longrightarrow \begin{array}{c}>C\!\!-\!\!-\!\!C< \\ | \quad\;\; | \\ >C\!\cdot \quad \cdot C<\end{array} \longrightarrow \begin{array}{c}>C\!\!-\!\!-\!\!C< \\ | \quad\;\; | \\ >C\!\!-\!\!-\!\!C<\end{array} \qquad (11)$$

$$R_3SiCX{=}CH_2 + CF_2{=}CYZ \longrightarrow \begin{array}{c} H_2 \quad\;\; F_2 \\ \boxed{} \\ R_3Si \quad\; YZ \\ X \end{array} \qquad (12)$$

X = H (mainly), Cl
Y = F (mainly), Cl
Z = Cl (mainly), F

Reaction takes place at about 200°C and a free radical inhibitor is often added to improve yields which, on the whole, are quite good. Cyclization occurs in the sense shown in reaction (12). When geometric isomers are possible, for example, if X = Z = Cl, these are produced in approximately equal amounts. However, one isomer predominates when R = X = Z = Cl and Y = F (80).* It seems that the silyl groups activate the double bond with respect to addition (179), possibly by stabilizing the radical intermediate.

The related [2 + 2] addition of an acetylene to an olefin to give a cyclobutene is known for fluorocarbon systems (e.g., 161). However, attempts to react fluoroalkynyl derivatives of metals such as $(C_2H_5)_3GeC{\equiv}CCF_3$ with fluoroolefins have not been successful (81).

2. The [2 + 4] Reaction

This cycloaddition reaction, the Diels–Alder reaction, is a thermal symmetry-allowed addition (218). It is a commonly studied reaction of fluorocarbons (e.g., 3, 193) and has been used to prepare hydrocarbon derivatives of metals and metalloids (e.g., 32, 128); however, only a few fluoroalicyclic derivatives of germanium and silicon have been obtained by this route (145, 184).

* The cyclization of $CH_2{=}CHCl$ with $CF_2{=}CFH$ is apparently stereospecific (181).

[Reaction (13): diphenyl-substituted silacyclopentadiene + CF₃C≡CCF₃ → bicyclic adduct] (13)

[Reaction (14): Cl₃SiCH=CH₂ + hexachlorofluorocyclopentadiene → bicyclic adduct] (14)

The germanium analog of the product of reaction (13) can probably be prepared by the same method. Reaction (14) is known for a variety of vinylsilanes and best yields are obtained if an Si–H bond is present.

C. Hydride Additions

1. Additions without Elimination

Addition of hydrides to fluoroolefins has been a fruitful source of fluorocarbon derivatives of many elements (5, 28, 114, 208). In the case of fluoroalicyclic olefins the product is usually that which would be obtained from an "addition–elimination" sequence (q.v.). However, some of the adducts are stable. Thus, trimethylsilane, -germane, and -stannane add to fluorocyclobutenes to yield cyclobutanes (82, 136).

$$(CH_3)_3MH + \text{[fluorocyclobutene]} \longrightarrow \text{[adduct]} \quad (15)$$

M = Ge; X = F, Cl
M = Si, Sn; X = F

Some thiols react similarly (65, 186), as do pentacarbonylmanganese and pentacarbonylrhenium hydrides (16, 50).

$$\text{[hexafluorocyclobutene]} + (CO)_5MH \longrightarrow \text{[adduct]} \quad (16)$$

M = Mn, Re

The nuclear magnetic resonance spectra of the adducts of reaction (15) suggest that cis hydride addition occurs (82). However, pentacarbonylmanganese hydride apparently adds trans to hexafluorocyclopentadiene (121).

$(CO)_5MnH$ + [cyclopentene-F6] ⟶ [1,2-adduct with (CO)5Mn, F2, H, F, F2, F] + [1,4-adduct with (CO)5Mn, F2, FH, F, F] (17)

The 1,4-adduct is the major product of this reaction at −78°, but it is unstable and rearranges to the 1,2-adduct and cyclopent-1-enylmanganese derivatives on warming to 20°.

[(CO)5Mn-cyclopentenyl FH intermediate] ⟶

[1,2-adduct] + [(CO)5Mn-HF-F2-F2-F] + [(CO)5Mn-F2-HF-F2-F] (18)

In the overall reaction the yield of pentacarbonyl-σ-(5H-hexafluoro-cyclopent-1-enyl)manganese is much lower (∼10%) than that of the other two products of reaction (18).

2. Additions with Elimination

This route has been a major source of fluoroalicyclic derivatives as summarized in the following equations (*39, 40, 44, 56, 65–67, 69, 75, 82, 124, 136, 165, 168, 185, 186, 200, 201*).

$$(CH_3)_2AsH + XC\underset{(CF_2)_n}{=\!=\!=}CY \longrightarrow (CH_3)_2AsC\underset{(CF_2)_n}{=\!=\!=}CY + HX \quad (19)$$

Y = H, Cl, F, C_2H_5, As $(CH_3)_2$; X = Cl, F

$$2(C_6H_5)_2PH + FC\underset{(CF_2)_n}{=\!=\!=}CF \longrightarrow (C_6H_5)_2PC\underset{(CF_2)_n}{=\!=\!=}CP(C_6H_5)_2 + 2HF \quad (20)$$

$$R_3MH + ClC\underset{(CF_2)_n}{=\!=\!=}CCl \longrightarrow (CH_3)_3MC\underset{(CF_2)_n}{=\!=\!=}CCl + HCl \quad (21)$$

M = Si, Ge

CH_3SH + [cyclopentene F2-F2-F2-H-Cl] ⟶ [CH3S, H, Cl cyclopentane F2-F2-F2-F] + HF (22)

$$\text{RSH} + \underset{F\quad F}{\overset{F_2\quad F_2}{\square}} \longrightarrow \underset{RS\quad SR}{\overset{F_2\quad F_2}{\square}} + 2\text{HF} \quad (23)$$

$$trans\text{-}[(C_2H_5)_3P]_2Pt(Cl)H + \underset{F\quad F}{\overset{F_2\quad F_2}{\square}} \longrightarrow$$

$$\underset{trans\text{-}[(C_2H_5)_3P]_2Pt(Cl)\quad F}{\overset{F_2\quad F_2}{\square}} + \text{HF} \quad (24)$$

In the reactions of thiols, silanes, and germanes [reactions (21)–(23)] both 1,1-adducts and HX eliminated products are obtained suggesting that the former are precursors of the latter. In some cases elimination of hydrogen fluoride is preferred over hydrogen chloride so vinylic substitution need not be the end result as is shown in reaction (22). It should also be pointed out that elimination of metal halides can also occur and in the case of trimethylgermane the resulting reduced cyclobutene apparently reacts further (82).

$$(CH_3)_3GeH + \underset{Cl\quad Cl}{\overset{F_2\quad F_2}{\square}} \longrightarrow (CH_3)_3Ge\underset{Cl\quad Cl}{\overset{F_2\quad\quad F_2}{\square}}H \quad (25)$$

$$\swarrow \qquad \searrow$$

$$\underset{(CH_3)_3Ge\quad Cl}{\overset{F_2\quad F_2}{\square}} + \text{HCl} \qquad \underset{H\quad Cl}{\overset{F_2\quad F_2}{\square}} + (CH_3)_3GeCl$$

$$\underset{(CH_3)_3Ge\quad H}{\overset{F_2\quad F_2}{\square}} \longleftarrow \underset{(CH_3)_3Ge\quad HCl}{\overset{F_2\quad\quad F_2}{\square}} \overset{(CH_3)_3GeH}{\longleftarrow}$$

All the germanium-containing products were isolated in this investigation. The rate of the arsine reactions [reaction (19)] depends on the electronegativity of X and Y, but no intermediates were detected (66, 69). The product reacts further with dimethylarsine to give a ditertiary arsine when X = Y = F and n = 2, 3, or 4 (40). Diphenylphosphine displaces both vinylic fluorine atoms [reaction (20)] to give the ditertiary phosphine (*67, 75, 200*), but monosubstitution is found when vinylic chlorine atoms are present and if the reaction is carried out in DMF (*200, 201*). Dialkylphosphines have a greater tendency to give monosubstituted products (*65*). Mixed derivatives are available by reacting the arsine products of reaction (19) (Y = F) with diphenylphosphine (*40, 56*). The ease of these substitution reactions decreases as the ring size increases, a result which may be related to ring strain (*200, 202*).

When the ring has a methoxy substituent, reaction with a thiol or a secondary amine gives a cyclobutenone (*67, 102*).

$$\text{CH}_3\text{SH} + \underset{\underset{\text{Cl}}{}\underset{\text{OCH}_3}{}}{\overset{\overset{\text{F}_2}{}\overset{\text{F}_2}{}}{\square}} \longrightarrow \underset{\underset{\text{CH}_3\text{S}}{}\underset{\text{Cl}}{}}{\overset{\overset{\text{F}_2}{}\overset{}{}\text{O}}{\square}} \quad (26)$$

An addition–elimination sequence is believed to take place when fluorocyclobutenes containing a vinylic iodine atom are reduced by lithium aluminum hydride (*30*).

$$\underset{\text{X}\text{I}}{\overset{\text{F}_2\text{F}_2}{\square}} \longrightarrow \left[\underset{\text{X}\text{AlH}_2\text{I}}{\overset{\text{F}_2\text{F}_2}{\square}}\right]^{-} + \text{H}_2 \quad (27)$$

$$\downarrow \underset{\text{X}\text{I}}{\overset{\text{F}_2\text{F}_2}{\square}}$$

$$\left[\underset{\text{X}\text{AlHI}_2}{\overset{\text{F}_2\text{F}_2}{\square}}\right]^{-} + \underset{\text{X}\text{H}}{\overset{\text{F}_2\text{F}_2}{\square}}$$

The first intermediate is formed by attack of the eliminated hydrogen iodide on the initially formed $R_f\text{AlH}_3^-$.

Reactions such as reaction (24) (*44*) involving transition metal hydrides have been little investigated. It seems that manganese pentacarbonyl hydride does not react with perfluorocyclopentene or the products of reaction (18) (*121*).

D. METAL–FLUORIDE ADDITIONS

The addition of metal fluorides to fluoroolefins and acetylenes is a useful source of fluorocarbon derivatives (*103, 170*), but the reaction has had limited application to cyclic fluoroolefins (*104*).

E. Metal–Alkyl and Metal–Aryl Additions

In the general fluorocarbon field only compounds with Mn–R and Pt–R bonds (R = alkyl or aryl group) have been extensively investigated with respect to their addition reactions with olefins and acetylenes (*43*, *167*, *217*). Two additions to hexafluoro(Dewar)benzene have been described (*16*).

R = CH$_3$, C$_6$H$_5$

Lithium derivatives have been postulated as the intermediates in the reaction of lithium alkyls with fluoroolefins or diradicals (e.g., *35*, *47*, *199*). (See also reaction (*57*) in Section III.)

Similar additions followed by elimination of metal halide are believed to be involved in the reduction of cyclic fluoroolefins with lithium aluminum hydride (e.g., *47*) and alkali metal alkoxides (e.g., *46*). Thus, these reactions can be regarded as involving unstable aluminum and alkali metal fluoroalicyclic derivatives.

F. Reactions with Compounds Containing Metal–Metal Bonds

In this section the phrase "metal–metal bond" is used to include such diverse reagents as $(CH_3)_2As$–$As(CH_3)_2$, C_6H_5SCu, and $(\pi\text{-}C_5H_5)Fe(CO)_2Na$.

Tetramethyldiarsine readily adds to fluoroolefins and fluorocarbon-acetylenes—a reaction that is a useful source of chelating ditertiary arsines (61, 77, 86). However, when cyclic fluoroolefins are used the product is that which would be expected from an addition–elimination sequence (69).

$$(CH_3)_2As\text{–}As(CH_3)_2 \xrightarrow{CF_2=CFCF_3} (CH_3)_2AsCF_2CF(CF_3)As(CH_3)_2$$

$(CH_3)_2As\text{–}As(CH_3)_2 +$ [cyclic $C_4F_6X_2$] $\longrightarrow (CH_3)_2As$–[cyclic C_4F_6X] $+ XAs(CH_3)_2$ (31)

X = Cl, F

Hexamethylditin does not react with fluorocyclobutenes (82), but the Sn–Mn bond in trimethyltin(pentacarbonyl)manganese is more easily cleaved (15). Trimethyltin fluoride is eliminated in a reaction analogous to reaction (31).

$(CH_3)_3Sn\text{–}Mn(CO)_5 +$ [cyclic C_4F_6] $\longrightarrow (CH_3)_3SnF + (CO)_5Mn$–[cyclic C_4F_5] (32)

A saturated product is obtained from the reaction of the same tin–manganese compound with hexafluoro(Dewar)benzene (16).

$(CH_3)_3Sn\text{–}Mn(CO)_3 +$ [hexafluoro(Dewar)benzene] $\longrightarrow (CO)_5Mn$–[HF-substituted cyclic structure] (33)

When perfluorocyclobutene and bis(trimethylsilyl)mercury are irradiated, the products are trimethylsilyl fluoride, mercury, and cyclobutenyl derivatives of silicon (*122, 123*).

$$(CH_3)_3Si-HgSi(CH_3)_3 + \underset{F\quad F}{\overset{F_2\quad F_2}{\square}} \longrightarrow \left[\underset{F}{\overset{F_2\quad F_2}{\square}}\; (CH_3)_3Si\text{—}\square\text{—}(F)HgSi(CH_3)_3 \right]$$

(34)

$$\underset{(CH_3)_3Si\quad Si(CH_3)_3}{\overset{F_2\quad F_2}{\square}} \xleftarrow{((CH_3)_3Si)_2Hg} \underset{F\quad Si(CH_3)_3}{\overset{F_2\quad F_2}{\square}}$$

These reactions are believed to proceed through adducts like the one shown. Similar intermediates have been detected in additions to acylic fluoroolefins.

Reactions involving metal carbonyl anions have been the most fruitful means of obtaining fluoroalicyclic derivatives of the transition metals. The tetrahydrofuran (THF) solutions of the anion are treated with the fluoroolefin and monosubstitution takes place in good yield (*6, 16, 26, 50, 148*).

$$[M]_2 + Na/Hg \xrightarrow{THF} [M]Na$$

$$[M]Na + XC\underset{\underset{(CF_2)_n}{\frown}}{=\!=}CX \longrightarrow [M]C\underset{\underset{(CF_2)_n}{\frown}}{=\!=}CX + NaX \quad (35)$$

$$[M]Na + \underset{F\;F\;F}{\overset{F\;F\;F}{\square\!\square}} \longrightarrow \underset{[M]\;F\;F}{\overset{F\;F\;F}{\square\!\square}} + NaF \quad (36)$$

M = (π-C$_5$H$_5$)Fe(CO)$_2$, (CO)$_5$Mn, (CO)$_5$Re, (C$_6$H$_5$)$_3$P(CO)$_4$Mn

Disubstitution does not occur. In one investigation (*26*) the saturated

1,2-tetrachlorocyclobutane was used; the product was the same as from reaction (35), but the yields were much lower.

Although the sodium derivative of di-n-butylphosphite failed to react with 1-chloroheptafluorocyclopentene at 20°C (*124*) an arsino-Grignard reagent gave the expected product when treated with perfluorocyclobutene. This is only of academic interest since the hydride precursor to the Grignard reagent also reacts to give the same product [cf. reaction (19)] (*69*).

$$(CH_3)_2AsH + CH_3MgI \longrightarrow (CH_3)_2AsMgI + CH_4$$

$$(CH_3)_2AsMgI + \underset{F\ \ \ \ F}{\overset{F_2\ \ \ \ F_2}{\square}} \longrightarrow \underset{(CH_3)_2As\ \ \ \ F}{\overset{F_2\ \ \ \ F_2}{\square}} + MgIF \quad (37)$$

Copper derivatives RSCu (R = C_6H_5, $C_6H_5CH_2$) have been used to prepare disubstituted fluoroalicyclic derivatives from 1,2-dichlorocycloolefins (*120*).

Clark and co-workers (*15*) have suggested that the reaction of hexafluorobutyne-2 with trimethyltin(pentacarbonyl)manganese gives a product which can be regarded as the adduct of the Sn–Mn compound with the dimer of the butyne [a cyclobutadiene, cf. reaction (49)].

$$(CH_3)_3Sn-Mn(CO)_5 + CF_3C\equiv CCF_3 \longrightarrow \underset{(CH_3)_3Sn\ \ \ \ (CF_3)Mn(CO)_5}{\overset{CF_3\ \ \ \ CF_3}{\underset{CF_3}{\square}}} \quad (38)$$

However, it is possible that an adduct with the butyne is first formed which then undergoes a [2 + 2] cycloaddition with more butyne.

G. Oxidative Addition Reactions

In the transition metal field the oxidative addition of, say, alkyl halides to derivatives of metals with a d^8 configuration has been exhaustively studied (*49*). However, there is only one report (*173*) of a reaction of this type leading to a fluoroalicyclic derivative. Fluoroolefins usually act as two electron donors or dimerize to give metallocyclopentane rings as is also shown in reaction (39).

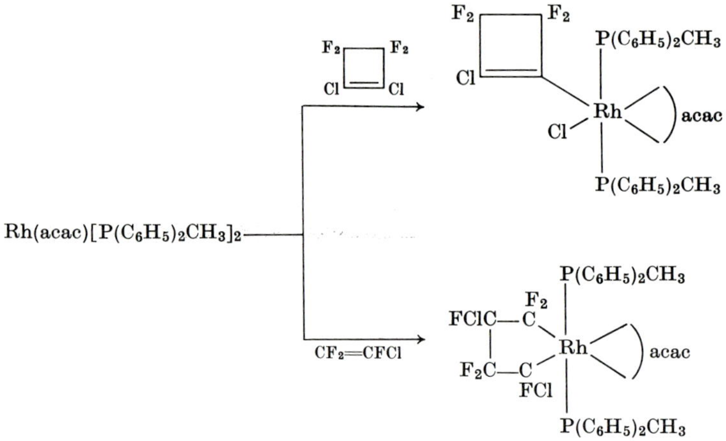

(39)

Triethylphosphite interacts with hexafluorocyclobutene to give a five-coordinate phosphorus compound, which on heating undergoes an Arbusov reaction to the phosphonate (155).

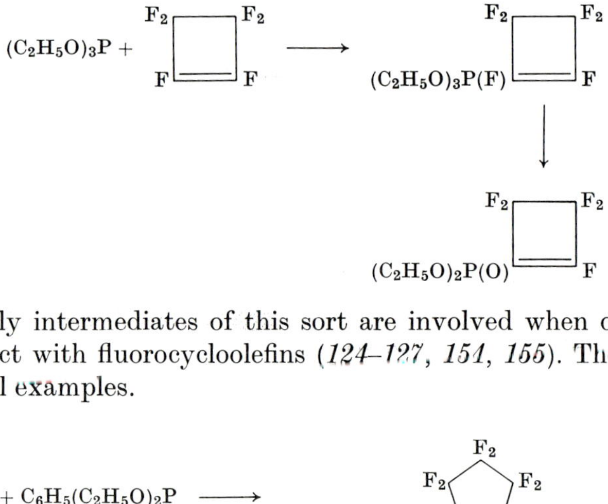

(40)

Presumably intermediates of this sort are involved when other phosphites react with fluorocycloolefins (124–127, 154, 155). The following are typical examples.

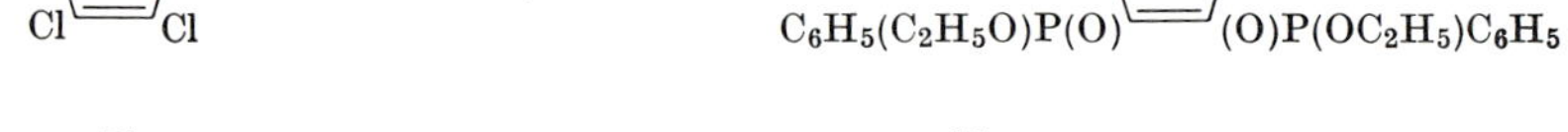

(41)

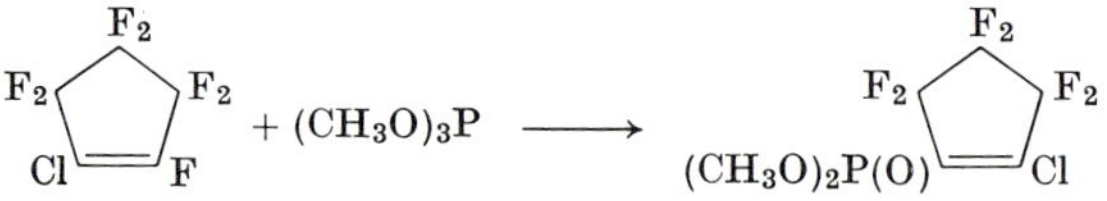

(42)

Frank (124–127) found that disubstitution occurs when 1,2-dichloroolefins are involved, but that monophosphonates can be obtained from

1-chloro-2-fluoroolefins. The vinylic chlorine atom of the product of reaction (42) can be replaced by further treatment with phosphite.

A 1:1 adduct which can be formulated as a nonclassical dipolar species is formed when triphenylphosphine and perfluorocyclobutene interact (*202*).

$$(C_6H_5)_3P + \underset{F\ \ \ \ F}{\overset{F_2\ \ \ \ F_2}{\square}} \longrightarrow \underset{\underset{(C_6H_5)_3}{P^{\oplus}}}{\overset{F_2\ \ \ \ \ \ \ F_2}{\underset{F\ \ \ \ \ \ \ F}{\boxed{\ominus}}}} \quad (43)$$

The intermediate is hydrolyzed to a phosphobetaine. Other alicyclic fluoroolefins react with tertiary phosphines in acetic acid/water solution to give phosphobetaines directly, presumably via an initially formed 1,1-adduct.

$$C_4H_9(C_6H_5)_2P + \underset{Cl\ \ \ Cl}{\overset{F_2}{\underset{F_2\ \ \ F_2}{\triangle}}} \longrightarrow C_4H_9(C_6H_5)_2\overset{\oplus}{P}\underset{O}{\overset{O\ \ F_2}{\underset{\ominus}{\pentagon}}}F_2 \quad (44)$$

H. Direct Reaction with a Metal

Apart from vinylic Grignard reagents, magnesium derivatives are not well known in acyclic fluorocarbon chemistry (*5, 114*). However, a number of fluoroalicyclic Grignard reagents derived from bicyclic systems are accessible by direct reaction of magnesium with the appropriate halogen derivative (*33–36*).

$$\underset{X = Cl, Br, I}{\text{(bicyclic F structure with X)}} \xrightarrow{Mg} \text{(bicyclic F structure with MgX)} \quad (45)$$

Saturated analogs of the products of reaction (45) are known, as are derivatives with hydrogen at the other bridgehead position. A saturated bicyclic derivative with two MgBr groups, one at either bridgehead position, has been reported.

Although the reaction of trifluoroiodomethane with mercury and other elements was one of the first to be studied in this field of chemistry (*5, 114*), the corresponding fluoroalicyclic derivatives have been little investigated in this respect. Park and his co-workers (*178*) have obtained uncharacterized products from the reaction of $\overline{IC{=}CClCF_2CF_2}$ with mercury. Bis(undecafluorobicyclo[2,2,1]heptan-1-yl)mercury is obtained when 1-iodoundecafluorobicyclo[2,2,1]heptane and mercury react under the influence of ultraviolet light (35).

$$\text{(bicyclic-I)} \xrightarrow[h\nu]{Hg} \text{(bicyclic)}_2\text{Hg} \tag{46}$$

I. Exchange Reactions

These have been used to prepare magnesium and lithium fluoroalicyclic derivatives. Halogen exchange affords cyclobutenyl and cyclopentenyl derivatives (179, 180, 182, 203). Electropositive halogens

$$\text{Cl—C}_4\text{F}_4\text{—Br} + C_2H_5MgBr \longrightarrow \text{Cl—C}_4\text{F}_4\text{—MgBr} + C_2H_5Br \tag{47}$$

normally need to be present for these reactions, otherwise substitution occurs.* However, a recent report (177) describes how even 1,2-dichlorofluorocycloolefins are lithiated by n-butyllithium.

Hydrogen exchange with methyllithium affords lithium derivatives of cyclobutene, cyclopentene, and cyclohexene (37). This method also yields lithium derivatives of the bicyclic systems studied by Tatlow and co-workers (33–35, 199).

$$\text{(bicyclic-Y,H)} \xrightarrow{CH_3Li} \text{(bicyclic-Y,Li)} \tag{48}$$

Y = F, H, CF$_3$

* Presumably substitution takes place via an addition–elimination process which would involve the formation of magnesium fluoroalicyclic intermediates (see Section II, E).

An unstable lithium derivative is probably involved as an intermediate in the synthesis of the unstable tetrakis(trifluoromethyl)-cyclobutadiene (*169*). The cyclobutadiene was not isolated, but its existence was inferred from the products actually obtained. For example,

(49)

J. Preparation by Modification of Existing Fluoroalicyclic Derivatives

Park and co-workers (*179*) established the direction of their [2 + 2] cycloaddition reactions (Section II,B,1) by a series of interconversions, some of which are shown in the scheme below.

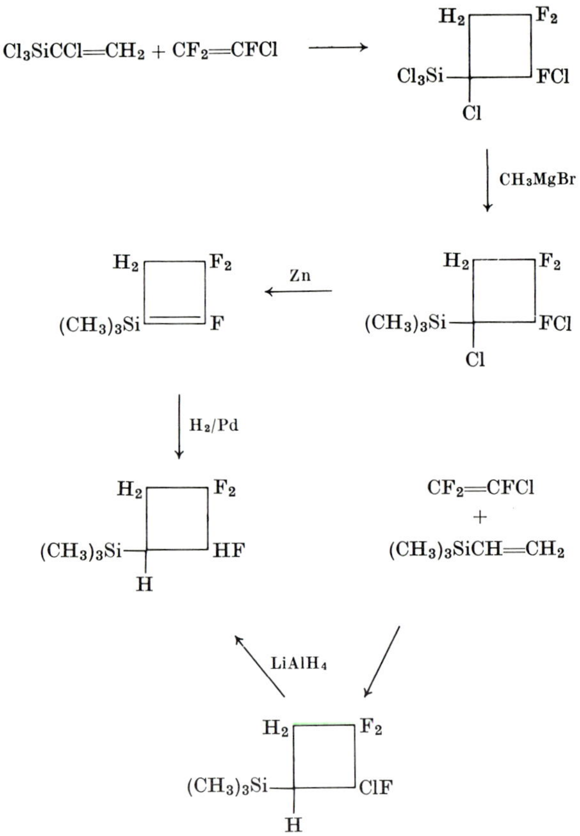

$$Cl_3SiCCl=CH_2 + CF_2=CFCl \longrightarrow \underset{Cl}{Cl_3Si}\overset{H_2\ \ \ \ F_2}{\underset{}{\square}}FCl \qquad (50)$$

$$\downarrow CH_3MgBr$$

$$(CH_3)_3Si\overset{H_2\ \ \ \ F_2}{\square}F \xleftarrow{Zn} \underset{Cl}{(CH_3)_3Si}\overset{H_2\ \ \ \ F_2}{\square}FCl$$

$$\downarrow H_2/Pd$$

$$(CH_3)_3Si\overset{H_2\ \ \ \ F_2}{\underset{H}{\square}}HF \qquad\qquad \begin{array}{c}CF_2=CFCl\\+\\(CH_3)_3SiCH=CH_2\end{array}$$

$$\underset{\nwarrow}{}\ LiAlH_4\ \underset{\nearrow}{}$$

$$(CH_3)_3Si\overset{H_2\ \ \ \ F_2}{\underset{H}{\square}}ClF$$

Base-assisted elimination of hydrogen chloride from a silylcyclobutane has also yielded a cyclobutene (*144*).

Frank (*124–127*) found that the phosphonate products of reactions such as reactions (41) and (42) also undergo a number of reactions which have synthetic potential [reaction (50a)]. The initial cleavage by PCl_5 was developed in order to obtain monophosphonates. These, as shown above [reaction (41)], are also available by treating phosphites with 1-chloro-2-fluorocycloolefins.

$(C_2H_5O)_2P(O)\overset{F_2}{\underset{}{\bigotimes}}(O)P(OC_2H_5)_2$ $\xrightarrow{PCl_5}$ $Cl\overset{F_2}{\underset{}{\bigotimes}}(O)PCl_2$ (50a)

$\downarrow$ HCl $\qquad\qquad\qquad$ HCl $\updownarrow$ C_2H_5OH

$(HO)_2P(O)\overset{F_2}{\underset{}{\bigotimes}}P(O)(OH)_2$ $\qquad\qquad$ $Cl\overset{F_2}{\underset{}{\bigotimes}}(O)P(OC_2H_5)_2$

$\downarrow$ POCl$_3$/PCl$_5$

$Cl_2P(O)\overset{F_2}{\underset{}{\bigotimes}}(O)PCl_2$ $\xrightarrow{CF_3CH_2OH}$ $(CF_3CH_2O)_2P(O)\overset{F_2}{\underset{}{\bigotimes}}(O)P(OCH_2CF_3)_2$

As previously described [reaction (44)], phosphobetaines are obtained from the hydrolysis of some phosphorus compounds (202). The butenone (3) seems to be one of the products obtained when tetraphenyldiphosphine is heated with perfluorocyclobutene (65).

(3) $(C_6H_5)_2P$-cyclobutanone-F, F$_2$

(4) $(CO)_5Mn$-cyclopentenone-F, F$_2$ (via HF)

(5) $Cl-Pt$ with $P(C_2H_5)_3$ ligands and furan-F,F,F,F

A similar hydrolysis product (4) can be isolated from the reaction described by reactions (17) and (18) (121). Cherwinski and Clark (39) have found that heating trans-$[(C_2H_5)_3P]_2PtCl(\overline{C=CFCF_2CF_2})$ with water in the presence of silicon tetrafluoride produces the cationic carbonyl salt trans-$\{[(C_2H_5)_3P]_2PtCl(CO)\}^+SiF_5^-$ (42) together with a compound of formula trans-$[(C_2H_5)_3P]_2PtCl(C_4F_3O)$. The latter could be a butenone analogous to (3), but spectroscopic evidence favors the structure (5).

Oxidation of bis-sulfides with potassium permanganate affords sulfones in good yield (120).

$$RSC\overset{(CF_2)_n}{=\!=\!=}CSR \longrightarrow RSO_2C\overset{(CF_2)_n}{=\!=\!=}CSR \qquad (51)$$

$n = 2, 3, 4$; R = aliphatic or aromatic

III. Other Chemical Properties

On the whole, the properties of the fluoroalicyclic derivatives resemble those of their better known acyclic analogs (*5, 114*). Thus, hydrolysis usually results in liberation of the fluorocarbon group (e.g., *6, 87*) (see the preceding Section II, for some exceptions).

$$(CH_3)_3Sn\text{-cyclopropene-}CF_3,F_2 \xrightarrow{OH^-} H\text{-cyclopropene-}CF_3,F_2 \quad (52)$$

$$(CO)_5Mn\text{-cyclopentene-}F_2,F_2,F_2,F \xrightarrow{H_2SO_4} \text{cyclopentene-}F_2,F_2,F_2,H,F \quad (53)$$

As previously mentioned in connection with the preparation of cyclopropenes (Section II,A), trimethyl(trifluoromethyl)tin acts as a difluorocarbene source when heated, the other product being trimethyltin fluoride (*45*). Although other acyclic fluorocarbon–tin derivatives appear to be more stable (*85*), the 1,1-adduct of trimethyltin hydride and perfluorocyclobutene slowly decomposes at 20° as follows (*82*):

$$(CH_3)_3Sn\text{-cyclobutane-}F_2,F_2,F_\alpha,F_\beta,H \longrightarrow (CH_3)_3SnF + \text{cyclobutene-}F_2,F_2,H,F \quad (54)$$

The initial hydride addition is believed to take place cis to the double bond, thus the mechanism of decomposition is possibly abstraction of F_α by the tin atom to form a carbene followed by hydrogen migration to yield the cyclobutene. Both α and β elimination have been proposed for similar reactions involving silicon compounds (e.g., *12, 95*) and in the reactions described by reactions (54) and (25) no distinction between these two possibilities can be made at the moment. A β elimination is believed to take place when one of the products of reaction (18) is heated. The resulting unstable diene can be trapped with CF_3NO, otherwise it dimerizes (*121*).

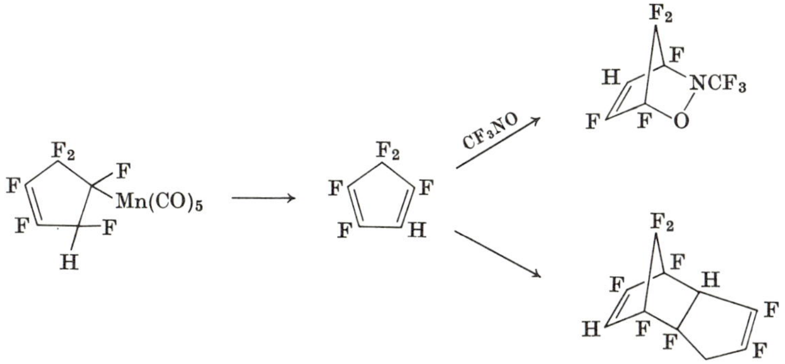

FLUOROALICYCLIC DERIVATIVES OF METALS AND METALLOIDS

(55)

The thermal stability of other fluoroalicyclic derivatives of Groups IV and V and the transition metals is quite high and even the cyclopropenes are stable. Apart from the mercury compounds, the derivatives of other elements exist only in solution and most are unstable. Here again there is a rough parallel with their acyclic analogs. The lithium and magnesium compounds, in particular, have a tendency to eliminate metal halide (*34, 35, 37, 199, 203*), often with the transient existence of cyclohexyne and bridgehead olefin intermediates (both may be equally well represented as diradicals). A related sequence (49) has previously been mentioned.

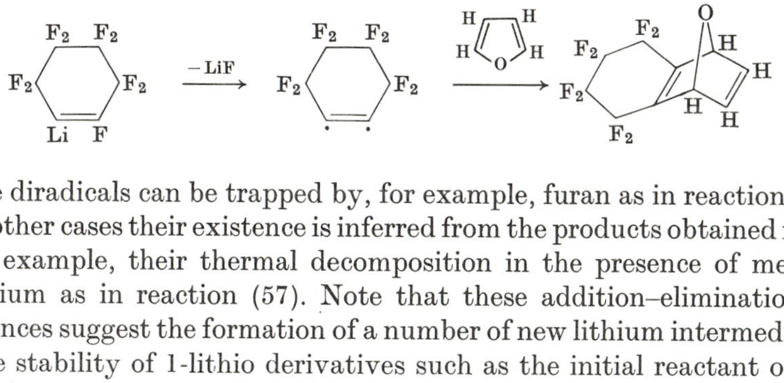

(56)

The diradicals can be trapped by, for example, furan as in reaction (56). In other cases their existence is inferred from the products obtained from, for example, their thermal decomposition in the presence of methyllithium as in reaction (57). Note that these addition–elimination sequences suggest the formation of a number of new lithium intermediates. The stability of 1-lithio derivatives such as the initial reactant of (57) is a function of the group at the other bridgehead position. This stability is in the order $F < H < CF_3$, which is not the order of acidity found in the $1H$ species (*199*).

Apart from these eliminations the lithium and magnesium fluoroalicyclic derivatives react normally with the usual reagents such as aldehydes, carbon dioxide, and halogens (*34, 35, 177, 199*) and, thus, offer a very convenient route to other derivatives. Many of these reactions are not possible in hydrocarbon chemistry. In some cases reaction of 1*Li*-4*H*-decafluorobicyclo[2.2.1]heptane leads to 1,4-disubstituted products. This is believed to be due to lithium–halogen and lithium–hydrogen exchange (*34*).

(57)

The manganese pentacarbonyl derivatives $(CO)_5MnC=CF(CF_2)_nCF_2$ ($n = 2$, 3) react with triphenylphosphine to give both cis and trans isomers of $[(C_6H_5)_3P](CO)_4MnC=CF(CF_2)_nCF_2$ (*6, 148*), whereas reaction of the perfluoroolefins with $[(C_6H_5)_3P](CO)_4Mn^-$ yields only the trans compounds.

IV. Physical Properties

The transition metal derivatives are all solids; those of the main group elements are mainly liquids. The only physical properties of the compounds which have been studied to any extent are their infrared and nuclear magnetic resonance (NMR) spectra which have proved to be useful in characterizing the derivatives.

The infrared spectra of the cyclopropenes $R_nM\overset{\frown}{C=C(R_f)C}F_2$ show a strong band in the region 1758–1733 cm^{-1}. This can be assigned to $\nu(C=C)$; however, it involves considerable contributions from normal skeletal modes and substituent vibrations (48, 216). When the cyclopropene has CF_3 groups in the 3-position, $\nu(C=C)$ is increased. For example $(CH_3)_3Ge\overset{\frown}{C=C(CF_3)\overset{\frown}{C}}(CF_3)_2$ absorbs at 1837 cm^{-1} (88) and the bistrimethylsilyl compound $(CH_3)_3Si\overset{\frown}{C=C(Si(CH_3)_3)\overset{\frown}{C}}(CF_3)_2$ [reaction (9)] has one of the highest frequencies noted for this absorption (79), in spite of the observation (211) that electropositive groups in the vinylic position lower the frequency.

The monosubstituted derivatives $R_nM\overset{\frown}{C=CFCF_2C}F_2$ have $\nu(C=C)$ in the region 1623–1662 cm^{-1} and this does not change much with increase in ring size. The same is true for the chloro compounds $R_nM\overset{\frown}{C=CCl(CF_2)_nC}F_2$ which have $\nu(C=C)$ at a lower frequency (~1575 cm^{-1}). The unsubstituted perfluorocycloolefins have $\nu(C=C)$ at higher frequencies, e.g., $F\overset{\frown}{C=CF(CF_2)_2C}F_2$ absorbs at 1754 cm^{-1}. Thus, the presence of an absorption at 1745 cm^{-1} in one of the products of reaction (18) is good evidence for the proposed 1,2-addition.

Symmetrically substituted compounds such as the diphosphonates and the ditertiary arsines and phosphines show little double-bond absorption. The intensity of the absorption increases somewhat when the substitution is unsymmetrical (40).

Little can be said about other regions of the infrared spectra. The most intense features are the C–F vibrations which, within a particular series of compounds, have a characteristic pattern. The presence of a vinylic fluorine atom is usually indicated by a strong absorption at the high-energy end (~1370 cm^{-1}) of the C–F stretching region.

The NMR spectra of all the derivatives are as expected and can often be interpreted on a first-order basis. Exceptions to this are the silylcyclobutanes. Cyclobutanes can be either bent or planar (e.g., 38, 163, 219). As a result of a study of the NMR spectra of fluorocyclobutanes Lambert and Roberts (156) suggested that two types of conformational equilibria are possible in solution at 30°, planar ⇌ bent and bent ⇌ bent, but in both cases bulky groups preferentially occupied "equatorial" positions. The NMR spectra of other fluorocyclobutanes have also been interpreted in terms of conformational equilibria (e.g., *116, 138, 174, 181*). The spectra of the silylcyclobutanes (**6**) and (**7**) have been studied with the aid of heteronuclear decoupling (*80*).

If certain assumptions are made, for example, that $^3J_{HF}$ couplings follow a Karplus type of dependence, it follows that (**7**) exists predominantly in the conformer having the bulky trimethylsilyl group equatorial. The trimethylsilyl methyl groups apparently couple with the fluorine atoms in the ring only when the trimethylsilyl and fluorine groups are cis to each other. However, when this information is used to establish the preferred conformation of (**6**) an axial preference of the trimethylsilyl group is indicated. These results indicate the difficulty of interpreting the spectra of cyclobutanes and although the NMR spectra of the 1,1-adducts of reaction (15) indicate that addition is cis to the double bond, this is by no means certain (*82*).

V. Coordination Complexes

The complexing abilities of ditertiary arsines and phosphines such as (**8**) (diars) and (**9**) (diphos) have been much investigated (*17, 31, 100, 137, 162*). However, until recently analogs with unsaturated bridging groups such as (**10**), which could conceivably also use their double bond in complex formation, were unknown and unstudied. Several groups are now working in this area (*1, 10, 117, 118, 151, 159*). The fluoroalicyclic bridged ditertiary arsines and phosphines (**11**) and their unsymmetrical analogs (**12**) whose preparations have been described in Section II,C,2

(11)

(11a) L = (CH$_3$)$_2$As, $n = 2$, f$_4$fars
(11b) L = (CH$_3$)$_2$As, $n = 3$, f$_6$fars
(11c) L = (CH$_3$)$_2$As, $n = 4$, f$_8$fars
(11d) L = (C$_6$H$_5$)$_2$P, $n = 2$, f$_4$fos
(11e) L = (C$_6$H$_5$)$_2$P, $n = 3$, f$_6$fos
(11f) L = (C$_6$H$_5$)$_2$P, $n = 4$, f$_8$fos

(12)

(12a) $n = 2$, f$_4$AsP
(12b) $n = 3$, f$_6$AsP
(12c) $n = 4$, f$_8$AsP

also have an unsaturated bridging group. Other interesting features of these molecules are the electronegative bridging groups which could enhance any π-acceptor properties of the ligands (54, 100),* and the possibility of varying the "bite" of the ligand, i.e., the direct donor-to-donor distance, by varying the ring size.

The fluorocarbon bridged ligands (11) and (12) interact with metal halides and metal carbonyls to give a variety of products in which the ligand is monoligate, biligate, triligate, or even rearranged, and these complexes are described in the next sections. It should be pointed out that where more extensive studies have been made the products actually isolated are very dependent on the reaction conditions. Thus, the reaction of f$_4$AsP (12a) with Fe(CO)$_5$ at 80° yields f$_4$AsPFe(CO)$_4$, at 150° f$_4$AsPFe$_2$(CO)$_6$, and under ultraviolet irradiation f$_4$AsPFe$_2$(CO)$_8$. Reaction of the same ligand with Fe$_3$(CO)$_{12}$ yields as major products f$_4$AsPFe$_3$(CO)$_{10}$, f$_4$AsPFe$_3$(CO)$_9$, and f$_4$AsPFe$_2$(CO)$_6$, or (f$_4$AsP)$_2$Fe$_2$(CO)$_4$ and (f$_4$AsP)$_2$Fe(CO)$_3$ depending on the ratio of reactants (40).

A. MONOLIGATE COMPLEXES

In spite of the presence of two potential donor atoms on the fluorocarbon-bridged ligands (11) and (12) and their similarity to the hydrocarbon-bridged analogs (8) and (9), there is a surprising tendency of some of the former class to yield complexes in which they are monoligate.

* This makes the assumption that in a coordination complex the bond between the phosphorus (or arsenic) atom and the metal has two components: (i) a sigma component arising from donation of the lone pair on the phosphorus (or arsenic) to a suitable acceptor metal orbital and (ii) a synergic back donation from filled metal orbitals (of mainly d character) into vacant orbitals of mainly d character on the donor atom. This back-bonding concept is a useful model for rationalizing a number of results, although it is by no means universally accepted (e.g., 2, 210, 220).

This is especially true for derivatives containing the cyclobutene ring and may be an indication that the bite of the ligand, which would be expected to be at maximum for the small ring and for compounds with $(CH_3)_2As$ groups because of the greater exocyclic angles and As–C bond distances, is such as to discourage chelation. However, as will be pointed out below, the bite of a ligand is very dependent on the type of complex formed and, hence, the geometry cannot be the only reason for the reluctance of, say, $f_4$fars (**11a**) to chelate.

When a petroleum ether solution of $f_4$fars and $Fe(CO)_5$ is irradiated with ultraviolet light, a yellow complex $f_4$farsFe(CO)$_4$ is obtained (74). Mild reaction conditions also afford $f_4$fosFe(CO)$_4$, f_4AsPFe(CO)$_4$, and in low yield $(f_4AsP)_2Fe(CO)_3$ and f_6AsPFe(CO)$_4$ (40, 72, 74). The (L–L)Fe-(CO)$_4$* derivatives have infrared spectra similar to the three-band pattern reported for compounds such as $(C_6H_5)_3PFe(CO)_4$ (53), which is indicative of apical substitution in a trigonal bipyramid. The parent pentacarbonyl has this trigonal-bipyramidal structure (8, 101). In the cases where (L–L) = $f_4$fos, f_4AsP, and f_6AsP, the asymmetry of the ligand results in an overall C_s symmetry for the molecule and an extra band appears presumably because of splitting of the E mode in a C_{3v} spectrum (74). In the case of f_4AsPFe(CO)$_4$ it is apparent that it is the phosphine end of the molecule that is coordinated, since the NMR spectrum of the complex shows that the chemical shift of the $(CH_3)_2$As moiety is virtually unaltered on complexing. The free ligand has an absorption at 1.33 ppm and the complex at 1.25 ppm.† If coordination through the arsenic occurred the resonance would shift downfield by ~0.5 ppm (10, 74, 176). This is seen in the spectrum of $f_4$farsFe(CO)$_4$, where the free ligand absorbs at 1.36 ppm and the complex at 1.44 ppm [the "free" $(CH_3)_2$As group] and 1.90 ppm. Similar considerations indicate that f_6AsPFe(CO)$_4$ is P-bonded as is $(f_4AsP)_2Fe(CO)_3$. The infrared spectrum of the latter compound can be interpreted in terms of a trigonal-bipyramidal structure with two apical phosphine substituents (40, 53).

A recent X-ray study of f_4AsPFe(CO)$_4$ (109) confirms that its structure is as shown in (**13**) with the ligand coordinated through the phosphorus atom. There is no significant difference between axial and equatorial iron–carbon bond lengths [1.74(2) Å] and the iron atom is dis-

---

* In this account (L–L) represents a ditertiary arsine or phosphine such as (**11**) and, when appropriate, a mixed ligand such as (**12**).

† In this chapter NMR chemical shifts are given in ppm downfield from internal TMS. Mössbauer parameters are derived from spectra run at 80°K. Isomer shifts, δ, and quadrupole splittings, Δ, are quoted in mm sec^{-1}. Isomer shifts for iron compounds are given relative to sodium nitroprusside.

placed 0.025 Å from the plane of the equatorial groups toward the axial carbonyl group.

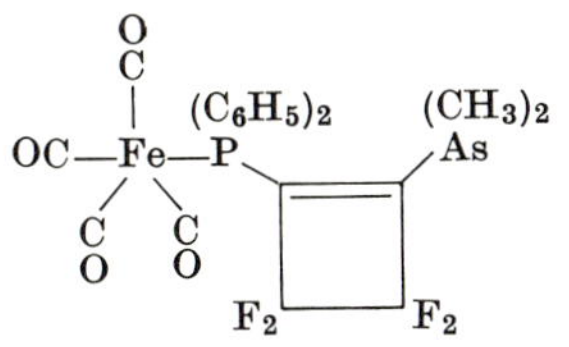

$$\text{(13)}$$
OC—Fe—P(C₆H₅)₂—[C₄F₄ ring]—As(CH₃)₂ with carbonyl and CO groups

(13)

The Mössbauer spectra of these and related five-coordinate iron complexes consist of a widely split quadrupole doublet as expected for a single iron atom in a noncubic field (*134, 215*). At 80°K the isomer shifts of these compounds lie in the narrow range of $\delta = 0.16$–0.23 mm sec^{-1} and the quadrupole splittings, Δ, lie in the range 2.12–2.83 mm sec^{-1}. The compounds with $(CH_3)_2As$ groups coordinating show greater shifts than those containing $(C_6H_5)_2P$ indicating that phosphorus is a better π acceptor than arsenic. As a result the spectrum of $f_4AsPFe_2(CO)_8$, (**18**), shows two doublets (Fe^A, $\delta = 0.21$, $\Delta = 2.83$; Fe^B, $\delta = 0.19$, $\Delta = 2.18$ mm sec^{-1}) and the one with the greater isomer shift, Fe^A, can be assigned to the iron atom bonded to the arsenic atom (*40*).

Triosmium dodecarbonyl (**14**) (M = Os), like its ruthenium analog (*164*), has a triangular arrangement of osmium atoms with each metal atom approximately octahedrally coordinated by four carbonyl groups and the other two osmium atoms (*51*). The stability of the $M_3(CO)_{12}$ skeleton increases in the series M = Fe < Ru < Os,* and a number of derivatives are known in which a carbonyl group has been replaced

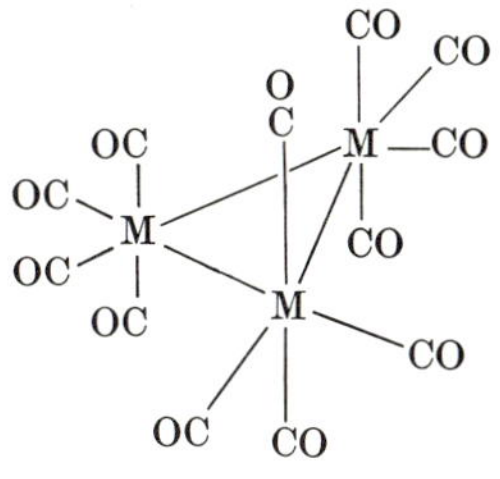

(14)

by a tertiary phosphine or arsine without rupture of the Os$_3$ skeleton (*97*). Both yellow $f_4fosOs_3(CO)_{11}$ and $f_8fosOs_3(CO)_{11}$ are obtained

* $Fe_3(CO)_{12}$ has a different structure with bridging carbonyl groups (**20**) (*212*).

under vigorous conditions by displacement of one, presumably equatorial, carbonyl group (55).

The ditertiary phosphine diphos (**9**) reacts with $(\pi\text{-}C_5H_5)Fe(CO)_2Sn(CH_3)_3$ under ultraviolet irradiation with displacement of both carbonyl groups (152). Under similar conditions f₆fos displaces only one carbonyl group from the same compound and its triphenyltin analog (84).

$$(\pi\text{-}C_5H_5)Fe(CO)_2\text{-}SnR_3 + f_6fos \xrightarrow{h\nu} (\pi\text{-}C_5H_5)Fe(CO)(f_6fos)\text{-}SnR_3 \quad (58)$$

R = CH₃, C₆H₅

The solid state structure of the trimethyltin compound is shown in Fig. 1

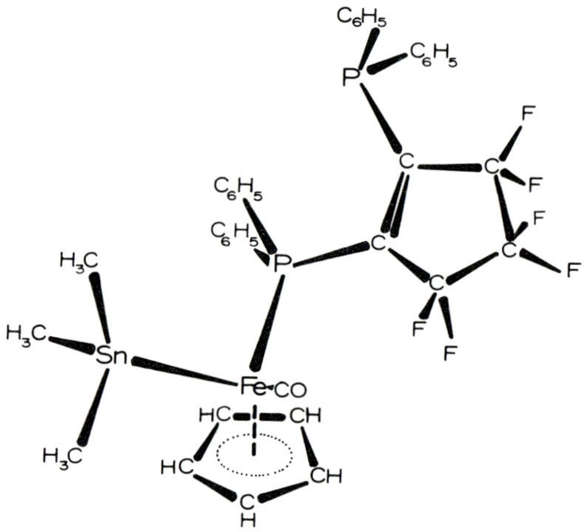

FIG. 1. The structure of $(\pi\text{-}C_5H_5)Fe(CO)(f_6fos)Sn(CH_3)_3$. (112).

(112), although rotamers appear to be present in solutions of this type of derivative (152) [see also (23, 147)]. The ligand is clearly monodentate with a P---P separation of 3.60(1) Å. The fluorocarbon ring is in an envelope conformation and the coordination is approximately tetrahedral round the tin atom, and octahedral round the iron assuming that the π-C₅H₅ ring occupies three coordination sites. The Fe–Sn distance of 2.562(4) Å is not significantly different from the sum of the covalent radii, although assigning radii to metals in low oxidation states is a hazardous undertaking, and the conclusion can be made that there is little π-character to the metal–metal bond. This same conclusion has been reached from X-ray studies on the parent molecules $(\pi\text{-}C_5H_5)Fe(CO)_2SnR_3$ (29, 132, 133) and from spectroscopic studies on their derivatives (*85*) (see also *119*). It should be pointed out that many workers favor a π-bonded model for metal–metal bonds especially those involving the lighter Group IV elements (e.g. *13*).

B. BILIGATE MONOMETALLIC (CHELATE) COMPLEXES

1. *Chelate Complexes Derived from the Group VI Hexacarbonyls*

The reaction of the Group VI hexacarbonyls, or their derivatives, with ditertiary arsines and phosphines to give the compounds $(L-L)M(CO)_4$ has been the subject of many investigations (*31, 100*). The fluorocarbon-bridged ligands also give complexes of the same stoichiometry (*59, 68, 70*). The complete series $f_n fosM(CO)_4$ is known for $n = 2$, 4, or 6 and M = Cr, Mo, or W in addition to $f_4 farsM(CO)_4$, where M = Cr or Mo. As usual, the tungsten compounds require more forcing conditions for their preparation. An interesting feature of the complexes $f_6 fosM(CO)_4$ is their ability to form solvates of formula $f_6 fosM(CO)_4 \cdot \frac{1}{2}S$, where S = benzene, cyclohexane, or chloroform, and where the solvent is quite strongly held. One of the main reasons for preparing these compounds was to study their carbonyl infrared spectra and the following conclusions have been made.

(*i*) Since the spectra of $diphosM(CO)_4$ and cis-$(C_6H_5)_2PCH=CHP$-$(C_6H_5)_2M(CO)_4$ are similar, the electronegative $>C=C<$ group with its sp^2-hybridized carbon atoms has little effect on the donor–acceptor properties of the ligand.

(*ii*) Since the spectra of $f_n fosM(CO)_4$ are approximately the same for a given M, differences in properties of the $f_n fos$ ligands are mainly due to geometric rather than electronic effects.

(*iii*) Since the spectra of the $f_n fos$ complexes have bands at higher wavenumber than those of the diphos complexes, the electronegative bridging groups cause the phosphorus atoms to be better π acceptors (*or worse σ donors*).

In addition, the spectra of the $f_6 fos$ complexes have one band more than expected for a molecule with C_{2v} symmetry. This has been explained as follows (*70*). The complex $f_6 fosM(CO)_4$ contains two five-membered rings which will be puckered (see Figs. 1 and 2), and it is possible that the combination of these two rings would give rise to two distinct isomers ["chair" and "boat" (*105*)] which could have a long enough lifetime in solution to be detected by infrared spectroscopy. The lifetime of any one isomer must be short since the ^{19}F NMR spectrum of the complex is almost identical with that of the free ligand.* An "extra" band is also seen in the spectrum of $f_6 fosFe(CO)_3$ (*74*).

* The NMR spectra indicate that the ligands are chelated.

2. Chelate Complexes Derived from Manganese and Rhenium Carbonyls

Dimanganese decacarbonyl and f_4fos interact to give a complex which is probably best formulated as $(CO)_5Mn-Mn(CO)_3f_4fos$ with the ligand chelated to one manganese atom (59). However, an alternative formulation involving ligand cleavage is possible (cf. Fig. 11 and Section V,E). The chelated carbonyl halide derivatives $f_4farsM(CO)_3X$ have also been synthesized. In the case of the iodides these were obtained by warming the biligate bimetallic complexes $f_4fars[M(CO)_3I]_2$ (see Section V,C,2). This reaction involves considerable redistribution of ligands and is rather surprising in view of the usual reluctance of f_4fars to chelate.

3. Chelate Complexes Derived from Iron and Ruthenium Carbonyls

Chelate complexes of formula $(L-L)Fe(CO)_3$ $[(L-L) = f_4fos, f_6fos, f_8fos, f_4AsP, f_6AsP]$ are obtained from most reactions involving iron carbonyls and the fluorocarbon-bridged ligands (40, 72, 74). Similar compounds are obtained from diars and diphos (73). The NMR spectra of these complexes indicate that the symmetry of the ligands is unaltered on complexing which suggests a C_{2v} structure with the ligand occupying two equatorial positions of a trigonal bipyramid. Thus, the L–Fe–L angle would be about 120° on a rigid model. It is apparent from the structures that have been determined that this angle should be closer to 90° (105, 139). Indeed the structure of $f_6fosFe(NO)_2$ (**16**) (139) is considerably distorted from one with tetrahedral angles round the iron atom (vide infra). Thus, geometrical requirements seem to favor aplical-equatorial substitution, and this is the structure found in an X-ray investigation of $diarsFe(CO)_3$ (22). If this were the structure of $diarsFe(CO)_3$ in solution then two arsenic–methyl environments would be seen in the NMR spectrum. However, the spectrum is a singlet in the As–CH_3 region down to −80°C which suggests that this molecule and the related $(L-L)Fe(CO)_3$ compounds are undergoing positional exchange (14, 172), probably by the "turnstile" mechanism (209), where the ligands (L–L) rotate so as to exchange axial and equatorial arsenic atoms.

It is significant that f_4fars does not form the chelate complex $f_4farsFe(CO)_3$ even though the closely related ligands (**15**) ($R = CF_3$) and

$(CH_3)_2As\overset{R}{\underset{}{\diagup}}C=C\overset{R}{\underset{As(CH_3)_2}{\diagdown}}$

(**15**)

(**16**)

f_4AsP do (40, 61). This suggests that geometric factors may be operating especially since in the phosphine series the ease of formation of $f_n fosFe(CO)_3$ increases as n increases (using yields obtained under comparable conditions as indicators) (72).

The corresponding complexes of ruthenium are known for f_4fos and f_8fos. They have similar properties to the iron analogs, but are less stable (71). Osmium analogs are unknown (55).

Reaction of f_4fars with iron dicarbonyl dinitrosyl results in displacement of only one carbonyl group even though other ligands give the chelate $f_n fosFe(NO)_2$ complexes ($n = 4$, 6, or 8) in good yield (58). The Mössbauer spectra of these Fe^{-2}, d^{10} derivatives, show little variation in $\delta(Fe)$ with ligand, and point charge calculations indicate that the ligand strength of $f_n fos$ is approximately the same as triphenylphosphine (58). The crystal structure of $f_6 fosFe(NO)_2$ shows some interesting features (16) (139). The ligand is in an envelope conformation, although here again this is not seen in the solution ^{19}F NMR spectrum. The nitrosyl groups are linear, but the P–Fe–P angle is only 87°, whereas the N–Fe–N angle is 125°. Thus, the ligand seems to prefer to chelate with a P–Fe–P angle of ~90°. The bite of the ligand is 3.084(3) Å which will be discussed later. Mingos and Ibers (171) have recently reported the structure of the isoelectronic ion $[(C_6H_5)_3P)_2Ir(NO)_2]^+$. Here there is a large angle (154°) between *nonlinear* nitrosyl groups.

The "folded book" structure of $Co_2(CO)_8$ (204) is also found in the iron complexes $[Fe(CO)_3SR]_2$ (91, 149). The compound with $R = CH_3$ exists as two isomers syn and anti depending on the orientation of the $S-CH_3$ groups in the bridge. Under mild conditions $f_n fos$ ($n = 4$, 8) displaces two carbonyl groups from the syn/anti mixture to yield products which have two different iron environments (57). The Mössbauer spectra indicate that one iron atom is little changed from the parent molecule, but the other is increased in isomer shift considerably (Table I), so it is probable that the ditertiary phosphines are chelated. The chemical shift difference between the two S-methyl resonances in the NMR spectrum indicates that the products have the anti S-methyl arrangement. The structure proposed for the f_4fos derivative is (17).

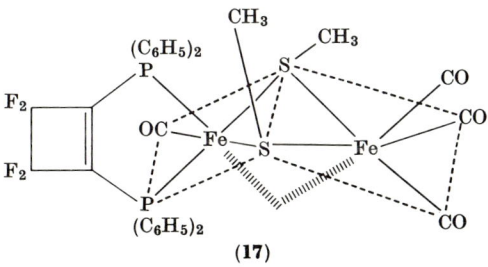

(17)

TABLE I

Spectroscopic Data for some $[Fe(CO)_3SCH_3]_2$ Derivatives[a]

Compound	Mössbauer		NMR
	$\delta(Fe)$	$\Delta(Fe)$	$\delta(S-CH_3)$
anti-$[Fe(CO)_3SCH_3]_2$	0.29	0.88	1.61, 2.12
anti-$[Fe_2(CO)_5(SCH_3)_2P(C_6H_5)(CH_3)_2]$	0.28	1.15	1.12, 1.80
	0.29	1.70	
$f_4fos[Fe(CO)_2SCH_3]_2$[b]	0.33	1.01	—
$f_4fars[Fe(CO)_2SCH_3]_2$[b]	0.36	1.08	1.57, 1.71
$f_4fos[Fe(CO)_2SCH_3]_2$[c]	0.28	1.14	0.85, 1.80
	0.40	0.47	—

[a] See footnote on p. 348.
[b] The ligand is bridging.
[c] The ligand is chelating.

Equatorial-axial substitution is more likely because monosubstitution by $C_6H_5P(CH_3)_2$ leads to a compound apparently with a plane of symmetry since the NMR spectrum shows that the P-methyl groups are equivalent (57). The carbonyl infrared spectra of these $f_n$fos derivatives are similar to that of the diphos complex which is also believed to be chelated (96).

4. *Chelate Complexes Derived from* $[Rh(CO)_2Cl]_2$ *and Other Metal Halides*

Reaction of the ditertiary phosphines $f_n fos(n = 4, 6)$ with the chlorine-bridged dimer $[Rh(CO)_2Cl]_2$ yields a product which can be regarded as the result of unsymmetrical cleavage of the dimer (83). A 1:4 ratio of reactants affords $[(f_n fos)_2Rh]^+Cl^-$ and in this case $n = 4, 6,$ or 8.

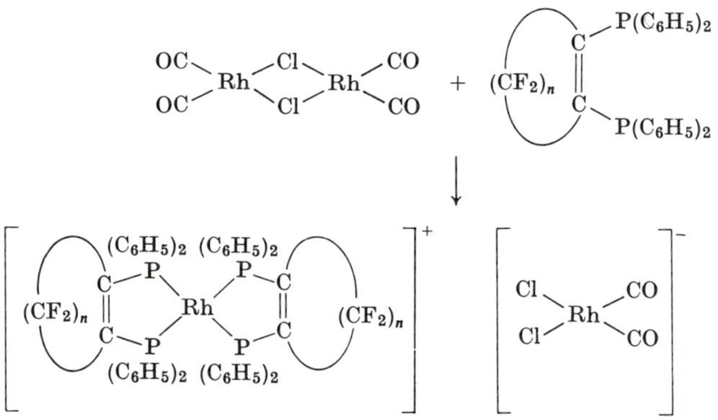

(59)

Similar products are obtained by reaction of other hydrocarbon-bridged ditertiary arsines and phosphines such as (**10**) (*159*), but diphos behaves differently (*142, 192*). The structures of the anion and cation products of reaction (59) were first proposed on the basis of their spectroscopic properties, and were later verified by an X-ray investigation (*105*). The structure of the cation [(f$_6$fos)$_2$Rh]$^+$ is shown in Fig. 2. Like the anion the cation has approximately the square-planar geometry around the central atom expected for a Rh^I (d^8) derivative. The cation is actually twisted slightly so that it is not superimposable on its mirror image. This twisting probably occurs to accommodate the anion which lies perpendicular to the plane of the cation with the two cis chlorine atoms closer to two trans phosphorus atoms in the cation. A similar structure has been found for the cation [(diphos)$_2$Rh]$^+$ (*135*).

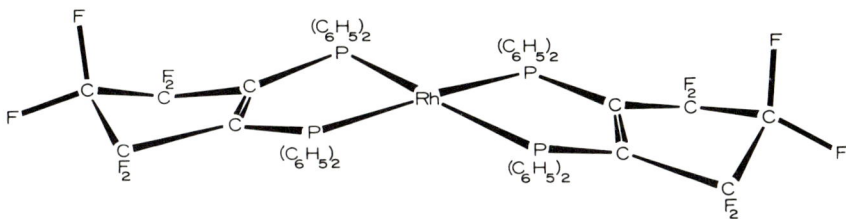

FIG. 2. The structure of [(f$_6$fos)$_2$Rh]$^+$ (*105*).

The bite of the ligand in Fig. 2 is 3.111(8) Å which is considerably shorter than that found when the same ligand is monodentate (*112*) (Fig. 1) but is similar to that found for f$_6$fosFe(NO)$_2$, 3.084(3) Å (*139*). These results show that the bite of a ligand can be very much dependent on the bonding situation and, hence, caution is needed when the concept is being used to rationalize chemical results.

One of the features of Rh^I derivatives is their ability to undergo oxidative addition (*49*). The electronegative bridging groups of [(f$_n$fos)$_2$Rh]$^+$ seem to inhibit this reaction and hydrogen chloride has been successfully added only to [(f$_4$fos)$_2$Rh]$^+$ (*83*).

Hydrocarbon-bridged ligands such as diars (**8**) have the ability to form chelate complexes with a wide variety of metals in a variety of oxidation states (*31, 100, 137, 162*). In preliminary investigations (*68, 69, 99*) it has been found that the fluorocarbon-bridged ligands are reluctant to react with salts of the first-row transition metals. Although compounds of formula diarsMX$_2$ (M = Zn, Cd, or Hg) are well established (*157*), only the HgCl$_2$ derivative is known for f$_4$fars and f$_4$fos. Other chelate complexes of the heavier metals have been isolated including

$f_4$farsPdCl$_2$ and (f$_4$fars)$_2$RhCl$_3$. The reaction of both f$_4$fos and f$_4$fars with Pt^{II} derivatives is complex and seems to resemble the behavior of the related ligand (15) (R = H) which gives bridged complexes (9).

C. Biligate Bimetallic Complexes

In these complexes the ditertiary arsine or phosphine is acting in a bridging bis-monoligate fashion.

1. Bridging Two Atoms not Otherwise Connected

When a solution of f$_4$fars and an iron carbonyl is irradiated with ultra-violet light, one of the products is f$_4$farsFe$_2$(CO)$_8$ (72, 74). Similar conditions produce f$_4$AsPFe$_2$(CO)$_8$ (40) and the analogous complexes of diars and diphos are known (73). There are numerous examples in the literature where it has been postulated that a ligand bridges two metal carbonyl moieties and this is the confirmed structure of diars[(π-CH$_3$C$_5$H$_4$)Mn(CO)$_3$]$_2$ (11). The spectroscopic properties of these Fe$_2$(CO)$_8$ derivatives are compatible with bridged structures similar to that shown for f$_4$AsPFe$_2$(CO)$_8$ (18) with substitution taking place in the apical position. Again these complexes are almost certainly not rigid in solution (see also Sections V,A and B above).

When f$_4$fars is treated with [Rh(CO)$_2$Cl]$_2$, a red complex is obtained (83) which has been assigned the structure (19) on the basis of its spectroscopic properties and its similarity to the previously known compounds

[(C$_6$H$_5$)$_2$ECH$_2$E(C$_6$H$_5$)$_2$Rh(CO)Cl]$_2$ (158, 159). It is worth noting that f$_n$fos ligands give chelate complexes under the same conditions as in reaction (59). Thus, changing the donor atoms is sufficient to change the type of product. It had been suggested (159) that the length of the bridging chain in the ligands $(C_6H_5)_2P(CH_2)_nP(C_6H_5)_2$ determines whether the complexes would be bridged or chelated.

2. Bridging Two Atoms Otherwise Connected

Most reactions of $Fe_3(CO)_{12}$ involve rupture of the basic skeleton (**20**)

(**20**)   (**21**)

(212).* However, some monodentate ligands have been found to displace one or more carbonyl groups from the basic skeleton. Examples include $(C_6H_5)_3PFe_3(CO)_{11}$ (4, 92, 93) and $(C_6H_5P(CH_3)_2)_3Fe_3(CO)_9$ (166), where the ligands are equatorial, i.e., they lie in the Fe_3 plane. Irradiation of a mixture of $f_4$fars and $Fe_3(CO)_{12}$ yields $f_4$fars$Fe_3(CO)_{10}$ in low yield (72, 76). The spectroscopic properties of this molecule, singlet ^1H and ^{19}F NMR resonances, and simple infrared spectrum in the carbonyl stretching region indicate a symmetrical structure. The Mössbauer spectrum shown in Fig. 3 is very similar to that of $Fe_3(CO)_{12}$. The centre peak associated with the unique Fe^A atom (**20**) (approximately octahedrally coordinated and therefore unsplit) is little changed in the complex. The two outer peaks arising because of quadrupole splitting of the absorption due to the two equivalent Fe^B atoms are slightly shifted to more positive velocities suggesting that the ligand has displaced a terminal carbonyl group from each Fe^B. This structure (**21**) was subsequently confirmed by an X-ray investigation (188). The $f_4$fars ligand in the solid state is not quite coplanar with the Fe_3 triangle, but presumably it is in solution to account for the spectroscopic properties. Otherwise, the Fe_3 fragment is little altered from that of the parent $Fe_3(CO)_{12}$. It should be noted that bands in the bridging carbonyl stretching region of the infrared spectrum of $f_4$fars$Fe_3(CO)_{10}$ are weak, just as are those of $Fe_3(CO)_{12}$ itself. Thus,

* The structure in solution is subject to debate (76, 183, 212).

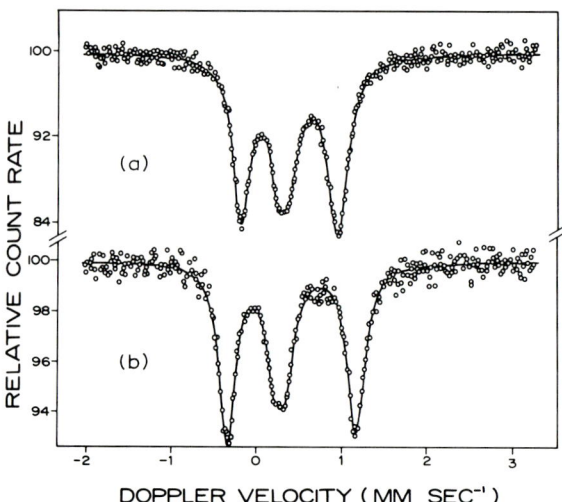

Fig. 3. The Mössbauer spectra at 80°K of (a) $Fe_3(CO)_{12}$ and (b) $f_4farsFe_3(CO)_{10}$.

it is not necessary to postulate a different structure in solution for $Fe_3(CO)_{12}$ to explain the low intensity. The only similar complex so far isolated is $f_4AsPFe_3(CO)_{10}$, which is formed in high yield (87%) when f_4AsP and $Fe_3(CO)_{12}$ are refluxed in cyclohexane solution for 45 min (40). Its spectroscopic properties are very similar to those of $f_4farsFe_3(CO)_{10}$. There is probably a connection between the formation of a (L–L)$Fe_3(CO)_{10}$ derivative and the reluctance of (L–L) to form the chelate complexes (L–L)$Fe(CO)_3$ (see Section V,B).

The structures of $Ru_3(CO)_{12}$ and $Os_3(CO)_{12}$ contain no bridging carbonyl groups (14) and their stability is greater than that of $Fe_3(CO)_{12}$. Consequently it is not surprising to find that both f_4fars and f_4fos react with $Ru_3(CO)_{12}$ to afford red (L–L)$Ru_3(CO)_{10}$ complexes on ultraviolet irradiation (71). More vigorous conditions (cyclohexane reflux) give (L–L)$_2Ru_3(CO)_8$. It may be significant that f_6fos which seems to be a better chelating ligand, does not form any Ru_3 complexes. No complex was isolated when f_4fars was treated with $Os_3(CO)_{12}$ presumably because of the vigorous reaction conditions needed; however, f_4fos gave $f_4fosOs_3(CO)_{10}$ and $(f_4fos)_2Os_3(CO)_8$ (55). The carbonyl infrared spectra of the Os_3 complexes are very similar to those of the corresponding Ru_3 derivatives and, hence, they are probably isostructural. The spectroscopic properties of $f_4farsRu_3(CO)_{10}$ suggested that the molecule has high symmetry and a ligand-bridged structure was proposed which was subsequently verified by X-ray studies (189). A similar bridged structure was proposed and found for $(f_4\text{fars})_2\text{Ru}_3(\text{CO})_8$ (*190*) (Fig. 4). The ruthenium triangle is maintained in both Ru$_3$ structures, but equatorial substitution by the arsine distorts the molecule considerably in the solid. The solution spectroscopic properties indicate higher symmetry and, thus, the molecule is either flexing in solution or the distortion is due to crystal packing. It is interesting that in both structures the Ru–Ru bonds which involve ruthenium atoms with the greatest number of carbonyl groups are shorter than the others. Thus, in $f_4\text{farsRu}_3(\text{CO})_{10}$ there are two bonds of length 2.831(3) Å and one

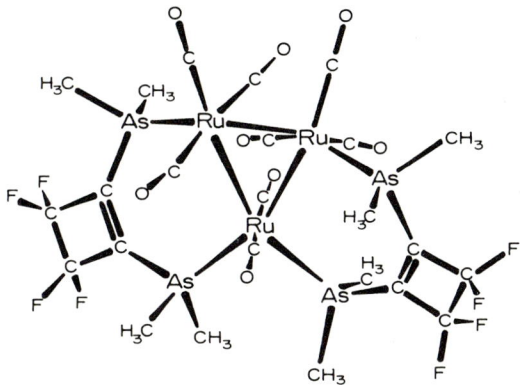

FIG. 4. The structure of $(f_4\text{fars})_2\text{Ru}_3(\text{CO})_8$ (*190*).

(the bridged one) of length 2.856(6) Å, while in the $\text{Ru}_3(\text{CO})_8$ derivative one Ru–Ru bond is 2.785(4) Å and the other two bridged ones are 2.853(3) Å. This can be ascribed to replacing a carbonyl group by a poorer π acceptor.

It is well established that dicobalt octacarbonyl exists in solution as an equilibrium mixture (approximately 1 : 1) of carbonyl bridged and nonbridged forms (*18, 175*). The bridged structure is found in the solid state (*204*). Many tertiary phosphine and arsine derivatives show this same tendency (*31*); however, some, e.g., $[(C_4H_9)_3\text{PCo(CO)}_3]_2$, exist in solution and the solid state in the nonbridged form (*146*). The ligands $f_4\text{fars}$, $f_6\text{AsP}$, and $f_n\text{fos}$, ($n = 4$, 6, or 8) react easily with dicobalt octacarbonyl to displace two moles of carbon monoxide and give complexes of formula $(L-L)\text{Co}_2(\text{CO})_6$ (*56*). The carbonyl infrared spectra of these complexes are simple, and show the presence of bridging carbonyl groups. Two types of spectra are obtained. In the $f_4\text{fars}$, $f_6\text{AsP}$, and $f_4\text{fos}$ derivatives the two frequencies associated with the bridging carbonyl groups are separated by ~56 cm^{-1}; in the f$_6$fos and f$_8$fos derivatives the frequency difference is much smaller (~12 cm^{-1}). These results lead to the suggestion that the f$_4$fars-type complexes were formed by displacement of two terminal carbonyl groups; one from each cobalt atom as shown in Fig. 5. Indeed, this is the X-ray determined structure of f$_4$farsCo$_2$(CO)$_6$ (*140*), and the same structure is assumed for the f$_4$fos and f$_6$AsP analogs. The f$_6$fos and f$_8$fos derivatives are believed to have a similar bridged structure, only here the ligand is situated trans to the Co–Co bond so that the bridging and terminal carbonyl groups are cis to the ligand which would account for the similarity of the asymmetric and symmetric bridging carbonyl stretching frequencies.

Similar bridged products are obtained when the acetylene-bridged complex Co$_2$(CO)$_6$C$_6$H$_5$C≡CH is treated with f$_4$fars and f$_4$fos.

Under xylene reflux f$_4$fos and f$_4$fars react with the syn/anti mixture of [Fe(CO)$_3$SCH$_3$]$_2$ to give (L–L)[Fe(CO)$_2$SCH$_3$]$_2$ (*57*). The Mössbauer spectra of the products (see Table 1) indicate that the two iron atoms are equivalent and that they are monosubstituted. The NMR spectrum shows that the As–CH$_3$ groups of the f$_4$fars complex are inequivalent and the S-methyl groups have chemical shifts characteristic of the syn isomer. Therefore, the bridged structure (**22**) is proposed for these molecules whose infrared spectra are very similar to that of (C$_6$H$_5$)$_2$PCH$_2$P(C$_6$H$_5$)$_2$[Fe(CO)$_2$SCH$_3$]$_2$, which is also believed to be bridged (*96*).

(**22**)

It is interesting to note that f$_4$fos[Fe(CO)$_2$SCH$_3$]$_2$ also exists as a chelated isomer (see Section V,B) and this is one of the two instances where a ligand gives isomers whose structure depends on whether it is bridged or chelated. The other example is afforded by the compound (f$_4$AsP)$_2$Fe$_2$(CO)$_4$ (*40*), which is a derivative of f$_4$AsPFe$_2$(CO)$_6$ (cf. Fig. 9 and Section V,D) in which the second ligand bridges the two Fe atoms or is chelated to one iron atom (Fe^A). The bridged isomer is obtained by reacting excess f$_4$AsP with Fe$_3$(CO)$_{12}$ or by irradiating a mixture of f$_4$AsPFe$_2$(CO)$_6$ and f$_4$AsP. The latter method also gives the chelated isomer (40, 108). A related complex (f$_4$AsP)(f$_6$AsP)Fe$_2$(CO)$_4$, obtained from f$_6$AsPFe$_2$(CO)$_6$, seems to exist only in the bridged form.

The cobalt clusters RCCo$_3$(CO)$_9$ (R = CH$_3$, CF$_3$), which are not carbonyl bridged in the solid state (205), react with f$_4$fars to give the substitution product f$_4$farsCH$_3$CCo$_3$(CO)$_7$ and the rearranged product f$_4$farsCo$_4$(CO)$_8$ (56), respectively (the fate of the fluorocarbon residue is

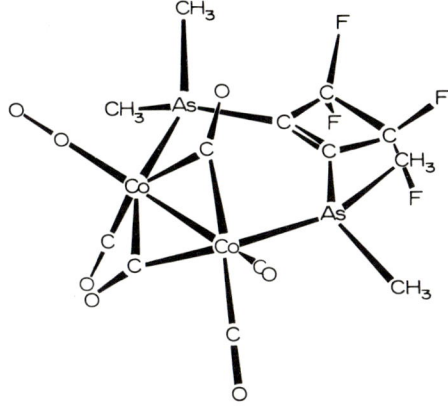

FIG. 5. The structure of f$_4$farsCo$_2$(CO)$_6$ (140).

unknown). The bridging carbonyl groups seen in the solution infrared spectrum are present in the solid state structure of f$_4$farsCH$_3$CCo$_3$(CO)$_7$ (Fig. 6) (106). However, the bridging carbonyl groups present in solutions of the Co$_4$(CO)$_8$ complex are *absent* in the solid state (Fig. 7) (107). The skeleton of the latter structure is similar to that of Ir$_4$(CO)$_{12}$ (214), but both Co$_4$(CO)$_{12}$ and Rh$_4$(CO)$_{12}$ are carbonyl-bridged in the solid state (52, 213). Some terminal carbonyl groups of (f$_4$fars)$_2$Co$_4$(CO)$_8$ are weakly interacting with adjacent cobalt atoms. This feature has been found in a number of structures determined recently (e.g., 41).

In the known structure of CH$_3$CCo$_3$(CO)$_8$P(C$_6$H$_5$)$_3$ (25) the phosphine has displaced an equatorial carbonyl group (with respect to the Co$_3$ plane) and the complex is unbridged like the parent CH$_3$CCo$_3$(CO)$_9$. However, bridging carbonyl groups are present in solutions of this phosphine derivative. In the structure shown in Fig. 6 the ligand has displaced two carbonyl groups from axial positions.

Under mild conditions f$_4$fars, f$_6$fars, and f$_4$fos react with M$_2$(CO)$_{10}$ (M = Mn, Re) to give bridged derivatives such as f$_4$farsMn$_2$(CO)$_8$ (*59, 78*). Other hydrocarbon-bridged bidentate ligands apparently afford chelate complexes (L–L)M(CO)$_3$ or (L–L)M$_2$(CO)$_8$ (*141, 191*). The spectroscopic properties of the f$_4$fars derivative are in agreement with the solid state structure (Fig. 8) (*60*). Here the Mn–Mn bond is longer than in

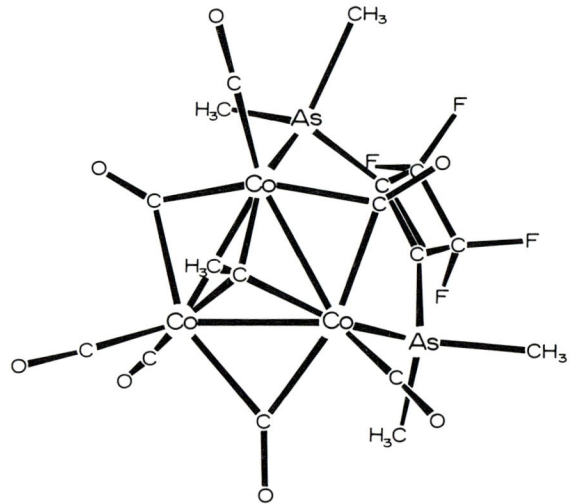

FIG. 6. The structure of f$_4$farsCH$_3$CCo$_3$(CO)$_7$ (*106*).

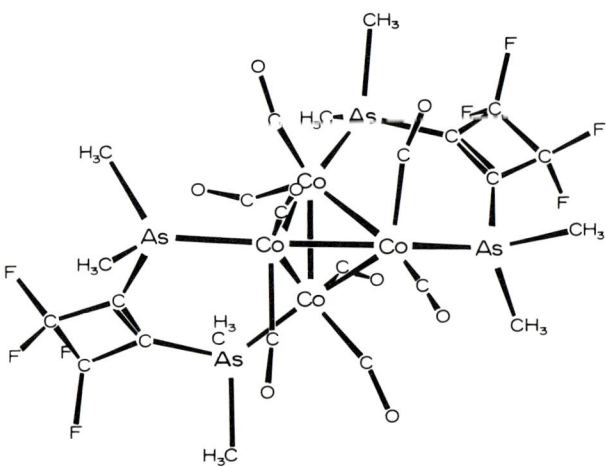

FIG. 7. The structure of (f$_4$fars)$_2$Co$_4$(CO)$_8$ (*107*).

the parent carbonyl (*90*). The carbonyl groups are staggered. Reaction of the bridged (L–L)Mn$_2$(CO)$_8$ (*59, 78*) derivatives with iodine results in easy cleavage of the M–M bond and the bridged derivatives (L–L)-[Mn(CO)$_4$I]$_2$ are obtained. The structure has been confirmed in the case of the f$_4$fars compound; the two Mn(CO)$_4$I moieties which are coordinated to separate –As(CH$_3$)$_2$ groups lie on opposite sides of the cyclobutene plane. The iodine atoms are cis to the arsenic (*60*).

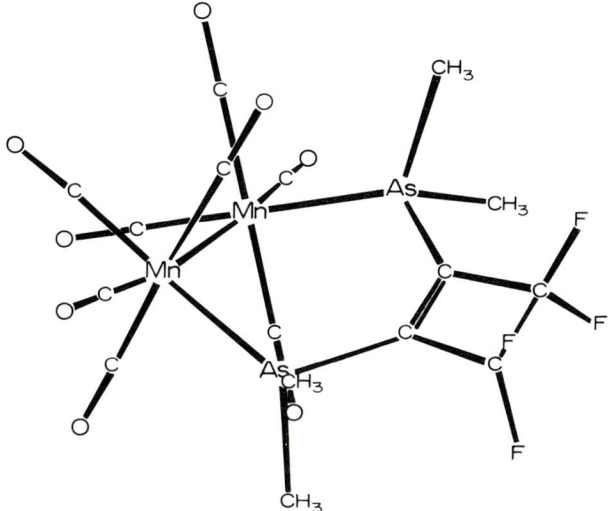

FIG. 8. The structure of f$_4$farsMn$_2$(CO)$_8$ (*60*).

D. TRILIGATE BIMETALLIC COMPLEXES

Ligands such as (**10**) and (**15**) with unsaturated bridging groups can, in principle, act as triligate groups by donating electron pairs from the two Group V atoms and the double bond. In practice this is realized only when the ligand is fluorocarbon-bridged, although not necessarily fluoroalicyclic since the complex (L–L)Fe$_2$(CO)$_6$ where (L–L) = (**15**) (R = CF$_3$) is of this type (*61*). The first complex in this class, f$_4$farsFe$_2$(CO)$_6$, had its crystal structure determined by accident [it was believed to be f$_4$farsFe(CO)$_3$ (*99*)]. The result is seen in Fig. 9 (*113*). One iron atom Fe^A is approximately octahedrally coordinated by two arsenic atoms and three carbonyl groups; the other by the double bond of the cyclobutene group, by three carbonyl groups, and by Fe^A which is acting as a donor. Thus, the coordination around Fe^B can be regarded as either a distorted

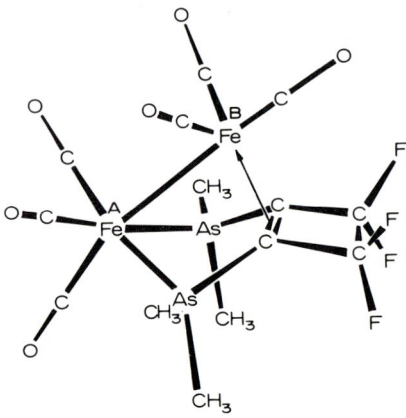

FIG. 9. The structure of f$_4$farsFe$_2$(CO)$_6$ (*113*).

trigonal bipyramid with the double bond occupying one site or a distorted octahedron with the two carbon atoms occupying two sites.* In any case the symmetry of Fe^B is lower than that of Fe^A.

This compound was eventually isolated and characterized from the reaction of f$_4$fars with Fe$_3$(CO)$_{12}$ (*75*), which has proven to be the best way of preparing the other complexes (L–L)Fe$_2$(CO)$_6$ [(L–L) = f$_4$fos, f$_6$fos (*75*), f$_4$AsP, f$_6$AsP (*40*), f$_6$fars, f$_8$fars (*64*)]. It is probably significant that in f$_4$farsFe$_2$(CO)$_6$ (Fig. 9) the bonding of the double bond of the cyclobutene ring to Fe^B allows the bite of the ligand to decrease enough for both arsenic atoms to coordinate to one iron atom (cf. the non-existence of f$_4$farsFe(CO)$_3$, Section V,B). A feature of the preparation of the f$_n$fos and f$_n$AsP compounds is that the yields decrease as the ring size is increased (compounds with $n = 8$ are unknown), which was taken as an indication that as the ring became less strained there was less tendency for the double bond to coordinate to Fe^B. However, the preparation of f$_6$fars and f$_8$fars derivatives in good yield (*64*) suggests that ring strain is not the only factor and that possibly steric hindrance plays some part. Similar compounds are obtained from Ru$_3$(CO)$_{12}$ (*71*) for f$_4$fars, f$_4$fos, and f$_6$fos and from Os$_3$(CO)$_{12}$ (*55*) for f$_4$fos.

Of all the types of complexes formed between the Group VIII carbonyls and the fluorocarbon ligands the series f$_4$fosM$_2$(CO)$_6$ is the only complete one for M = Fe, Ru, or Os. Their ease of preparation decreases in the order M = Fe > Ru > Os, which is expected in view of the stability of the parent M$_3$(CO)$_{12}$ species. The spectroscopic properties

---

* See the footnote on p. 323.

of all these compounds are very similar and, hence, they all have the same basic structure (55).

The iron complexes have been most investigated and it has been found that the Mössbauer spectrum of, for example, $f_4farsFe_2(CO)_6$, consists of two doublets which can be assigned mainly on the basis of symmetry around the Fe atoms. The parameters are listed in Table II

TABLE II

Mössbauer Parameters for $(L-L)Fe_2(CO)_6$ and Related Complexes[a]

Complexes	δ(Fe)	Δ(Fe)	Assignment[b]
$f_4farsFe_2(CO)_6$	0.28	0.64	Fe^A
	0.32	1.44	Fe^B
$f_4fosFe_2(CO)_6$	0.23	0.66	Fe^A
	0.32	1.30	Fe^B
$f_4AsPFe_2(CO)_6$	0.27	0.83	Fe^A
	0.31	1.45	Fe^B
$[(C_6H_5)_3P]f_4AsPFe_2(CO)_5$	0.36	0.56	Fe^A
	0.30	1.38	Fe^B
$(f_4AsP)^c(f_4AsP)Fe_2(CO)_4$	0.36	1.21	Fe^A
	0.36	1.21	Fe^B
$(f_4AsP)^d(f_4AsP)Fe_2(CO)_4$	0.50	0.61	Fe^A
	0.28	1.07	Fe^B

[a] See footnote on p. 348.
[b] See Fig. 9.
[c] The ligand is bridging.
[d] The ligand is chelating (108).

(40, 75). Note that Fe^A, the more symmetrical, has a lower quadrupole splitting than Fe^B, and that the isomer shift of Fe^A is less than that of Fe^B indicating that the s electron density at Fe^A is greater than Fe^B. Also note that replacement of a $(CH_3)_2As$ group by a $(C_6H_5)_2P$ group causes a decrease in isomer shift at Fe^A, presumably due to the better π-acceptor properties of the phosphorus-containing moiety, and that when f_4fos is replaced by f_4AsP only Fe^A changes.

Also listed in Table II are parameters for some derivatives of $(L-L)Fe_2(CO)_6$. These are obtained as follows, where $(L-L)'$ can be bridging or chelating.

$$(L-L)Fe_2(CO)_6 + (L-L)' \xrightarrow[h\nu]{acetone} (L-L)'(L-L)Fe_2(CO)_4 \quad (60)$$

Monosubstitution by $(C_6H_5)_3P$ apparently takes place on Fe^A as judged by the increase in $\delta(Fe^A)$. When (L–L)′ is a bridging ligand, only a two line spectrum is obtained and the isomer shifts of both Fe^A and Fe^B are thus essentially the same although increased over those of the unsubstituted complex.

E. DECOMPOSITION PRODUCTS OF COORDINATION COMPLEXES

When $f_4farsFe_3(CO)_{10}$ (21) is heated in cyclohexane solution, the major product is a compound of formula $f_4farsFe_3(CO)_9$. On further

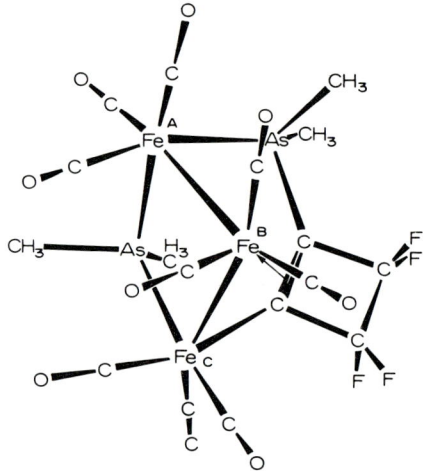

FIG. 10. The structure of $f_4farsFe_3(CO)_9$ (111).

heating, this loses its fluorocarbon group to yield $As_2(CH_3)_2CH_2Fe_3(CO)_9$ (76). The spectroscopic properties of the first $Fe_3(CO)_9$ derivative indicate that the molecule has little symmetry. The X-ray determined structure shown in Fig. 10 (111) indicates that an $As(CH_3)_2$ moiety has been displaced from the ligand. The structure consists of four dissimilar iron atoms and one arsenic atom which form a distorted square plane. One Fe^A–Fe^B bond is very long, 2.917(5) Å. The cyclobutene "double bond" of length 1.44 Å is situated 2.00 Å from Fe^B. In $f_4farsFe_2(CO)_6$ (Fig. 9) the double bond has a length of 1.51 Å and is situated 1.90 Å from Fe^B, implying a stronger bond in this case. Rearrangement of the ligand also occurs when $f_4AsPFe_3(CO)_{10}$ is heated to give the stable complex $f_4AsPFe_3(CO)_9$ (40). The carbonyl infrared spectrum of this $Fe_3(CO)_9$

derivative is very similar to that of the f₄fars complex suggesting a similar structure. This has now been confirmed by an X-ray investigation (*108a*) and the structure is essentially the same as in Fig. 9 except that a $(C_6H_5)_2P$ group replaces the $(CH_3)_2As$ group bonded to the cyclobutene ring.

Displacement of an $As(CH_3)_2$ group also occurs when $f_4farsMn_2(CO)_8$ (Fig. 8) is refluxed in xylene solution. The isomeric product was first believed to have the ligand chelating rather than bridging; however, the solid state structure is shown in Fig. 11. The two manganese atoms

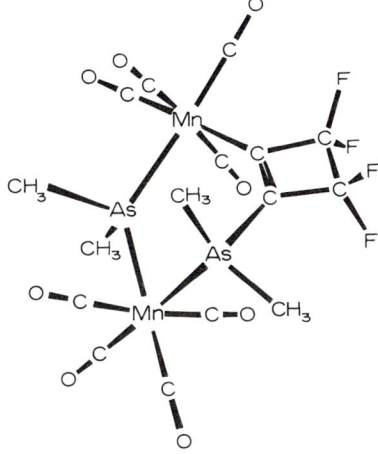

Fig. 11. The structure of the isomer of $f_4farsMn_2(CO)_8$ (cf. Fig. 8) (*110*).

are approximately octahedrally coordinated and bridged by the displaced $As(CH_3)_2$ moiety (*60, 78, 110*) and again a transition metal–fluorocarbon bond is formed. Both isomers of the rhenium analogs $(L-L)Re_2(CO)_8$ [(L–L) = f₄fars, f₆fars] have also been isolated (*60, 78*).

Perhaps the most remarkable ligand transformation occurs when solutions containing $f_4farsCo_2(CO)_6$ (Fig. 5) are heated. One unstable product has the formula $f_4farsCo_2(CO)_5$ and is believed to have a structure analogous to that of $f_4farsFe_2(CO)_6$ (*56*); another, obtained in low yield has the empirical formula $(f_4fars)_2Co_4(CO)_9$ (2H?). The structure is shown in Fig. 12 (*110*). The molecule has a twofold axis. Again $(CH_3)_2As$ groups have been displaced, but this time the cyclobutene rings have united to form a bicyclobutyl system. The two carbonyl-bridged cobalt atoms are apparently one electron short of a "closed shell" configuration and it is possible that the sixth coordination position around each is occupied by a hydrogen atom.

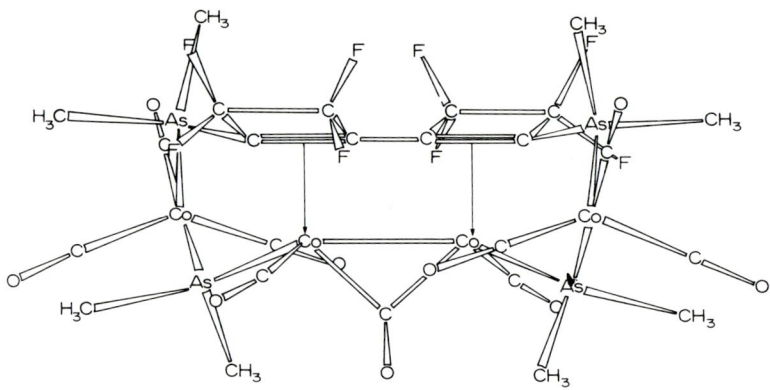

FIG. 12. The structure of $(f_4fars)_2Co_4(CO)_9$ (2H?) (110).

There are few examples of hydrocarbon ligands undergoing this type of modification on complex formation. However, this is not to say that they do not occur. Bosnich and co-workers (19) have described that a compound originally reported as $[(diars)_3Ni][ClO_4]_2$ is, in fact, $[(diars)(triars)Ni][ClO_4]_2$ {triars = $CH_3As[C_6H_4As(CH_3)_2\text{-}o]_2$}, where rearrangement of diars has occurred. Reaction of diars with iron carbonyls also seems to result in fragmentation of the ligand (73).

Acknowledgments

The author is indebted to his many graduate students, postdoctoral fellows, and colleagues in Vancouver who have contributed so significantly to much of the work described in this review. The financial support of the National Research Council of Canada is gratefully acknowledged. The figures which were drawn by Beatrix Krizsan are reproduced with permission of Dr. J. Trotter and Dr. F. W. B. Einstein and the Chemical Society, London, the American Chemical Society, and the editors of *Inorganica Chemica Acta*, and the *Canadian Journal of Chemistry*.

References

1. Aguiar, A. M., Mague, J. T., Aguiar, H. J., Archibald, T. G., and Prejean, G., *J. Org. Chem.* **33**, 1681 (1968).
2. Allen, F. H., and Pidcock, A., *J. Chem. Soc. A* 2700 (1968).
3. Anderson, L. P., Feast, W. J., and Musgrave, W. K. R., *J. Chem. Soc. C* 211 (1969).
4. Angelici, R. J., and Siefert, E. E., *Inorg. Chem.* **5**, 1457 (1966).
5. Banks, R. E., "Fluorocarbons and Their Derivatives." Oldbourne Press, London, 1964.
6. Banks, R. E., Haszeldine, R. N., Lappin, M., and Lever, A. B. P., *J. Organometal. Chem.* **29**, 427 (1971).

7. Bartlett, P. D., Montgomery, L. K., and Seidel, B., *J. Amer. Chem. Soc.* **86**, 616 (1964).
8. Beagley, B., Cruickshank, D. W. J., Pinder, P. M., Robiette, A. G., and Sheldrick, G. M., *Acta Crystallogr. Sect. B* **25**, 737 (1969).
9. Bennett, M. A., Erskine, G. J., and Wild, J. D., *Inorg. Chim. Acta* **2**, 379 (1969).
10. Bennett, M. A., and Wild, J. D., *J. Chem. Soc. A* 545 (1971).
11. Bennett, M. J., and Mason, R., *Proc. Chem. Soc.* 395 (1964).
12. Bell, T. N., Haszeldine, R. N., Newlands, M. J., and Plumb, J. B., *J. Chem. Soc.* 2107 (1965).
13. Berry, A. D., Corey, E. R., Hagen, A. P., MacDiarmid, A. G., Saalfeld, F. E., and Wayland, B. B., *J. Amer. Chem. Soc.* **92**, 1940 (1970).
14. Berry, R. S., *J. Chem. Phys.* **32**, 933 (1960).
15. Bichler, R. E. J., Booth, M. R., and Clark, H. C., *Inorg. Nucl. Chem. Lett.* **3**, 71 (1967).
16. Booth, B. L., Haszeldine, R. N., and Tucker, N. I., *J. Organometal. Chem.* **11**, P5 (1968).
17. Booth, G., *Advan. Inorg. Chem. Radiochem.* **6**, 1 (1964).
18. Bor, G., *Spectrochim. Acta* **19**, 2065 (1963).
19. Bosnich, B., Nyholm, R. S., Pauling, P. J., and Tobe, M. L., *J. Amer. Chem. Soc.* **90**, 4741 (1968).
20. Breslow, R., Eicher, T., Krebs, A., Peterson, R. A., and Posner, J., *J. Amer. Chem. Soc.* **87**, 1320 (1965).
21. British Patent 760,201 (1956).
22. Brown, D. A., and Bushnell, G. W., *Acta Crystallogr.* **22**, 296 (1967).
23. Brown, D. A., Lyons, H. J., and Manning, A. R., *Inorg. Chim. Acta* **4**, 428 (1970).
24. Browning, J., Green, M., and Stone, F. G. A., *J. Chem. Soc. A* 453 (1971).
25. Brice, M. D., Penfold, B. R., Robinson, W. T., and Taylor, S. R., *Inorg. Chem.* **9**, 362 (1970).
26. Bruce, M. I., Jolly, P. W., and Stone, F. G. A., *J. Chem. Soc. A* 1602 (1966).
27. Bruce, M. I., and Stone, F. G. A., in "Preparative Inorganic Reactions," (W. L. Jolly, ed.) Vol. 4, p. 177. Wiley, New York, 1968.
28. Bruce, M. I., and Stone, F. G. A., *Angew. Chem. Int. Ed. Engl.* **7**, 747 (1968).
29. Bryan, R. F., *J. Chem. Soc. A* 192 (1967).
30. Burton, D. J., and Mettille, F. J., *Inorg. Nucl. Chem. Lett.* **4**, 9 (1968).
31. Calderazzo, F., Ercoli, R., and Natta, G., in "Organic Synthesis Via Metal Carbonyls" (I. Wender and P. Pino, eds.) Vol. 1, p. 1. Wiley (Interscience), New York, 1967.
32. Callot, H. J., and Benezra, C., *Chem. Commun.* 485 (1970).
33. Campbell, S. F., Leach, J. M., Stephens, R., and Tatlow, J. C., *Tetrahedron Lett.* 4269 (1967).
34. Campbell, S. F., Leach, J. M., Stephens, R., Tatlow, J. C., and Wood, K. N., *J. Fluorine Chem.* **1**, 103 (1971/72).
35. Campbell, S. F., Leach, J. M., Stephens, R., and Tatlow, J. C., *J. Fluorine Chem.* **1**, 85 (1971/72).
36. Campbell, S. F., Stephens, R., and Tatlow, J. C., *Tetrahedron* **21**, 2997 (1965).
37. Campbell, S. F., Stephens, R., and Tatlow, J. C., *Chem. Commun.* 151 (1967).
38. Chang, C. H., Porter, R. F., and Bauer, S. H., *J. Mol. Struct.* **7**, 89 (1971).
39. Cherwinski, W. J., and Clark, H. C., *J. Organometal. Chem.* **29**, 451 (1971).

40. Chia, L. S., M.Sc. Thesis, University of British Columbia, 1971.
41. Churchill, M. R., and Veidis, M. V., *Chem. Commun.* 529 (1970).
42. Clark, H. C., Corfield, P. W. R., Dixon, K. R., and Ibers, J. A., *J. Amer. Chem. Soc.* **89**, 3360 (1967).
43. Clark, H. C., and Puddephatt, R. J., *Inorg. Chem.* **9**, 2670 (1970).
44. Clark, H. C., and Tsang, W. S., *J. Amer. Chem. Soc.* **89**, 529 (1967).
45. Clark, H. C., and Willis, C. J., *J. Amer. Chem. Soc.* **82**, 1888 (1960).
46. Clayton, A. B., Collins, D., Stephens, R., and Tatlow, J. C., *J. Chem. Soc. C* 1177 (1971).
47. Clayton, A. B., Feast, W. J., Sayers, D. R., Stephens, R., and Tatlow, J. C., *J. Chem. Soc. C* 1183 (1971).
48. Closs, G. L., *Advan. Alicyclic Chem.* **1**, 53 (1966).
49. Collman, J. P., and Roper, W. R., *Advan. Organometal. Chem.* **7**, 53 (1968).
50. Cook, D. J., Green, M., Mayne, N., and Stone, F. G. A., *J. Chem. Soc. A* 1771 (1968).
51. Corey, E. R., and Dahl, L. F., *Inorg. Chem.* **1**, 521 (1964).
52. Corradini, P., and Sirigu, A., *Ric. Sci.* **36**, 188 (1966).
53. Cotton, F. A., and Parish, R. V., *J. Chem. Soc.* 1440 (1960).
54. Cotton, F. A., and Wilkinson, G., "Advanced Inorganic Chemistry," 2nd ed. Wiley (Interscience), New York, 1966.
55. Crow, J. P., and Cullen, W. R., *Inorg. Chem.* **10**, 1529 (1971).
56. Crow, J. P., and Cullen, W. R., *Inorg. Chem.* **10**, 2165 (1971).
57. Crow, J. P., and Cullen, W. R., *Can. J. Chem.* **48**, 2948 (1971).
58. Crow, J. P., Cullen, W. R., Herring, F. G., Sams, J. R., and Tapping, R. L., *Inorg. Chem.* **10**, 1616 (1971).
59. Crow, J. P., Cullen, W. R., and Hou, F. L., *Inorg. Chem.* in press.
60. Crow, J. P., Cullen, W. R., Hou, F. L., Chan, L. Y. Y., and Einstein, F. W. B., *Chem. Commun.* 1229 (1971).
61. Crow, J. P., Cullen, W. R., Sams, J. R., and Ward, J. E. H., *J. Organometal. Chem.* **22**, C 29 (1970).
62. Cullen, W. R., *Fluorine Chem. Rev.* **3**, 73 (1969).
63. Cullen, W. R., and Bruce, M. I., *Fluorine Chem. Rev.* **4**, 79 (1969).
64. Cullen, W. R., and Chia, L. S., unpublished results.
65. Cullen, W. R., Dawson, D. S., and Dhaliwal, P. S., *Can. J. Chem.* **45**, 683 (1967).
66. Cullen, W. R., and Dhaliwal, P. S., *Can. J. Chem.* **45**, 719 (1967).
67. Cullen, W. R., Dhaliwal, P. S., and Stewart, C. J., *Inorg. Chem.* **6**, 2256 (1967).
68. Cullen, W. R., Dhaliwal, P. S., and Stewart, C. J., unpublished results.
69. Cullen, W. R., Dhaliwal, P. S., and Styan, G. E., *J. Organometal. Chem.* **6**, 364 (1966).
70. Cullen, W. R., Dong, D. F., and Thompson, J. A. J., *Can. J. Chem.* **47**, 4671 (1969).
71. Cullen, W. R., and Harbourne, D. A., *Inorg. Chem.* **9**, 1839 (1970).
72. Cullen, W. R., and Harbourne, D. A., unpublished results.
73. Cullen, W. R., and Harbourne, D. A., *Can. J. Chem.* **47**, 3371 (1969).
74. Cullen, W. R., Harbourne, D. A., Liengme, B. V., and Sams, J. R., *Inorg. Chem.* **8**, 1464 (1969).
75. Cullen, W. R., Harbourne, D. A., Liengme, B. V., and Sams, J. R., *Inorg. Chem.* **8**, 95 (1969).

76. Cullen, W. R., Harbourne, D. A., Liengme, B. V., and Sams, J. R., *Inorg. Chem.* **9**, 702 (1970).
77. Cullen, W. R., and Hota, N. K., *Can. J. Chem.* **42**, 1123 (1964).
78. Cullen, W. R., and Hou, F. L., unpublished results.
79. Cullen, W. R., and Hou, F. L., *Can. J. Chem.* **49**, 2749 (1971).
80. Cullen, W. R., Johnson, R. N., and Hall, L. D., unpublished work.
81. Cullen, W. R., and Leeder, W. R., *Inorg. Chem.* **5**, 1004 (1966).
82. Cullen, W. R., and Styan, G. E., *J. Organometal. Chem.* **6**, 633 (1966).
83. Cullen, W. R., and Thompson, J. A. J., *Can. J. Chem.* **48**, 1730 (1970).
84. Cullen, W. R., Sams, J. R., and Thompson, J. A. J., *Inorg. Chem.* **10**, 843 (1971).
85. Cullen, W. R., Sams, J. R., and Waldman, M. C., *Inorg. Chem.* **9**, 1682 (1970).
86. Cullen, W. R., and Styan, G. E., *J. Organometal. Chem.* **4**, 151 (1965).
87. Cullen, W. R., and Waldman, M. C., *J. Fluorine Chem.* **1**, 151 (1971/72).
88. Cullen, W. R., and Waldman, M. C., *Can. J. Chem.* **48**, 1885 (1970).
89. Cullen, W. R., and Waldman, M. C., *J. Fluorine Chem.* **1**, 41 (1971/72).
90. Dahl, L. F., and Rundle, R. E., *Acta Crystallogr.* **16**, 419 (1963).
91. Dahl, L. F., and Wei, C. H., *Inorg. Chem.* **2**, 328 (1963).
92. Dahm, D. J., and Jacobson, R. A., *J. Amer. Chem. Soc.* **90**, 5106 (1968).
93. Dahm, D. J., and Jacobson, R. A., *Chem. Commun.* 496 (1966).
94. Davidsohn, W., and Henry, M. C., *J. Organometal. Chem.* **5**, 29 (1966).
95. Davidson, I. M. T., Eaborn, C., and Lilly, M. N., *J. Chem. Soc.* 2624 (1964).
96. de Beer, J. A., Haines, R. J., Greatrex, R., and Greenwood, N. N., *J. Organometal. Chem.* **27**, C33 (1971).
97. Deeming, A. J., Johnson, B. F. G., and Lewis, J., *J. Chem. Soc. A* 897 (1970).
98. Dehmlow, E. V., *J. Organometal. Chem.* **6**, 296 (1966).
99. Dhaliwal, P. S., Ph.D. Thesis, University of British Columbia, 1966.
100. Dobson, G. R., Stolz, I. W., and Sheline, R. K., *Advan. Inorg. Chem. Radiochem.* **8**, 1 (1966).
101. Donohue, J., and Caron, A., *J. Phys. Chem.* **70**, 603 (1966).
102. Dreier, F., Duncan, W., and Mill, T., *Tetrahedron Lett.* 1951 (1964).
103. Dyatkin, B. L., Mochalina, E. P., and Knunyants, I. L., *Fluorine Chem. Rev.* **3**, 45 (1969).
104. Dyatkin, B. L., Sterlin, S. R., Martynov, B. I., and Knunyants, I. L., *Abstr. 6th Int. Symp. Fluorine Chem. Durham*, p. A31 (1971).
105. Einstein, F. W. B., and Hampton, C. R. S. M., *Can. J. Chem.* **49**, 1901 (1971).
106. Einstein, F. W. B., and Jones, R. D. G., *Inorg. Chem.* **11**, 395 (1972).
107. Einstein, F. W. B., and Jones, R. D. G., *J. Chem. Soc. A* 3359 (1971).
108. Einstein, F. W. B., and Jones, R. D. G., *Inorg. Chem.*, in press.
108a. Einstein, F. W. B., and Jones, R. D. G., unpublished results.
109. Einstein, F. W. B., and Jones, R. D. G., *J. Chem. Soc. Dalton* 442 (1972).
110. Einstein, F. W. B., Jones, R. D. G., MacGregor, A. C., and Cullen, W. R., unpublished results.
111. Einstein, F. W. B., Pilotti, A. M., and Restivo, R., *Inorg. Chem.* **10**, 1947 (1971).
112. Einstein, F. W. B., and Restivo, R., *Inorg. Chim. Acta* **5**, 501 (1971).
113. Einstein, F. W. B., and Trotter, J., *J. Chem. Soc. A* 824 (1967).
114. Eleméus, H. J., "The Chemistry of Fluorine and its Compounds," Academic Press, New York, 1969.

115. Eméleus, H. J., and Haszeldine, R. N., *J. Chem. Soc.* 2948, 2953 (1949).
116. Ernst, R. R., *Mol. Phys.* **16**, 241 (1969).
117. Erskine, G. J., *Can. J. Chem.* **47**, 2699 (1969).
118. Feltham, R. D., Metzger, H. G., and Silverthorn, W., *Inorg. Chem.* **7**, 2003 (1968).
119. Fenton, D. E., and Zuckerman, J. J., *J. Amer. Chem. Soc.* **90**, 6226 (1968).
120. Ferretti, A., and Tesi, G., *Chem. Ind. (London)* 1987 (1964).
121. Fields, R., Green, M., Harrison, T., Haszeldine, R. N., Jones, A., and Lever, A. B. P., *J. Chem. Soc. A* 49 (1970).
122. Fields, R., Haszeldine, R. N., and Hubbard, A. F., *Chem. Commun.* 647 (1970).
123. Fields, R., Haszeldine, R. N., Hubbard, A. F., and Palmer, P. J., *Abstr. 6th Int. Symp. Fluorine Chem. Durham*, p. A15 (1971).
124. Frank, A. W., *J. Org. Chem.* **31**, 1917 (1966).
125. Frank, A. W., *J. Org. Chem.* **30**, 3663 (1965).
126. Frank, A. W., *J. Org. Chem.* **31**, 1521 (1966).
127. Frank, A. W., *J. Org. Chem.* **31**, 1920 (1966).
128. Freeburger, M. E., and Spialter, L., *J. Org. Chem.* **35**, 652 (1970).
129. Gale, D. M., Middleton, W. J., and Krespan, C. G., *J. Amer. Chem. Soc.* **87**, 657 (1965).
130. Green, M. L. H., "Organometallic Compounds", Vol. 2. Methuen, London, 1968.
131. Green, M., Osborn, R. B. L., Rest, A. J., and Stone, F. G. A., *J. Chem. Soc. A* 2525 (1968).
132. Greene, P. T., and Bryan, R. F., *J. Chem. Soc. A* 2261 (1970).
133. Greene, P. T., and Bryan, R. F., *J. Chem. Soc. A* 1696 (1970).
134. Greenwood, N. N., *Chem. Brit.* **3**, 56 (1967).
135. Hall, M. C., Kilbourn, B. T., Taylor, K. A., *J. Chem. Soc. A* 2539 (1970).
136. Haluska, L. A., U.S. Patent 2,800,494 (1957).
137. Harris, C. M., and Livingston, S. E., in "Chelating Agents and Metal Chelates" (F. P. Dwyer and D. P. Meller, eds.), p. 130. Academic Press, New York, 1964.
138. Harris, R. K., and Robinson, V. J., *J. Mag. Res.* **1**, 362 (1969).
139. Harrison, W., and Trotter, J., *J. Chem. Soc. A* 1542 (1971).
140. Harrison, W., and Trotter, J., *J. Chem. Soc. A* 1607 (1971).
141. Hieber, W., and Freyer, W., *Chem. Ber.* **93**, 462 (1960).
142. Hieber, W., and Kummer, R., *Chem. Ber.* **100**, 148 (1967).
143. Hine, J., "Divalent Carbon." Ronald Press, New York, 1964.
144. Holbrook, G. W., Gordon, A. F., and Pierce, O. R., *J. Amer. Chem. Soc.* **82**, 825 (1960).
145. Hota, N. K., and Willis, C. J., *J. Organometal. Chem.* **15**, 89 (1968).
146. Ibers, J. A., *J. Organometal. Chem.* **14**, 423 (1968).
147. Jetz, W., and Graham, W. A. G., *J. Amer. Chem. Soc.* **89**, 2773 (1967).
148. Jolly, P. W., Bruce, M. I., and Stone, F. G. A., *J. Chem. Soc.* 5830 (1965).
149. King, R. B., *J. Amer. Chem. Soc.* **84**, 2460 (1962).
150. King, R. B., and Egger, C. A., *Inorg. Chim. Acta* **2**, 33 (1968).
151. King, R. B., Kapoor, R. N., and Pannell, K. H., *J. Organometal. Chem.* **20**, 187 (1969).
152. King, R. B., and Pannell, K. H., *Inorg. Chem.* **7**, 1510 (1968).
153. Kirmse, W., "Carbene Chemistry." Academic Press, New York, 1964.

154. Knunyants, I. L., Pervova, E. Ya., and Tyuleneva, V. V., *Dokl. Akad. Nauk SSSR* **129**, 576 (1959).
155. Knunyants, I. L., Tyuleneva, V. V., Pervova, E. Ya., and Sterlin, R. N., *Izv. Akad. Nauk SSSR Ser. Khim.* 1797 (1964).
156. Lambert, J. B., and Roberts, J. D., *J. Amer. Chem. Soc.* **87**, 3884, 3891 (1965).
157. Lewis, J., Nyholm, R. S., and Phillips, D. J., *J. Chem. Soc.* 2177 (1962).
158. Mague, J. T., *Inorg. Chem.* **8**, 1975 (1969).
159. Mague, J. T., and Mitchener, J. P., *Inorg. Chem.* **8**, 119 (1969).
160. Mague, J. T., and Wilkinson, G., *Inorg. Chem.* **7**, 542 (1968).
161. Mahler, W., *J. Amer. Chem. Soc.* **84**, 4600 (1962).
162. Manuel, T. A., *Advan. Organometal. Chem.* **3**, 181 (1965).
163. Margulis, T. N., *J. Amer. Chem. Soc.* **93**, 2193 (1971).
164. Mason, R., and Rae, A. I. M., *J. Chem. Soc.* A 778 (1968).
165. McBee, E. T., Turner, J. J., Morton, C. J., and Stefani, A. P., *J. Org. Chem.* **30**, 3698 (1965).
166. McDonald, W. S., Moss, J. R., Raper, G., Shaw, B. L., Greatrex, R., and Greenwood, N. N., *Chem. Commun.* 1295 (1969).
167. Mitchell, C. M., and Stone, F. G. A., *Proc. 4th Int. Conf. Organometal. Chem. Bristol*, E5 (1969).
168. Mill, T., Rodin, J. O., Silverstein, R. M., and Woolf, C., *J. Org. Chem.* **28**, 836 (1963).
169. Miller, W. T., Hummel, R. J., and Pelosi, L. F., *Abstr. 6th Int. Symp. Fluorine Chem., Durham, 1971*, p. A30 (1971).
170. Miller, W. T., Snider, R. H., and Hummel, R. J., *J. Amer. Chem. Soc.* **91**, 6533 (1969).
171. Mingos, D. M. P., and Ibers, J. A., *Inorg. Chem.* **9**, 1105 (1970).
172. Muetterties, E. L., *Accounts Chem. Res.* **3**, 266 (1970).
173. Mukhedkar, A. J., Mukhedkar, V. A., Green, M., and Stone, F. G. A., *J. Chem. Soc. A* 3166 (1970).
174. Newmark, R. A., Apai, G. R., and Michael, R. O., *J. Mag. Res.* **1**, 418 (1969).
175. Noack, K., *Spectrochim. Acta* **19**, 1925 (1963).
176. Nyholm, R. S., Snow, M. R., and Stiddard, M. H. B., *J. Chem. Soc.* 6564 (1965).
177. Park, J. D., Bertino, C. D., and Nakata, B. T., *J. Org. Chem.* **34**, 1490 (1969).
178. Park, J. D., Choi, S. K., and Romine, H. E., *J. Org. Chem.* **34**, 2521 (1969).
179. Park, J. D., Groves, J. D., and Lacher, J. R., *J. Org. Chem.* **25**, 1628 (1960).
180. Park, J. D., and McMurtry, R. J., *Tetrahedron Lett.* 1301 (1967).
181. Park, J. D., Michael, R. O., and Newmark, R. A., *J. Org. Chem.* **34**, 2525 (1969).
182. Park, J. D., and Nakata, B. T., *Abstr. 154th Meeting Amer. Chem. Soc., Chicago, 1967*.
183. Poliakoff, M., and Turner, J. J., *J. Chem. Soc. A* 654 (1971).
184. Ponomarenko, V. A., and Snegova, A. D., *Izv. Akad. Nauk SSSR Otd. Khim. Nauk* 135 (1960).
185. Pruett, R. L., Barr, J. T., Rapp, K. E., Bahner, C. T., Gibson, J. D., and Lafferty, R. H., *J. Amer. Chem. Soc.* **72**, 3646 (1950).
186. Rapp, K. E., Pruett, R. L., Barr, J. T., Bahner, C. T., Gibson, J. D., and Lafferty, R. H., *J. Amer. Chem. Soc.* **72**, 3642 (1950).
187. Roberts, J. D., and Sharts, C. M., *Org. Reactions* **12**, 1 (1962).

188. Roberts, P. J., Penfold, B. R., and Trotter, J., *Inorg. Chem.* **9**, 2137 (1970).
189. Roberts, P. J., and Trotter, J., *J. Chem. Soc. A* 1479 (1971).
190. Roberts, P. J., and Trotter, J., *J. Chem. Soc. A* 3246 (1970).
191. Sacco, A., *Gazz. Chim. Ital.* **93**, 698 (1963).
192. Sacco, A., and Ugo, R., *J. Chem. Soc.* 3274 (1964).
193. Sargeant, P. B., *J. Amer. Chem. Soc.* **91**, 3061 (1969).
194. Seyferth, D., *Organometal. Chem. Rev.* **134**, 242 (1968).
195. Seyferth, D., and Damrauer, R., *J. Org. Chem.* **31**, 1660 (1966).
196. Seyferth, D., and Darragh, K. V., *J. Org. Chem.* **35**, 1297 (1970).
197. Seyferth, D., Jula, T. F., Dertouzos, H., and Pereyre, M., *J. Organometal. Chem.* **11**, 63 (1968).
198. Seyferth, D., Dertouzos, H., Suzuki, R., Mui, J. Y. P., *J. Org. Chem.* **32**, 2980 (1967).
199. Stephens, R., Tatlow, J. C., and Wood, K. N., *Abstr. 6th Int. Symp. Fluorine Chem., Durham*, p. A1 (1971).
200. Stockel, R. F., *Can. J. Chem.* **47**, 867 (1969).
201. Stockel, R. F., *Can. J. Chem.* **46**, 2625 (1968).
202. Stockel, R. F., Megson, F., and Beachem, M. T., *J. Org. Chem.* **33**, 4395 (1968).
203. Sullivan, R., Lacher, J. R., and Park, J. D., *J. Org. Chem.* **29**, 3664 (1964).
204. Sumner, G. G., Klug, H. P., and Alexander, L. E., *Acta Crystallogr.* **17**, 732 (1964).
205. Sutton, P. W., and Dahl, L. F., *J. Amer. Chem. Soc.* **89**, 261 (1967).
206. Tarrant, P., Johnson, R. W., and Brey, W. S., *J. Org. Chem.* **27**, 602 (1962).
207. Tarrant, P., and Oliver, W. H., *J. Org. Chem.* **31**, 1143 (1966).
208. Treichel, P. M., and Stone, F. G. A., *Advan. Organometal. Chem.* **1**, 143 (1964).
209. Ugi, I., Marquarding, D., Klusacek, H., Gillespie, P., and Ramirez, F., *Accounts Chem. Res.* **4**, 288 (1971).
210. Venanzi, L. M., *Chem. Brit.* **4**, 162 (1968).
211. Waldman, M. C., Ph.D. Thesis, University of British Columbia (1968).
212. Wei, C. H., and Dahl, L. F., *J. Amer. Chem. Soc.* **91**, 1351 (1969).
213. Wei, C. H., and Dahl, L. F., *J. Amer. Chem. Soc.* **88**, 1821 (1966).
214. Wei, C. H., Wilkes, G. R., and Dahl, L. F., *J. Amer. Chem. Soc.* **89**, 4792 (1967).
215. Wertheim, G. K., "Mössbauer Effect: Principles and Applications." Academic Press, New York, 1964.
216. Wilcox, C. F., and Craig, R. R., *J. Amer. Chem. Soc.* **83**, 3866 (1961).
217. Wilford, J. B., Treichel, P. M., and Stone, F. G. A., *Proc. Chem. Soc.* 218 (1963).
218. Woodward, R. B., and Hoffman, R., *Angew. Chem. Int. Ed. Engl.* **8**, 781 (1969).
219. Wright, J. S., and Salem, L., *Chem. Commun.* 1370 (1969).
220. Zumdahl, S. S., and Drago, R. S., *J. Amer. Chem. Soc.* **90**, 6669 (1968).

THE SULFUR NITRIDES

H. G. Heal
Queen's University of Belfast, Belfast, Northern Ireland

I. Introduction 375
II. Thiazyl, SN, and Its Polymers 376
   A. Sulfur Nitride, SN 376
   B. Tetrasulfur Tetranitride, S_4N_4 377
   C. Disulfur Dinitride, S_2N_2 393
   D. Polythiazyl, $(SN)_x$ 395
III. Tetrasulfur Dinitride, S_4N_2 396
IV. Saturated Sulfur–Nitrogen Frameworks, Coupled and Fused Rings, and Polymers 399
   A. General 399
   B. The Coupled-Ring Nitrides, $S_7N-S_x-NS_7$ 400
   C. The Fused-Ring Nitride, $S_{11}N_2$ 403
   D. Other Fused-Ring Nitrides 406
   E. Polymeric Saturated Sulfur Nitrides 407
V. Conclusion 408
References 409

I. Introduction

Although tetrasulfur tetranitride, S_4N_4, was first prepared about 1835 (53, 54), the chemistry of the sulfur nitrides progressed slowly until the middle of the twentieth century. In 1940 the structure of S_4N_4 was not known and could not even be reliably postulated. Of the eight other sulfur nitrides described as individual chemical species in the present review, only two were known, and these had not been fully characterized. Since then, there have been great advances and the whole subject has been put on a systematic footing. Some key contributions have come from the schools of Donohue, who correctly determined the structure of S_4N_4; Becke-Goehring and Meuwsen, whose wide-ranging and meticulous experimental work and original thinking have extended our knowledge in many directions; and Weiss, whose introduction of chromatography to this area of chemistry has proved to be one of the most significant of the many advances made by him. At present, several research groups in Europe, North America, nd Asia are working on sulfur nitrides, and the subject is still in a lively state, though the effort is on a much smaller scale than in more fashionable fields of inorganic chemistry.

There has been no review devoted entirely to sulfur nitrides since 1959 (8), but the subject has been covered in three more recent reviews of broader scope (14, 16b, 57). "Gmelin's Handbuch" (47) covers the literature up to 1961.

In the present review, the literature has been examined up to September 1971. The main emphasis is on recent work, and an attempt is made to point out promising lines for future research.

II. Thiazyl, SN, and Its Polymers

A. SULFUR NITRIDE, SN

Sulfur nitride, SN, or thiazyl monomer, is a radical with one unpaired electron. In contrast to its homolog, nitric oxide, it polymerizes so readily that it cannot be isolated as a monomeric solid or liquid and has only a transient existence in the gaseous phase. Nevertheless, much is known about its formation and properties, mainly from its emission spectrum.

The band spectrum of SN was first observed in 1933 in the light emitted by an electric discharge through a mixture of nitrogen and sulfur vapor, and was recognized as coming from SN by its resemblance to the spectrum of NO (34). There have been several subsequent investigations of the SN spectrum, in the course of which emissions involving three electronically excited states ($^2\Sigma$, $^2\Delta$, and $^2\Pi$), as well as the ground state ($^2\Pi$), have been observed (52, 67, 94). The ESR spectrum of SN has been described (26).

SN is formed by the passage of an electric discharge through sulfur vapor and nitrogen (34, 67, 81) or SF_6 and nitrogen (94), from the reaction of "active nitrogen" with sulfur vapor (17), H_2S (52, 67, 88), or SCl_2 (52), or from flash photolysis of a mixture of COS and NF_3 (75).

The bond length of SN, 149.7 pm, calculated from its spectroscopic moment of inertia (118), is shorter by 24–28 pm than the sum of single-bond covalent radii, which indicates a bond order between 2 and 3. This parallels the situation for NO and is as expected for the electronic structure of both molecules, viz., three bonding MO's fully occupied and an odd electron in an antibonding MO. The dissociation energy of this strong S–N bond is estimated from spectroscopic data at 463 kJ (87). Nevertheless, the compound is endothermic and unstable with respect to its elements, like the other sulfur nitrides; ΔH^0_f is 281 kJ (87). The ionization potential has been estimated as 950 kJ from these thermochemical values coupled with the experimental electron-impact appearance potential of SN^+ from NSF (87).

The experimental dipole moment of SN is 1.8 ± 0.02 Debye units in either the ground ($^2\Pi_{1/2}$) or first excited ($^2\Pi_{3/2}$) states (87), corresponding to "24% ionic character."

Ab initio calculations of several properties of this molecule have been carried out (87), using the matrix Hartree–Fock method and the experimental internuclear distance. Good results were obtained for the dipole moment and the ionization potential, but the calculated binding energy (a small difference between two large quantities) was seriously in error.

The polymerization of SN has not been studied in detail, but a deposit of S_4N_4 has been observed in the exit pumping system of the apparatus used for the reaction of active nitrogen with H_2S or SCl_2 (52).

Salts of the SN$^+$ ion, analogous to the nitrosonium salts, have recently been prepared (46a) by the reaction of thiazyl fluoride, NSF, with arsenic and antimony pentafluorides.

B. Tetrasulfur Tetranitride, S_4N_4

This is the best known nitride of sulfur, and has formed the starting point of most investigations in sulfur–nitrogen chemistry. One cannot work for long with sulfur–nitrogen compounds without encountering S_4N_4, for it is formed as a by-product of many reactions.

1. Preparation

The most convenient method of preparation of S_4N_4 is still the traditional one, from a sulfur chloride and ammonia in an inert solvent, usually carbon tetrachloride. This preparation has recently been investigated (110) with the object of maximizing yields; the findings largely supersede earlier scattered information on the effects of procedural variations. The oxidation state of sulfur in S_4N_4 is +3, so, in principle, it would be desirable to start from a sulfur(III) chloride if one existed. Since none does exist, the use of a CCl_4 solution of S_2Cl_2, saturated with chlorine, is recommended. It is not certain that the chlorine actually raises the oxidation state of the sulfur; the reaction of S_2Cl_2 with chlorine to give SCl_2 is known to be slow in absence of a catalyst. However, saturation with chlorine doubles the yield of S_4N_4. Excess ammonia is passed into the mixture, keeping the temperature between 20° and 50°. At 0° the yield of S_4N_4 is 65% lower; there is a 40% increase in the yield of the other main sulfur–nitrogen compound formed, S_7NH. The S_4N_4 is mostly precipitated in the reaction mixture together with ammonium chloride, from which it can be extracted with dioxane to give a yield of 28% based on sulfur.

S_4N_4 explodes when subjected to percussion, friction, or sudden heating. The above preparation is safe, but care is needed with the crystalline end product. Its sensitivity to shock or heat is stated to increase with purity (*110*). No serious accidents with S_4N_4 have been reported, but a research worker in our laboratory was badly cut on the hand when 0.5 gm of the compound which she was crushing with a glass rod in a small beaker detonated violently. The mishandling of quantities of a few grams could cause devastating explosions and endanger life. Obviously one should not leave stocks of S_4N_4 lying about or prepare more than is needed for the purpose in hand.

The course of this S_4N_4 synthesis is now partly understood (*13, 15*). Whether the starting material is S_2Cl_2 or SCl_2, the volatile and rather unstable NSCl monomer is formed when the flow of ammonia starts; it can be driven out of the reaction mixture by a stream of nitrogen and identified by its infrared spectrum (*15*). NSCl is known to react with S_2Cl_2 (*73*) according to the equation

$$2\,NSCl + S_2Cl_2 \rightarrow [S_3N_2Cl]^+Cl^- + SCl_2$$

giving the red-brown crystalline compound $[S_3N_2Cl]^+Cl^-$; in the reaction mixture this reaction goes much more quickly than the competing trimerization of NSCl. The $[S_3N_2Cl]^+Cl^-$ has not been isolated from the mixture, as it transforms rapidly in the presence of excess sulfur chlorides and ammonia to the next intermediate, the golden-yellow crystalline thiotrithiazyl chloride, $[S_4N_3]^+Cl^-$. This precipitates and can be isolated in 45% yield (based on sulfur) by interrupting the ammonia stream at the right stage. Finally, as has been shown by separate experiments (*13*), the $[S_4N_3]^+Cl^-$ reacts with more ammonia to give S_4N_4.

A variant of this synthesis, handy for small-scale preparations but giving poor percentage yields, is to pass S_2Cl_2 vapor through pelleted ammonium chloride at 160° (*65*).

An adaptation of Herring's "azide synthesis" of phosphonitrilics is worth mentioning for the principle involved, although it has not yet been developed into a reliable S_4N_4 synthesis (*45*). S_2Cl_2 is added to a suspension of lithium azide in an inert solvent. Nitrogen is evolved. The reactions postulated are as follows:

$$4LiN_3 + 2S_2Cl_2 \rightarrow 2S_2(N_3)_2 + 4LiCl$$
$$\text{(unstable)}$$
$$2S_2(N_3)_2 \rightarrow 4SN + 4N_2$$
$$4SN \rightarrow S_4N_4$$

In the experiments reported, the S_4N_4 was not isolated, but underwent its known reaction with more S_2Cl_2 (Section II, B, 10) to give $[S_4N_3]^+Cl^-$ in 60% yield based on S_2Cl_2.

2. General Description

Tetrasulfur tetranitride forms monoclinic (24, 29), strongly birefringent (105) crystals of density 2.20 to 2.23 gm/ml at room temperature (47), which are usually twinned. It is orange-yellow at room temperature, becoming almost colorless at −190°, orange-red at 100°, and red at higher temperatures. The melting points of well-crystallized specimens range from 178° to 187° (110). Since the compound in solution in molten sulfur decomposes completely between 130° and 190° (101), the melting point is probably affected by incipient decomposition and, consequently, by the rate of heating. S_4N_4 cannot readily be vaporized even in a high vacuum at room temperature, but it sublimes below 130° at 0.1 mm Hg (78). No vapor pressures have been reported, but various qualitative observations (47) suggest that the vapor pressure reaches 1 mm Hg somewhere between 80° and 130°. It is insoluble in, and not quickly affected by, water, but soluble in a wide range of organic solvents, in most cases without decomposition. Quantitative solubility data are available for four solvents (Table I). In addition, it is stated (47) to be

TABLE I
SOLUBILITIES OF TETRASULFUR TETRANITRIDE

Solvent	Temperature (°C)	Solubility (moles of S_4N_4 per kg of solvent)
Carbon disulfide	0	0.0160
	10	0.0295
	20	0.0404
	30	0.0568
	46.25	0.0810
Benzene	0	0.0140
	10	0.0263
	20	0.0389
	30	0.0536
	40	0.0686
	50	0.0847
	60	0.1056
Ethanol	0	0.0044
	10	0.0057
	20	0.0072
	30	0.0087
	40	0.0102
	50	0.0113

appreciably soluble in carbon tetrachloride, chloroform, aniline, benzaldehyde, formic acid, acetic acid pyridine, quinoline, thiophene, and naphthalene, and to be "almost insoluble" in nitromethane, acetonitrile, nitrobenzene, and n-pentane. In the laboratory S_4N_4 is easily recognized

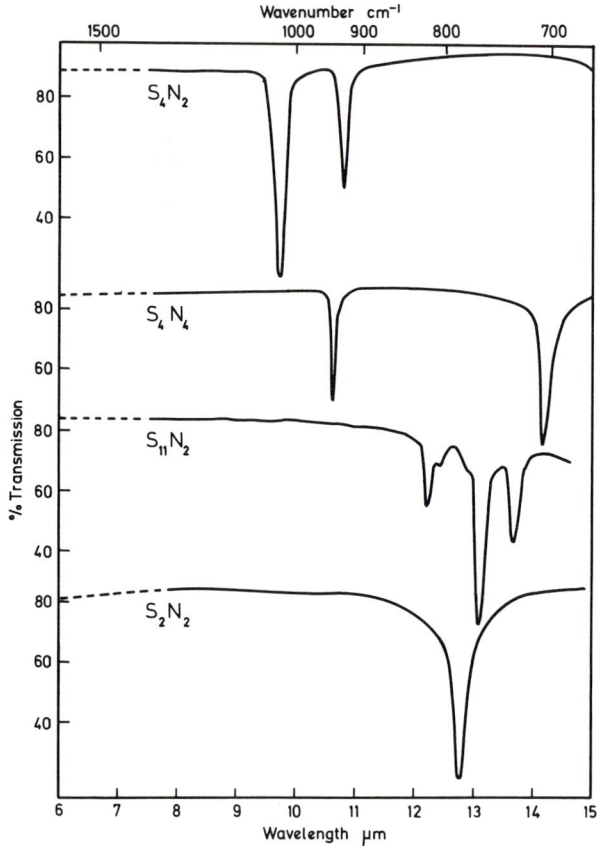

FIG. 1. Infrared spectra of CS_2 solutions of sulfur nitrides. 2–5% Solutions in CS_2; 0.2-mm cell. Dotted lines in regions of strong solvent absorption.

by its characteristic infrared spectrum in CS_2 solution, with two strong bands at 705 and 938 cm^{-1} (Fig. 1). Its X-ray powder diffraction pattern has also been published (56).

3. Molecular Structure

The molecular formula S_4N_4 has been established by analysis and cryoscopic and ebullioscopic determinations of the molecular weight

(1, 99). Diffraction studies of the vapor (71) and crystal (104) show the molecule to have the shape of Fig. 2 in both states of aggregation, with D_{2d} ($\bar{4}$2m) symmetry. The nitrogen atoms form a square and the sulfur atoms a bisphenoid. The S–N bond lengths are all the same, 161.6 ± 1.0 pm (108), and correspond to about a 1.65 bond order (27a). The distances S-1 to S-3 and S-2 to S-4 are 258 pm, longer than a S–S single bond (205 pm), but shorter than the sum of van der Waals' radii (about 330 pm). This suggests weak S–S bonding, supported by MO calculations (108). A molecule with the shape shown in Fig. 2 should have a zero dipole

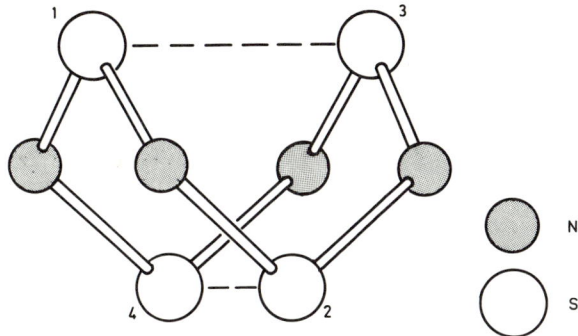

FIG. 2. Molecular structure of S_4N_4. After Sharma and Donohue (104).

moment. Nonzero values, 0.52 and 0.72 D, have, however, been calculated from the experimental dielectric constants and refractive indices of solutions in benzene and CS_2, respectively (96). These values may be spurious, arising from failure to correct adequately for the atomic polarization, which is likely to be abnormally high in this molecule (74, 96). The infrared and Raman spectra of S_4N_4, in the solid state and in solution, have been explained in terms of the structure given in Fig. 2 (19, 55), with assignments shown in Table II. Fourteen vibrational fundamentals are expected for this structure, of which two (those of A_2 species) would be inactive in both the infrared and Raman spectra, while seven (species B_2 and E) would be active in the infrared. A valence force field has been calculated (19) in good agreement with the observed frequencies.

The structure of S_4N_4 does now seem to be virtually settled, although an alternative to that shown in Fig. 2, with the sulfur and nitrogen atoms interchanged and N–N bonds present, was being discussed as recently as 1966 (108). MO calculations (108) show that this alternative

TABLE II

Fundamental Vibrations of the Molecule of Tetrasulfur Tetranitride[a]

Wavenumber (cm^{-1})	Infrared or Raman spectra[b]	Assignment
		A_1
716	Raman only, polarized (2)	S–N
529.7	Raman only[c]	SSN
213	Raman only, polarized (10)	S–S
		A_2
	Inactive in both modes	S–N
	Inactive in both modes	SSN
		B_1
888	Raman only (2)	S–N
615	Raman only (3)	SSN
		B_2
705	Infrared (vs) and Raman	S–N
564	Infrared (m) and Raman (2)	SSN
177.5	Infrared (m) and Raman, depolarized (6)	S–S
		E
938	Infrared (s) and Raman (2)	S–N
766	Infrared (vw) and Raman	S–N
519.3	Infrared (w) and Raman	SSN
314	Infrared (m) and Raman (1.5)	SSN

[a] From Bragin and Evans.

[b] Numbers in parentheses represent relative Raman intensities. Letters in parentheses represent infrared intensities (very weak, weak, medium, strong, or very strong).

[c] So far not actually observed in the Raman spectrum, but only as a weak band in the infrared spectrum of the solid.

should be considerably less stable than the structure given in Fig. 2. Moreover, the chemical behavior of S_4N_4 is always compatible with Fig. 2 and sometimes inconsistent with the alternative (inasmuch as hydrazine is not formed by hydrolysis or reduction).

4. Electronic Structure

The facts relating to the electronic structure of S_4N_4 are as follows. All S–N bonds have the same length, corresponding to a bond order of about 1.65 (27a). The bond distances (Section II,B,3) suggest weak S–S bonding. The electronic absorption spectrum shows poorly resolved vibronic bands of medium intensity at about 420 and 327 nm, which are responsible for the color; a very strong band at 257 nm; and a strong band beyond 185 nm with a shoulder at 204 nm (20). S_4N_4 is diamagnetic, with $\chi = -102 \times 10^{-6}$ mole^{-1}; this value exceeds the value for $S_4(NH)_4$ and is considered evidence of a diamagnetic ring current (20). The ^{14}N NMR spectrum of S_4N_4 (74) consists of a single line at 485 ± 20 ppm upfield from aqueous nitrite ion; a value nearer to typical values for singly bonded nitrogen than for doubly bonded nitrogen, although the bond lengths, and Lewis structures (I) to (III), point to a degree of multiple bonding.

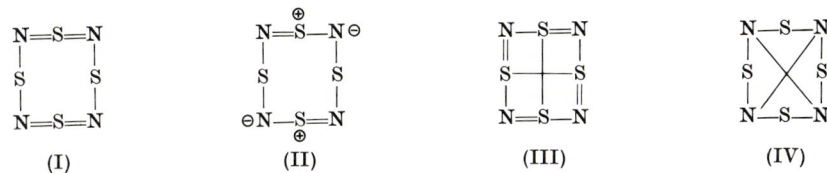

There have been three theoretical models of the electronic structure, all with delocalized π electrons as a feature. In the earliest and simplest, the "electron-on-a-sphere" model of 1962 (27a), these electrons were assumed to move in a sphere of uniform potential coterminous with the molecule. Subsequently, two fuller MO treatments have been published (20, 108). The interpretation of the experimental data in terms of these models is complicated and not entirely settled (74). Reference should be made to the original papers for details. The more recent MO treatments do, however, support the evidence of the bond lengths that there is appreciable S–S bonding but negligible N–N bonding; in other words, structure (III) contributes but (IV) does not. The main ultraviolet band at 257 nm may well represent an $n \rightarrow \pi^*$ transition into a π orbital which the NMR data suggest must be spherically symmetrical (74).

5. Thermochemistry

Although the S–N bonds in S_4N_4 are quite strong (~307 kJ; the exact value depends on the strength assumed for the S–S bonds), the compound has a positive enthalpy of formation from its elements. This, of course, is a consequence of the high dissociation energy of the nitrogen molecule. ΔH^0_f has recently been determined by decomposition of the nitride to its elements in a calorimeter [5]; it is given as 460 ± 8 kJ/mole. This replaces a value of 536 kJ determined in the last century.

6. General Reactions

The reactions of S_4N_4 are very diverse. It is convenient to classify them along conventional lines as thermal decomposition and dissociation, oxidation, reduction, reactions with electrophiles, reactions with nucleophiles, and addition reactions.

The eight-membered ring of S_4N_4 is not to be compared to organic rings which persist essentially unchanged through a variety of substitutions, couplings, and ring fusions. It is better to think of it as a labile polymer of four thiazyl, SN, units, which are individually fairly stable, but are as likely to appear in the reaction products of S_4N_4 in groups of two, three, or five, as in fours.

7. Thermal Decomposition and Dissociation

When heated to 130° (*51*) or higher at ordinary pressures, S_4N_4 decomposes wholly to nitrogen and sulfur, usually with an explosion. Differential thermal analysis of samples diluted with a large excess of octasulfur (*102*) shows that under these conditions decomposition proceeds quietly, though exothermically, between 120° and 190°.

When the vapor of S_4N_4 at low pressure is passed through a region heated to 300°, S_2N_2 is formed (Section II,C). There is some accompanying decomposition of S_4N_4 to its elements, and if steps are not taken to remove the sulfur so formed, S_4N_2 (Section III) is produced by reaction of this sulfur with S_4N_4 or S_2N_2.

8. Oxidation, including Halogenation

S_4N_4 is little affected by air, dry or moist, at room temperature, but it burns vigorously when slowly heated. Rapid heating in air causes detonation (*47*). It reacts with ozone and hydrogen peroxide, but the reactions are not understood (*47*). Nitrogen dioxide in carbon tetrachloride or carbon disulfide oxidizes S_4N_4 to nitrosyl disulfate, $(NO)_2S_2O_7$ (*77*). Chloramine-T in acidified dioxane converts it quantitatively to sulfuric acid and ammonia (*79*).

The direct reaction of S_4N_4 with fluorine has not been investigated, but electrochemical fluorination breaks the molecule down completely, giving NF_3 and SF_6 as major products (*33*). Milder fluorination can be effected by metal fluorides in high oxidation states. Under the mildest conditions, when a CCl_4 solution of the nitride is warmed with silver(II) fluoride and then allowed to cool, colorless needles of tetrathiazyl tetrafluoride (V) crystallize out (*46*). This compound contains the S_4N_4 ring intact, but flattened (*114*). It depolymerizes to thiazyl fluoride

```
F—S=N—S—F          F\   N   /F           Cl\   N   /Cl
 |      ||           S ⁄ ‖ \ S              S ⁄ ‖ \ S
 N      N            |   N   |              |   N   |
 ||     |            N≤   ⁄                 N≤   ⁄
F—S—N=S—F              \S                      \S
                        |                        |
                        F                        Cl

   (V)                 (VI)                    (VII)
```

monomer, NSF, when heated to 300° *in vacuo* (*44*). NSF, a colorless gas boiling at 0.4°, can also be made in good yield by fluorination of S_4N_4 under more vigorous conditions, viz., refluxing with HgF_2 in carbon tetrachloride (*46*). The cyclic trimer, $N_3S_3F_3$ (VI), also exists, but it is made by fluorination of $N_3S_3Cl_3$ (VII), not directly from S_4N_4 (*46*). $N_3S_3Cl_3$ (m.p. 162.5°) results from the action of chlorine on a solution or suspension of S_4N_4 in CS_2 or CCl_4 (*100*). In this standard preparation, the $N_3S_3Cl_3$ only crystallizes hours after the chlorine has been passed into the vessel. This is because, as recently shown (*82*), the primary chlorination product is not $N_3S_3Cl_3$, but $N_4S_4Cl_4$, the chlorine analog of (V). The $N_4S_4Cl_4$ has not been isolated; it decomposes according to the equation

$$N_4S_4Cl_4 \rightarrow N_3S_3Cl_3 + NSCl$$

with a half-reaction time of about 1 hr in CS_2 at room temperature. $N_3S_3Cl_3$ can be depolymerized to gaseous NSCl *in vacuo* at 110°, whereas at room temperature NSCl trimerizes easily (*43, 82*). These reactions of the chlorine and fluorine compounds illustrate the lack of kinetic stability of the polythiazyl ring systems and the importance of thermodynamics in determining the reaction products under practical conditions.

Bromine and S_4N_4 in CS_2 react to give bronze crystals, insoluble in nonpolar solvents, which until recently were thought to be poly(thiazyl bromide), $(NSBr)_x$. In our own investigations, however, we have not been able to get analyses in good agreement with this formula, and we suspect that the bronze compound may really be $S_3N_2Br_2$. It gives a conducting solution in nitromethane, which $N_3S_3Cl_3$ does not. There is infrared evidence of the formation of NSBr as a soluble intermediate in the bromination reaction (*84*).

When S_4N_4 reacts at room temperature with bistrifluoromethyl nitroxide, $(CF_3)_2NO$, a white crystalline compound is formed with the tetrameric formula $[NSON(CF_3)_2]_4$; X-ray diffraction shows it to have a structure like (V), with the S_4N_4 ring intact although altered in shape (*32a*).

S_4N_4 dissolves in concentrated sulfuric acid to give an orange solution. At the same time it is oxidized in a complex reaction (*66, 70*). One of the early products is $S_2N_2^+$, the identity of which has been established by ESR spectroscopy. The end products include sulfur dioxide and sulfamic acid.

9. Reduction

The MO calculations on S_4N_4 (Section II, B, 4) show that there should be low-lying vacant orbitals; the presence of these is confirmed by the ability of the molecule to add an electron. Electrolysis in tetrahydrofuran solution, or treatment with potassium in the same solvent, gives the paramagnetic species $S_4N_4^-$ (76), recognized by its ESR spectrum and stable only below −25°. Potassium in dimethoxyethane at room temperature reduces S_4N_4 to a series of colored ions, two of them paramagnetic, which were formerly thought to be the result of simple electron addition to S_4N_4, but must now be attributed to unidentified decomposition products (27, 76). There is not yet convincing evidence for the direct addition of more than one electron to the S_4N_4 molecule. However, an orange-red sodium salt thought to contain the anion $S_4N_4^{4-}$ has been obtained from tetrasulfur tetraimide (VIII) by the reaction

$$S_4(NH)_4 + 4[Ph_3C]^-Na^+ \xrightarrow[\text{dioxane}]{\text{ether-}} 4Ph_3CH + [S_4N_4]^{4-}Na_4^+$$

and the lithium salt can be obtained similarly using butyllithium (4a).

S_4N_4 can easily be reduced to sulfur imides. In the fast reaction with tin(II) chloride in boiling methanol/benzene (51), a hydrogen atom is simply added to each nitrogen of the ring. The product (VIII) is a colorless crystalline compound melting at 156°, with a puckered ring molecule very like that of octasulfur (98). By contrast, the slow reduction of S_4N_4 with hydrazine adsorbed on silica gel in benzene at 46° (39–41) gives a mixture of cyclic sulfur imides $S_{8-n}(NH)_n$ (n = 1, 2, 3, or 4) as well as sulfur. Examples of these imides are shown in formulas (VIII) to (XII) (109).

```
HN—S—NH          S—S—NH          S—S—NH
 |    |           |    |           |    |
 S    S           S    S           S    S
 |    |           |    |           |    |
HN—S—NH          S—S—S           S—S—NH
  (VIII)           (IX)             (X)

S—S—NH           S—S—NH                    +
 |    |           |    |          ⎡  S—S   ⎤
 S    S           S    S          ⎢ ╱    ╲ ⎥
 |    |           |    |          ⎢N      N⎥
S—NH—S           HN—S—S           ⎢ ╲    ╱ ⎥
  (XI)            (XII)           ⎣  S—N—S ⎦
                                    (XIII)
```

Hydrogen iodide in CCl_4 completely reduces S_4N_4 to H_2S and ammonia (*80*).

10. Reactions with Electrophiles

The reactions of S_4N_4 with Lewis acids are a topic of current interest. Surprisingly, however, little attention has been given to the hydrogen ion. A dark red precipitate thought to be $S_4N_4 \cdot HCl$ is formed (*72*) initially from the nitride and HCl in carbon tetrachloride, and reacts with more HCl according to the equation

$$S_4N_4 \cdot HCl + 3HCl \longrightarrow [S_4N_3]^+Cl^- + NH_4Cl + Cl_2$$

The reactions with HBr (*72*) and HF (*93*) are similar in their end results.

TABLE III

Adducts of Tetrasulfur Tetranitride with Main Group Halides and Sulfur Trioxide

S_4N_4 plus:	Reference	S_4N_4 plus:	Reference
BF_3	*116*	$2SbI_3$	*97*
BCl_3	*92, 116*	$2SO_3$	*51*
$2BCl_3$	*92*	$4SO_3$	*51*
BBr_3	*116*	$Se_2Cl_2(?)$	*115*
$\frac{1}{2}SnCl_4$	*116*	$SeCl_4$	*92*
$\frac{1}{2}SnBr_4$	*4a*	$SeCl_4 \cdot SO_3$	*92*
$2SbF_5$	*97*	$TeCl_4$	*92*
$4SbF_5$	*30*	$TeCl_4 \cdot BCl_3$	*92*
$SbCl_5$	*92, 116*	$TeCl_4 \cdot SbCl_5$	*92*
$2SbCl_5$	*92*	$TeCl_4 \cdot SO_3$	*92*
$2SbBr_3$	*97*		

The halides of the ion $[S_4N_3]^+$ (XIII) (thiotrithiazyl halides) are well characterized crystalline salts (*57*). S_4N_4 probably acts as a base toward H^+ in these reactions, but a more direct investigation of its Brønsted base character in the absence of complications is obviously needed.

With the halides of the main-group elements, in inert solvents, S_4N_4 often gives donor–acceptor complexes in which nitrogen is the donor atom (Table III). The adduct $S_4N_4 \cdot BF_3$ is obtained (*116*) as dark burgundy-red crystals by passing BF_3 into a CH_2Cl_2 slurry of S_4N_4. In its crystal structure (*32*) one nitrogen of the S_4N_4 ring is coordinated to boron, and the ring is flatter than in S_4N_4, with the four sulfur atoms coplanar (cf. Fig. 3). The orange-red $S_4N_4 \cdot BCl_3$, formed similarly, is believed, on infrared evidence, to be similarly constituted (*116*). $S_4N_4 \cdot BBr_3$ exists (*116*), but has not been obtained pure. Following the usual order of Lewis acid strengths toward donor nitrogen, the BF_3 adduct is less stable than the BCl_3 adduct, losing BF_3 easily on warming or *in vacuo*, with an enthalpy of dissociation of only 62.8 kJ/mole, and being converted to $S_4N_4 \cdot BCl_3$ by treatment with BCl_3. The red, crystalline, and rather stable $S_4N_4 \cdot SbCl_5$, formed by bringing the components together in CCl_4, $CHCl_3$, or CH_2Cl_2 (*116*), has been shown by X-ray diffraction (*86*) to have the structure given in Fig. 3, with one donor

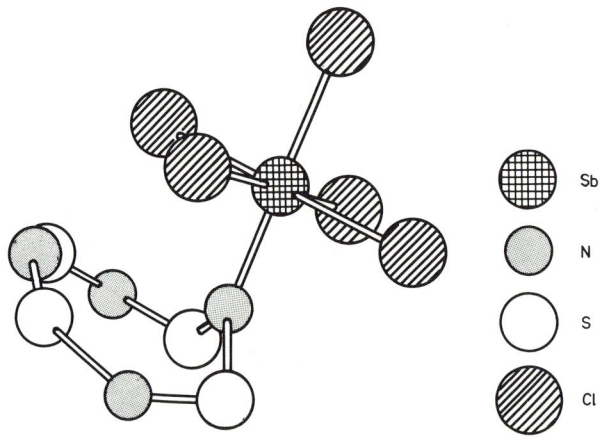

FIG. 3. Molecular structure of $S_4N_4 \cdot SbCl_5$. After Neubauer and Weiss (*86*).

nitrogen atom and the ring flattened. There has been some discussion whether the S_4N_4 molecule can add molecules of Lewis acid to more than one of its nitrogen atoms. There is no clear structural proof that it can, although such behavior would not be surprising in view of the proven behavior of S_2N_2 as a bidentate ligand (Section II,C,4). Diadducts of S_4N_4 with BCl_3 or $SbCl_5$ are reported to be formed in liquid SO_2 at $-40°$ (*92*), but their infrared spectra point to the ionic structures $[S_4N_4 \cdot BCl_2]^+[BCl_4]^-$ and $[S_4N_4 \cdot SbCl_4]^+[SbCl_6]^-$, with unidentate S_4N_4. The mixed adduct $S_4N_4 \cdot BCl_3 \cdot SbCl_5$ is likewise believed (*116*) on infrared evidence to be $[S_4N_4 \cdot BCl_2]^+[SbCl_6]^-$, and the mixed adduct $S_4N_4 \cdot BCl_3 \cdot SO_3$ has been formulated as $[S_4N_4 \cdot BCl_2]^+[SO_3Cl]^-$ on the evidence of infrared and conductance data (*92*). Adducts with two or four molecules of SO_3 (*51*) or SbF_5 (*30*) have been reported, but considering the tendency of SO_3 and SbF_5 to polymerize, these adducts may well consist of clusters of Lewis acid molecules coordinated to one nitrogen atom of the S_4N_4 ring. 1:1 Complexes with $SeCl_4$ and $TeCl_4$ have been given (92) an ionic formulation $[S_4N_4 \cdot MCl_3]^+Cl^-$, and they yield adducts formulated ionically as $[S_4N_4 \cdot MCl_3]^+[ACl]^-$ (A = BCl_3, $SbCl_5$, or SO_3) with a second Lewis acid. But the evidence cited for these formulations (conductance data in acetonitrile) does not establish the constitution in the solid state, and X-ray crystallographic studies are needed.

An important reaction of S_4N_4 with a Lewis acid is the standard preparation of thiotrithiazyl chloride [cation (XIII)]. It is tempting to

$$3S_4N_4 + 2S_2Cl_2 \xrightarrow[CCl_4]{\text{boiling}} 4[S_4N_3]^+Cl^-$$
(90% yield)

postulate addition of S_4N_4 to S_2Cl_2 as the first step in this reaction. However, tracer results with ^{35}S are not easily reconciled with this view (51), and another mechanism, involving the dissociation of S_4N_4 to $2S_2N_2$, has been suggested. The corresponding reaction with Se_2Cl_2 is more complicated (2, 38). In CCl_4, the products are thiotrithiazyl chloride and selenium. In thionyl chloride, a yellow compound, recrystallizable from formic acid, is produced which has the empirical formula $SeS_2N_2Cl_2$ and which may be the hexachloroselenate of the selenotrithiazyl cation $[SeS_3N_3]^+$, analogous in structure to (XIII).

Many reactions of S_4N_4 with the compounds of transition and post-transition metals give rise to complexes in which there are S–N ligands, but probably not intact molecules of S_4N_4. These have been well reviewed (112) and only the main results will be described here. Some of the reactions have been carried out in inert solvents and seem relatively simple. With iron, cobalt, and nickel carbonyls in benzene, for example, S_4N_4 gives strongly colored crystalline compounds $M(SN)_4$, of unknown constitution, but monomeric in some solvents. $Mo(CO)_6$ gives the explosive $Mo(SN)_5CO$ (117). The ligands in these complexes are probably unidentate SN or bidentate S_2N_2. The same holds for the complexes $CuCl_2 \cdot S_2N_2$ and $CuBr_2S_2N_2$, obtained from S_4N_4 and the halide in dimethylformamide. In the other reactions of complex formation described in the literature, the solvents participate chemically in obscure ways. The nickel, cobalt, and palladium complexes (XIV), (XV), and (XVI), obtained by heating methanol or dimethylformamide solutions of the metal halide with S_4N_4, are colored, crystalline compounds with definite melting points (112), the structures of which have been well established crystallographically and otherwise. The hydrogen atoms in the NH groups must come from the solvent; their presence was recognized by infrared spectroscopy and their substitution reactions have been studied.

$$M \begin{bmatrix} NH-S \\ | \\ S=N \end{bmatrix}_2 \qquad M \begin{bmatrix} S-S \\ | \\ S=N \end{bmatrix}_2 \qquad \begin{array}{c} S-S \\ | \quad M \quad | \\ N=S \end{array} \begin{array}{c} NH-S \\ | \\ S=N \end{array}$$

(XIV)   (XV)   (XVI)

The reactions of S_4N_4 with various metal halides in thionyl chloride have recently been investigated (4). They are complicated reactions, because S_4N_4 is known to react with thionyl chloride in the absence of other Lewis acids to give thiotrithiazyl chloride [cation (XIII)] and thiodithiazyl dioxide (XVII) (2). The products usually have the stoichiometry of adducts of SN or S_2N_2, but most of them are insoluble in nonpolar solvents and probably polymeric. Exceptionally, the reaction with $AlCl_3$ gives a yellow, soluble, crystalline salt $[S_5N_5]^+[AlCl_4]^-$, the cation of which is a heart-shaped nearly planar ring (3).

11. Reactions with Nucleophiles

S_4N_4 and S_2N_2 (Section II,C) both dissolve in liquid ammonia to give (after a few minutes standing in the case of S_4N_4) the same red solution, which on evaporation leaves a cinnabar-red solid formulated as $(SN)_2 \cdot NH_3$ or $S_4N_4 \cdot 2NH_3$ or $H_3N_3S_2$ (6, 7, 85, 119). Its molecular weight has not been reported. The suggestion that it is an acid with the structure $H-N=S=N-S-NH_2$ has been made (6) on the evidence of its sodium salts. Titration of the ammonia solution with triphenylmethylsodium gives a brown monosodium salt and then a yellow, explosive trisodium salt, probably because the imide hydrogen and the two amide hydrogens are successively replaced. Only one H atom readily ionizes in ammonia solution, however, and the compound behaves as a 1:1 electrolyte (119). $H_3N_3S_2$ might be expected to behave as a chelating ligand, and if it did so, would give six-membered chelate rings with metal ions. Actually, however, it reacts with "b" class metal ions to give complexes which seem to be derived, not from $H_3N_3S_2$ itself, but from another ligand $H_2N_2S_2$ resulting from disproportionation.

$$2H_3N_3S_2 \rightarrow H_2N_2S_2 + S(NH)_2$$

Addition of lead nitrate or iodide, for example, to the ammonia solution of $H_3N_3S_2$ precipitates green $Pb(NS)_2 \cdot NH_3$, which has been shown by X-ray diffraction to contain the chelate structure (XVIII) (112). The drive for the postulated disproportionation probably comes from the superior stability of five-membered chelate rings and the preference of

$$\text{}^+\text{S}\overset{\overset{\text{N}}{\diagdown}}{\underset{\text{O}^-}{|}}\text{S}\overset{\overset{\text{N}}{\diagdown}}{\underset{\text{O}^-}{|}}\text{S}^+$$

(XVII)

$$\text{H}_3\text{N}\rightarrow\overset{\overset{\text{N}=\text{S}}{\diagup}}{\underset{\underset{\text{S}-\text{N}}{\diagdown}}{\text{Pb}}}$$

(XVIII)

"b" ions for sulfur over nitrogen as a ligand atom. There is evidence for the other product of the disproportionation, $S(NH)_2$; a "salt" of this compound, $HgN_2S \cdot NH_3$, precipitates when mercury(II) iodide is added to the filtrate from precipitation of the lead salt. Thallium, silver, and copper salts can be made in the same manner as the lead salt(XVIII). The formulation usually given for the copper and silver salts, $M(NS)_2$ indicates a +2 oxidation state for the metal, although the copper salt can only be made from Cu(I). It has been suggested (112) that these salts are really MS_2N_2H derived from the +1 oxidation state. Air oxidation of a pyridine solution of the compound $(SN)_2 \cdot NH_3$ gives rise to the ammonium salt of the anion $[S_4N_5O]^-$, tetrasulfur tetranitride oxide imide (16a, 105a), a yellow, well-crystallized compound forming stable solutions in polar solvents. The structure of this anion is not known; its infrared spectrum is complex and suggests that a $\text{\textgreater}S{=}O$ group is present.

The hydrolysis of S_4N_4 has been studied under heterogeneous (48) and homogeneous (80) conditions. The time required in a homogeneous, aqueous–organic solvent medium containing 0.5 N alkali is about 2 hr at room temperature. Reaction in acid solutions is generally slower, but the rate increases with acidity. The nitrogen is always completely converted to ammonia (48), confirming the physical–structural deduction that there are no N–N bonds in the molecule and justifying the usual assignment of oxidation number −3 to nitrogen and +3 to sulfur. On this view of the sulfur oxidation state, reasonable mechanisms have been suggested to explain the complex mixture of sulfur oxidation numbers found in the end products of hydrolysis (18, 80, 95). In the presence of excess sulfite ion, the overall reaction of "hydrolysis" becomes relatively simple, but it is not clear whether the thiophilic sulfite ion attacks the

$$S_4N_4 + 6H_2O + 2H_2SO_3 \rightarrow 2H_2S_3O_6 + 4NH_3$$

S_4N_4 ring itself or its primary hydrolysis products (80). Hydrolysis of S_4N_4 with aqueous ammonia (104a) gives similar products and also sulfamate, probably formed via trithionate.

The thiophiles, cyanide ion and triphenylphosphine, attack S_4N_4 in dimethylformamide solution (35). One sulfur atom is removed from the molecule in each case. The product from cyanide attack is thought to be a derivative of the dimer of cyanamide in which the hydrogens have been replaced by S_3N_3 rings (XIX) (35). It has not been obtained pure.

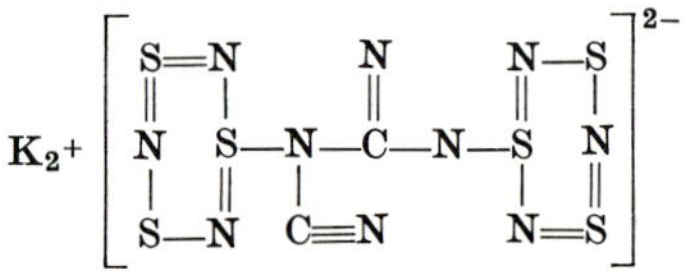

(XIX) 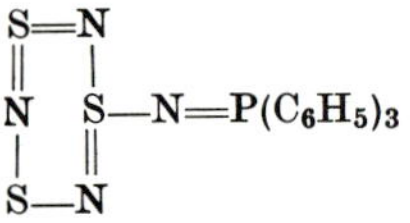 (XX)

The product from triphenylphosphine (35, 68) forms red monoclinic crystals, and X-ray diffraction (64) shows it to have the structure (XX). The corresponding cyclohexyl compound has also been made (68). Phenyldichlorophosphine probably attacks the S_4N_4 ring similarly, but an excess of the phosphine was used in the reported experiments, leading to total breakdown of the ring with the formation of phosphonitrilic chlorides and phenyldichlorophosphine sulfide (36).

Nucleophilic ring opening of S_4N_4 by arylmagnesium bromides gives rise to an interesting series of stable, colored, well-crystallized compounds, Ar–S–N=S=N–S–Ar (113), which deserve further investigation.

12. Additions

Some of the Lewis formulations of S_4N_4 [e.g., (II) and (III)] show the molecule with alternate S=N bonds. It is natural to ask whether S_4N_4 would behave as an "inorganic diene" and add to dienophiles in reactions of the Diels–Alder type. It does, in fact, readily form the colorless, crystalline adducts, $S_4N_4 \cdot 4C_5H_6$ (with cyclopentadiene), $S_4N_4 \cdot 2C_7H_{10}$ (with norbornene), and $S_4N_4 \cdot 2C_7H_8$ (with norbornadiene) (16, 16b). The structures of the adducts have not been determined, but by analogy with organic Diels–Alder reactions the structure (XXI) has been suggested for the norbornene adduct.

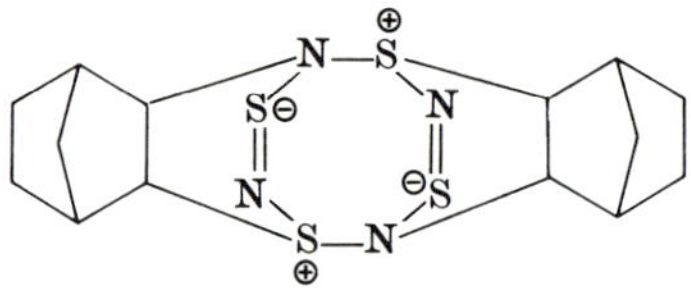

(XXI)

C. Disulfur Dinitride, S_2N_2

1. Preparation

Disulfur dinitride, S_2N_2, the dimer of thiazyl, can be made in good yield by passing the vapor of S_4N_4 at a pressure of about 1 mm Hg or less through silver wool at 300° (11, 19, 21, 89). Better yields and safer working conditions are claimed if the operation is carried out at lower temperatures and pressures (31). The silver is converted to silver sulfide. Its main function is to remove sulfur generated by the decomposition of S_4N_4 to its elements; if this were not done, the product would be contaminated with too much S_4N_2 formed from this sulfur and S_4N_4 or S_2N_2. The silver sulfide may catalyze the depolymerization of S_4N_4.

2. Description

S_2N_2 forms large, colorless crystals which can be sublimed at 10^{-2} mm Hg at room temperature (50). It is insoluble in water, but soluble in benzene, carbon tetrachloride, ether, acetone, and especially tetrahydrofuran and dioxane (50); it can be recrystallized from ether. It has a repulsive smell and is the least stable of the thiazyl polymers, detonating with friction, shock, or heating above 30°. It polymerizes rapidly at room temperature in the solid state or in solution (19).

3. Molecular Structure

The cryoscopic molecular weight in benzene corresponds to the formula S_2N_2 (50). The infrared spectra of the solid, vapor, and solution indicate a planar, nearly square ring structure with alternating sulfur and nitrogen atoms and D_{2h} symmetry (19, 111), but the instability of the compound has so far prevented a structure determination by X-ray or electron diffraction.

4. Reactions

As S_2N_2 is not very easy to prepare or purify, its reactions have not been as extensively investigated as those of S_4N_4.

Rapid and quantitative polymerization to S_4N_4 takes place when traces of alkalis or potassium cyanide are added to solutions of S_2N_2 in organic solvents (21). However, when dry, purified S_2N_2 is stored for 30 days in an evacuated desiccator, nearly 100% conversion to polythiazyl $(SN)_x$ (Section II,D) is observed. In moist air, S_2N_2 changes quickly to a mixture of S_4N_4, hydrolysis products, and a little $(SN)_x$ (31, 50). When heated to 250° in a sealed tube, it decomposes quantitatively to its elements (89).

The reactions of S_2N_2 with Lewis acids are a subject of current interest. Adducts of S_2N_2 can be formed, but in some circumstances adducts of S_4N_4 or $(SN)_x$ are obtained. When S_2N_2 is gradually added to $SbCl_5$ in dichloromethane at room temperature (*89*), the following sequence of reactions takes place:

$$S_2N_2 + 2SbCl_5 \rightarrow S_2N_2(SbCl_5)_2 \text{ (yellow crystalline precipitate)}$$

$$S_2N_2 + S_2N_2(SbCl_5)_2 \text{ (dissolves)} \rightarrow S_2N_2 \cdot SbCl_5 \text{ (orange, crystallizes on evaporation)}$$

$$S_2N_2 \cdot SbCl_5 + S_2N_2 \rightarrow S_4N_4 \cdot SbCl_5 \text{ (orange-red precipitate)}$$

Any further S_2N_2 added is quickly changed to S_4N_4, apparently because of catalysis by some unidentified component of the mixture. The adduct $S_2N_2(SbCl_5)_2$ has been shown by X-ray crystallography to have the structure (XXII), with two donor nitrogen atoms and a planar S_2N_2 ring almost the same in dimensions as the free S_2N_2 molecule (*91*).

$$Cl_5Sb \leftarrow N \underset{S}{\overset{S}{\diamond}} N \rightarrow SbCl_5$$

(XXII)

Mono- and diadducts are also formed with BCl_3 at $-78°$ (*90*). The diadduct easily loses BCl_3. The monoadduct readily polymerizes to a brown solid $[(S_2N_2)(BCl_3)]_x$, which almost certainly does not contain intact S_2N_2 rings. It resembles, and may be identical with, the adduct of $(SN)_x$ with BCl_3 (Section II, D, 4). With excess BF_3 (*90*), S_2N_2 gives only the known $S_4N_4 \cdot BF_3$ (Section II, B, 10). This may be a result of the weakness of BF_3 as a Lewis acid; even in the presence of excess BF_3, there would be a relatively high concentration of free S_2N_2 in equilibrium with the presumed primary adduct $S_2N_2 \cdot BF_3$, permitting the secondary reaction

$$S_2N_2 + S_2N_2 \cdot BF_3 \rightarrow S_4N_4 \cdot BF_3$$

to take place quickly.

The nucleophilically catalyzed dimerization of S_2N_2 has been mentioned above. There is little further information on its behavior with nucleophiles. With liquid ammonia (*6, 7, 85*) and aqueous alkali (*50*) it gives the same products as S_4N_4.

D. Polythiazyl, $(SN)_x$

1. Preparation

This compound, of special interest because it conducts electricity, was first observed in 1910 as a blue film with bronzy reflex, accompanying the pyrolysis of S_4N_4 vapor (25). We now know that it originates from the polymerization of S_2N_2. The only way to make it in quantity is to store S_2N_2 in an evacuated desiccator, or in a dry, inert atmosphere, for about 30 days at room temperature (31, 50). According to the size of the S_2N_2 crystals used, the $(SN)_x$ crystals in the resulting mass may be very small or up to 3 mm long. They look like brass and have a fibrous makeup.

2. Molecular Structure

The molecular structure of polythiazyl is unknown. It is sometimes assumed to consist of zigzag chains of alternating sulfur and nitrogen atoms. Perhaps this assumption should be questioned, in view of an observation (31) that the compound can be sublimed almost unchanged at 100° and 10^{-3} mm Hg, giving in the process no S_4N_4 and very little S_2N_2. It is diamagnetic, with susceptibility equal to the sum of the Pascal constants (50), which suggests, but does not prove, that ring currents and rings are absent. It has not been possible so far to form good single crystals for X-ray diffraction. Nevertheless, some progress is being made with X-ray and electron diffraction studies (18a, 28, 31, 50); the dimensions of the unit cell have been determined, and it is known to contain four SN radicals (18a).

3. Electrical Properties

The resistivity of pelleted samples is of the order of 10^{-2} ohm-cm, and drops with rising temperature (28, 50, 69). Although this suggests semiconduction, the plot of log of resistivity against $1/T$ is not linear (28, 31), and its slope gives very low activation energies for conduction, <0.02 eV. The Seebeck coefficient shows conduction to be mainly by electrons, and the Hall effect is undetectable, showing that the current carriers have very low mobility (69). Although $(SN)_x$ is sometimes called a semiconductor, it has been suggested (69) that the conduction may really be metallic, the effect of rising temperature being to increase the mobility of the carriers rather than their number. The optical spectrum contains intense bands peaking at 1.77 eV and about 4 eV (8, 31). Neither of these could correspond to the <0.02 eV activation of the "dark" conductance. It has been impractical, however, to examine the material for photoconductivity in these bands because of its low "dark" resistance (28).

Obviously, we must still admit to considerable ignorance of the molecular and electronic structure of polythiazyl.

4. Chemical Behavior

Polythiazyl is not explosive. It ignites in air near 130°. It is insoluble in all ordinary solvents, but can be hydrolyzed with concentrated sodium hydroxide solution to ammonia, sulfite, and thiosulfate (50). It is slowly hydrolyzed in moist air (31) and forms an adduct $[(S_2N_2)(BCl_3)]_x$ on standing at 20° with BCl_3 (90).

III. Tetrasulfur Dinitride, S_4N_2

Traces of this nitride, recognizable by its dark red color, are formed in many reactions involving sulfur–nitrogen compounds. S_4N_2 was first properly characterized in 1951 (78), although it had been known, in an impure state, since 1896.

1. Preparation

S_4N_2 is best prepared by heating S_4N_4 with sulfur to 100°–120° in CS_2 solution in an autoclave (12). The mechanism of this synthesis is complicated and little understood. The solvent must participate, because much thiocyanogen polymer, $(CNS)_x$, is formed; also, experiments with ^{35}S have shown that 30–35% of the sulfur in the S_4N_2 produced comes from the solvent (8). The use of an autoclave can be avoided. S_4N_2 is formed, with loss of nitrogen, when a benzene (or better, xylene) solution of S_4N_4 is refluxed for some hours (8, 78); but yields are smaller, and the method has not been fully investigated. Another very interesting preparation (78) not requiring an autoclave

$$Hg_5(NS)_8 + 4S_2Cl_2 \xrightarrow[20°]{CS_2} Hg_2Cl_2 + 3HgCl_2 + 4S_4N_2$$

(42% yield)

employs one of the mercury salts of $S_4(NH)_4$ (VIII). The idea of this preparation could be further exploited. There are many other salts and complexes of "b" class metals (Section II,B) with sulfur–nitrogen ligands which would probably react with chlorosulfanes to give sulfur nitrides. It is conceivable that new sulfur nitrides, including perhaps isomers of S_4N_2 (below), might be formed in this way. Such reactions would probably go at low temperatures and might therefore be made to yield, by kinetic control, products which are thermodynamically unstable.

2. Isomerism

Physical and chemical evidence described below show that S_4N_2 has the structure (XXIII). Many hypothetical isomers of S_4N_2 can be

```
      (+2)
       S
(0)S⁄   ⁄S(0)       S⁄S⁄S              S⁄N⁄S
  |    |           |   |              |   ||
  N⁄   ⁄N          N⁄   ⁄S             S⁄   ⁄S
    S≠              N                    N
   (+4)
  (XXIII)           (XXIV)             (XXV)
```

written, such as those represented by formulas (XXIV) and (XXV). None, however, is known. They are probably thermodynamically unstable with respect to (XXIII), for, as shown by chromatography (83), only one form of S_4N_2 is present in the products of the autoclave preparation. The same holds true for the products of the $Hg_5(NS)_8$ method (above); the reason in this case may be thermodynamic or structural.

3. General Description

Tetrasulfur dinitride forms opaque, red–gray needles which melt at 23° to a dark red liquid looking like bromine (8, 83). In the pure state it decomposes in a few hours at room temperature, and decomposition becomes explosive at 100° (8). It keeps much better in solution in CS_2. The vapor pressures have not been measured, but the nitride can easily be sublimed at room temperature at about 1 mm Hg or lower pressures. Vacuum sublimation, or chromatography in CS_2 solution on silica gel, are good ways of purifying it (83). It is insoluble in and slowly hydrolyzed by water. It dissolves readily without decomposition in many organic solvents, including CS_2, benzene, chloroform, ether, acetone, ethanol, formic acid, nitrobenzene, and hexane (49, 78).

4. Molecular Structure

The molecular formula S_4N_2 has been established by analysis and cryoscopic determination of the molecular weight in benzene (78). Determination of the molecular structure has been more difficult. An astonishing number of Lewis structures can be written for the formula S_4N_2. Excluding unlikely features such as three-membered rings and adjacent charges of the same sign, the number is of the order of sixty. Only a few of these have ever been discussed in the literature. There are, unfortunately, no X-ray crystallographic data on S_4N_2, and the instability of the crystals would make an X-ray study difficult. However, the following physical evidence makes it possible to rule out the great majority of hypothetical structures at once, and then to select (XXIII) as the most likely structure among the few remaining contenders (83). First, the ^{14}N NMR spectrum contains a single resonance 105 ppm upfield from aqueous nitrate ion, indicating that the two nitrogen atoms are equivalent and doubly bonded. Most of the hypothetical structures have either nonequivalent nitrogen atoms or nitrogen atoms at bridgeheads (i.e., singly bonded), so these need not be further considered. Second, the dipole moment, 1.74 ± 0.28 Debye units, is unmistakably nonzero, and there are coincidences between Raman and infrared spectral frequencies, so that centrosymmetric structures [including planar and chair form (XXV)] can be ruled out. Third, linear structures with equivalent nitrogen atoms cannot be reconciled with the mass spectral fragmentation pattern (83). This leaves only (XXIII), (XXIV), and boat form (XXV) to be discussed. There is no frequency in the vibrational spectrum high enough for an N=N bond. Hence (XXIV) can be ruled out. Finally, there are four fundamental S–N stretching vibrations (83), the number expected for (XXIII); boat form (XXV) would give only three. The physical evidence, then, is consistent with (XXIII). This structure is also supported by chemical evidence. Hydrolysis (49, 78) converts all the nitrogen in the compound to ammonia; no hydrazine or molecular nitrogen is formed as would perhaps be expected from (XXIV). Detailed studies of the hydrolysis products in the presence and absence of HSO_3^- (49) have shown that the sulfur in S_4N_2 behaves as if it were in three oxidation states. This can be understood in terms of

$$S_4N_2 \rightarrow S(+4) + S(+2) + 2S(0) + 2N(-3)$$

(XXIII), by assigning reasonable oxidation states to the different sulfur atoms as shown; but it would be difficult to reconcile with (XXV), in which all the sulfur atoms must have the same oxidation number.

It is noteworthy that S_4N_2 has an entirely different structure from N_2O_4. N_2O_4 is in equilibrium with the paramagnetic NO_2. S_4N_2, in contrast, gives no evidence of dissociation to radicals. It is diamagnetic, with susceptibility near the sum of the Pascal constants (49).

5. Chemical Behavior

Apart from the hydrolysis just described, the chemistry of S_4N_2 has not been adequately investigated.

S_4N_2 appears to be less strongly basic than S_4N_4 or S_2N_2, undergoing no reaction with BCl_3 in CS_2 solution at room temperature (84). In contrast to organic nitrogen bases such as pyridine it does not hydrogen-bond to phenol. It does, however, combine with $SbCl_5$ to give a moisture-sensitive compound with approximately the composition of a 1:1 adduct. At the same time, $SbCl_5$ oxidatively destroys some of the S_4N_2, giving rise to $S_4N_4 \cdot SbCl_5$ (Section II, B, 10) and $[S_4N_3]^+[SbCl_6]^-$.

Its sensitivity to oxidative ring opening is also apparent in its reactions with chlorine and bromine, which still have to be fully worked out (84). Simple halogenation products containing the intact S_4N_2 ring do not appear to be formed; reaction with chlorine produces instead S_4N_4, $[S_4N_3]^+Cl^-$, and $[S_6N_4]^{2+}Cl_2^-$.

Like other sulfur nitrides, S_4N_2 is easily reduced by HI in anhydrous

$$S_4N_2 + 6H^+ + 6I^- \rightarrow 3I_2 + 4S + 2NH_3$$

formic acid (49). Reduction with hydrogen and palladium, potassium borohydride, sodium dithionite, lithium aluminum hydride, or hydrazine (84) gives a mixture of cyclic sulfur imides with eight-membered rings, including compounds (VIII) to (XII). A literature report (37) of the formation of cyclo-1,3-$S_4(NH)_2$ by reduction of S_4N_2 with tin(II) chloride could not be confirmed, repetition of the experiment giving mainly (IX) (84).

IV. Saturated Sulfur–Nitrogen Frameworks, Coupled and Fused Rings, and Polymers

A. GENERAL

The analogy between the sulfanes and the alkanes has often been pointed out, and there is a similar analogy between the cycloalkanes and the various sulfur ring molecules S_x ($x = 6, 7, 8, 9, 10, 12,$ or 18) present in different crystalline sulfur allotropes. But sulfur is always divalent in polysulfur chains, and branching seems not to occur. Thus, the branched chains, fused or coupled rings, and cages of organic chemistry can have no analogs among structures composed wholly of sulfur atoms. The situation changes if a few tervalent heteroatoms, such as nitrogen, are introduced as branching points (62). An infinite variety of sulfur–nitrogen frameworks can then be postulated, with corresponding structural features to known carbon frameworks. These sulfur nitrides differ from those so far discussed in being "formally saturated"; i.e., they can be written with all single bonds. It is for the experimenter to find out which of them are stable under practical conditions. Only a small fraction of the vast range of conceivable structures has actually been realized. Four coupled-ring nitrides of the homologous series $S_7N-S_x-NS_7$ have been characterized. One fused-ring nitride, $S_{11}N_2$, is very stable and has been extensively studied. In addition, there are incompletely characterized sulfur–nitrogen polymers—some regular and others with more random structures.

B. The Coupled-Ring Nitrides, $S_7N-S_x-NS_7$

1. Preparation

The members of this series (formula XXVI) with $x = 1, 2, 3,$ and 5 have been isolated. Those with $x = 1$ ($S_{15}N_2$) and $x = 2$ ($S_{16}N_2$) were

$$\begin{array}{c} S-S-S \\ S \diagup \quad \diagdown N-S_x-N \diagup \quad \diagdown S \\ S-S-S \qquad\qquad S-S-S \end{array}$$

(XXVI)

the first saturated sulfur nitrides to be discovered. Their synthesis has set the pattern for all later investigations in the field.

All these compounds are prepared by the condensation of two molecules of heptasulfur imide, S_7NH (IX) with a chlorosulfane S_xCl_2.

$$2S_7NH + S_xCl_2 \xrightarrow{-2HCl} S_7N-S_x-NS_7$$

Heptasulfur imide, a stable crystalline compound, is the simplest of a series of imides (57) exemplified by formulas (VIII) to (XII), all derived, in principle, from octasulfur, and all with puckered-ring molecules (109) like octasulfur. It can easily be prepared in quantity from ammonia and S_2Cl_2 in a polar solvent (58). Its condensation with chlorosulfanes is slow when the reactants are simply mixed in a solvent such as CS_2; with pyridine present, however, it reacts almost instantaneously at room temperature.

The nitrides of this group with $x = 1$ and $x = 2$ were prepared in 1959 (9, 10), and those with $x = 3$ and $x = 5$ more recently (42).

2. Purification; Molecular Exclusion Chromatography

The easiest of these nitrides to prepare pure and in high yield is $S_{17}N_2$ ($x = 3$). When prepared as described above, $S_{15}N_2$ and to a lesser extent $S_{16}N_2$ are always contaminated with uncrystallizable sulfur nitrides, which may be other members of the same homologous series, formed in consequence of the disproportionation of the chlorosulfane during the reaction. The purification by crystallization of products thus contaminated is frustrating, because the impurities often separate as a second liquid phase (a red oil) into which the desired compound partitions very strongly. Adsorption chromatography on silica gel is also unsatisfactory because all the compounds are rather weakly adsorbed and tend to move with the solvent front. The stronger adsorbent, activated alumina, decomposes the sulfur–nitrogen rings. Vapor phase chromatography is out of the question because of the low vapor pressure and thermal instability of the compounds. A breakthrough in this purification problem has recently been achieved with the use of molecular exclusion chromatography on an inert polystyrene gel (63). Besides its application to the compounds under discussion, this advance has opened up the whole field of saturated sulfur–nitrogen frameworks to systematic investigation. Its advantages include the ability to handle thermodynamically unstable compounds with the least possible risk of rearrangement; the ability to separate nonpolar or weakly polar compounds not amenable to adsorption chromatography provided their molecules differ in size; and the fact that the elution volume of a sought-for compound can be accurately predicted by means of its simple relationship to molecular weight (63).

3. Molecular Structure

No structure determinations on the solid compounds by physical methods have been reported. The method of synthesis leaves little room for doubt that (XXVI) correctly represents the structural formulas. From the formation of S_7NH (IX) by reaction of $S_{15}N_2$ and $S_{16}N_2$ with piperidine (9), it has been argued that these nitrides must contain intact S_7N rings, but this is not proof, since S_7NH is also formed by the reduction of S_4N_4, $[S_4N_3]^+Cl^-$, and S_4N_2.

With all members of this homologous series except the first, a question arises as to the structure and conformation of the $-S_x-$ chain. The dipole moments (101) so far measured (in CS_2) are (Debye units): $S_{15}N_2$ ($x = 1$) 0.79; $S_{16}N_2$ ($x = 2$) 1.46; and $S_{17}N_2$ ($x = 3$) 0.88. To begin with, these values for $S_{16}N_2$ and $S_{17}N_2$ rule out branched chains, for a sulfur "side-chain" $>S \rightarrow S$ would introduce a relatively large moment. It seems reasonable to assume that the dihedral angles in the $-S_x-$ chains in $S_{16}N_2$ and $S_{17}N_2$ are near the usual values for polysulfur structures (90°–105°). If so, it follows from simple reasoning (101), which will not be given here, that $S_{17}N_2$ must exist in solution as a mixture of conformers.

4. Properties and Reactions

$S_{15}N_2$ (m.p. 137°), $S_{16}N_2$ (m.p. 122°), and $S_{17}N_2$ (m.p. 97°) are yellow, crystalline solids, stable in moist air for periods of a few days. Over many months they slowly decompose, giving CS_2-insoluble polymeric material. $S_{19}N_2$ ($x = 5$) could not be crystallized (101), but gave satisfactory analyses and molecular weight values, and has been proved chromatographically to be a single substance.

All these nitrides are fairly soluble in CS_2, but only slightly soluble in other nonpolar solvents. It is a curious property of these and saturated

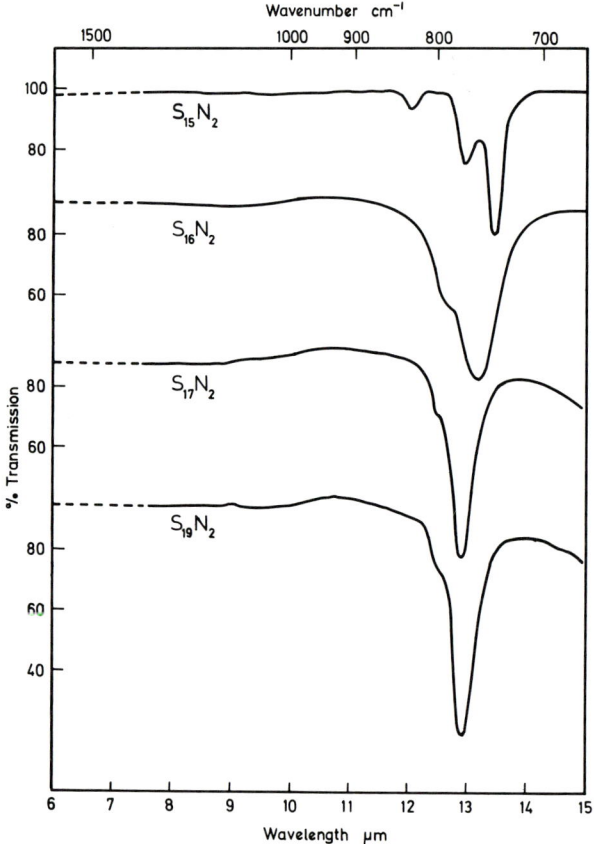

FIG. 4. Infrared spectra of nitrides of the homologous series $S_7N-S_x-NS_7$: 2–4% solutions in CS_2; 0.2-mm cell. Dotted lines in regions of strong solvent absorption.

polymeric sulfur nitrides (Section IV, E) that they are precipitated from solution in CS_2 by adding quite small amounts of CCl_4.

The infrared spectra of CS_2 solutions of these nitrides show a strong S–N stretch at 750–775 cm^{-1}. In $S_{19}N_2$ the nitrogen atoms are in a nearly symmetrical environment and the band is solitary and nearly symmetrical. On moving down the homologous series to $S_{15}N_2$, a shoulder, becoming a satellite band, develops as the nitrogen environment becomes less symmetrical (Fig. 4). This regularity would be a useful tool for investigating the structures of saturated sulfur–nitrogen polymers (Section IV, E).

On heating for a few minutes just above their melting points, $S_{15}N_2$, $S_{16}N_2$, and $S_{17}N_2$ decompose, giving S_4N_4, which subsequently breaks down to sulfur and nitrogen (103). The largest yield of S_4N_4 is obtained from $S_{15}N_2$, which alone of the three compounds has in its structure the –S–N–S–N– configuration present in S_4N_4.

C. THE FUSED-RING NITRIDE, $S_{11}N_2$

1. General

This compound, with structure (XXVII) (see also Fig. 5) is one of the most interesting recent developments in the chemistry of sulfur nitrides.

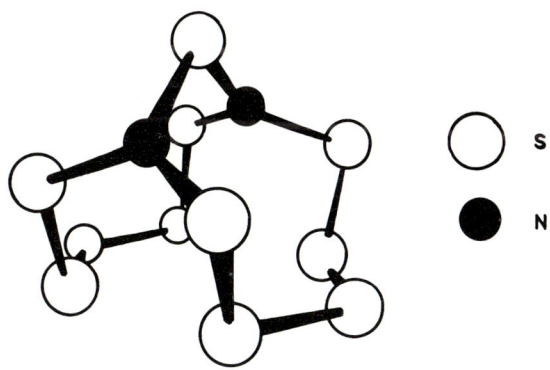

FIG. 5. Molecular structure of $S_{11}N_2$.

Its existence was expected, since molecular models show that two rings with the crown conformation of octasulfur can be fused at the 1,3-positions with little strain (Fig. 5). Experience has now shown $S_{11}N_2$ to be stable and easily handled. Its decomposition, which starts at about

145°, is exothermic but not explosive, at any rate with the small quantities hitherto made. It is not sensitive to shock or friction.

$$
\begin{array}{c}
\text{S—S—N—S—S} \\
\text{|\quad\quad|\quad\quad|} \\
\text{S\quad\;\;S\quad\;\;S} \\
\text{|\quad\quad|\quad\quad|} \\
\text{S—S—N—S—S}
\end{array}
$$

(XXVII)

2. Preparation

The best route to $S_{11}N_2$ is a double condensation of 1,3-hexasulfur diimide (X) with an equimolar quantity of S_5Cl_2 (*60, 63*), in the presence of pyridine. This diimide isomer is formed in rather small yield from the S_2Cl_2–ammonia reaction (*58*) and from the reduction of S_4N_4 with hydrazine (*40*). It is difficult to separate it from its 1,5-isomer; the separation is easier when one prepares it from S_4N_4 and hydrazine, for this method gives relatively little of the 1,5-compound. There are indications that the crude mixture of 1,3- and 1,5-isomers may work nearly as well as pure 1,3-$S_6(NH)_2$ in the preparation of $S_{11}N_2$ (*63*), so separation may not be necessary.

The reaction between the 1,3-diimide and S_5Cl_2 is a polycondensation between two bifunctional molecules, to which the standard theory of polycondensations applies. The products to be expected are linear polymers and ring oligomers of the general formula $[-(S_6N_2)-S_5-]_x$. $S_{11}N_2$ (XXVII) can be thought of as a ring oligomer with $x = 1$. Primary ring closure should be promoted, relative to chain growth, by mixing the reactants under Ruggli–Ziegler dilution conditions. Experimentally, however, the dilution technique gave only a 9% yield of $S_{11}N_2$ from the initial work-up, the remainder of the product being a sticky mixture of polymers. Much more $S_{11}N_2$ was obtained by refluxing the polymers for a few hours with CS_2 (*63*). The polymer has evidently a thermodynamic tendency to eliminate molecules of $S_{11}N_2$. Consequently, the dilution technique is not really needed. It is more convenient and almost as productive to mix the reactants quickly and then decompose the resulting polymer by refluxing it with CS_2.

3. Description

$S_{11}N_2$ forms amber-colored crystals. The color of pure samples is very pale, but does seem to be intrinsic. There are two crystalline forms, both monoclinic and both with four molecules in the unit cell (*61*). Slow evaporation of CS_2 solutions gives unsymmetrical octahedra of the α form; rapid evaporation gives thin platelets of the β form. The transition temperature, deduced from the crossing of the solubility curves, is 21° approximately; below this temperature the α form is stable and less soluble than the β form. At 21°, the solubility of both forms is 2.8 gm/100 gm CS_2. α-$S_{11}N_2$ is only very slightly soluble in other common solvents, for example (gm per 100 gm solvent at 20°), ether 0.03, hexane 0.02, and chloroform 0.09. The α crystals melt at 149°–150° when heated quickly in a capillary, but this figure is not very reproducible because of the polymorphic transition and incipient decomposition near the melting point. ΔH for the α→β transition near room temperature is approximately 10 kJ (deduced from the difference in slopes of the solubility curves).

Neither form is quickly affected by storage in moist air at room temperature, but a few β crystals have been observed to become opaque, presumably because of transition to the α form. CS_2 solutions keep well in the dark in absence of oxygen, but when evaporated in air and light they give some sticky, insoluble polymeric matter.

In the laboratory, $S_{11}N_2$ is easily recognized by its characteristic infrared spectrum in the S–N stretching region (Fig. 1).

4. Molecular Structure

X-Ray crystallographic studies on the α-form of $S_{11}N_2$ show that it has the structure of Fig. 5 (42a). Other physical evidence (10) is consistent with this structure, in which both rings have essentially the shape of the S_8 ring. The mass spectrum shows the molecular ion $S_{11}N_2^+$ and expected fragments. The vibrational spectrum is complicated and difficult to interpret, but the number of infrared-active S–N fundamentals and the number of infrared–Raman coincidences in the S–N stretching region seem consistent with the C_{2v} symmetry of Fig. 5. In CS_2 solution, α- and β-$S_{11}N_2$ give identical infrared spectra, although there are minor differences in their KBr disk spectra; hence, the difference between the forms is probably in the mode of packing in the crystal, not in the structure of the molecules themselves. The dipole moment in CS_2 solution (0.46 Debye) is nonzero as expected from Fig. 5.

5. Bonding and Chemical Behavior

The X-ray studies show that the three bonds from each nitrogen atom in α-$S_{11}N_2$ are coplanar. This implies sp^2 hybridization and suggests that the nitrogen lone pairs are used in $p\pi$–$d\pi$ bonding to sulfur, and so unavailable to confer basic properties on the molecule.

Several observations confirm this. In the first place, $S_{11}N_2$ is not strongly basic. It is insoluble in, and not affected by, concentrated sulfuric acid at room temperature, and it does not combine with BCl_3 at room temperature in CS_2 solution. Second, the dipole moment is much lower than would be expected from two lone pairs in partial reinforcement. Third, the average S–N bond energy term is 248 kJ, about right for a bond order of 1.33, which is what would be expected if the p_z electrons of each sp^2-hybridized nitrogen were fully utilized in $p\pi$–$d\pi$ bonding to its three neighbor sulfur atoms. This bond energy term has been deduced from data for the exothermic decomposition of $S_{11}N_2$ ($\Delta H = -262$ kJ) in solution in molten sulfur in a differential thermal analysis apparatus (*102*). Finally, the electronic absorption spectrum of $S_{11}N_2$ in hexane is remarkably like that of S_8, and contains no additional band attributable to an $n \to \pi^*$ transition of lone-pair electrons on nitrogen (*101, 102*).

6. *Other Modes of Formation*

$S_{11}N_2$ is formed in condensation reactions of sulfur imides from which it would not be expected on obvious mechanistic grounds. Good yields are obtained from (X) and S_7Cl_2, poorer yields from (X) and S_3Cl_2. Still more surprisingly, it can be obtained in moderate yield from (XII) (with its imide groups apparently in the wrong relative positions) and S_7Cl_2. These facts are a warning against pressing the analogy between S–N chemistry and organic chemistry too far, and a challenge to the experimenter to try out "wild" ideas.

D. OTHER FUSED-RING NITRIDES

Models of other fused-ring nitrides such as S_9N_2 (XXVIII) and $S_{13}N_2$ (XXIX) can be made up with reasonable bond angles and dihedral angles. Attempts to make (XXVIII) and (XXIX) from (X) and S_3Cl_2 or S_7Cl_2, respectively, have however failed, probably because of kinetic obstacles which may eventually be overcome by using other methods of condensation (*63*). Just as S_6 and S_{10} are thermodynamically unstable relative to S_8, so compounds such as (XXVIII) and (XXIX) are likely to be unstable with respect to $S_{11}N_2$, and their preparation will call for reaction at low temperatures followed by mild methods of purification such as molecular exclusion chromatography (Section IV, B, 2).

The unknown compound (XXX) may well be thermodynamically very stable. Attempts to make it by condensing (VIII) with S_5Cl_2 or (X) with SCl_2 have not so far succeeded (101).

E. POLYMERIC SATURATED SULFUR NITRIDES

In all condensation reactions of the bifunctional hexasulfur diimides (X, XI, and XII) with chlorosulfanes S_xCl_2, polymers of the general formula $[-(S_6N_2)-S_x-]_y$ are formed in quantity as tarry red materials.

When equimolar quantities of diimide and chlorosulfane are used, the polymers contain a negligible concentration of end groups and are true sulfur nitrides. Part of the polymer so obtained from (X) is soluble in CS_2, and, from its molecular weight, appears to be a mixture of dimer and trimer, presumably cyclic. The rest is insoluble and presumably consists of linear high polymers. Both kinds are labile at or near room temperature, each giving rise to some of the other when refluxed with CS_2 (63). As already mentioned, they eliminate $S_{11}N_2$ (Section IV, C, 2) at the same time.

Condensation of S_2Cl_2 with two moles of (X), (XI), or (XII) gives linear oligomers with NH end groups (59). These compounds are not strictly sulfur nitrides, but approximate thereto as the chains get longer. Complex possibilities of isomerism arise among them, which are worth discussing because they show some of the problems to be expected in the experimental study of large, complex saturated sulfur–nitrogen frameworks. The rings of (X), for example, can couple cis or trans to each other through a sulfur chain; the resulting isomers have been separated chromatographically and shown to transform to an equilibrium mixture with a half-reaction time of about 3 hr at room temperature. The rings of (XI) are optically active and give diastereoisomeric coupling products through a sulfur chain; the diastereoisomers have been separated chromatographically and shown to epimerize with a half-reaction time of, again, about 3 hr. Further, polysulfur chains exhibit conformational isomerism, because, with dihedral angles of about 90°, each successive sulfur atom can be added to a chain in a "right-handed" or a "left-handed" way (107). The energy barriers between conformations are low enough (107) to permit rapid isomerization at room temperature. Considering all these points, it is to be expected that a typical polysulfur nitride built up of a combination of rings, chains, and cages would be formed as a mixture of isomers with very similar gross properties, that would be very difficult to separate and would equilibrate among themselves within minutes to hours at room temperature. In some cases, too, there will be a labile equilibrium between different degrees of polymerization. It is not surprising, then, that the polymeric nitrides so far described (*22, 23, 63*), even the dimers and trimers, are uncrystallizable tars.

$$\begin{array}{c} S\!-\!S\!-\!S \\ / \qquad \backslash \\ S \qquad N\!-\!Hg\!-\!N \qquad S \\ \backslash \qquad / \\ S\!-\!S\!-\!S \qquad\qquad S\!-\!S\!-\!S \end{array}$$

(XXXI)

Besides the condensation reactions already described, there is a second route to polymeric saturated sulfur nitrides. The mercury(II) "salt" of S_7NH (XXXI) is treated with iodine in CS_2 solution (*22, 23*). The sequence of reactions is probably

$$Hg(S_7N)_2 + I_2 \rightarrow HgI_2 + 2S_7N\cdot \text{ (unstable)}$$

$$2S_7N\cdot \rightarrow \text{polymer}$$

The polymer, a CS_2-soluble red tar, is of unknown structure and has a molecular weight in the range of 2000–3000. It can be separated by chromatography on silica gel into fractions with varying elemental compositions, infrared spectra, and molecular weights. On standing for a few weeks it eliminates some S_8 and produces a polymer insoluble in CS_2. This may well prove to be typical of large, saturated polysulfur nitride molecules; as discussed above, they are labile at room temperature and therefore (cf. plastic sulfur) will gradually eliminate thermodynamically stable decomposition products such as S_8 and perhaps $S_{11}N_2$.

V. Conclusion

There is obviously a great deal more worthwhile work to be done in the field of sulfur nitrides. This includes studies of their behavior as bases; structural studies on the nitrides themselves, their adducts, and metal derivatives; and investigations of the stability and transformations of saturated polysulfur nitrides. One promising area of research is the use of S–N compounds of metals for the attempted synthesis of new sulfur nitrides (an aspect of the general topic of "reactions of coordinated ligands"). The recently investigated anions of the sulfur imides (*106*) are also attractive starting materials for new syntheses. The relatively new experimental technique of preparative-scale molecular exclusion chromatography opens up many opportunities for the experimenter.

REFERENCES

1. Andreocci, A., *Z. Anorg. Chem.* **14**, 246 (1896).
2. Banister, A. J., and Padley, J. S., *J. Chem. Soc. A* 1437 (1967).
3. Banister, A. J., Dainty, P. J., Hazell, A. C., Hazell, R. G., and Lomborg, J. G., *Chem. Commun.* 1187 (1969).
4. Banister, A. J., and Padley, J. S., *J. Chem. Soc. A* 658 (1969).
4a. Banister, A. J., and Younger, D., *J. Inorg. Nucl. Chem.* **32**, 3763 (1970).
5. Barker, C. K., Cordes, A. W., and Margrave, J. L., *J. Phys. Chem.* **69**, 334 (1965).
6. Becke-Goehring, M., and Schwartz, R., *Z. Anorg. Allgem. Chem.* **296**, 3 (1958).
7. Becke-Goehring, M., *Advan. Inorg. Chem. Radiochem.* **2**, 169 (1960).
8. Becke-Goehring, M., *Progr. Inorg. Chem.* **1**, 207 (1959).
9. Becke-Goehring, M., Jenne, H., and Rekalic, V., *Chem. Ber.* **92**, 855 (1959).
10. Becke-Goehring, M., Jenne, H., and Rekalic, V., *Chem. Ber.* **92**, 1237 (1959).
11. Becke-Goehring, M., *Inorg. Syn.* **6**, 123 (1960).
12. Becke-Goehring, M., *Inorg. Syn.* **6**, 128 (1960).
13. Becke-Goehring, M., and Latscha, H. P., *Z. Anorg. Allgem. Chem.* **333**, 181 (1964).
14. Becke-Goehring, M., and Fluck, E., in "Developments in Inorganic Nitrogen Chemistry" (C. B. Colburn, ed.), Vol. I, pp. 150–240. Elsevier, Amsterdam, 1966.
15. Becke-Goehring, M., and Schläfer, D., *Z. Naturforsch. B* **21**, 492 (1966).
16. Becke-Goehring, M., and Schläfer, D., *Z. Anorg. Allgem. Chem.* **356**, 234 (1968).
16a. Becke-Goehring, M., and Erhardt, K., *Naturwissenschaften* **56**, 415 (1969).
16b. Becke-Goehring, M., *Inorg. Macromol. Rev.* **1**, 17 (1970).
17. Bett, J. A. S., and Winkler, C. A., *J. Phys. Chem.* **68**, 2501 (1964).
18. Blasius, E., and Wagner, H., *J. Chromatog.* **26**, 549 (1967).
18a. Boudeulle, M., Douillard, A., Michel, P., and Vallet, G., *C. R. Acad. Sci. Ser. C* **272**, 2137 (1971).
19. Bragin, J., and Evans, M. V., *J. Chem. Phys.* **51**, 268 (1969).
20. Braterman, P. S., *J. Chem. Soc.* 2297 (1965).
21. Brauer, G., "Handbook of Preparative Inorganic Chemistry," Vol. 1, p. 409. Academic Press, New York, 1963.
22. Buckley, J., and Heal, H. G., *J. Inorg. Nuclear Chem.*, **25**, 321 (1963).
23. Buckley, J., M.Sc. Thesis, Queen's University of Belfast, 1964.
24. Buerger, M. J., *Amer. Mineral.* **21**, 575 (1936).
25. Burt, F. P., *J. Chem. Soc.*, 1171 (1910).
26. Carrington, A., *Proc. Roy. Soc. Ser. A* **302**, 291 (1967).
27. Chapman, D., and Massey, A. G., *Trans. Faraday Soc.* **58**, 1291 (1962).
27a. Chapman, D., and Waddington, T. C., *Trans. Faraday Soc.* **58**, 1679 (1962).
28. Chapman, D., Warn, R. J., Fitzgerald, A. G., and Yoffe, A. D., *Trans. Faraday Soc.* **60**, 294 (1964).

29. Clark, D., *J. Chem. Soc.* 1615 (1952).
30. Cohen, B., Hooper, T. R., Hugill, D., and Peacock, R. D., *Nature (London)* **207**, 748 (1965).
31. Douillard, A., May, J.-F., and Vallet, G., *C. R. Acad. Sci. Ser. C* **269**, 212 (1969).
32. Drew, M. G. B., Templeton, D. H., and Zalkin, A., *Inorg. Chem.* **6**, 1906 (1967).
32a. Eméleus, H. J., Forder, R. A., Poulet, R. J., and Sheldrick, G. M., *Chem. Commun.* 1483 (1970).
33. Engelbrecht, A., Mayer, E., and Pupp, C., *Monatsh. Chem.* **95**, 633 (1964).
34. Fowler, A., and Bakker, C. J., *Proc. Roy. Soc., Ser. A* **136**, 28 (1932).
35. Fluck, E., Becke-Goehring, M., and Dehoust, G., *Z. Anorg. Allgem. Chem.* **312**, 60 (1961).
36. Fluck, E., and Reinisch, R. M., *Z. Anorg. Allgem. Chem.* **328**, 165 (1964).
37. Garcia-Fernandez, H., *Bull. Soc. Chim. Fr.* 760 (1959).
38. Garcia-Fernandez, H., *C. R. Acad. Sci.* **252**, 411 (1961).
39. Garcia-Fernandez, H., *C. R. Acad. Sci.*, **260**, 1183 (1965).
40. Garcia-Fernandez, H., *C. R. Acad. Sci.* **260**, 6107 (1965).
41. Garcia-Fernandez, H., and Heal, H. G., *C. R. Acad. Sci. Ser. C* **266**, 1449 (1968).
42. Garcia-Fernandez, H., Heal, H. G., and Shahid, M. S., *C. R. Acad. Sci. Ser. C* **272**, 60 (1971).
42a. Garcia-Fernandez, H., and Messager, M., to be published.
43. Glemser, O., and Perl, H., *Naturwissenschaften* **48**, 620 (1961).
44. Glemser, O., *Angew. Chem. Int. Ed. Engl.* **2**, 532 (1963).
45. Glemser, O., Haas, A., and Reinke, H., *Z. Naturforsch. B* **20**, 809 (1965).
46. Glemser, O., *Prep. Inorg. React.* **1**, 227 (1964).
46a. Glemser, O., and Koch, W., *Angew. Chem. Int. Ed. Engl.* **10**, 127 (1971).
47. "Gmelins Handbuch der Anorganischen Chemie," Schwefel, Part B, Section 3. Verlag Chemie, Weinheim, 1963.
48. Goehring, M., *Chem. Ber.* **80**, 110 (1947).
49. Goehring, M., Herb, H., and Wissemeier, H., *Z. Anorg. Allgem. Chem.* **267**, 238 (1952).
50. Goehring, M., and Voigt, D., *Z. Anorg. Allgem. Chem.* **285**, 181 (1956).
51. Goehring, M., "Ergebnisse und Probleme der Chemie der Schwefelstickstoffverbindungen." Akademie-Verlag, Berlin, 1957.
52. Goudmand, P., and Dessaux, O., *J. Chim. Phys.* **64**, 135 (1967).
53. Gregory, W., *J. Pharm.* **21**, 315 (1835).
54. Gregory, W., *J. Pharm.* **22**, 301 (1835).
55. Griffith, W. P., and Rutt, K. J., *J. Chem. Soc. A* 2331 (1968).
56. Hamada, S., Takanashi, A., and Shirai, T., *Bull. Chem. Soc. Jap.* **44**, 1433 (1971).
57. Heal, H. G., *in* "Inorganic Sulfur Chemistry" (G. Nickless, ed.), pp. 459–508. Elsevier, Amsterdam, 1968.
58. Heal, H. G., and Kane, J., *Inorg. Syn.* **11**, 184 (1968).
59. Heal, H. G., and Kane J., *J. Polym. Sci. Part C* 3491 (1968).
60. Heal, H. G., Shahid, M. S., and Garcia-Fernandez, H., *Chem. Commun.* 1063 (1969).
61. Heal, H. G., Shahid, M. S., and Garcia-Fernandez, H., *C. R. Acad. Sci. Ser. C* **269**, 1543 (1969).
62. Heal, H. G., *Int. J. Sulfur Chem. C* **1**, 27 (1971).

63. Heal, H. G., Shahid, M. S., and Garcia-Fernandez, H., *J. Chem. Soc. A* 3846 (1971).
64. Holt, E. M., and Holt, S. L., *Chem. Commun.* 1704 (1970).
65. Jolly, W. L., and Maguire, K. D., *Inorg. Syn.* **9**, 102 (1967).
66. Jolly, W. L., and Lipp, S. A., *Inorg. Chem.* **10**, 33 (1971).
67. Joshi, K. C., *Z. Phys. Chem. (Frankfurt)* **55**, 173 (1967).
68. Kraus, H. L., and Jung, H., *Z. Naturforsch. B* **16**, 624 (1961).
69. Kronick, P. L., Kaye, H., Chapman, E. F., Mainthia, S. B., and Labes, M. M., *J. Chem. Phys.* **36**, 2235 (1962).
70. Lipp, S. A., Chang, J. J., and Jolly, W. L., *Inorg. Chem.* **9**, 1970 (1970).
71. Lu, C.-S., and Donohue, J., *J. Amer. Chem. Soc.* **66**, 818 (1944).
72. MacDiarmid, A. G., *J. Amer. Chem. Soc.* **78**, 3871 (1956).
73. Maguire, K. D., Smith, J. J., and Jolly, W. L., *Chem. Ind. (London)* 1589 (1963).
74. Mason, J., *J. Chem. Soc. A* 1567 (1969).
75. McGrath, W. D., and Morrow, T., *Nature (London)* **212**, 5063 (1966).
76. Meinzer, R. A., Pratt, D. W., and Myers, R. J., *J. Amer. Chem. Soc.* **91**, 6623 (1969).
77. Meuwsen, A., and Krüger, S., *Z. Anorg. Allgem. Chem.* **236**, 221 (1938).
78. Meuwsen, A., *Z. Anorg. Allgem. Chem.* **266**, 251 (1951).
79. Nair, C. G. R., and Murthy, A. R. V., *Chem. Ind. (London)* 1539 (1962).
80. Nair, C. G. R., and Murthy, A. R. V., *J. Inorg. Nucl. Chem.* **25**, 453 (1963).
81. Narasimham, N. A., and Subramanian, T. K. B., *J. Mol. Spectrosc.* **29**, 294 (1969).
82. Nelson, J., and Heal, H. G., *Inorg. Nucl. Chem. Lett.* **6**, 429 (1970).
83. Nelson, J., and Heal, H. G., *J. Chem. Soc. A* 136 (1971).
84. Nelson, J., and Heal, H. G., unpublished observations.
85. Nelson, J. T., and Lagowski, J. J., *Inorg. Chem.* **6**, 1292 (1967).
86. Neubauer, D., and Weiss, J., *Z. Anorg. Allgem. Chem.* **303**, 28 (1960).
87. O'Hare, P. A. G., *J. Chem. Phys.* **52**, 2992 (1970).
88. Pannetier, G., Goudmand, P., Dessaux, O., and Tavernier, N., *C. R. Acad. Sci.* **255**, 91 (1962).
89. Patton, R. L., and Jolly, W. L., *Inorg. Chem.* **8**, 1389 (1969).
90. Patton, R. L., and Jolly, W. L., *Inorg. Chem.* **8**, 1392 (1969).
91. Patton, R. L., and Raymond, K. N., *Inorg. Chem.* **8**, 2426 (1969).
92. Paul, R. C., Arora, C. L., Kishore, J., and Malhotra, K. C., *Aust. J. Chem.* **24**, 1637 (1971).
93. Peacock, R. D., *Chem. Abstr.* **67**, 7581 (1967).
94. Peyron, M., and Lam Than My, *J. Chim. Phys. Physiochim. Biol.* **64**, 129 (1967).
95. Roesky, H. W., Glemser, O., and Hoff, A., *Chem. Ber.* **101**, 1215 (1968).
96. Rogers, M. T., and Gross, K. J., *J. Amer. Chem. Soc.* **74**, 5294 (1952).
97. Rötgers, K., Ph.D. Dissertation, Berlin, 1907.
98. Sass, R. L., and Donohue, J., *Acta Crystallog.* **11**, 497 (1958).
99. Schenk, R., *Ann. Chem.* **290**, 171 (1896).
100. Schroeder, H., and Glemser, O., *Z. Anorg. Allgem. Chem.* **298**, 78 (1959).
101. Shahid, M. S., Ph.D. Thesis, Queen's University of Belfast, 1971.
102. Shahid, M. S., Heal, H. G., and Garcia-Fernandez, H., *J. Inorg. Nucl. Chem.*, in press.
103. Shahid, M. S., Heal, H. G., and Garcia-Fernandez, H., *J. Inorg. Nucl. Chem.* **33**, 4364 (1971).

104. Sharma, B.D., and Donohue, J., *Acta Crystallog.* **16**, 891 (1963).
104a. Shieh, M.-C., Katabe, K., and Okabe, T., *Bull. Chem. Soc. Jap.* **43**, 3449 (1970).
105. Smith, G. F. H., *Mineral. Mag.* **16**, 97 (1913).
105a. Steudel, R., *Z. Naturforsch.* B **24**, 934 (1969).
106. Tingle, E. M., and Olsen, F. P., *Inorg. Chem.* **8**, 1741 (1969).
107. Tuinstra, F., "Structural Aspects of the Allotropy of Sulphur and the Other Divalent Elements." Waltman, Delft, 1967.
108. Turner, A. G., and Mortimer, F. S., *Inorg. Chem.* **5**, 906 (1966).
109. Van de Grampel, J. C., and Vos, A., *Acta Crystallog.* **25**, 611 (1969).
110. Villena-Blanco, M., and Jolly, W. L., *Inorg. Syn.* **9**, 98 (1967).
111. Warn, J. R. W., and Chapman, D., *Spectrochim. Acta* **22**, 1371 (1966).
112. Weiss, J., *Fortschr. Chem. Forsch.* **5**, 635 (1966).
113. Weiss, J., and Piechaczek, H., *Z. Naturforsch.* B **18**, 1139 (1963).
114. Wiegers, G. A., and Vos. A., *Proc. Chem. Soc.* 387 (1962).
115. Wölbling, H. *Z. Anorg. Allgem. Chem.* **57**, 286 (1908).
116. Wynne, K. J., and Jolly, W. L., *Inorg. Chem.* **6**, 107 (1967).
117. Wynne, K. J., and Jolly, W. L., *J. Inorg. Nucl. Chem.* **30**, 2851 (1968).
118. Zeeman, P. B., *Can. J. Phys.* **29**, 174 (1951).
119. Zipp, A. P., and Evers, E. C., *Inorg. Chem.* **8**, 1746 (1969).

AUTHOR INDEX

Numbers in parentheses are reference numbers and indicate that an author's work is referred to although his name is not cited in the text. Numbers in italics show the page on which the complete reference is listed.

A

Abeledo, C., 220(246), *249*
Ablov, A. V., 169(1–3), 177(1, 3), 180(3), 181(1, 3), *241, 242*
Abraham, M. H., 266(35), *316*
Abrahams, S. C., 10(111), 13(111), *53, 56*
Adams, C. J., 24(1a), 32(1a), *53*
Adams, D. M., 24(2), *53*
Adams, R. W., 274(67), 284(85, 86), 287(85), 288(85, 104, 105), 289(85, 86, 108), *317, 318*
Adloff, J. P., 100(249), 101(249), *249*
Agrawal, M. M., 262(18), 291(117), 294, *316, 319*
Agron, P. A., 6(34, 118), 7(3), *53, 54, 57*
Aguiar, A. M., 346(1), *368*
Aguiar, H. J., 346(1), *368*
Akhtar, M., 33(147), 35(147), *58*
Aksnes, O., 38(4), 40(4), *53*
Albers, H., 294, *319*
Alcock, N. W., 10(5), 13(5), 14(6), 33(8), 35(8), 36(8), 39(7), 40(7), *53*, 135(4), *242*
Aleksandrov, A. Yu., 70(5), 103(5, 6, 8, 9), 106(9), 136(8), 141(6, 7), 142(5, 6, 7, 9), *242*
Alexander, L. E., 353(204), 359(204), *374*
Ali, K. M., 124(10), 143(10), 144(10), 145(10), 148(10), 149(10), 151(10), 152, 154(10), 155(10), *242*
Allen, F. H., 347(2), *368*
Allen, H. C., 284(90), 307(90), *318*
Almin, K. E., 30(9), 34(9), *53*
Alsdorf, H., 261(15), 262(17), 295(15), *316*
Altena, D., 51(36b), *54*
Alti, G. De., 177(11, 12), *242*

Alyea, E. C., 213(13), 214(13), *242*, 284(88), 285(88), 286(95), 287(101, 102), 300(178, 179, 183), 304(183), 308(179, 183, 223), 309(183), 315(179), *318, 320, 322*
Amma, E. L., 51(130a), *57*
Anderson, L. P., 327(3), *368*
Andreocci, A., 381(1), *409*
Andrews, L. J., 5(10), 47(10), *53*
Andrianov, K. A., 297, *320*
Angelici, R. J., 357(4), *368*
Anisimov, K. N., 122(364), 123(275, 364), 143(275), 159(275), 160(275, 364), 161(275, 364), *250, 252*
Apai, G. R., 346(174), *373*
Araujo, F. T. de, 102(15), *242*
Archer, E. M., 11(11), 12(12), 14(11), *53*, 196(16), *242*
Archibald, T. G., 346(1), *368*
Arora, C. L., 387(92), 388(92), 389(92), *411*
Arora, M., 264(29), 266, *316, 317*
Artman, J. O., 67(17), *242*
Ashby, E. C., 300, *321*
Attridge, C. J., 108(18), *242*
Atwood, J. L., 311, 312(230), *322*
Aubke, F., 140(572), 142(572), 145(574), 146(574), 147(574), 148(574), 149(574), 150(574), 152(574), 163(572, 573), *258*
Axtmann, R. C., 214(19, 320, 321), 218(19, 320, 321), 219(320, 321), *242, 251*
Aylett, B. J., 312(231), *322*
Aynsley, E. E., 24(13), *53*

B

Bäcklund, S., 32(37, 38), 35(38), *54*
Bagnall, K. W., 44(14), *54*

Bahner, C. T., 328(186), 329(185, 186), *373*
Bailey, M., 19(152), 20(152), *58*
Bains, M. S., 261, 270(53), 271, *316, 317*
Baird, H. W., 17(149), *58*
Baker, W. A., 95(482), 169(135), 173(135), 178(135), 180(135), 181(135), 211 (480, 481), 213(480, 481), 217(135), 219(480), 220(135, 480), *246, 255, 256*
Bakker, C. J., 376(34), *410*
Balizs, A., 225(179), 226(179), *247*
Ballhausen, C. J., 224(154), *246*
Bally, R., 30(15), 34(15), *54*
Bancroft, G. M., 64(568), 68(44), 70(121), 71(36, 44), 73, 77(29, 39, 44), 78(44), 79, 80, 82(121), 84, 88(29), 94(30, 31, 32), 98(20, 21, 23–26, 40), 99(24, 26, 40), 100(24, 27), 102(38), 122(29), 123(29), 127, 128(29), 129, 133(29), 159(44), 162(28), 167, 168(34, 39, 44), 169(34, 37, 39), 170, 171(36), 172(34, 36), 173(34, 35), 174(39, 45), 175(34, 36, 44), 176(34), 177(37, 44), 178, 179(44), 180(39, 44), 181(44, 45), 182(39, 44), 183(44), 184(22, 37), 185(22), 187(37), 210(41), 211(32, 42, 43, 399), 212(32, 41, 42, 399), 213(32, 42, 43), 214(43), 216(121), 219(44), 227(33), 230(44), *242, 243, 245, 253, 258*
Banister, A. J., 386(4a), 387(4a), 389(2), 390(2–4), *409*
Banks, R. E., 324(5), 328(5), 334(6), 337(5), 342(5, 6), 344(6), *368*
Baranovskii, V. I., 161(46), *243*
Barber, M., 162, 163(47), *243*
Bargeron, C. B., 96(48), *243*
Barker, C. K., *409*
Barr, J. T., 328(186), 329(185), 329(185, 186), *373*
Barraclough, C. G., 288(105), 289(108), *318*
Bartlett, N., 7, 8(142), 10(142), 33(147), 35(147), *54, 56, 58*
Bartlett, P. D., 327(7), *369*
Bartley, W. G., 260(9), *316*
Bartunik, H. D., 91(49), 219(49), 227(49), 228(49), 229(49), 230(49), *243*
Basi, J. S., 286(95), 287(101, 102), 299(168, 169), 300(168, 169, 179), 308(169, 179), 315(179), *318, 320*
Batwara, J. M., 291(113), 296, *319, 320*
Bauer, S. H., 345(38), *369*
Baxter, R. M., 51(100a), *56*
Beachem, M. T., 331(202), 337(202), 341(202), *374*
Beagley, B., 312(213), *322*, 348(8), *369*
Beard, G. B., 90(496), 232(496), 233(496), 234(496), *256*
Bearden, A. J., 90(50), *243*
Beattie, I. R., 114(51), *243*
Beauchamp, A. L., 236(52), *243*
Becke-Goehring, M., 376(8, 14, 16b), 378(13, 15), 390(6, 7), 391(16a), 392(16, 16b, 35), 393(11), 394(6, 7), 395(8), 396(8, 12), 397(8), 400(9, 10), 401(9), 405(10), *409, 410*
Becker, K. A., 30(17, 18), *54*
Beckett, R., 220(53), 221(53), *243*
Bell, N. A., 305(216), *321*
Bell, T. N., 342(12), *369*
Belozerskii, G. N., 169(1), 177(1), 181(1), *241*
Benezra, C., 327(32), *369*
Bennett, M. A., 346(10), 348(10), 356(9), *369*
Bennett, M. J., 236(52), *243*, 356(11), *369*
Bensey, F. D., 9(30, 31), *54*
Bergen, A. Van den, 211(54), 213(54), 214(54), *243*
Bernstein, J. L., 10(111), 13(111), *53, 56*
Berrett, R. R., 77(55), 167, 168(55), 180(55), 181(55), 211(56), 212(56), 213(56), 220(56), *243*
Berry, A. D., 351(13), *369*
Berry, R. S., 352(14), *369*
Bersin, T., 293, 294, *319*
Bersuker, I. B., 177(2), *241*
Bertino, C. D., 338(177), 343(177), *373*
Bett, J. A. S., *409*
Bichler, R. E. J., 332(15), 335(15), *369*
Bigotto, A., 177(11, 12), *242*
Bilevich, K. A., 112(57), *243*
Bindal, S. R., 271(61), *317*
Bindschadler, E., 292(123, 124), 294(138), *319*
Birchall, T., 90(62), 93(102, 104), 112(103), 212(58), 217(58), 220(61), 221

AUTHOR INDEX

(59, 61), 223(60), 224(60), 225(60), 232(62, 63), 233(63), *243*, *244*, *245*
Bird, S. R. A., 85(64), 88(64), 106(65), 112(65), 117(64), 118(64), 122(64), 123(64), 128(64), 129(64), 159(64), 160(64, 65), 161(64), 162, 235(63), *244*
Biryukov, B. P., *244*
Bishop, E., 284(85), 287(85), 288(85), 289(85), *318*
Bjorvatten, T., 31(19), 34(19), *54*
Blackman, A. G., 227(67), *244*
Blasius, E., 391(18), *409*
Bledsoe, W., 194(570), 197(570), 199(570), *258*
Bloembergen, N., 191(68), *244*
Blom, E. A., 244(69), *244*
Blume, D., 292(124), 294(138), *319*
Boer, F. P., *257*
Böttcher, B., 295(143), 312(143), *319*
Bokii, N. G., 116(70), 117(70), *244*
Bolef, D. I., 191(408), *253*
Bolhuis, F. V., 10(52a), 12(20), 13(52a), 15(20), *54*, *55*
Bondi, A., 3, 4, *54*
Bonnett, R., 299(175), *320*
Booth, B. L., 328(16), 332(16), 333(16), 334(16), *369*
Booth, G., 346(17), *369*
Booth, M. R., 332(15), 335(15), *369*
Booth, R., 237(547), 238(548), *257*
Bor, G., 359(18), *369*
Borgen, B., 37(22), 38(22), *54*
Borshagovskii, B. V., 123(275), 143(275), 159(275), 160(275), 161(275), *250*
Bosnich, B., *369*
Boswijk, K. H., 11(88), 12(23, 24), 15(23, 88, 153), 16(23, 24), *54*, *56*, *58*
Boudeulle, M., 395(18a), *409*
Bowen, L. H., 90(71, 522), 232(71, 72, 388, 389, 522), 233(72, 389), 234(72, 389, 522), 235(72, 389, 522), 236(72, 389, 522), *244*, *253*, *257*
Boylan, M. J., 95(73), *244*
Boyle, A. J. F., 62(75), 202(74), 203(74), 204, *244*
Bradley, D. C., 213(13), 214(13), *242*, *259*(1, 3, 4), 260(1, 4, 7, 8), 262(19), 265(1), 266, 267, 271, 272(1, 3, 46), 275(70), 276, 277, 279, 280, 281(83),
282(70, 75, 80, 84), 283(75), 284(88, 89), 285, 286, 287(73, 92, 95, 98, 101, 102), 288(103), 290, 291, 292, 293, 294(135, 136), 295(7), 296, 297, 299(3, 166, 168–175), 300(183), 301(166, 170, 172, 190, 191), 302(199, 201, 203–206), 303, 304(183), 305(173, 176, 207, 214), 306, 307, 308(169, 176, 179, 183, 221–223), 309(183), 314, 315(179, 221), *316*, *317*, *318*, *319*, *320*, *321*, *322*
Brady, P. R., 91(76), 168(76), 173(76), 176(76), 177(76), 178(76), 179(76), *244*
Braekken, H., 32(25), *54*
Brändén, C. I., 147(77), *244*
Bragin, J., 381(19), 393(19), *409*
Braibanti, A., 50(25a), 51(25b), *54*
Braterman, P. S., 383(20), *409*
Brauch, K. F., 169(237), 180(237), 181(237), *249*
Brauer, G., 393(21), *409*
Brauman, J. I., 301(197), *321*
Bray, P. J., 192(105), *245*
Breslow, R., 324(20), *369*
Brey, W. S., 327(206), *374*
Brice, M. D., 361(25), *369*
Bringeland, R., 39(54), *54*
Britton, D., 37, 38(27, 134), 39(27, 66, 67, 133, 135), 40(66, 67, 135), 41, 53(114a, 154a), *54*, *55*, *57*, *58*, 112(506), 137(506), *256*
Brown, C. A., 295, *319*
Brown, D. A., 287(100), *318*, 350(23), 352(22), *369*
Brown, I. D., 238(78, 162), *244*, *247*
Brown, L. M., 291, *319*
Brown, R. N., 9(28), 10(28), *54*
Brown, T. L., 112(558), *258*, 261, *316*
Browning, J., 323(24), *369*
Brubaker, C. H., 285(93, 94), 289, *318*
Bruce, M. I., 324(28, 63), 328(28), 334(26, 148), 344(148), *369*, *370*, *372*
Brunge, O., 147(315), *251*
Bryan, R. F., 142(296), 152(79), *244*, *250*, *253*, 350(29, 132, 133), *369*, *372*
Bryce-Smith, D., 266, *316*
Bryden, J. H., 19(29), 20(29), *54*
Bryuchova, E. V., 141(81), *244*
Bryukhanov, V. A., 103(84), 136(83, 84), 141(84), 143(84), 152(513), 203(82),

232(513), 233(513), 237(85, 513), 238(85, 513), *244, 256*
Buchanan, D. N. E., 94(560), *258*
Buckley, A. N., 211(54, 86, 87), 213(54, 86, 87), 214(54), *243, 244*
Buckley, J., 408(22, 23), *409*
Bürger, H., 259(5), 300, 301, 302, 304, 305, 306, 309(181), 312, *316, 320, 321*
Buerger, M. J., 379(24), *409*
Bukshpan, S., 91(90), 191, 192(89), 193 88–93), 198, 199(91, 93), 200, 201(90, 92), *244, 245*
Bulliner, P. A., 7(142), 8(142), 10(142), *58*
Bullock, R. J., 91(272), 100(272), *250*
Bunbury, D. St. P., 202(74), 203(74), 204 (74), *244*
Burbank, R. D., 7(32, 33, 108), 9(30, 31), *54, 56*
Burbridge, C. D., 216(94–97), 217(94, 96, 97), 218(97), *245*
Burger, K., 90(376), 232(377), *253*
Burns, G. R., *245*
Burns, J. H., 6(34, 35), *54*
Burns, R. G., 98(23–26, 40), 99(24, 26, 40), 100(24, 27), *242, 243*
Burt, F. P., 395(25), *409*
Burton, D. J., *369*
Burton, J. R., 266(42), *317*
Bushnell, G. W., 352(22), *369*
Buss, B., 20(36), 24, 51(36a,b), *54*
Butler, K. D., 77(29), 85(29), 88(29), 122(29), 123(29), 127(29), 128(29), 129(29), 133(29), 162(28), *242*
Buyrn, A. B., 238(99), *245*
Byström, A., 32(37, 38), 35(38), *54*

C

Calderazzo, F., 346(31), 351(31), 355(31), 359(31), *369*
Caldwell, E. W., 294(136), *319*
Callot, H. J., 327(32), *369*
Cameron, J. A., 102(100), 103(100), *245*
Camerman, A., 310(225), *322*
Camerman, N., 29(39), 30, 34(39), 39(40), 40(40), *54*
Campbell, I. G. M., 234(101), *245*
Campbell, S. F., 332(35), 337(33–36), 338(33–35, 37), 343(34, 35, 37), 344 (34), *369*
Cardin, D. J., 301(194), *321*
Caron, A., 348(101), *371*
Carpenter, G. B., 12(41), 15(41), 37(91), *54, 56*
Carrai, G., 30(42), 34(42), 53(42), *54*
Carrington, A., 376(26), *409*
Carter, V. B., *54*
Carter, W., 93(104), 112(103), *245*
Carty, A. J., 93(102, 104), 112(104), 145, 147, 148(103), 150(103), 152(103), *245*
Casabella, P. A., 192(105), *245*
Cassidy, J. E., 205(106), 206(106), *245*
Caughlan, C. N., 272(65), 274, 277, 280 (81), *317, 318*
Chadha, S. L., 270(56), *317*
Chakravarti, B. N., 292(125), *319*
Chamberlain, M. M., 295, *319*
Champion, A. R., 96(107), 97(107), *245*
Chan, L. Y. Y., 51(42b), 52(42c), *54, 55,* 362(60), 363(60), 367(60), *370*
Chandler, L., 88(569), 122(569), 123(569), 133(569), 159(569), 160(569), 161 (569), *258*
Chandra, G., 301(192), 302(198, 200), *321*
Chandra, S., 64(345), 119(327), 125(327), 165(327–329), *251, 252*
Chang, C. H., 345(38), *369*
Chang, J. J., 385(70), *411*
Chapman, D., 381(27a), 382(27a), 383 (27a), 386(27), 393(111), 395(28), *409, 412*
Chapman, E. F., 395(69), *411*
Charalambous, J., 302(206), *321*
Charlton, J. S., 91(108), 227(108), 228 (108), 229(108), 230(108), *245*
Chatt, J., 167, *245*
Chatterjee, A. K., 291(118–120), 292(125, 126, 128), *319*
Chatterjee, Amar K., 292(128), *319*
Chatterjee, Amiya K., 292(128), *319*
Cheng, H. S., 137, 141(112), 142(112), 150 (330), 151(330), 152(112, 113), 154 (112), 155(113, 330), 165(112), *245, 251*
Cherwinski, W. J., 329(39), 341, *369*
Chia, L. S., 329(40), 330(40), 331(40), 345(40), 347(40), 348(40), 349(40), 352(40), 353(40), 356(40), 358(40),

AUTHOR INDEX

360(40), 361(40), 364(40, 64), 365(40), 366(40), *370*
Chien, J. C. W., 302(203), *321*
Chisholm, M. H., 276(73), 285(91), 286(95), 287(73, 95, 102), 299(169, 173), 300(169, 173, 176, 177, 179), 303(207), 304(173, 176, 207, 208), 305(173, 176, 207), 306(173, 176), 307, 308(169, 176, 179), 309(173), 315 (179), *317, 318, 320, 321*
Chivers, T., 105(114, 115), 106(115), 107 (115), 108(114, 115), 122(114, 115), 153(114, 115), 158(114, 115), *245*
Choi, S. K., 337(178), *373*
Chow, Y. M., 33(43), 41(44), 53(114a), *55, 57,* 140(116), *245*
Christe, K. O., 50(118a), *57,* 200(117), *245*
Christensen, A. T., 37(45), 38(45), *55*
Christofferson, G. D., 22(46), 28, *55*
Churchill, M. R., 361(41), *370*
Ciuffarin, E., 301(197), *321*
Clark, A. H., 312, *322*
Clark, D., 379(29), *410*
Clark, H. C., 114(118), *245,* 324(43, 45), 329(39, 44), 331(44), 332(43), 333 (15), 335, 341, 342(45), *369, 370*
Clark, J. R., 99(436), *254*
Clark, M. G., 70(121), 71(122), 72, 77, 80(122), 81(122), 82(121), 83(122), 85, 86(122), 87(122), 88, 110(122), 112(122), 114(122), 116(122), 117 (122), 119(122), 120, 121(122), 127 (122), 128(122), 129(122), 134(122), 216(121), *245*
Clark, R. J. H., 114(123), 207, *245*
Clausen, C. A., 91(128, 130), 115(128), 150, 151(127), 154(127), 155(127), 185(129), 186(129), 187(473), 210 (125), 211, 212(125), 213(125), *245, 246, 255*
Clayton, A. B., 332(46, 47), *370*
Clifford, A. F., 91(197), *248*
Closs, G. L., 345(48), *370*
Coates, G. E., 264, 265, 266, 299, 305, *316, 317, 320, 321*
Cody, V., 115(131), 138(131), 158(131), *246*
Cohen, B., 387(30), 388(30), *410*
Cohen, S. G., 227(427), *254*

Colapietro, M., 30(47), 31, 34(47), *55*
Collins, D., 332(46), *370*
Collins, R. L., 64(136), 65(136), 70(132, 136), 71(406), 85(406), 93(206), 164 (206), 169(135), 173(135), 178(135), 180(135), 181(135), 208, 209, 210(406), 217(135), 220(135), 222(133, 134), 224 (132, 133), *246, 248, 253*
Collman, J. P., 335(49), 355(49), *370*
Cook, D. J., 328(50), 334(50), *370*
Copperthwaite, R. G., 213(13), 214(13), *242,* 300(183, 186), 304(183), 308(183, 186, 222, 223), 309(183, 224), *320, 321, 322*
Cordes, A. W., 21(48, 151), 25, 26, *55, 58, 409*
Cordey Hayes, M., 103(138, 141), 106 (141), 107(138, 140), 110(141), 111 (141), 119(141), 222(138, 140), 124 (141), 138(141), 139(141), 141(138, 141), 144(141), 154(139), 155(137), 160(139), 202(137), 203(137), *246*
Corey, E. R., 115(131), 119(426), 128(426), 138(131), 158(131), *246, 254,* 349(51), 351(13), *369, 370*
Corfield, P. W. R., *370*
Cornwell, C. D., 192(571), 193(571), *258*
Corradini, P., 361(52), *370*
Costa, G., 177(12), *242*
Costa, N. L., 168(142), 177(142), 178 (142), 179(142), 180(142), 220(142), *246*
Costan, C. J., 101(352), *252*
Cotton, F. A., 93(143), 236(52), *243, 246,* 347(54), 348(53), *370*
Cotton, S. A., 308(222), *322*
Cowan, M., 192(144), *246*
Cox, M., 95(145), 219(145), 220(145), 221(145), *246*
Craig, P. P., 100(541), 101(541), *257*
Craig, R. R., 345(216), *374*
Cromer, D. T., 14(49, 116), 36(123), *55, 57,* 201(394), *253*
Crow, J. P., 209(146), 210(146), 223(146), 225(146), *246,* 329(56), 331(56), 333(61), 350(55), 351(59), 352(59), 353(55, 57, 58, 61), 354(57), 358(55), 359(56), 360(57), 361(56), 362(59, 60), 363(59–61), 364(55), 365(55), 367(56, 60), *370*

AUTHOR INDEX

Crowe, R. W., 277(77), *318*
Cruickshank, D. W. J., 312(231), *322*, 348(8), *369*
Cullen, W. R., 105(150), 106(150), 108(150, 151), 122(150), 133(149), 160(149), 161(149), 209(146), 210(146), 222(147, 149), 223(146), 224(147, 149), 225(146), 235(148), *246*, 324(62, 63, 81, 87, 89), 326(79, 88), 327(80, 81), 328(65), 329(56, 65–67, 69, 75, 82), 330(66, 69, 82), 331(56, 65, 67, 75), 333(61, 77, 82, 86), 335(69), 341(65), 342(82, 85, 87), 345(79, 88), 346(80, 82), 348(72, 74), 350(55, 84), 351(59, 68, 70, 74, 85), 352(59, 72–74), 353(55, 57, 58, 61, 71, 72), 354(57, 83), 355(68, 69, 83), 356(72–74, 83), 357(72, 76), 358(55, 71), 359(56), 360(57), 361(56), 362(59, 60, 78), 363(59–61, 78), 364(55, 64, 71, 75), 365(55, 75), 366(76), 367(56, 60, 78, 110), 368(110), *370, 371*
Cunningham, D., 124(10), 133(152), 143(10), 144(10), 145(10), 147, 148(10, 152), 149(10, 152), 150(152), 151(10, 152), 152(10), 154(10), 155(10, 152), *242, 246*, 287(100), *318*
Curran, C., 110(417), 112(457), 124(417, 418, 420, 421, 455), 125(417, 418, 420, 421, 455), 129, 130, 140(418), 141(418), 143, 144(418), 145(457), 147(457), 149(433, 455), 150(433, 455), 151(433, 455), 152(421, 457), 153(417), 154(421, 433, 455), 155(417, 418, 421, 433, 457), 157(418, 421), 158(421) 203(194), 205(194), 206(194), 207(194), *248, 254, 255*
Cushen, D. W., 32(50), *55*

D

Da Costa, M. I., 192(153), 195(153), *246*
Dähne, W., *252*
Dahl, J. P., 224(154), *246*
Dahl, L. F., 92(557), 93(557), *258*, 263(22), *316*, 349(91, 212), 353(91), 357(212), 361(205, 213, 214), 363(90), *370, 371, 374*
Dahm, D. J., 357(92, 93), *371*
Dailey, B. P., 187(540), *257*
Dainty, P. J., 390(3), *409*
Dale, B., 162(28), *242*
Dale, B. W., 168(156), 169(156), 172(157), 175(156), 177(156), 178(156), 180(156), 181(156), 225, 226(155), *246*
Dalton, R. F., 145(159), *246*
Dam, C. F., 227(491), *256*
Damrauer, R., 324(195), *374*
Danon, J., 62(160), 91(470), 168(142, 161), 172(161), 176(160), 177(142), 178(142), 179(142), 180(142), 185(470), 187(470), 220(145), *246, 247, 255*
Darken, J., 95(145), 219(145), 220(145), 221(145), *246*
Darragh, K. V., 325(196), *374*
Das, A. K., 238(162), *247*
Das, T. P., 187(163), *247*
Dasent, W. E., 13(51), *55*
Davidsohn, W., 326(94), *371*
Davidson, I. M. T., 342(95), *371*
Davies, A. G., 33(52), 35(52), 36, *55*, 114(123), 120(167), 124(168), 125(168), 130(166), 131(168), 140(167, 168), 141(167), 143(168), 144(168), 145, 150(168), 155(168), 157(168), 161(169), 206(165, 166), *245, 247*
Davis, G. L., 263(22), *316*
Dawson, D. S., 328(65), 329(65), 331(65), 341(65), *370*
Dear, R. E. A., 263, *316*
de Beer, J. A., 354(96), 360(96), *371*
de Boer, B. G., 7(142), 8(142), 10(142), *58*
de Boer, E., 90(177), 102(178), *247*
de Boer, J. L., 10(52a), 13(52a), *55*
Debye, N. W. G., 110(170), 113(171), 124(171), 135, 136(170), 137, 141(171), 143(171), 146(171), *247*
Deeming, A. J., 349(97), *371*
Deeney, F. A., 95(73), *244*
Deer, W. A., 98(172), *247*
Dehmelt, H. G., 192(173), 193(493), 196(493), *247, 256*
Dehmlow, E. V., 324(98), *371*
Dehn, J. T., 224(174), *247*
Dehoust, G., 392(35), *410*
Delmaldé, A., 52(52b), *55*
Delyagin, N. N., 70(5), 103(5, 6, 84), 136(83, 84), 141(6, 84), 142(5, 6), 143(84), 203(82), *242, 244*

DePasquali, G., 91(310), 96(435), 97 (435), 101(352), 189(310), 190(310), 191(310), 192(310), 193(310), *251, 252, 254*
Dermer, D. C., 299(165), *320*
Dertouzos, H., 325(197, 198), *374*
Dessaux, O., 376(52, 88), 377(52), *410, 411*
Deutsch, M., 294(137), *319*
DeVoe, J. R., 60(175), *247*
Devooght, J., 111(176), *247*
deVries, J. L. K. F., 90(177), 102(178), *247*
DeWaard, H., 91(310), 189(310), 190 (310), 191(310), 192(310), 193(310), *251*
d'Eye, R. W. M., 44(14), *54*
Dezsi, E., 102(180), 225(179), 226(179), *247*
Dhaliwal, P. S., 328(65), 329(65–67, 69), 330(66, 69), 331(65, 67), 333(69), 335(69), 341(65), 351(68), 355(68, 69, 99), 363(99), *370, 371*
Dharmawardena, K. G., 94(30, 31, 32), 211(32), 213(32), *242, 243*
Di Lorenzo, J. V., 236(508), *256*
Dixon, K. R., *370*
Djatkina, M. E., 224(514), *256*
Dobson, G. R., 346(100), 347(100), 351 (100), 355(100), *371*
Dodd, R. E., 24(13), *53*
Dollase, W. A., 37(53), *55,* 98(181), 99 (181), *247*
Domenicano, A., 30(47), 31(47), 34(47), *55*
Donaldson, J. D., 85(64), 88(64), 106(65), 112(65), 117(64), 118(64), 122(64), 123(64), 124(10), 128(64), 129(64), 133(152), 143(10), 144(10), 145(10), 147(152), 148(10, 152), 149(10, 152), 150(152), 151(10, 152), 152(10), 154 (10), 155(10, 152), 159(64), 160(64, 65), 161(64), 162(65), 201(188), 202, 203, 204, 205, 206(106, 165, 166, 184–187, 191, 193), 207(106, 184–187), *242, 244, 245, 246, 247*
Dong, D. F., 351(70), *370*
Donohue, J., 27(54), *55,* 348(101), *371, 381,* 386(98), *411, 412*
Dorfman, Ya. G., 141(7), 142(7), *242*

Doskey, M. A., 203(194), 205(194), 206 (194), 207(194), *248*
Dosser, R. J., 95(195), 169(195), 180(195), *248*
Douillard, A., 393(31), 395(18a, 31), 396 (31), *409, 410*
Downs, A. J., 24(1a), 32(1a), *53*
Drago, R. S., 109, 141(338), 146(388), *252,* 347(220), *374*
Dreier, F., 331(102), *371*
Drew, M. G. B., 387(32), *410*
Drickamer, H. G., 96, 97(107, 435), 101 (352), *243, 245, 248, 249, 252, 253, 254*
Drummond, I., 223(60), 224(60), 225(60), *244*
Duax, W. L., 50(129a), *57*
Dufresne, A., 102(15), *242*
Dulaney, G. W., 91(197), *248*
Duncan, J. F., 60(198), 91(76), 168(76), 173(76), 176(76), 177(76), 178(76), 179(76), 216(284), 217(284), 218(199, 284), *244, 248, 250*
Duncan, W., 331(102), *371*
Dundon, R. W., 98(200), *248*
Dunn, P., 272, *317*
Dyatkin, B. L., 331(103, 104), *371*
Dzevitskii, B. E., 155(534), 161(46), *243, 257*

E

Eaborn, C., 342(95), *371*
Edwards, A. J., 9, 10(57, 62), 20(56), 21(58, 59, 60), 23, 24(58, 59, 60, 61), 25, 31, 32, 35(55, 63), 44(62), *55*
Edwards, C., 202(74), 203(74), 204(74), *244*
Edwards, P. R., 95(195), 168(156), 169 (156, 195), 172(157), 175(156), 177 (156), 178(156), 180(156, 195), 181 (156), 210(201), 212(201), 216(202), 217(202), 218(202), 225(155), 226 (155), *246, 248*
Edwards, W. T., 93(143), *246*
Efratry, A., 93(102), *245*
Eger, C., 50(129a), *57*
Egger, C. A., *372*
Ehrlich, B. S., 91(205), 190(203), 193 (203), 194(204, 205), 196, 197(205), 198, 199(203), *248*

Eicher, T., 324(20), *369*
Eilbeck, W. J., 95(195), 169(195), 180 (195), *248*
Einstein, F., 7(16), 22(64), 27, *54, 55*
Einstein, F. W. B., 51(42b), 52(42c), *54, 55*, 348(109), 350(112), 351(105), 352(105), 355(105, 112), 361(106–108), 362(60, 106, 107), 363(60, 113), 364(113), 365(108), 366(111), 367(60, 108a, 110), 368(110), *370, 371*
Ellis, I. A., 312(231), *322*
Ellison, R. D., 6(35), *54*
Emeléus, H. J., 324(114, 115), 328(114), 337(114), 342(114), *371, 372*, 385(32a), *410*
Emerson, G. F., 93(206), 164(206), *248*
Emerson, K., 37(65), 38(65), 39(66, 67), 40(66, 67), 41, *55*
Emsley, J. W., 305(216), *321*
Engel, G., 44(68), *55*
Engelbrecht, A., 384(33), *410*
Ensling J., 111(207), 116, 123(207), 133(207), 146(207), 158(207), *248*
Epstein, L. M., 124(210), 125(210), 169(209), 178(209), 180(209), 181(209), 212(211), 220(113), 222(208), 226(212), 227(212), *248*
Ercoli, R., 346(31), 351(31), 355(31), 359(31), *369*
Erhardt, K., 391(16a), *409*
Erich, U., 227(214), *248*
Erickson, N. E., 62(216), 91(218, 219), 92(218), 93(218), 94(309), 117(217), 118(217), 119(217), 125(217), 141(217), 169(215), 180(215), 181(215), 211(480), 213(480), 217(220), 219(480), 220(480), 222(218), 238, 239(219), *248, 251, 255*
Ernst, R. R., 346(116), *372*
Erskine, G. J., 346(117), 356(9), *369, 372*
Evans, B. J., 98(221), *248*
Evans, M. V., 381(19), 393(19), *409*
Evers, E. C., 390(119), *412*

F

Fairhall, A. W., 91(218), 92(218), 93(218), 222(218), *248*
Faktor, M. M., 266, 290, 291(110), *317, 318*

Faltens, M. O., 91(222), 227(222), 228(222), 229(222), 230, 231, *248*
Farmery, K., 93(223, 224), 222(224), 223(224), *248*
Feast, W. J., 327(3), 332(47), *368, 370*
Fehler, F., 38(69), 40(69), *55*
Feikema, Y. D., 10(70), 14, *55*
Feltham, R. D., 346(118), *372*
Fenger, J., 100(225), 101(225), *248*
Fenton, D. E., 110(170), 122(226, 227), 133, 135(170), 136(170), 137(170), 160(226, 227), 161(227), 162, *247, 248*, 351(119), *372*
Fernandez, V. P., 268(47), 269(47), 270(47), *317*
Fernelius, W. C., 299(165), *320*
Ferretti, A., 335(120), 341(120), *372*
Fetter, N. R., 305(215), *321*
Fieggen, W., 267, 268, 270, *317*
Fields, R., 331(121), 334(122, 123), 342(121), *372*
Filmore, E. J., 203(183), 204(183), 205(183), *247*
Finger, L. W., 99(228, 229), *248*
Fischer, D. C., 96(230), *248*
Fischer, K. F., 99(231), *248*
Fisher, K. J., 259(3), 272(3), 299(174, 175), 305(214), 309, *316, 320, 321*
Fishwick, A. H., 264(24), 265, 305(217), *316, 321*
Fitzgerald, A. G., 395(28), *409*
Fitzsimmons, B. W., 77(55, 234), 94(309), 95(145), 110(234), 111(207), 13(234), 114, 116(207), 117(232), 118(232), 123(207, 234), 124(232–234), 125(234), 126(233), 129, 133(207), 140(234), 142(234), 143(234), 144(234), 146(207), 156(234), 157(233, 234), 158(207), 167, 168(55, 312), 180(55, 312), 181(55, 312), 211(56), 212(56), 213(13, 56), 214(13), 219(145), 220(56, 145), 221(145), *242, 243, 246, 248, 249, 251*, 308(223), *322*
Flinn, P. A., 63(381), 152(381), 202, 203(381), *253*
Fluck, E., 60(235), 90(236), 91(238, 239), 168(238, 239), 169(236, 237), 176(238, 239), 177(236, 238), 178(238), 179(238), 180(236–239), 181(236–238),

220(238, 239), *249*, 376(14), 392(35, 36), *409*, *410*
Flynn, J. J., *257*
Ford, B. F. E., 121, 135(240, 241), 136(240, 241), 139, 140(242, 572), 142(242, 572), 145, 163, 164(240, 241), 165(240), *249*, *258*
Forder, R. A., 33(71), 35(71), 53(72), *55*, 112(243), 137(243), 140(244), *249*, 385(32a), *410*
Forker, C., 305(213), 312(213), *321*
Forrester, J. D., 10(145), 14(145), *58*
Forster, D., 210(245), *249*
Foss, O., 19(74), 20(74), 22(75, 76, 77), 27(75, 77), 28, 29, 38(4), 39(26), 40 (4), 49, *53*, *54*, *55*
Foster, G., 286, *318*
Fowler, A., 376(34), *410*
Fox, W. B., 263(21), *316*
Fraga, E. F. R., 192(153), 195(153), *246*
Frank, A. W., 329(124), 335(124), 336, 340, *372*
Frank, E., 220(246), *249*
Frauenfelder, H., 60(247), 62(247), *249*
Frazer, M. J., 124(10), 133(152), 143(10), 144(10), 145(10), 147(152), 148(10, 152), 149(10, 152), 150(152), 151(10, 152), 152(10), 154(10), 155(10, 152), *242*, *246*
Fredericks, R. J., 263(21), *316*
Freeburger, M. E., 327(128), *372*
Freeman, A. J., 62(555), *257*
Freeman, J. H., 44(14), *54*
French, D., 11(107), 15(107), *56*
Freyer, W., 362(141), *372*
Friedt, J. M., 100(248, 249), 101(248, 249), *249*
Fröhlich, F., 227(214), *248*
Fung, S. C., 96(196, 250, 251), *248*, *249*

G

Gabriel, J. R., 64(252), *249*
Galasso, V., 177(11, 12), *242*
Gale, D. M., 326(129), *372*
Gallagher, P. K., 94(253), 95(253), *249*
Gancedo, R. R., 140(254), *249*
Ganorkar, M. C., 302(201), *321*
Garcia-Fernandez, H., 386(39–41), 388 (38), 399(37), 400(42), 401(63), 403 (103), 404(40, 60, 61, 63), 405(42a, 61), 406(63, 64, 102), 407(63), 408 (63), *410*, *411*
Garg, A. N., 182(255), 211(256), 212(256), *249*
Garrett, B. S., 11(78), 14(78), *55*
Garrod, R. E. B., 71(36), 102(38), 139, 168 (34), 169(34, 37), 170(36), 174(34, 36), 176(34), 177(37), 171(36), 172 (34, 36), 173(34, 35), 184(37), 187 (37), 227(33), *243*, *249*
Garrou, P. E., 90(71), 232(71), *244*
Gassenheimer, B., I11(258), 122(258), 137 (258), 138(258), 140(258), 141(258), 143(258), 158(258), *249*
Gay, R. S., 109(342), *252*
Geissler, H., 114(378), *253*
Geller, S., 37(79), 38(79), *55*
George, T. A., 302, *321*
Gerding, H., 236(430), *254*, 268(50), *317*
Gerloch, M., 211(259–261), *249*
Ghose, S., 11(80), 56(80), *56*, 98(221, 263, 264, 503), 99, 100(503), *248*, *249*, *256*
Ghotra, J. S., 300(184), *320*
Gibb, T. C., 60(270), 70(265), 91(266), 103(269), 104(269), 117, 203(271), 204 (271), 216(268), 217(268), 218(268), 222(267), 223(267), 224(267), 227 (270), 237, 238(266), 239(266), *249*, *250*
Gibson, J. D., 328(186), 329(185, 186), *373*
Gibson, J. F., 308(222), *322*
Gielen, M., 111(176), *247*
Gilbert, E. E., 263(21), *316*
Gillard, R. D., 182(434), *254*
Gillespie, P., 352(209), *374*
Gillespie, R. J., 6(81), *56*
Gilman, H., 292, 294, 299(122), *319*
Gitlitz, M. H., 299(171), 302(199), 303 (171), 304, 306, *320*, *321*
Glass, W. K., 287(100), *318*
Glemser, O., 377(46a), 378(45), 384(46), 385(43, 44, 46, 100), 391(95), *410*, *411*
Glentworth, P., 91(272), 100(272), *250*
Glockling, F., 299, *320*
Goehring, M., 384(51), 386(51), 387(51), 388(51), 389(51), 391(48), 393(50),

394(50), 395(50), 396(50), 397(49), 398(49), 399(49), *410*
Goel, P. L., 188(255), 211(256), 212(256), *249*
Goel, R. G., 139(242), 142(242), 163(242), *249*
Goldanskii, V. I., 60(274, 277), 90(280, 532), 102(424), 103(84, 279), 105(278), 106(368), 107(368), 112(57, 278), 122(368), 123(375), 136(83, 84), 138(279), 141(81, 84), 142(279), 143(84, 275), 145, 148(278), 150(273), 153(273, 276), 155(278), 159(275), 160(275), 161(275), 169(1, 3), 177, 180(3), 181(1, 3), 222(532), 223(533), *241, 242, 243, 244, 250, 252, 254, 257*
Golding, R. M., 60(198), 69(281, 283), 208, 216(284), 217(284), 218, 219(281, 282), *248, 250*
Goldstein, C., 191(89), 192(89), 193(89, 90), 195(285), 200(89, 90), 201(90), *244, 250*
Goldstein, L., 32(96), 35(96), *56*
Goldstein, M., 147(286), 149(286), 155(286), *250*
Golovanov, I. B., 261, *316*
Good, M. L., 91(128, 130), 115(128), 150, 151(127), 154(127), 155(127), 185(129), 186(129), 187(473), 210(125), 211, 212(125), 213(125), *245, 246, 255*
Goodgame, D. M. L., 210(245), *249*
Goodgame, M., 216(97), 217(97), 218(97), *245*
Goodman, B. A., 85(288), 88(288), 117, 118(288–291), 119(288–291), 121, 122(288), 123(288), 124(291), 125(291), 128(288), 129, 131(291), 133(288), 138(290), 141(289, 290), 157(291), 159(288), 160(288), 161(288), 203(271), 204(271), *250*
Gordon, A. F., 327(144), 340(144), *372*
Gordon, G., 284(90), 307(90), *318*
Gordy, W., 188(292), 192(144), 193(143), 196(143), *246, 250, 256*
Gorobchenko, V. D., 225(179), 226(179), *247*
Gorodinskii, G. M., 153(276), *250*
Goscinny, Y., 88(331), 102(331), 122(331), 123(331), 160(331), 161(331), *251*

Gosling, K., 312(229), *322*
Gottardi, G., 30(42, 82), 34(42, 82), 53(42), *54, 56*
Goubeau, J., 305(213), 312(213), *321*
Goudmand, P., 376(52, 88), 377(52), *410, 411*
Gouterman, M., 226(575), *258*
Graham, W. A. G., 109(342), *252*, 350(147), *372*
Grdenic, D., 201(363), *252*
Greatrex, R., 85(288), 88(288), 91(266), 92(293), 93(223, 224, 295), 117(288), 118(288), 119(288), 121(288), 122(288), 123(288), 128(288), 129(288), 133(288), 159(288), 160(288), 161(288), 185(294), 186(294), 187(294), 193(293), 222(224, 267), 223(224, 267), 237(266), 238(266), 239(266), *248, 249, 250*, 354(96), 357(166), 360(96), *371, 373*
Green, M., 323(24, 131), 328(50), 331(121), 334(50), 335(173), 342(121), *369, 370, 372, 373*
Green, M. L. H., 323(130), *372*
Greene, P. T., 142(296), *250*, 350(132, 133), *372*
Greenwood, N. N., 24(83), *56*, 60, 85(288), 88(288), 89(297), 91(266), 92(293), 93(223, 224, 295), 103(269), 104(269, 300), 105(300), 112(299, 300), 117, 118(288–291), 119(288–291), 121(288), 122(288), 123(288), 124(291), 125(291), 128(288), 129(288, 291), 131(291), 133(288), 138(290), 140(300), 141(289, 290), 144(300), 149(300), 151(299), 153(299), 154(300), 155(300), 157(291), 159(288), 160(288), 161(288), 185(294), 186(294), 187(294), 193(293), 203(271), 303, 204(271), 205(303), 206(303), 207(303), 216(268), 217(268), 218(268), 220(61), 221(61), 222(224, 267), 223, 224(267), 227(270), 237(266), 238(266, 301, 302), 239(266), *244, 248, 249, 250, 251*, 349(134), 354(96), 357(166), 360(96), *371, 372, 373*
Gregory, N. W., 211(304, 484), *251, 256*
Gregory, W., 375(54, 55), *410*
Griffith, W. P., 381(55), *410*

Grigor'ev, A. I., 264(30), 265(30), 266 (30), *316*
Grodzins, L., 238(99), *245*
Gross, K. J., 381(96), *411*
Groth, P., 10(84), 13(84), *56*
Grover, J. E., 100(305), *251*
Groves, J. D., 327(179), 338(179), 340 (179), *373*
Grützmacher, H. F., 295(146), 296(146), *320*
Gruntvig, F., 51(84a), *56*
Grushko, Yu. S., 192(419), *254*
Gütlich, P., 94(308, 309), 111(207), 116 (207), 123(207), 133(207), 146(207), 158(207), 227(214), *248, 251*
Gukasyan, S. E., 232(307), 235, 237(306), 238(306), *251*
Gulein, S. P., 223(533), *257*
Gut, R., 294, *319*

H

Haaland, A., 312, *322*
Haas, A., 378(45), *410*
Haas, H., 185(517), *256*
Hach, R. J., 11(107), 15(107), *56*
Hafemeister, D. W., 91(310), 189(310), 190, 191, 192(310), 193(310), *251*
Hafner, S. S., 98(221, 264, 311, 549, 550), 99, *249, 251, 257*
Hagen, A. P., 351(13), *369*
Hagihara, M., 286, *318*
Hahn, E. L., 187(163), *247*
Haines, R. J., 354(96), 360(96), *371*
Haiwaidy, F. I., 20(56), 23, *55*
Hall, D., 168(312), 180(312), 181(312), *251*
Hall, H. E., 62(75), *244*
Hall, J. B., 33(85), *56*
Hall, L. D., 327(80), 346(80), *371*
Hall, L. H., 211(313, 519), 212(313, 519), *251, 257*
Hall, M. C., 355(135), *372*
Halsey, M. J., 94(314), *251*
Haluska, L. A., 328(136), 329(136), *372*
Hamada, S., 380(56), *410*
Hamilton, W. C., 33(138), *58*, 112(507), 114(507), 140(507), 236(509), *256*

Hammons, J. H., 301(197), *321*
Hampton, C. R. S. M., 351(105), 352 (105), 355(105), *371*
Hansen, F., 51(85a), *56*
Hansson, A., 147(315), *251*
Harbourne, D. A., 222(147), 224(147), *246*, 329(75), 331(75), 348(72, 74), 351(74), 352(72-74), 353(71, 72), 356 (72-74), 357(72, 76), 358(71), 364(71, 75), 365(75), 366(76), *370, 371*
Harder, B., 293(133), *319*
Harmon, K. M., 116(316), 152(316), 154 (316), 155(316), *251*
Harpold, M., 101(422), *254*
Harris, C. M., 346(137), 355(137), *372*
Harris, R. K., 346(138), *372*
Harrison, P. G., 90(317), 107(317, 460), 109(460), 137, 138(318), *251, 255*
Harrison, T., 331(121), 342(121), *372*
Harrison, W., 352(139), 353(139), 355 (139), 360(140), 361(140), *372*
Hart, F. A., 300(184), *320*
Hartmann, H., 185(517), *256*
Hartwell, G. E., 261, *316*
Hashimoto, F., 183(500), *256*
Hassel, O., 5(86), 10(84), 13(84), 18(86), 29(86), 37, 38(22), 47, *54*, *56*
Hassellbach, K. M., 94(308), 111(207), 116(207), 123(207), 133(207), 146 (207), 158(207), *248, 251*
Haszeldine, R. N., 324(115), 328(16), 331(121), 332(16), 333(16), 334(6, 16, 122, 123), 342(6, 12, 121), 344(6), *368, 369, 372*
Hauge, S., 19(74, 87), 20(74, 87), 23, 53 (87a,b), *55, 56*
Havinga, E. E., 11(88), 15, 15(153), *56, 58*
Hayter, R. G., 167, 222(332), *245, 251*
Hayward, G. C., 24(89), *56*
Hazell, A. C., 38(90), *56*, 390(3), *409*
Hazell, R. G., 51(85a, 126f), *56, 57*, 390(3), *409*
Hazony, Y., 61(319), 100(322), 165(329), 178(319), 191(443), 196(443), 214(19, 320, 321), 218, 219(320, 321), *242*, *251, 254*
Heal, H. G., 376(57), 384(102), 385(82, 84), 386(41), 387(57), 397(83), 398 (83, 84), 399(62, 84), 400(42, 57, 58), 401(63), 403(103), 404(58, 60, 61, 63),

406(63, 64, 102), 407(59, 63), 408(22, 63), *409, 410, 411*
Heath, C. E., 299(173), 300(173), 305 (173), 306(173), 309(173), 310, 314 (234), *320, 322*
Heath, G. A., 220(53), 221(53), *243*
Heiart, R. B., 37(91), 38(91), *56*
Helmholtz, L., 11(131), 14(131), *57*
Hendra, P. J., 24(89), *56*
Henry, C., 326(94), *371*
Herb, H., 397(49), 398(49), 399(49), *410*
Herber, R. H., 60(324), 88(331), 101(322), 102(331), 103(334), 104(334), 109 (338), 111(258, 336), 112, 113(336), 119(323, 327), 121, 122(258, 331, 336), 123(331), 124(323), 125(327, 336), 132 (336), 136(323, 336), 137, 138(258, 336, 383), 139(326, 335, 520), 140 (258), 141(112, 258, 327, 336, 338, 383, 531), 142, 143(258), 144(323), 145 (336), 146(338), 149(530), 150(330), 151(330), 152, 153(383), 154(112, 383), 155(113, 330, 530), 157(323), 158(258), 160(324, 331), 161(324, 331), 162(323), 163(531), 164(383), 165, 193(91), 199(91), 209(333), 221, 222(332, 333, 561), 223(325, 561), 224(333, 561), *244, 245, 249, 251, 252, 253, 257, 258*
Hermodsson, Y., 21(92, 93, 94), 25(92, 93), 26, *56*, 147(337), *251*
Herring, F. G., 209(146), 210(146), 223 (146), 225(146), *246*, 353(58), *370*
Heslop, J. A., 266(34), *316*
Hesse, L. L., 116(316), 152(316), 154 (316), 155(316), *251*
Hieber, W., 355(142), 362(141), *372*
Hill, H. A. O., 201(488), 219(487–489), 220(485, 489), 221(489), 225(486), *256*
Hill, J., 194(570), 197(570), 199(570), *258*
Hill, J. C., 109(338), 141(338), 146(338), *252*
Hillier, I. H., 49(95), *56*
Hillyer, M. J., 281(83), 282(84), 285(92), 286(92), 303, *318*
Hine, J., 324(143), *372*
Hinsperger, T., 145(103), 147(103), 148 (103), 150(103), 152(103), *245*
Hinze, J., 203, *252*
Hoard, J. L., 32(96), 35(96), *56*

Hochstrasser, R. M., 235(148), *246*
Hodgson, D. J., 8, 10(97), *56*
Hodgson, W., 277(77), *318*
Hoff, A., 391(95), *411*
Hoffman, R., 327(218), *374*
Hoffmann, K., 295(146), 296(146), *320*
Hofmann, W., 201(341), *252*
Hogben, M. G., 109(342), *252*
Hogg, C. S., 98(343), *252*
Holbrook, G. W., 327(144), 340(144), *372*
Holding, A. F. LeC., 85(64), 88(64), 112 (65), 117(64), 118(64), 122(64), 123 (64), 128(64), 129(64), 159(64), 160 (64), 161(64), *244*
Holloway, C. E., 275(70), 277, 279, 280 (70, 80), 282(70, 75, 80), 285(75), 306 (220), 307, *317, 318, 321*
Holt, E. M., 392(64), *411*
Holt, S. L., 392(64), *411*
Hooper, T. R., 387(30), 388(30), *410*
Hope, H., 22(98, 99), 27(98, 99), *56*
Hoppe, R., *252*
Hoppe, W., 37(100), *56*
Hordvik, A., 51(84a, 100a), *56*
Hoskins, B. F., 220(53), 221(53), *243*
Hota, N. K., 327(145), 333(77), *371, 372*
Hou, F. L., 326(79), 345(79), 351(59), 352(59), 362(59, 60, 78), 363(59, 60, 78), 367(60, 78), *370, 371*
Howie, R. A., 98(25, 172), *242, 247*
Hoy, G. R., 64(345), *252*
Huang, H. H., 108(346), *252*
Huang, S.-R., 266(42), *317*
Hubbard, R. F., 334(122, 123), *372*
Hudson, A., 168(347), 169(347), 175(347), 181(347), 226(347), *252*
Hudswell, F., 293(133), *319*
Hüfner, S., 217(372), 225(372), 226(372), *253*
Huggins, D. K., 263(21), *316*
Hugill, D., 387(30), 388(30), *410*
Hui, K. M., 108(346), *252*
Hull, G. W., 225(563), 227(563), *258*
Hulme, R., 29, 30(102), 32(50), 34(101), 52(101a), *55, 56*, 165(348), *252*
Hummel, R. J., 331(170), 339(169), *373*
Hurley, J. W., Jr., 214(19, 321), 218(19, 321), 219(321), *242, 251*
Hursthouse, M. B., 280(82), 299(173), 300(173, 185, 187), 301(190, 191), 305

(173), 306(173), 308(221), 309(173), 310, 314(233, 234), *318, 320, 321, 322*
Husebye, S., 19, 20(103, 104), 22(75), 23, 27(75), 28, *55, 56*

I

Iannarella, L., 168(161), 172(161), *247*
Ibers, J. A., 8, 10(97, 105), 14(105), *56*, 272(64), 276, *317*, 353(171), 359(146), *370, 372, 373*
Ichiba, S., 148(349), 193(350), 200, *252*
Ingalls, R., 66(351), 68(351), 69(351), 100(352), 101(352), 208, 214(351), 215(351), 216(351), 218(351), *252*
Iofa, B. Z., 152(513), 232(513), 233(513), 237(85, 306, 513), 238(85, 306, 513), *244, 251, 253, 256*
Isaacs, N. W., 121(353), 164(353), *252*
Ito, T., 18(106), 20(106), *56*

J

Jabs, G. A., 295(145), *319*
Jaccarino, V., 63(553), 189(354), 195(354), *252, 257*
Jackson, F., 69(283), 208(283), *250*
Jacobson, R. A., 32(130), 35(130), *57*, 357(92, 93), *371*
Jaffé, H. H., 204, *252*
Jain, S., 302(206), *321*
James, W. J., 11(107), 15(107), *56*
Jamieson, P. B., *53*
Janssen, M. J., 135(355), 165(355), *252*
Jaura, K. L., 117(291), 118(291), 119(291), 124(291), 125(291), 129(291), 131(291), 157(291), *250*
Jelen, A., 203(184), 205(106), 206(106, 184), 207(184), *245, 247*
Jenkins, A. D., 301(193), 302(200, 202), *321*
Jenne, H., 400(9, 10), 401(9), 405(10), *409*
Jensen, G., 52(107a), *56*
Jesson, J. P., 95(356), 96, *252*
Jetz, W., 350(147), *372*
Johnson, B. F. G., 349(97), *371*
Johnson, C. E., 60(357), 70(438), 95(195), 117, 118(439), 119, 120(439), 122(439), 124(439), 138(439), 144(439), 168(156), 169(156, 195), 172(157), 174(439), 175(156), 177(156), 178(156), 180(156, 195), 181(156), 201(488), 210(201), 212(201), 213(13), 216(202), 218(202), 219(487–489), 220(485, 489), 221(489), 225(155, 158, 486), 226(155), *242, 246, 248, 252, 254, 256*, 308(223), *322*
Johnson, C. K., 7(3), *53*
Johnson, R. N., 327(80), 346(80), *371*
Johnson, R. W., 327(206), *374*
Jolly, P. W., 334(26, 148), 344(148), *369, 372*
Jolly, W. L., 377(110), 378(65, 110), 379(110), 385(66, 70), 387(116), 388(116), 389(117), 393(89), 394(89, 90), 396(90), *411, 412*
Jones, A., 331(121), 342(121), *372*
Jones, C. H. W., 100(358), *252*
Jones, G. R., 7(32, 33, 108), 9(57), 10(57), 21(58, 59, 60), 24(58, 59, 60), 25, *54, 55, 56*
Jones, K., 145(159), *246*, 298, 301(162), 304(162), *320*
Jones, M. T., *252*
Jones, R. D. G., 348(109), 361(106–108), 362(106, 107), 365(108), 367(108a, 110), 368(110), *371*
Jones, R. G., 292(122–124), 294(138), 299(122), *319*
Josephson, B. D., 61(360), *252*
Joshi, K. C., 376(67), *411*
Jula, T. F., 325(197), *374*
Jung, H., 392(68), *411*
Jung, P., 91(361), 193(361), 237(361), 238(361), 239(361), *252*

K

Kacsoh, L., 100(102), *245*
Kaindl, G., 91(49, 362, 471), 185(362), 219(49), 227(49), 228(49), 229(49), 230(49), *243, 252, 255*
Kakos, G. A., 274(68), 287, *317, 318*
Kalvius, G. M., 90(496), 232(496), 233(496), 234(496), *256*
Kamenar, B., 201(363), *252*
Kane, J., 400(58), 404(58), 407(59), *410*
Kaplan, M., 91(205), 102(425), 190(203),

193(203), 194(204, 205), 196, 197 (205), 198, 199(203), *248, 254*
Kaplan, S. F., 51(108a), *56*
Kapoor, R. N., 291, 292(127, 129), 296(148), *318, 319, 320,* 346(151), *372*
Karasev, A. N., 237(306), 238(306), *251*
Karasyov, A. N., 88(364), 122(364), 123(364), 160(364), 161(364), *252*
Karle, I. L., 15(109), *56*
Karmas, G., 292(122–124), 299(122), *319*
Karraker, D. G., 293, *319*
Karyagin, S. V., 153(276), 165(365), *250, 252*
Kasai, N., 112(366), 137(366), *252*
Kaspi, P., 185(294), 186(294), 187(294), *250*
Katabe, K., 391(104a), *412*
Katz, W., 277(77), *318*
Kawasaki, A., 271(59), *317*
Kay, M. I., 51(108a), *56*
Kaye, H., 395(69), *411*
Keefer, R. M., 5(10), 47(10), *53*
Kelly, B. P., 220(53), 221(53), *243*
Kemmitt, R. D. W., 107(140), 122(140), *246*
Kennard, C. H. L., 121(353), 164(353), *252*
Keppie, S. A., 106(65), 112(65), 160(65), 162(65), *244,* 301(194), *321*
Kerler, W., 91(238), 168(238), 176(238), 177(238), 178(238), 180(238), 181(238), 220(238), *249*
Keszthelyi, L., 102(100, 180), 103(100), *245, 247*
Ketelaar, J. A. A., 37(110), 38(110), *56*
Keve, E. T., 10(111), 13, *56*
Khotsyanova, T. L., 12(112), 16(112), 17(112, 114), *57, 58*
Khrapov, V. V., 60(277), 90(280), 102(424), 103(279), 105(278), 106(367, 368), 107(368), 112(57, 278), 122(367), 136(367), 138(279, 367), 141(81, 367), 142(279, 367), 144(367), 145(424), 148(278), 153(276), 155(278), 169(1), 177(1), 181(1), *241, 242, 243, 250, 252, 254*
Kilbourn, B. T., 355(135), *372*
Kilner, M., 93(223, 224), 222(224), 223(224), *248*
King, J. G., 189(354), 195(354), *252*

King, R. B., 209(333), 221, 222(333), 224(333), *251,* 346(151), 350(151), 353(149), *372*
Kinsella, E., 304(209), 306(209), *321*
Kirmse, W., 324(153), *372*
Kishi, I., 271(59), *317*
Kishore, J., 387(92) 388(92), 389(92), *411*
Kistner, O. C., 59(369), *252*
Kjekshus, A., 11(140), 14(140), 41(140), *58*
Klein, M. P., 101(422, 423), *254*
Kleinschmidt, D. C., 268, 269, *317*
Klemann, L. P., 116(316), 152(316), 154(316), 155(316), *251*
Klinsberg, E., 18(113), *57*
Klug, H. P., 353(204), 359(204), *374*
Klusacek, H., 352(209), *374*
Knight, J., 93(270), *252*
Knox, K., 200(475), *255*
Knudsen, J. M., 102(15), *242*
Knunyants, I. L., 331(103, 104), 336(154, 155), *371, 373*
Kobayashi, H., 226(575), *258*
Kobelt, D., 51(113a), *57*
Koch, W., 377(46a), *410*
Kocher, C. W., 116(316), 132, 152(316), 154(316), 155(316), *251, 258*
Kocheshkov, K. A., 261(12, 13), 266(37), *316, 317*
Kojima, S., 192(371), *252*
Kokoszka, G. F., 284(90), 307, *318*
Kolobova, N. E., 88(364), 122(364), 123(275, 364), 143(275), 159(275), 160(275, 364), 161(275, 364), *244, 250, 252*
Konig, E., 95(373–375), 96, 150(373), 217(372), 225(372, 375), 226(372, 375), *253*
Konnert, J. H., 37, 38(114), 53(114a), *57*
Kono, H., 181(501), *256*
Korecz, L., 90(376), *253*
Korytko, L. A., 103(84), 136(83, 84), 141(84), 143(84), 153(276), *244, 250*
Koster, P. B., 12(20), 15(20), *54*
Kostikas, A., 102(515), 103(515), *256*
Kostyanovskii, R. G., 106(368), 107(368), 122(368), *252*
Kothekar, V., 152(513), 232(377, 513), 233(513), 237(513), 238(513), *253, 256*
Kovar, R., 300, *321*

AUTHOR INDEX

Kracht, D., 15(154), 41(154), 44, 45, 48, 49, *58*
Krastinak, W., 294(137), *319*
Kraus, H. L., 392(68), *411*
Kravtsov, D. N., 102(424), 145(424), *254*
Krebs, A., 324(20), *369*
Krebs, B., 20(36), 24, 51(36a,b), *54*
Krebs Larsen, F., 51(114b), *57*
Kremer, S., 95(373), 96(373), 150(373), *253*
Krespan, C. G., 326(129), *372*
Kriegsmann, V. H., 114(378, 379), *253*
Krizhanskii, L. M., 106(456, 494), 109(494), 142(380), 153(276), *250, 253, 255, 256*
Kronick, P. L., 395(69), *411*
Krüger, A. G., 289, *318*
Krüger, S., 384(77), *411*
Kruse, F. H., 22(115), 28(115), *57*
Kruse, W., 302(203), *321*
Kubo, M., 194(502), 200(502), *256*
Kühr, H., 261(15), 295(15), *316*
Kuhn, P., 91(239), 168(239), 176(239), 179(239), 180(239), 220(239), *249*
Kummer, R., 355(142), *372*
Kundig, W., 101(422, 423), *254*
Kuo, F.-T., 266(42), *317*
Kurkjian, C. R., 94(253), 95(253), *249*
Kuzmin, R. N., 203(82), *244*

L

Labes, M. M., 395(69), *411*
Lacher, J. R., 324(203), 327(179), 338(179, 203), 340(179), 343(203), *373, 374*
Lafferty, R. H., 328(186), 329(185, 186), *373*
Lagowski, J. J., 394(85), *411*
Lambert, J. B., 345, *373*
Lambert, J. L., 211(313, 519), 212(313, 519), *251, 257*
Lam Than My, 376(94), *411*
Landman, D. A., 227(67), *244*
Lang, G., 176, 219(432), *254*
LaPlaca, S. J., 236(509), *256*
Lappert, M. F., 106(65), 112(65), 160(65), 162(65), *244*, 259(6), 284, 298, 300(178), 301, 302, 304(162), *316, 318, 320, 321*
Lappin, M., 334(6), 342(6), 344(6), *368*
Large, N. R., 91(272), 100(272), *250*
Larkworthy, L. G., 95(145), 219(145), 220(145), 221(145), *246*
Larson, A. C., 14(49, 116), 36(123), *55, 57*, 201(394), *253*
Latscha, H. P., 378(13, 15), *409*
Lauder, A., 264(27), *316*
Leach, J. M., 332(35), 337(33–35), 338(33–35), 343(34, 35), 344(34), *369*
Leciejecwitz, J., 18, 20(117), *57*
Leeder, W. R., 327(81), *371*
Lees, J. K., 63(381), 152(381), 202, 203(381), *253*
Lehmann, M. S., 51(114b), *57*
Lejeune, S., 111(176), *247*
Lenné, H. U., 37(100), *56*
Lependina, O. L., 141(7), 142(7), *242*
Lesikar, A. V., 222(382), *253*
Leung, K. L., 137(383), 138(383), 139(520), 141(383), 142(383), 152(383), 153(383), 154(383), 164(383), 165(383), *253, 257*
Lever, A. B. P., 225(385), *253*, 331(121), 334(6), 342(6, 121), 344(6), *368, 372*
Levey, R. P., 227(490), *256*
Levy, H. A., 6(34, 35, 118), 7(3), *54, 57*
Lewis, G. K., 96(196, 385, 386), *248, 253*
Lewis, J., 211(259, 387), *249, 253*, 349(97), 355(157), *371, 373*
Libbey, E. T., 77(39), 168(39), 169(39), 174(39), 180(39), 182(39), *243*
Lichtenstein, D., 101(423), *254*
Liengme, B. V., 135(240, 241), 136(240), 140(240), 145(240), 164(240), 165(240), 222(147), 224(147), *246, 249*, 329(75), 331(75), 348(74), 351(74), 352(74), 357(76), 364(75), 365(75), 366(76), *370, 371*
Lilly, M. N., 342(95), *371*
Lima, G. C. de, 102(15), *242*
Lind, M. D., 50(118a), *57*
Lindqvist, I., 18(121), 20(121), 21(119), 25(119), 32(120), *57*
Linke, K. H., 38(59), 40(69), *55*
Lipp, S. A., 385(66, 70), *411*
Little, R., 24(13), *53*
Livingston, C. E., 346(137), 355(137), *372*
Lock, P. J., 24(2), *53*
Lomborg, J. G., 390(3), *409*

Long, G. G., 90(71), 232(71, 72, 388, 389), 233(72, 389), 234(72, 389), 235(72, 389), 236(72, 389), *244, 253*
Long, G. J., 211(481), 213(481), *255*
Long, G. L., 215(390), 216(390), 217(390), 218(390), *253*
Long, T. V., 107(460), 109(460), *255*
Lorimer, J. W., 297(158, 159), *320*
Lu, C.-S., 381(71), *411*
Lucken, E. A. C., 67, 80(391), *253*
Ludwig, G. W., 193(392), 194(392), *253*
Luijken, J. G. A., 135(355), 165(355), *252*
Lukashevich, I. I., 225(179), 226(179), *247*
Lure, B. G., 192(419), *254*
Lurio, A., 227(67), *244*
Lutz, H. D., 265(31, 32), 266(31, 32), *316*
Lynch, C. T., 291(114), *319*
Lynch, T. R., 51(121a), *57*
Lyons, H. J., 350(23), *369*

M

Mabbs, F. E., 211(259–261, 387), *249, 253*, 306(220), 307(220), *321*
McBee, E. T., 329(165), *373*
McClure, D. S., 73, *253*
McCullough, J. D., 18(122), 19(29), 20(29, 122), 22(46, 99, 115, 156), 27(156), 28(46, 115), *54, 55, 56, 57, 58*
MacDiarmid, A. G., 351(13), *369*, 387(72), *411*
McDonald, R. R., 36(123), *57*, 201(394), *253*
McDonald, W. S., 38(124), 39(124), 40, *57*, 357(166), *373*
McEwen, W. E., 236(509), *256*
McFarlane, W., 145(169), 161(169), *247*
MacGillivray, C. H., 10(125), 13(125), *57*
McGrady, M. M., 158(395), *253*
McGrath, W. D., 376(75), *411*
MacGregor, A. C., 367(110), 368(110), *371*
Machado, A. A. S. C., 95, 96, *250*
MacKay, K. M., 236(396), *253*
McKinley, S. V., 116(316), 152(316), 154(316), 155(316), *251*
McLaughlin, G. M., 312(229), *322*

McMurtry, R. J., 338(180), *373*
McQuillan, G. P., 114(51), *243*
McRae, V. M., 7(126), 8, 10(126), *57*
McWhinnie, W. R., 110(465), 124(465), 125(465), 129(465), 153(465), 156(465), 164(465), 169(397), 180(397), *253, 255*
Maddock, A. G., 71(36, 122), 77(122), 80(122), 83(122), 85(122), 86(122), 87(122), 91(219), 94(30, 31), 98(26, 40), 99(26, 40), 100(225, 398), 101(225), 109(400), 110(122), 112(122), 114(122), 115(400), 116(122), 117(122, 400, 402), 118(402), 119(122, 400), 120(122), 121(122), 122(400, 402), 124(402), 127(122), 128(122), 129(122), 132(401), 134(122), 135(400), 138(400), 140(254), 141(400), 142(400), 168(34), 169(34, 37), 170(36), 171(36), 172(34, 36), 173(34, 35), 174(34, 36), 176(34), 177(34), 184(37), 187(37), 210(41), 211(42, 43, 399), 212(41, 42, 399), 213(42, 43), 214(43), 227(33), 238, 239(219), *242, 243, 245, 248, 249, 253*
Madeja, K., 95(374, 375), 96, 217(372), 225(372, 375), 226(372, 375), *253*
Maeda, Y., 193(350), 200(350), *252*
Maguire, K. D., 378(65, 73), *411*
Mague, J. T., 323(160), 346(1, 159), 355(159), 356(158, 159), 357(159), *368, 373*
Mahler, J. E., 93(206), 164(206), *248*
Mahler, W., 325, 327(161), *373*
Mahon, C., *58*
Mainthia, S. B., 395(69), *411*
Makarov, E. F., 90(532), 103(84, 279), 105(278), 112(278), 123(275), 136(83, 84), 138(279), 141(84), 142(279), 143(84, 275), 148(278), 153(276), 155(278), 159(275), 160(275), 161(275), 169(1, 3), 177(1, 3), 180(3), 181(1, 3), 222(532), 223(533), *241, 242, 244, 250, 257*
Makhini, H. S., 270(56), *317*
Malathi, N., 212(403), *253*
Malhotra, K. C., 388(92), 389(92), *411*
Malmros, G., 52(126a), *57*
Mangia, A., 52(52b), *55*
Manning, A. R., 350(23), *369*

Manotti Lanfredi, A. M., 50(25a), 51 (25b, 126b), *54, 57*
Manuel, T. A., 346(162), 355(162), *373*
Maresca, L., 207(124), *245*
Margrave, J. L., *409*
Margulis, T. N., 345(163), *373*
Maroy, K., 22(76), 28(76), *55*
Marquarding, D., 352(209), *374*
Marsh, H. S., 90(50), *243*
Marsh, R. E., 22(115), 28(115), *57*
Martin, G. A., 292(122, 124), 294(138), 299(122), *319*
Martin, R. L., 220(53), 221(53), *243*, 277, 284, 287(85), 288(85, 104, 105), 289, *318*
Martineau, E., 232(63), 233(63), 235(63), *244*
Martynov, B. I., 331(104), *371*
Mason, J., 381(74), 383(74), *411*
Mason, R., 349(164), 356(11), *369, 373*
Massey, A. G., 386(27), *409*
Mastin, S. H., 52(126c), *57*
Matas, J., 169(404), 182(404), *253*
Mathur, V. K., 271(61), *317*
Matoni, V. A., 263(23), *316*
Matthews, C. K., 211(405), 212(405), 213(405), *253*
May, J.-F., 393(31), 395(31), 396(31), *410*
Mayer, E., 384(33), *410*
Mayne, N., 328(50), 334(50), *370*
Mays, M. J., 68(44), 71(36, 44), 73(44), 77(44), 78(44), 79(44), 80(44), 84(44), 93(270), 159(44), 167(44), 168(44), 169(37), 170(36, 44), 171(36), 172(36), 174(45), 175(36, 44), 177(37, 44), 178(44), 180(44), 181(44, 45), 182(44), 183(44), 184(37), 187(37), 219(44), 230(44), *243, 252*
Mazak, R. A., 71(406), 85(406), 208, 209, 210(406), *253*
Mazdiyasni, K. S., 291, *319*
Meads, R. E., 98(343), *252*
Medeiros, L. O., 211(399), 212(399), *253*
Meerwein, H., 293, 294, *319*
Megson, F., 331(202), 337(202), 341(202), *374*
Mehrotra, R. C., 259(2), 260(2), 262(18), 264, 266, 267, 268, 270, 271(60, 61), 272(2), 291, 294, 296, *316, 317, 318, 319, 320*

Mehrotra, R. K., 271(60), 291(113), *317, 319*
Mehta, M. L., 287(98), *318*
Meinzer, R. A., 386(76), *411*
Mellor, I. P., 51(121a), *57*
Melson, G. A., *253*
Menes, M., 191(408), *253*
Merrett, F. R., 225(563), 227(563), *258*
Messager, M., 405(42a), *410*
Mettile, F. J., *369*
Metzger, H. G., 346(118), *372*
Meuwsen, A., 379(78), 384(77), 396(78), 397(78), 398(78), *411*
Michael, R. O., 327(181), 346(174, 181), *373*
Michel, P., 395(18a), *409*
Middleton, W. J., 326(129), *372*
Migchelsen, T., 11(20), 15(20), *54*
Mihichuk, L., 145(103), 147(103), 148(103), 150(103), 152(103), *245*
Mijlhoff, F. C., 10(128), 13(128), *57*
Mill, T., 329(168), 331(102), *371, 373*
Milledge, H. J., 33(52), 35(52), 36(52), 55, 120(167), 140(167), 141(167), 145(167), *247*
Miller, W. T., 331(170), 339(169), *373*
Milne, J. B., 232(63), 233(63), 235(63), *244*
Mingos, D. M. P., 353(171), *373*
Minkin, V. I., 109(576), *258*
Mishima, M., 148(349), *252*
Misra, S. N., 291(111, 112, 116), *318, 319*
Misra, T. N., 291 (111, 116), *318, 319*
Mitchell, C. M., 332(167), *373*
Mitchener, J. P., 346(159), 355(159), 356(159), 357(159), *373*
Mitrofanov, K. P., 70(5), 103(5, 6, 8), 136(8), 141(6, 7), 142(5, 6, 7), *242*
Mochalina, E. P., 331(103), *371*
Moedritzer, K., 306, *321*
Mössbauer, R. L., 59, 91(49, 362, 470, 471), 185(362, 470), 186(470), 187(470), 219(49), 227(49), 228(49), 229(49), 230(49), *243, 252, 254, 255*
Mohana Rao, J. K., 51(126d,e), *57*
Mok, K. F., 91(76), 168(76), 173(76), 176(76), 177(76), 178(76), 179(76), 216(284), 217(284), 218(199, 284), *244, 248, 250*

Molnar, B., 102(180), 225(179), 226(179), *247*
Monaghan, J. J., 312(213), *322*
Montgomery, L. K., 327(7), *369*
Monty, M. A., 24(61), *55*
Moore, W. J., 202(409), *253*
Mootz, D., 295(143), 312(143), *319*
Morandi, G., 37(100), *56*
Morosin, B., 13(132), 51(108a), *56, 57*
Morrow, T., 376(75), *411*
Mortimer, F. S., 381(108), 383(108), *412*
Morton, C. J., 329(165), *373*
Moser, W., 205(106), 206(106), *245*
Moss, J. R., 357(166), *373*
Moss, R. H., 284(89), 285(89), 307(89), *318*
Moss, T. H., 225(410), 226(410), *254*
Mueller, R. F., 100(414), *254*
Muetterties, E. L., 352(172), *373*
Mui, J. Y. P., 325(198), *374*
Mukhedkar, A. J., 335(173), *373*
Mukhedkar, V. A., 335(173), *373*
Mulay, L. N., 224(174), *247*
Mullins, F. P., 124(416, 455), 125(415, 416, 455), 129(415, 455), 130(415, 416), 131(415, 416), 143(415), 144(415), 149(455), 150(455), 151(455), 154(455), *254, 255*
Mullins, M. A., 110(417), 112(457), 124(417, 418), 125(417, 418), 129(417, 418), 130, 140(418), 141(418), 143, 144(418), 145(457), 147(457), 148(457), 149(457), 152(457), 153(417), 155(457), 156(417, 418), 157(418), *254, 255*
Multani, R. K., 288(103), *318*
Murin, A. N., 192(419), *254*
Murray, K. S., 211(54, 86), 213(54, 86, 87), 214(54), *243, 244*
Murthy, A. R. V., 384(79), 387(80), 391(80), *411*
Musgrave, W. K. R., 327(3), *368*
Myers, R. J., 386(76), *411*

N

Nagy, G., 102(100), 103(100), *245*
Nahringbauer, G., 21(119), 25(119), *57*

Naik, D. V., 124(420, 421), 125(420, 421), 129(420, 421), 152(421), 154(421), 155(421), 157(421), 158(421), *254*
Nair, C. G. R., 384(79), 387(80), 391(80), *411*
Nakamura, D., 194(502), 200(502), *256*
Nakata, B. D., 338(177, 182), 343(177), *373*
Narasimham, N. A., 376(81), *411*
Nardelli, M., 52(52b), *55*
Nath, A., 101(422, 423), *254*
Natta, G., 346(31), 351(31), 355(31), 359(31), *369*
Neese, H. J., 259(5), 301, 302, 304, 306, *316*
Negita, H., 148(349), 193(350), 200(350), *252*
Nelson, J., 385(82, 84), 398(83, 84), 399(84), *411*
Nelson, J. T., 394(85), 397(83), *411*
Nelson, S. M., 95(73), *244*
Nelson, W. H., 277(76), 278, 281, *318*
Nesmeyanov, A. N., 102(424), 145(424), 223(533), *254, 257*
Neubauer, D., 388, *411*
Neuwirth, W., 91(238), 168(238), 176(238), 177(238), 178(238), 180(238), 181(238), 220(238), *249*
Newing, C. W., 300(185), 302(203–205), 308(221), *320, 321, 322*
Newlands, M. J., 342(12), *369*
Newmark, R. A., 327(181), 346(174, 181), *373*
Newnham, R. E., 30(139), 34(139), *58*
Ng, T. W., 93(102, 104), 112(104), *245*
Nibbering, N. M. M., 268(50), *317*
Nichols, A. J., 91(272), 100(272), *250*
Nichols, D. I., 91(108), 227(108), 228(108), 229(108), 230(108), *245*
Nicholson, D. G., 203(185–187), 205, 206(106, 185–187), 207(185–187), *245, 247*
Nielsen, B. R., 51(126f), *57*
Niggli, A., 32(120), *57*
Niki, E., 106(554), *257*
Noack, K., 359(175), *373*
Norton, D. A., 50(129a), *57*
Novoselova, A. V., 266(39, 40), *317*
Nowik, I., 227(427), *254*
Nozik, A. J., 102(425), *254*

AUTHOR INDEX

Nyburg, S. C., 51(121a), 52(126g), *57*
Nyholm, R. S., 348(176), 355(157), *369, 373*

O

O'Brien, R. J., 114(118), *245*
O'Connor, J. E., 119(426), 128(426), *254*
Odar, S., 94(309), *251*
Ofer, S., 227(427), *254*
Ogata, Y., 271(59), *317*
Ogawa, S., 192(371), *252*
O'Hare, P. A. G., 376(87), 377(87), *411*
Ohlberg, S. A., 37(127), 38(127), 40(127), *57*
Okabe, T., 391(104a), *412*
Okawara, R., 112(366), 114(428), 137(366), *252, 254*
Okazaki, A., 201(429), *254*
Okhlobystin, O. Yu., 103(8, 9), 106(9), 136(8), 142(9, 380), *242, 253*
Olie, K., 10(128), 13(128), 50(127a), *57*, 236(430), *254*
Oliver, J. G., 268, 270(54), 271(54), 272(54), 294(54), *317*
Oliver, W. H., 327(207), *374*
Olsen, F. P., 409(106), *412*
Olson, D. H., 201(499), *256*
Olthof-Hazekamp, R., 10(52a), 13(52a), *55*
Onaka, S., 88(431), 122(431), 123(431), 133, 159(431), 160(431), 161(431), 162, *254*
Ong, W. K., 210(41), 211(42), 212(41, 42), 213(42), *243*
Oosterhuis, W. T., 176, 219(432), *254*
Opalenko, A. A., 237(85), 238(85), *244*
O'Rourke, M., 149(433), 150(433), 151(433), 154(433), 155(433), *254*
Orville, P. M., 100(305), *251*
Osborn, J. A., 182(434, 537), *254, 257*
Osborn, R. B. L., 323(131), *372*
Osipoua, O. P., *244*
Oteng, R., 201(188), 203(189), *247*
Owusu, A. A., 124(233), 126(233), 129(233), 157(233), 211(56), 212(56), 213(56), 220(56), *243, 249*
Ozin, G. A., 52(126g), *57*

P

Padley, J. S., 389(2), 390(2, 4), *409*
Palenik, G. J., 93(104), 112(103), *245*
Palmer, P. J., 334(123), *372*
Pannell, K. H., 346(151), 350(151), *372*
Pannetier, G., 376(88), *411*
Panyushkin, V. N., 96(435), 97(435), *254*
Papike, J. J., 99(436), *254*
Parish, R. V., 60(437), 70(438), 71(440, 442), 77(440, 442), 84(442), 105, 106 (440), 107(440), 108(440), 109(440), 110(440), 111(440, 442), 113(440, 442), 114, 115(442), 116, 117, 118(439), 119, 120(439, 442), 122(439, 440, 442), 124 (439, 442), 132(442), 133(442), 135 (440), 138(439), 141(442), 143(442), 144(439), 145(440), 146(442), 151 (441), 152(440), 153(440, 441), 155 (442), 156(440, 442), 157(441), 158 (441, 442), 159(437), 174(439), *254*, 348(53), *370*
Parisi, G. I., 104(334), *251*
Park, J. D., 324(203), 327(179, 181), 337, 338(177, 179, 180, 182, 203), 340, 343 (177, 199), 346(181), *373, 374*
Pasternak, M., 91(445), 93(445), 101 (444), 191, 192(445, 446), 195(285), 196, 200(445), 238(445), 241(445), *250, 254, 255*
Patterson, D. O., 227(490), *256*
Patton, R. L., 393(89), 394(89-91), 396 (90), *411*
Paul, R. C., 270(56), 301(189), 306(189), *317, 321*, 387(92), 388(92), 389(92), *411*
Pauling, L., 3, *57*, 202(409), *253*
Pauling, P. J., *369*
Paulus, E. F., 51(113a), *57*
Pauson, P. L., 93(295), *250*
Peacock, R. D., 7(126), 8(126), 10(126), *57*, 107(140), 110(141), 111(141), 119 (141), 122(140), 124(141), 138(141), 139(141), 141(141), 144(141), *246*, 387 (30), 388(30), *410, 411*
Peck, D. N., 50(129a), *57*
Pelah, I., 102(447), 103(447), *255*
Pelizzi, G., 52(52b), *55*
Pellinghelli, M. A., 50(25a), 51(126b), *54, 57*

Pelosi, L. F., 339(169), *373*
Penfold, B. R., 121(69), *244*, 357(188), 361(25), *369*, *374*
Pereyre, M., 325(197), *374*
Perkins, P. G., 112(299), 151(299), 153(299), *250*
Perlow, G. J., 91(451, 452, 453, 454), 101, 189(452), 191, 192(452), 193(452), 196, 198(452), 238(454), 240, 241(449, 453), *255*
Perlow, M. R., 91(451, 452, 453), 101, 189(452), 191, 192(452), 193(452), 196, 198(452), 240(453), 241(453), *255*
Pervova, E. Ya., 336(154, 155), *373*
Peters, F. M., 305(215), *321*
Peterson, R. A., 324(20), *369*
Petrides, D., 102(515), 103(515), 124(455), 125(455), 129(455), 149(455), 150(455), 151(455), 154(455), *255, 256*
Petrov, A. A., 106(456, 494), 109(494), *255, 256*
Pettit, L. D., 38(124), 39(124), 40, *57*
Pettit, R., 93(206), 164(206), 169(135), 173(135), 178(135), 180(135), 181(135), 217(135), 220(135), 222(133, 134), 224(133), *246, 248*
Peyron, M., 376(94), *411*
Pfalzgraf, L. G., 276, *317*
Philip, J., 112(457), 145(457), 147(457), 148(457), 149(457), 152(457), 155(457), *255*
Phillips, D. J., 355(157), *373*
Pidcock, A., 347(2), *368*
Piechaczek, H., 392(113), *412*
Pierce, O. R., 327(144), 340(144), *372*
Pillinger, W. L., 91(458), 195(458), 227(458), *255*
Pilotti, A. M., 366(111), *371*
Pinajian, J. J., 101(352), *252*
Pinder, P. M., 348(8), *369*
Pischtschan, S., 114(379), *253*
Platt, R. H., 71(122, 440, 442), 77(122, 440, 442), 80(122), 83(122), 84(442), 85(122), 86(122), 87(122), 105, 106(440), 107(440), 108(440), 109(400, 440), 110(122, 440), 111(440, 442), 112(122), 113(440, 442), 114, 115(400, 442), 116, 117(122, 400, 402), 118(402), 119(122, 400, 440), 120(122, 442), 121(122), 122(400, 402, 440, 442), 124(402, 442), 127(122), 128(122), 129(122), 132(401, 442), 133(442), 134(122), 135(400, 440), 138(400), 139(257), 140(254), 141(400, 442), 142(400), 143(442), 145(440), 146(442), 151(441), 152(440), 153(440, 441), 155(442), 156(440, 442), 157(441), 158(441, 442), *245, 249, 253, 254*
Plieth, K., 30(17, 18), *54*
Plotnikova, M. V., 141(7), 142(7), *242*
Plumb, J. B., 342(12), *369*
Pocs, L., 102(180), *247*
Poder, C., 110(459), 135(459), 136(459), 137, 158(459), *255*
Poeth, T. P., 107(460), 109(460), *255*
Poh, B. L., 163(573), *258*
Polak, L. S., 70(5), 88(364), 103(5, 6, 8, 9), 106(9), 122(364), 123(364), 136(8), 141(6, 7), 142(5, 6, 7, 9), 160(364), 161(364), *242, 252*
Poliakoff, M., 102(38, 461), *243, 255*, 357(183), *373*
Poller, R. C., 110(465), 121, 124(463–465), 125(463, 465), 129, 136(464), 143(462), 144(462), 145, 153(462, 465), 156(462, 465), 157(462), 163, 164(462, 464, 466, 467, 468), 169(397), 180(397), *253, 255*
Polynova, T. W., 235(469), *255*
Pomerance, H., 227(491), *256*
Ponomarenko, V. A., 327(184), *373*
Popov, A. V., 142(380), *253*
Popovkin, B. A., 266(39, 40), *317*
Porai-Koshits, M. A., 235(469), *255*
Porter, R. F., 345(38), *369*
Porter, S. K., 32(130), 35(130), *57*
Posner, J., 324(20), *369*
Potzel, W., 91(49, 362, 470, 471), 185(362, 470), 186(470), 187(470), 219(49), 227(49), 228(49), 229(49), 230(49), *243, 252, 255*
Poulet, R. J., 385(32a), *410*
Powell, A. R., 182(537), *257*
Prados, R. A., 91(128, 130), 115(128), 185(129, 472), 186(129), 187(472, 473), *245, 246, 255*
Prater, B. E., 68(44), 71(36, 44), 73(44), 77(44), 78(44), 79(44), 80(44), 84(44), 158(44), 167(44), 168(44), 169(37), 170(36, 44), 171(36), 172(36), 174(45), 175(36, 44), 177(37, 44), 178(44), 180

AUTHOR INDEX

(44), 181(44, 45), 182(44), 183(44), 184(37), 187(37), 219(44), 230(44), *243*
Pratt, D. W., 386(76), *411*
Prejean, G., 346(1), *368*
Prevedorou-Demas, C., 297, *320*
Prince, R. H., 210(41), 211(42), 212(41, 42), 213(42), *243*
Priskunov, A. K., 261(12), *316*
Pritchard, A. M., 94(314), *251*
Prokai, B., 259(6), *316*
Prokof'ev, A. K., 106(368), 107(368), 122 (368), *252*, 311(227), *322*
Pruett, R. L., 328(186), 329(185, 186), *373*
Puddephatt, R. J., 114(123), *245*, 324(43), 332(43), *370*
Pudjaatmaka, A. H., 301(197), *321*
Pupp, C., 384(33), *410*
Puri, S. P., 212(403), *253*
Puxley, D. C., 33(52), 35(52), 36(52), *55*, 120(167), 140(167), 141(167), 145 (167), *247*

R

Rae, A. I. M., 349(164), *373*
Rake, A. T., 77(29), 85(29), 88(29), 122 (29), 123(29), 127(29), 128(29), 129 (29), 133(29), 162(29), *242*
Ramirez, F., 352(209), *374*
Randl, R. P., 211(43), 213(43), 214(43), *243*
Raper, G., 357(166), *373*
Rapp, K. E., 328(186), 329(185, 186), *373*
Rasmussen, S. E., 51(85a, 126f), *56, 57*
Raymond, K. N., 394(91), *411*
Reagan, W. J., 285(94), *318*
Reddy, J. M., 200(475), *255*
Redwood, M. E., 262(19), 263(20), 266 (34), *316*
Reichle, W. T., 111(336), 112(336), 113 (336), 121(336), 122(336), 125(336), 132(336), 136(336), 138(336), 141 (336), 145(336), *251*
Reiff, W. M., 208(479), 211(480, 481), 213(480, 481), 214(476, 478), 219(477, 480), 220(480), *255*
Reinisch, R. M., 392(36), *410*

Reinke, H., 378(45), *410*
Rekalic, V., 400(9, 10), 401(9), 405(10), *409*
Rendall, I. F., 299(175), 301(190, 191), *320, 321*
Renovitch, G. A., 95(482), *256*
Rentzeperis, P., *256*
Rest, A. J., 323(131), *372*
Restivo, R., 350(112), 355(112), 366(111), *371*
Rice, S. A., 49(95), *56*
Richards, A., 211(259, 387), *249, 253*
Richards, R. R., 211(484), *256*
Richards, S. M., 12(41), 15(41), *54*
Rickards, R., 201(488), 219(487–489), 220(485, 489), 221(489), 225(486), *256*
Ridley, D., 264(25, 26), 266(34), *316*
Ridley, D. R., 139(242), 142(242), 163 (242), *249*
Riess, J. G., 276, *317*
Rimmer, G. D., 107(140), 122(140), *246*
Roberts, J. D., 327(187), 345, *373*
Roberts, L. D., 227, *256*
Roberts, P. D., 266(36), *317*
Roberts, P. J., 357(188), 358(189), 359 (190), *374*
Robiette, A. G., 312(231), *322*, 348(8), *369*
Robin, M. B., 200, *255, 256*
Robinson, A. B., 225(410), 226(410), *254*
Robinson, H., 193(493), 196(493), *256*
Robinson, V. J., 346(138), *372*
Robinson, W. T., 121(69), *244*, 361(25), *369*
Rochev, V. Ya., 90(280), 102(424), 112 (57), 145(424), *243, 250, 254*
Rochow, E. G., 114(428), *254*
Rodesiler, P. F., 51(130a), *57*, 280(82), 314(233), *318, 322*
Rodin, J. O., 329(168), *373*
Römming, C., 5(86), 18(86), 29(86), 37, 38(22), 47, *54, 56*
Roesky, H. W., 391(95), *411*
Rötgers, K., 387(97), *411*
Rogers, K. A., 95(145), 221(145), *246*
Rogers, M. T., 11(131), 14(131), *57*, 381 (96), *411*
Rogozev, B. I., 106(456, 494), 109(494), 142(380), *253, 255, 256*
Rokhlina, E. M., 102(424), 145(424), *254*

Romers, C., 21(143), 24(143), *58*
Romine, H. E., 337(178), *373*
Roos, I. A. G., 220(53), 221(53), *243*
Roper, W. R., 335(49), 355(49), *370*
Rosenberg, E., 113(171), 124(171), 141 (171), 143(171), 146(171), *247*
Rosenzweig, A., 13(132), *57*
Rother, P., 91(495), 101(495), *256*
Ruby, S. L., 63(498, 510), 64(252), 90 (496), 91, 101(497), 102(447), 103 (447), 194(570), 196, 197(570), 199 (570), 231(498, 510), 232(496), 233, 234(496, 498), *249, 255, 256, 257, 258*
Ruddick, J. N. R., 104(300), 105(300), 110(465), 112(300), 121(463), 124(463–465), 125(463, 465), 129(463, 465), 136(464), 140(300), 143(462), 144 (300, 462), 145(464), 149(300), 153 (462, 465), 154(300), 155(300), 156 (462, 465), 157(462), 163(464), 164 (462, 464, 465), *250, 255*
Ruff, J. K., 298, *320*
Rukhadze, E. G., 90(532), 222(532), *257*
Rumbold, B. D., 211(86), 213(86), *244*
Rundle, R. E., 11(107), 15(107), *56*, 201(499), *256*, 363(90), *371*
Russell, D. R., 7(126), 8(126), 10(126), *57*
Russo, W. R., 277(76), 278, 281, *318*
Rutt, K. J., 381(55), *410*
Ryan, R. R., 52(126c), *57*
Rychnovsky, V., 39(133), *58*

S

Saad, M. A., 291(121), *319*
Saalfeld, F. E., 351(13), *369*
Sacco, A., 355(192), 362(191), *374*
Saegusa, T., 270(55), *317*
Sakai, H., 148(349), 193(350), 200(350), *252*
Salem, L., *374*
Sales, K. D., 213(13), 214(13), *242*, 284(89), 285(89), 307(89), 308(222, 223), 309(224), *318, 322*
Sams, J. R., 105(114, 115, 150), 106(115, 150), 107(115), 108(114, 115, 150), 110(459), 121, 122(114, 115, 150), 133, 135, 136(240, 241, 459), 137, 139(242, 257), 140(242, 572), 142(242, 572), 145(240, 556, 574), 146(556, 574), 147(574), 148(574), 149(574), 150 (574), 152(574), 153(114, 115), 158 (114, 115, 459, 556), 160(149), 161 (149), 163(242), 164(240, 241), 165 (240), 209(146), 210(146), 222(147, 149), 223(146), 224(147, 149), 225 (146), *245, 246, 249, 255, 257, 258*, 329 (75), 331(75), 333(61), 342(85), 348 (74), 350(84), 351(74, 85), 352(74), 353(58, 61), 357(76), 364(75), 365(75), 366(76), *370, 371*
Samulski, E. T., 293, *319*
Sanger, A. R., 284, 300(178), 301(195, 196), *318, 320, 321*
Sankla, B. S., 291(112), *318*
Sano, H., 88(431), 107(528), 109(528), 122(431, 528), 123(431), 133(431), 138(525), 141(528, 531), 142 (531), 143(528), 149(530), 155(530, 156(528), 159(431), 160(431), 161(431), 162, 163(529, 531), 181(501), 183(500), *254, 256, 257*
Sans, Th. Texeira, 304(210), *321*
Santoro, R. P., 30(139), 34(139), *58*
Sargeant, P. B., 327(193), *374*
Sarma, A. C., 91(266), 236(266), 238(302), *249, 251*
Sasaki, Y., 88(431), 122(431), 123(431), 133(431), 159(431), 160(431), 161 (431), 162(431), *254*
Sasane, H., 194(502), 200(502), *256*
Sass, R. L., 386(98), *411*
Satten, R. A., 189(354), 195(354), *252*
Sawada, H., 18(106), 20(106), *56*
Sawodny, W., 200(117), *245*, 300(182), 305(182), 306(182), *320, 321*
Saxena, S. K., 98(503), 99, 100(503), *256*
Sayers, D. R., *370*
Scaramuzza, L., 30(47), 31(47), 34(47), *55*
Schawlow, A. L., 37(79), 38(79), *55*
Schenk, R., 381(99), *411*
Schilt, A. A., 173(504), *256*
Schindler, F., 295(142), *319*
Schläfer, D., 378(15), 392(16), *409*
Schlemper, E. O., 33(138), 37, 38(134), 39(135, 136, 137), 40(135), 41(136,

137), *58,* 112(506, 507), 114(507), 119(505), 137(506), 140(507), 143 (505), 164(505), *256*
Schmidbaur, H., 295(141, 142), 296, 300 (152), *319, 320*
Schmidt, M., 296, *320*
Schneider, R. F., 236(508), *256*
Schroeder, H., 385(100), *411*
Schwartz, R., 390(6), 394(6), *409*
Seeley, N. J., 77(234), 110(234), 113(234), 114(234), 123(234), 124(233, 234), 125 (234), 126(233), 129(233), 140(234), 142(234), 143(234), 144(234), 156 (234), 157(233, 234), *249*
Segal, D. J., 30(139), 34(139), *58*
Seidel, B., 327(7), *369*
Selig, H., 101(497), *256*
Selte, K., 11(140), 14(140), 41(140), *58*
Selzter, H. E., 91(471), *255*
Semenov, S. I., 237(306), 238(306), *251, 253*
Semin, G. K., 141(81), *244*
Senior, B. J., 85(64), 88(64), 116(64), 122(64), 123(64), 124(10), 128(64), 129(64), 143(10), 144(10), 145(10), 148(10), 149(10), 151(10), 152(10), 154(10), 155(10), 159(64), 160(64), 161(64), 202(190), 203, 204, 205(187, 191), 206(187, 191, 193), 207(187), *242, 244, 247*
Senov, S. I., 152(513), 232(513), 233(513), 237(513), 238(513), *256*
Sergeev, V. P., 161(46), *243*
Seyferth, D., 324(195), 325(194, 196–198), *374*
Shahid, M. S., 379(101), 384(102), 400 (42), 401(63, 101), 402(101), 403(103), 404(60, 61, 63), 406(63, 64, 101, 102), 407(63, 101), 408(63), *410, 411*
Shamir, J., 193(90, 92), 200(90, 92), 201 (90, 92), *244*
Sharma, B. D., 381, *412*
Sharma, H. D., 145(103), 147(103), 148 (103), 150(103), 152(103), *245*
Sharma, K. K., 117(291), 118(291), 119 (291), 124(291), 125(291), 129(291), 131(291), 157(291), *250*
Sharts, C. M., 327(187), *373*
Shaw, B. L., 357(166), *373*

Shearer, H. M. M., 264(28), *316*
Sheldrick, G. M., 33(71), 35(71), 39(7), 40(7), *53, 55,* 112(243), 137(243), 140(244), *249,* 312(231), 315(235), *322,* 348(8), *369,* 385(32a), *410*
Sheldrick, W. S., 312(231), 315(235), *322*
Sheline, R. K., 346(100), 347(100), 351(100), 355(100), *371*
Shen, K., 236(509), *256*
Shenoy, G. K., 63(498, 510), 91, 194(570), 196, 197(570), 199(570), 231(498, 510), 234(498, 510), *256, 258*
Shieh, M.-C., 391(104a), *412*
Shigorin, D. N., 261(13), *316*
Shimauchi, A., 192(371), *252*
Shiner, V. J., 268, 269, 270(47), 271, *317*
Shiotani, A., 296, 300(152), *320*
Shirai, T., 380(56), *410*
Shirley, D. A., 91(222), 227, 228(222), 229(222), 230, 231, *248, 256*
Shpinel, V. S., 70(5), 88(364), 103(5, 6, 8, 9, 84), 106(9), 122(364), 123(364), 136(8, 83, 84), 141(6, 7, 84), 142(5–7, 9), 143(84), 152(513), 160(364), 161 (364), 203(82), 232(307, 513), 233 (513), 235, 237(85, 513), 238(85, 306, 513), *242, 244, 251, 252, 253, 256*
Shripkin, V. V., 123(275), 143(275), 159 (275), 160(275), 161(275), *250*
Shukla, P. N., 212(256), *249*
Shustorovich, E. M., 224(514), *256*
Siddal, T. H., 293(131), *319*
Siefert, E. E., 357(4), *368*
Siekierska, K. E., 100(225), 101(225), *248*
Sills, R. J. C., 9, 10(62), 31(63), 32, 35 (63), 44(62), *55*
Silverstein, R. O., 329(168), *373*
Silverthorn, W., 346(118), *372*
Silverthorn, W. E., 91(562), 186(562), *258*
Sim, G. A., 312(229), *322*
Simonov, A. P., 261, *316*
Simopoulos A., 102(515), 103(515), 191 (443), 196(443), *254, 256*
Simpson, W. B., 206(166), *247*
Sinn, E., 69(283), 208(283), *250*
Sirigu, A., 361(52), *370*
Skapski, A. C., 30(141), *58*
Sladky, F. O., 7(142), 8, 10(142), *58*
Slater, J. H., 168(312), 180(312), 181(312), *251*

Sletten, E., 51(100b), *56*
Sletten, J., 53(87b), *56*
Smail, W. R., 306(220), 307(220), *321*
Smallwood, R. J., 300(187), *321*
Smith, A. W., 77(234), 95(145), 110(234), 113(234), 114(234), 123(234), 124(233, 234), 125(234), 126(233), 129(233), 140(234), 142(234), 143(234), 144(234), 156(234), 157(233, 234), 219(145), 220(145), 221(145), *246, 249*
Smith, B. C., 292(127, 129), 296(148), *319, 320*
Smith, G. F. H., 379(105), *412*
Smith, H. M., 274(69), *317*
Smith, H. S., 277(77), *318*
Smith, J. D., 312(229), *322*
Smith, J. S., 291(114), *319*
Smith, L., 124(168), 125(168), 128(168), 131(168), 140(168), 143(168), 144(168), 145(169), 150(168), 155(168), 157(168), 161(169), *247*
Smith, P. J., 33(52), 35(52), 36(52), *55*, 60(516), 106(516), 111(516), 120(167), 124(168), 125(168), 128(168), 131(168), 136(516), 138(516), 140(167, 168), 141(167, 516), 142(516), 143(168), 144(168), 145(167, 169), 146(516), 150(168), 155(168), 157(168), 160(516), 161(169), 207(124), *245, 247, 256*
Smitskamp, C. C., 236(430), *254*
Snegova, A. D., 327(184), *373*
Snider, R. H., 331(170), *373*
Snow, M. R., 348(176), *373*
Snyder, R. E., 90(496), 232(496), 233(496), 234(496), *256*
Søtofte, I., 51(114b), *57*
Sonnino, T., 91(445), 93(445), 101(444), 191, 192(89, 153, 445), 193(89, 93), 195(153), 196, 198, 199(93), 200(89, 445), 238(445), 241(445), *244, 245, 246, 254, 255*
Soriano, J., 193(90, 92), 200(90, 92), 201(90, 92), *244*
Sorokin, P. P., 191(68), *244*
Sowerby, D. B., 236(396), *253*
Sparks, R. A., 22(46), 28(46), *55*
Spencer, C. B., 264(28), *316*
Spialter, L., 327(128), *372*
Spiess, H. W., 185(517), *256*

Spijkerman, J. J., 60(175), 211(313, 519), 212(313, 519), 227(518), *247, 251, 257*
Spillman, J. A., 121(463), 124(463), 125(463), 129(463), *255*
Spiro, T. G., 263(23), *316*
Sreenathan, B. R., 301(189), 306(189), *321*
Srivastava, R. C., 301(193), 302(200, 202), *321*
Stålhandske, C.-I., 52(142a), *58*
Stammreich, H., 304(210), *321*
Stapfer, C. H., 139, *251, 257*
Stefani, A. P., 329(165), *373*
Stefanini, F. P., 174(45), 181(45), *243*
Steichele, E., 217(372), 225(372), 226(372), *253*
Stein, L., 200(521), *257*
Stephens, R., 332(35, 46, 47, 199), 333(33–36), 338(33–35, 37), 342(34, 35, 37), 343(199), 344(34), *369, 370, 374*
Sterlin, R. N., 336(155), *372*
Sterlin, S. R., 331(104), *371*
Steudel, R., 391(105a), *412*
Stevens, J. G., 90(522), 232(72, 388, 389, 522), 233(72, 389, 522, 523), 234(72, 389, 522), 235(72, 389, 522), 236(72, 389, 522), *244, 253, 257*
Stevens, P. J., 179(524), 231(524), *257*
Stewart, C. J., 329(67), 331(67), 351(68), 355(68), *370*
Stewart, D. F., 7(16), *54*
Stewart, W. E., 293(131), *319*
Stiddard, M. H. B., 348(178), *373*
Stockel, R. F., 329(200, 201), 331(200–202), 337(202), 341(202), *374*
Stöckler, H. A., 107(528), 109(528), 111(336), 112(336), 113(336), 121(336), 122(336, 528), 125(336), 132(336), 136(336), 138(336, 525), 141(336, 528, 531), 142(531), 143(528), 145(336), 149(530), 155(530), 156(528), 162, 163(529, 531), 165, *251, 257*
Stokley, P. F., *253*
Stolz, I. W., 346(100), 347(100), 351(100), 355(100), *371*
Stone, A. J., 70(121), 82(121), 94(32), 100(27), 211(32, 42), 212(42), 213(32, 42), 216(121), *242, 243, 245*
Stone, F. G. A., 323(24, 131), 324(27, 28), 328(28, 50, 208), 332(167, 217),

AUTHOR INDEX

334(26, 50, 148), 335(173), 344(148), *369, 370, 372, 373, 374*
Stone, J. A., 91(458), 195(458), 227(458), *255*
Stork-Blaisse, B. A., 21(143), 24(143), *58*
Stranski, K. A., 30(17, 18), *54*
Straub, D. K., 124(210), 125(210), 212(211), 220(213), 226(212), 227(212), *248*
Straughan, B. P., 24(83), *56*, 238(301, 302), *250, 251*
Streitwieser, A., 301, *321*
Strens, R. G. J., 98(40), 99(40), *243*
Stroke, H. H., 189(354), 195(354), *252*
Struchkov, Yu. T., 17(144), *58*, 116(70), 117(70), *244*
Stucky, G. D., 311, 312(230), *322*
Stukan, R. A., 60(277), 90(532), 103(279), 105(278), 112(278), 123(275), 138(279), 142(279), 143(279), 148(278), 155(278), 159(275), 160(275), 161(275), 169(3), 177(3), 180(3), 181(3), 222(532), 223(533), *242, 250, 257*
Styan, G. E., 328(82), 329(69, 82), 330(69, 82), 333(69, 82, 86), 335(69), 342(82), 346(82), 355(69), *370, 371*
Subramanian, T. K. B., 376(81), *411*
Sukhoverkhov, V. F., 155(534), *257*
Sullivan, R., 324(203), 338(203), 343(203), *374*
Sumarokova, T. N., 105(278), 112(278), 148(278), 155(278), *250*
Sumner, G. G., 359(204), 353(204), *374*
Sunyar, A. W., 59(369), *252*
Sutin, N., 217(220), *248*
Sutton, P. W., 361(205), *374*
Suzdalev, I. P., 103(84), 136(83, 84), 141(84), 153(276), *244, 250*
Suzuki, R., 325(198), *374*
Swift, P., 162, 163(47), *243*
Szymanski, J. T., 30(102), 52(126g), *56, 57*

T

Taft, R. W., 108(535), *257*
Takanashi, A., 380(56), *410*
Talalaeva, T. V., 261(12, 13), 266, *316, 317*
Tapping, R. L., 209(146), 210(146), 223(146), 225(146), *246*, 353(58), *370*
Tarasenko, N. A., 311(227), *322*
Tarrant, P., 327(206, 207), *374*
Tatlow, J. C., 332(35, 46, 47, 199), 337(33–36), 338, 343(34, 35, 37, 199), 344(34), *369, 370, 374*
Tavernier, N., 376(88), *411*
Taylor, B., 124(464), 136(464), 145(464), 163(464), 164(464), *255*
Taylor, K. A., 355(135), *372*
Taylor, S. R., 361(25), *369*
Temkin, A. Ya., 141(7), 142(7), *242*
Templeton, D. H., 10(145), 14(145), 50(145a), *58*, 387(32), *410*
Templeton, L. K., 50(145a), *58*
Tesi, G., 335(120), 340(120), *372*
Thevarasa, M., 110(465), 124(465), 125(465), 129(465), 153(465), 156(465), 164(465), 169(397), 180(397), *253, 255*
Thirtle, J. R., 292(124), 294(138), *319*
Thomas, I. M., 296, 299(166, 167, 170, 172), 301(166, 170, 172), 303(166), *320*
Thomas, K., 182(537), *257*
Thompson, D. T., 222(267), 223(267), *249*
Thompson, J. A. J., 133(149), 160(149), 161(149), 222(149), 224(149), *246*, 350(84), 351(70), 354(83), 355(83), *370, 371*
Thompson, J. B., 100(538), *257*
Thomson, J. O., 227(490, 491), *256*
Thorp, T. L., 225(158), *246*
Timms, R. E., 33(8), 35(8), 36(8), *53*, 135(4), *242*
Timmick, A., 203(303), 205(303), 206(303), 207(303), *251*
Tingle, E. M., 409(106), *412*
Tinkham, M., 218(539), *257*
Tiripicchio, A., 50(25a), 51(25b, 126b), *54, 57*
Tiripicchio Camellini, M., 51(126b), *57*
Tjomsland, O., 22(77), 27(77), *55*
Tobe, M. L., *369*
Tobias, R. S., 158(395), *253*
Toley, D. L. B., 124(464), 136(464), 145(464), 163(464, 467), 164(464, 466, 467, 468), *255*
Townes, C. H., 187(540), *257*

Travis, J. C., 64(136), 65(136), 70(136), *246*
Treichel, P. M., 324(208), 328(208), 332(217), *374*
Tricker, M. J., 85(64), 88(64), 116(64), 122(64), 123(64), 128(64), 129(64), 159(64), 160(64), 161(64), 203(183), 204(183), 205(183), *244, 247*
Triftshauser, W., 91(361), 100(541), 101(541), 193(361), 237(361), 238(361), 239(361), *252, 257*
Tripathi, U. D., 291(113), *319*
Trofimenko, S., 95(356), 96(356), *252*
Trooster, J. M., 90(177), 102(178), *247*
Trotter, J., 7(16), 22(64), 27(64), 29(39), 30, 32(146, 148), 33, 34(39), 35(146, 147, 148), 39(40), 40(40), *54, 55, 58*, 114(118), 234(542), *245, 257*, 352(139), 353(139), 355(139), 357(188), 358(189), 359(190), 360(140), 361(140), 363(113), 364(113), *371, 372, 374*
Trozzolo, A. M., 225(563, 564), 226(564), 227(563, 564), *258*
Trukhtanov, V. A., 103(279), 105(278), 112(278), 138(279), 142(279), 148(278), 155(278), 169(1), 177(1), 181(1), *241, 250*
Tsai, J. H., *257*
Tsang, W. S., 329(44), 331(44), *370*
Tsareva, G. W., 261(12), *316*
Tsukada, K., 192(371), *252*
Tucker, N. I., 328(16), 332(16), 333(16), 334(16), *369*
Tuinstra, F., 407(107), *412*
Tullbane, R. J., 232(389), 233(389), 234(389), 235(389), 236(389), *253*
Turner, A. G., 381(108), 383(108), *412*
Turner, J. J., 102(38, 461), *243, 255*, 329(165), 357(183), *373*
Turova, N. Ya., 264(30), 265(30), 266(30, 39, 40), *316, 317*
Tyuleneva, V. V., 336(154, 155), *373*

U

Uchida, T., 200(544), *257*
Ueda, I., 201(429), *254*
Ugi, I., 352(209), *374*
Ugo, R., 355(192), *374*

Ulrich, S. E., 110(170), 135(170), 136(170), 137(170), *247*
Underhill, A. E., 95(195), 169(195), 180(195), *248*
Unisimov, K. N., *244*
Unland, M. L., 91(545), 237(545), 239(545), *257*
Unsworth, W. D., 146(286), 149(286), 155(286), *250*

V

Vaciago, A., 30(47), 31(47), 34(47), *55*
Valle, B. D., 90(62), 232(62, 63), 233(63), 235(63), *244*
Vallet, G., 393(31), 395(18a, 31), 396(31), *409, 410*
Van de Grampel, J. C., 386(109), 400(109), *412*
van der Heide, J., 12(23), 15(23), 16(23), *54*
Van Der Kerk, G. J. M., 135(355), 165(355), *252*
van Eck, C. L. P., 10(125), 13(125), *57*
van Schalkwyk, T. G. D., 12(12), *53*, 196(16), *242*
Van Wazer, J. R., 274(66), 278(66), 279, 280(66), 281(66), 302, *317*
Vaughan, P. A., 37(127), 38(127), 40(127), *57*
Vaughan, R. W., 96(107), 97(107), *245*
Veidis, M. V., 361(41), *370*
Venanzi, L. M, 347(210), *374*
Veshima, T., 270(55), *317*
Vidoni Tani, M. E., 52(52b), *55*
Viers, J. W., 17(149), *58*
Viguesnel, K., 266(42), *317*
Vilkov, L. V., 311, *322*
Villena-Blanco, M., 377(110), 378(110), 379(110), *412*
Violet, C. E., 237(546, 547), 238(548), 239(546), *257*
Virgo, D., 98(311, 549, 550), 99, *251, 257*
Viswamitra, M. A., 51(126e), *57*
Voigt, D., 393(50), 394(50), 395(50), 396(50), *410*
Vonk, C. G., 12(150), 16, *58*
Von Osten, H., 294(137), *319*
Vos, A., 10(52a, 70), 12(23), 13(52a), 14,

AUTHOR INDEX 439

15(23), 16(23), *54, 55,* 384(114), 386 (109), 400(109), *412*
Vucelic, M., 110(141), 111(141), 119(141), 124(141), 138(141), 139(141), 141 (141), 142(551), 144(141), *246, 257*

W

Waddington, T. C., 13(51), *55,* 381(27a), 382(27a), 383(27a), *409*
Wade, K., 168(312), 180(312), 181(312), *251*
Wagner, C. F., 225(565), *258*
Wagner, F. E., 91(362, 470, 471, 495, 552), 101(495), 185(362, 470), 186(470), 187 (470), *252, 255, 256, 257*
Wagner, H., 391(18), *409*
Wakefield, B. J., 266, *316*
Waldman, M. C., 105(150), 106(150), 108(150, 151), 122(150), *246,* 324(87, 89), 326(88), 342(85, 87), 345(88, 211), 351(85), *371, 374*
Walker, L. R., 63(553), *257*
Wall, D. H., 112(299), 153(299), *250*
Walter, L. S., 98(200), *248*
Wampler, D. L., 263(22), *316*
Wang, B.-C., 21(151), 26, *58*
Wannagat, U., 300, 305(182), 306(182), 309(181), *320*
Warburton, D., 98(311), 99(311), *251*
Ward, J. E. H., 333(61), 353(61), 363(61), *370*
Wardlaw, W., 260(9), 276(72), 279(72), 288(103), 291(118–121), 294, *316, 317, 318, 319*
Warn, J. R. W., 393(111), *412*
Warn, R. J., 395(28), *409*
Warren, J. L., 100(358), *252*
Watanabe, N., 106(554), *257*
Watenpaugh, K., 274(69), 280(81), *317, 318*
Watson, R. E., 62, *257*
Wayland, B. B., 295(145), *319,* 351(13), *369*
Webb, G. A., 227(214), *248*
Webster, D. E., 114(428), *254*
Wedd, R. W. J., 145(556), 146(556), 158(556), *257*

Wei, C. H., 92(557), 93(557), *258,* 349 (212), 353(91), 357(212), 361(213, 214), *371, 374*
Weickhardt, P. L., 220(53), 221(53), *243*
Weiher, J. F., 95(356), 96(356), *252*
Weingarten, H., 274(66), 278(66), 279, 280(66), 281(66), 302, *317*
Weiss, E., 261, 262(17), 295(146), 296(15, 146), *316, 320*
Weiss, J., 388, 389(112), 390(112), 391 (112), 392(113), *411, 412*
Welch, A. J., 300(185, 187), *320, 321*
Wells, A. F., 19(152), 20(152), *58*
Wells, W. L., 112(558), *258*
Wentworth, R. A. D., 285(93), *318*
Wertheim, G. K., 60(559), 63(553), 94(560), 209(333), 221, 222(333, 561), 223(561), 224(333, 561), *251, 257, 258,* 349(215), *374*
West, B. O., 211(54), 213(54), 214(54), *243*
West, K., 263(22), *316*
Westgren, A., 30(9), 34(9), *53*
Westlake, A. H., 279, 285(79), 293, *318*
Wheatley, P. J., 260, *316*
Whitfield, H. J., 168(347), 169(347), 175 (347), 181(347), 226(347), *252*
Whitley, A., 276(72), *317*
Whitney, D. L., 215(390), 216(390), 217 (390), 218(390), *253*
Whittaker, D., 268(47), 269(47), 270(47), 271, *317*
Wiberg, K. B., 286, *318*
Wicholas, M., 289, *318*
Wickman, H. H., 91(562), 102(515), 103 (515), 186(562), 225(563, 564, 565), 226(564), 227(563, 564), *256, 258*
Wiebenga, E. H., 11(88), 12(23, 24, 150), 15(23, 88, 153), 16, 41(154), 44, 45, 48, 49, *54, 56, 58*
Wiegers, G. A., 384(114), *412*
Wilcox, C. F., 345(216), *374*
Wild, J. D., 346(10), 348(10), 356(9), *369*
Wilford, J. B., 332(217), *374*
Wilhelmi, K.-A., 32(37, 38), 35(38), *54*
Wilhoit, R. C., 266, 268(51), *317*
Wilkes, G. R., *374*
Wilkinson, G., 182(434, 537), *254, 257,* 323(160), 347(54), *370, 373*

Willeford, B. R., 107(460), *255*
Williams, H. J., 225(563), 227(563), *258*
Williams, P. G. L., 64(568), *258*
Williamson, S. M., 10(145), 14(145), *58*
Williams, D. A., 272(63), *317*
Williams, D. E., 132, *258*
Williams, R. J. P., 168(156), 169(156), 172(157), 175(156), 177(156), 178(156), 180(156), 181(156), 218(202), 225(155, 158), 226(155), *246, 248*
Willis, C. J., 262(19), 263(20), *316,* 327(145), *370, 372*
Williston, C., 22(64), 27(64), *55*
Wilson, A. E., 24(83), *56*
Wilson, G. V. H., 211(86), 213(86, 87), *244*
Wilson, J. W., 268, *317*
Winkler, C. A., *409*
Winter, G., 274, 277, 284(85, 86), 287, 288(85, 104, 105), 289, *317, 318*
Wissemeier, H., 397(49), 398(49), 399(49), *410*
Witmore, W. G., 37(45), 38(45), *55*
Witt, J. R., 53(154a), *58*
Witters, R. D., 272(65), 277(65), *317*
Wölbling, H., 387(115), *412*
Wolf, A. P., 236(509), *256*
Wood, K. N., 332(199), 337(34), 338(34), 343(34, 199), 344(34), *369, 374*
Woodward, R. B., 327(218), *374*
Woolf, C., 329(168), *373*
Wooten, F., 238(548), *257*
Worrall, I. J., 268, 270(54), 271(54), 272(54), 294(54), *317*
Wright, D. A., 272(63), *317*
Wright, J. S., *374*
Wyckoff, R. G., 30(155), *58*
Wynne, K. J., 387(116), 388(116), 389(117), *412*
Wynter, C. I., 88(569), 122(569), 123(569), 133(569), 159(569), 160(569), 161(569), 194(570), 197(570), 199, *258*

X

Xavier, R. M., 168(142), 177(142), 178(142), 179(142), 180(142), 220(142), *246*

Y

Yamasaki, R. S., 192(571), 193(571), 286, *258, 318*
Yasuda, K., 112(366), 137(366), *252*
Yeats, P. A., 140(572), 142(572), 145(574), 146, 147, 148(574), 149(574), 150, 152(574), 163(572, 573), *258*
Yoeman, F. A., 292(123, 124), *319*
Yoffe, A. D., 395(28), *409*
Yoshida, H., 91(454), 101(454), 238(454), 240(454), *255*
Young, A. E., 152(316), 154(316), 155(316), *251*
Young, W. C., 236(396), *253*
Younger, D., 386(4a), 387(4a), *409*

Z

Zahn, U., 91(362, 470, 495, 552), 101(495), 185(362, 470), 186(470), 187(470), *252, 255, 256, 257*
Zakharova, G. N., 116(70), 117(70), *244*
Zakharova, M. Ya., *244*
Zalkin, A., 7(142), 8(142), 10(142, 145), 14(145), *58,* 387(32), *410*
Zaugorodnii, V. S., 106(456, 494), 109(494), *255, 256*
Zeeman, P. B., 376(118), *412*
Zeitler, V. A., 295, *319*
Zemcik, T., 169(404), 182(404), *253*
Zenina, G. V., 266(37), *317*
Zerner, M., 226, *258*
Zhdonov, Y. A., 109(576), *258*
Zinnius, A., 295(143), 312(143), *319*
Zipp, A. P., 390(119), *412*
Zobel, T., 32(148), 35(147), *58,* 234(542), *257*
Zuccaro, D. E., 22(156), 27(156), *58*
Zuckerman, J. J., 60(577), 90(50, 317), 107(317, 460), 110(170), 113(171), 122(226, 227), 124(171), 133, 135(170), 136(170), 137, 138(318), 141(171), 145(577), 146(171), 149(577), 159(226), 160(226, 227), 161(227), 162, *243, 247, 248, 251, 255, 258,* 351(119), *372*
Zumdahl, S. S., 347(220), *374*
Zussman, J., 98(172), *247*
Zwartsenberg, J. W., 37(110), 38(110), *56*

SUBJECT INDEX

A

Actinides, alkoxides of, preparation and properties, 290–293
Alkali metals, alkoxides of, structure and properties, 260–263
Alkaline earths, alkoxides of, preparation and properties, 264–266
Alkoxides, of metals, *see* Metal alkoxides
Aluminium alkoxides, preparation and properties, 266–272
Antimony compounds, secondary bonding by, 29–36
Arsenic compounds, secondary bonding by, 29–36, 52
Arsenic cyanides, secondary bonding in, 39, 40

B

Benzeneselinic acid, secondary bonding by, 19
Beryllium, alkoxides of, preparation and properties, 264–266
Bismuth compounds, secondary bonding by, 32–33, 52
Bonding, secondary, *see* Secondary bonding
Bromine compounds, secondary bonding in, 9–13, 50–51

C

Carbene and carbenoid additions, in preparation of fluoroalicyclic derivatives, 324–325
Chelates, from metal fluoroalicyclic derivatives, 351–356
Chlorine compounds, secondary bonding in, 9–13, 50–51
Coordination complexes, of fluorocyclic derivatives, 346–368
Coupled-ring nitrides
 molecular structure of, 401–6
 preparation of, 400
 properties and reactions of, 402–403
 purification of, 400–401
Cyanides, secondary bonding in, 36–41, 53

Cycloaddition reactions, in preparation of fluoroalicyclic derivatives, 327–328
Cyclopropenes
 preparation of, 326, 342
 physical properties of, 345
 stability of, 343

D

Dialkylamides, of metals, *see* Metal dialkylamides
Dimethylarsinodimethyldithioarsinate, secondary bonding by, 29–30
Disulfur dinitride
 description of, 393
 molecular structure of, 393
 preparation of, 393
 reactions of, 393–394

E

Elements, nonmetallic, secondary bonding to, 1–58
Exchange reactions, in preparation of fluoroalicyclic derivatives, 338–339

F

Fluorine, xenon compounds with, secondary bonding in, 6–9, 10
Fluoroalicyclic derivatives, of metals and metalloids, 323–374
 chemical properties of, 342–343
 coordination complexes, 346–368
  biligate bimetallic, 356–363
  chelate, 351–356
  decomposition products, 366–368
  monoligate 347–351
  triligate bimetallic, 363–366
 modification of, 340–341
 physical properties of, 345–346
 preparative methods for, 324–341
  carbene and carbenoid additions, 324–325
  from compounds with metal-metal bonds, 333–335

Fluoroalicyclic derivatives, preparative methods for—*cont.*
  cycloaddition reactions, 327–328
  direct reaction with metal, 337–338
  exchange reactions, 338–339
  hydride additions, 328–331
  metal-alkyl additions, 332
  metal–fluoride additions, 331–332
  oxidative addition reactions, 335–337
Frozen solutions, Mössbauer spectra in studies of, 101–103
Fused-ring nitrides
  bonding and chemical behaviour in, 405–406
  description of, 404–405
  general description of, 403–404
  miscellaneous types, 406–407
  molecular structure of, 405
  preparation of, 404

G

Gallium alkoxides, preparation and properties, 266–272
Germanium compounds, secondary bonding by, 33–36
Gold-197 compounds, Mössbauer spectra of, bonding and structure in, 227–231
Goldanskii—Karyagin effect, in Mössbauer activity of tin compounds, 165

H

Hafnium, alkoxides of, preparation and properties, 272–290
Halogen complexes, of tin(IV), Mössbauer data on, 135–150
Halogens
  iodine compounds with, secondary bonding in, 15–17
  secondary bonding to, 23–29, 31–33
Hexacarbonyls, chelate complexes from, 351
Hydrides, addition reactions of, in preparation of fluoroalicyclic derivatives, 328–331

I

Iodine compounds
  Mössbauer spectra of, bonding and structure in, 187–201
  secondary bonding to oxygen, 13–17

Iodomethanes, bonding parameters for, 199
Iridium compounds, Mössbauer spectra of bonding and structure in, 184–187
Iron carbonyls, chelates from, 352–354
Iron complexes, Mössbauer spectra of bonding, and structure in, 221–225
Iron compounds
  Mössbauer spectra of, bonding in, 166–184
  center shifts, 177–184
  electronic structure, 95–97, 208–210
  quadrupole splitting, 167–177

L

Lanthanides, alkoxides of, preparation and properties, 290–293

M

Magnesium alkoxides, preparation and properties, 264–266
Manganese carbonyls, chelates from, 352
Metal–metal bonds, compounds containing in preparation of fluoroalicyclic derivatives, 333–335
Metal alkoxides, 159–297
  of actinides, 290–293
  of alkali metals, 260–263
  of alkaline earths, 264–266
  of aluminium, 266–272
  of beryllium, 264–266
  double type, 293–294
  of gallium, 266–272
  of lanthanides, 290–293
  of magnesium, 264–266
  properties of, 260
  of transition metals, 272–290
  trialkylsilyloxides, 295–297
  of zinc, 264–266
Metal dialkylamides
  chemical properties of, 301–302
  infrared and Raman spectra of, 304–305
  metal–nitrogen bond of, 303–304
  NMR spectra of, 305–306
  physical properties of, 302–315
  paramagnetic, 306–309
    four-coordinated derivatives, 306–308
    three-coordinated derivatives, 308–309
  structures of, by X-ray and electron diffraction, 309–315

SUBJECT INDEX

Metalloids, fluoroalicyclic derivatives of, 323–374
Metals, fluoroalicyclic derivatives of, 323–374
Mössbauer effect, theory of, 60–61
  of oxidation states, 89–93
  in preparation of novel compounds, 100–101
  quadrupole splitting in, 63–71
    additivity model for, 71–88
    molecular orbital approaches, 73–88
    point charge formalism, 71–73
    sign of, 69–71
  of site populations in silicate minerals, 97–100
  in studies of electronic structure of iron compounds, 95–97
Mössbauer spectra, of inorganic compounds, 59–258
  of bonding and structure, 103–241
    in antimony-21 compounds, 231–236
    in gold-197 compounds, 227–231
    in iodine compounds, 187–201
    in iridium compounds, 184–187
    in iron complexes, 221–225
    in iron compounds, 166–184, 208–210, 214–221
    in ruthenium compounds, 184–187
    in tellurium-125 compounds, 236–238
    in tin compounds, 103–165, 201–207
    in xenon-129 compounds, 239–241
  of decomposition reactions, 94–95
  fingerprint uses of, 89–103
  in frozen solution studies, 101–103
  isomer and center shift in, 61–63

N

Niobium, alkoxides of, preparation and properties, 272–290
Nitrides, of sulfur, *see* Sulfur nitrides
Nitrogen, secondary bonding to, 29–30
Nonmetallic elements, secondary bonding to, 1–58

O

Organotin compounds, Mössbauer data on, 106–107
Oxidation, of inorganic compounds, Mössbauer spectra of, 89–93

Oxidative addition reactions, in preparation of fluoroalicyclic derivatives, 335–337
Oxygen
  bonding of iodine compounds to, 13–17
  secondary bonding to, 18–23, 30–31

P

Perfluoroalkoxides, of alkali metals, preparation, 263
Phenylarsine bis(diethyl dithiocarbamate), secondary bonding by, 30–31
Polythiazyl
  chemical properties of, 396
  electrical properties of, 395–396
  molecular structure of, 395
  preparation of, 395
Pyridinium tetrachloroantimonite, secondary bonding by, 32–33

R

Rhenium carbonyls, chelates from, 352
Ruthenium carbonyls, chelates from, 352–354
Ruthenium compounds, Mössbauer spectra of, bonding and structure in, 184–187

S

Secondary bonding, to nonmetallic elements, 1–58
  angles in, 41–42
  bridge bonds in, 5
  charge transfer adducts and, 5
  classification in, 5–6
  in cyanides, 36–41
  distances in, 44–46
  geometry of, 42–44
  significance of, 49–50
  theoretical aspects of, 2–6
    calculations, 48–49
  van der Waals distances in, 3–5
Selenium compounds, secondary bonding by, 18–29, 51–52
Selenium cyanides, secondary bonding in, 37–40
Selenium dioxide, secondary bonding in, 18
Selenium halides, secondary bonding by, 24
Selenous acid, secondary bonding by, 19

Silicate minerals, site populations in, from Mössbauer spectra, 97–100
Silyl cyanide, secondary bonding in, 40–41
Sulfur, secondary bonding to, 18–23, 30–31
Sulfur nitride, preparation and properties of, 376–377
Sulfur nitrides, 375–412
  coupled-ring type, 400–403
  disulfur dinitride, 393–394
  fused-ring type, 403–407
  polymeric saturated types, 407–408
  polythiazyl, 395–396
  SN, 376–377
  tetrasulfur dinitride, 396–399
  tetrasulfur tetranitride, 375–376, 377–392

T

Tantalum alkoxides, preparation and properties, 272–290
Tellurite, secondary bonding by, 18
Tellurium bis(dimethyldithiophosphate), secondary bonding by, 19
Tellurium catecholate, secondary bonding by, 18, 19
Tellurium compounds, secondary bonding in, 18–29, 51–52
Tellurium-125 compounds, Mössbauer spectra of, bonding and structure in, 236–238
Tellurium diethylxanthate, secondary bonding by, 19
Tellurium dioxide, secondary bonding in, 18
Tellurium tetrafluoride, secondary bonding by, 23–24
Tetrasulfur dinitride
  chemical behaviour of, 398–399
  general description of, 397
  isomerism of, 397
  molecular structure of, 397–398
  preparation of, 396
Tetrasulfur tetranitride
  addition reactions of, 392
  electronic structure of, 382–383
  general description of, 379–380
  general reactions of, 384
  historical aspects of, 375–376
  molecular structure of, 380–382
  oxidation of, 384–386
  preparation of, 375–376, 377–378
  reactions with electrophites, 387–390
  reactions with nucleophiles, 390–392
  reduction of, 386–387
  thermal decomposition and dissociation of, 384
  thermochemistry of, 383
Tin compounds
  bonding and structure of, Mössbauer spectra, 103–165
  center shifts, 150–162
  quadrupole splittings, 103–150
  temperature dependence, 162–165
  valence II, 201–207
  valence IV 103–165
  secondary bonding by, 33–36, 52–53
Titanium alkoxides, preparation and properties, 272–290
Transition metals alkoxides, preparation and properties, 272–290
Trialkylsilyloxides, of metals, preparation and properties, 295–297
Trifluoromethoxides, of alkali metals, preparation, 262–263
Trimethyl germyl cyanide, secondary bonding by, 41
Trithiourea tellurium hydrogen difluoride, secondary bonding by, 19–20
Tungsten alkoxides, preparation and properties, 272–290

V

Van der Waals distances in secondary bonding, 3–5
Vanadium, alkoxides of, preparation and properties, 272–290

X

Xenon-129 compounds, Mössbauer spectra of, bonding and structure in, 239–241
Xenon fluorides, secondary bonding in, 6–9, 10

Z

Zinc alkoxides, preparation and properties 264–266, 272–290

CONTENTS OF PREVIOUS VOLUMES

Volume I

Mechanisms of Redox Reactions of Simple Chemistry
  H. Taube

Compounds of Aromatic Ring Systems and Metals
  E. O. Fischer and H. P. Fritz

Recent Studies of the Boron Hydrides
  William N. Lipscomb

Lattice Energies and Their Significance in Inorganic Chemistry
  T. C. Waddington

Graphite Intercalation Compounds
  W. Rüdorff

The Szilard-Chalmers Reaction in Solids
  Garman Harbottle and Norman Sutin

Activation Analysis
  D. N. F. Atkins and A. A. Smales

The Phosphonitrilic Halides and Their Derivatives
  N. L. Paddock and H. T. Searle

The Sulfuric Acid Solvent System
  R. J. Gillespie and E. A. Robinson

AUTHOR INDEX—SUBJECT INDEX

Volume 2

Stereochemistry of Ionic Solids
  J. D. Dunitz and L. E. Orgel

Organometallic Compounds
  John Eisch and Henry Gilman

Fluorine-Containing Compounds of Sulfur
  George H. Cady

Amides and Imides of the Oxyacids of Sulfur
  Margot Becke-Goehring

Halides of the Actinide Elements
  Joseph J. Katz and Irving Sheft

Structures of Compounds Containing Chains of Sulfur Atoms
Olav Foss

Chemical Reactivity of the Boron Hydrides and Related Compounds
F. G. A. Stone

Mass Spectrometry in Nuclear Chemistry
H. G. Thode, C. C. McMullen, and K. Fritze

AUTHOR INDEX—SUBJECT INDEX

Volume 3

Mechanisms of Substitution Reactions of Metal Complexes
Fred Basolo and Ralph G. Pearson

Molecular Complexes of Halogens
L. J. Andrews and R. M. Keefer

Structures of Interhalogen Compounds and Polyhalides
E. H. Wiebenga, E. E. Havinga, and K. H. Boswijk

Kinetic Behavior of the Radiolysis Products of Water
Christiane Ferradini

The General, Selective, and Specific Formation of Complexes by Metallic Cations
G. Schwarzenbach

Atmospheric Activities and Dating Procedures
A. G. Maddock and E. H. Willis

Polyfluoroalkyl Derivatives of Metalloids and Nonmetals
R. E. Banks and R. N. Haszeldine

AUTHOR INDEX—SUBJECT INDEX

Volume 4

Condensed Phosphates and Arsenates
Erich Thilo

Olefin, Acetylene, and π-Allylic Complexes of Transition Metals
R. G. Guy and B. L. Shaw

Recent Advances in the Stereochemistry of Nickel, Palladium, and Platinum
J. R. Miller

The Chemistry of Polonium
  K. W. Bagnall

The Use of Nuclear Magnetic Resonance in Inorganic Chemistry
  E. L. Muetterties and W. D. Phillips

Oxide Melts
  J. D. Mackenzie

AUTHOR INDEX—SUBJECT INDEX

Volume 5

The Stabilization of Oxidation States of the Transition Metals
  R. S. Nyholm and M. L. Tobe

Oxides and Oxyfluorides of the Halogens
  M. Schmeisser and K. Brandle

The Chemistry of Gallium
  N. N. Greenwood

Chemical Effects of Nuclear Activation in Gases and Liquids
  I. G. Campbell

Gaseous Hydroxides
  O. Glemser and H. G. Wendlandt

The Borazines
  E. K. Mellon, Jr., and J. J. Lagowski

Decaborane-14 and Its Derivatives
  M. Frederick Hawthorne

The Structure and Reactivity of Organophosphorus Compounds
  R. F. Hudson

AUTHOR INDEX—SUBJECT INDEX

Volume 6

Complexes of the Transition Metals with Phosphines, Arsines, and Stibines
  G. Booth

Anhydrous Metal Nitrates
  C. C. Addison and N. Logan

Chemical Reactions in Electric Discharges
  Adli S. Kana'an and John L. Margrave

The Chemistry of Astatine
  A. H. W. Aten, Jr.

The Chemistry of Silicon–Nitrogen Compounds
  U. Wannagat

Peroxy Compounds of Transition Metals
  J. A. Connor and E. A. V. Ebsworth

The Direct Synthesis of Organosilicon Compounds
  J. J. Zuckerman

The Mössbauer Effect and Its Application in Chemistry
  E. Fluck

AUTHOR INDEX—SUBJECT INDEX

Volume 7

Halides of Phosphorus, Arsenic, Antimony, and Bismuth
  L. Kolditz

The Phthalocyanines
  A. B. P. Lever

Hydride Complexes of the Transition Metals
  M. L. H. Green and D. J. Jones

Reactions of Chelated Organic Ligands
  Quintus Fernando

Organoaluminum Compounds
  Roland Köster and Paul Binger

Carbosilanes
  G. Fritz, J. Grobe, and D. Kummer

AUTHOR INDEX—SUBJECT INDEX

Volume 8

Substitution Products of the Group VIB Metal Carbonyls
  Gerard R. Dobson, Ingo W. Stolz, and Raymond K. Sheline

Transition Metal Cyanides and Their Complexes
  B. M. Chadwick and A. G. Sharpe

Perchloric Acid
  G. S. Pearson

Neutron Diffraction and Its Application in Inorganic Chemistry
  G. E. Bacon

Nuclear Quadrupole Resonance and Its Application in Inorganic Chemistry
  Masaji Kubo and Daiyu Nakamura

The Chemistry of Complex Aluminohydrides
  E. C. Ashby

AUTHOR INDEX—SUBJECT INDEX

Volume 9

Liquid–Liquid Extraction of Metal Ions
  D. F. Peppard

Nitrides of Metals of the First Transition Series
  R. Juza

Pseudohalides of Group IIIB and IVB Elements
  M. F. Lappert and H. Pyszora

Stereoselectivity in Coordination Compounds
  J. H. Dunlop and R. D. Gillard

Heterocations
  A. A. Woolf

The Inorganic Chemistry of Tungsten
  R. V. Parish

AUTHOR INDEX—SUBJECT INDEX

Volume 10

The Halides of Boron
  A. G. Massey

Further Advances in the Study of Mechanisms of Redox Reactions
  A. G. Sykes

Mixed Valence Chemistry—A Survey and Classification
Melvin B. Robin and Peter Day

AUTHOR INDEX—SUBJECT INDEX—CUMULATIVE TOPICAL INDEX FOR VOLUMES 1–10

Volume 11

Technetium
K. V. Kotegov, O. N. Pavlov, and V. P. Shvedov

Transition Metal Complexes with Group IVB Elements
J. F. Young

Metal Carbides
William A. Frad

Silicon Hydrides and Their Derivatives
B. J. Aylett

Some General Aspects of Mercury Chemistry
H. L. Roberts

Alkyl Derivatives of the Group II Metals
B. J. Wakefield

AUTHOR INDEX—SUBJECT INDEX

Volume 12

Some Recent Preparative Chemistry of Protactinium
D. Brown

Vibrational Spectra of Transition Metal Carbonyl Complexes
Linda M. Haines and M. H. B. Stiddard

The Chemistry Complexes Containing 2,2'-Bipyridyl, 1,10-Phenanthroline, or 2,2', 6',2''-Terpyridyl as Ligands
W. R. McWhinnie and J. D. Miller

Olefin Complexes of the Transition Metals
H. W. Quinn and J. H. Tsai

Cis and Trans Effects in Cobalt(III) Complexes
J. M. Pratt and R. G. Thorp

AUTHOR INDEX—SUBJECT INDEX

Volume 13

Zirconium and Hafnium Chemistry
  E. M. Larsen

Electron Spin Resonance of Transition Metal Complexes
  B. A. Goodman and J. B. Raynor

Recent Progress in the Chemistry of Fluorophosphines
  John F. Nixon

Transition Metal Clusters with Π-Acid Ligands
  R. D. Johnston

AUTHOR INDEX—SUBJECT INDEX

Volume 14

The Phosphazotrihalides
  M. Bermann

Low Temperature Condensation of High Temperature Species as a Synthetic Method
  P. L. Timms

Transition Metal Complexes Containing Bidentate Phosphine Ligands
  W. Levason and C. A. McAuliffe

Beryllium Halides and Pseudohalides
  N. A. Bell

Sulfur–Nitrogen–Fluorine Compounds
  O. Glemser and R. Mews

AUTHOR INDEX—SUBJECT INDEX